PHYCOBIOTECHNOLOGY

Biodiversity and Biotechnology of Algae and
Algal Products for Food, Feed, and Fuel

Innovations in Biotechnology

PHYCOBIOTECHNOLOGY

Biodiversity and Biotechnology of Algae and
Algal Products for Food, Feed, and Fuel

Edited by
Jeyabalan Sangeetha, PhD
Devarajan Thangadurai, PhD
Sanniyasi Elumalai, PhD
Shivasharana Chandrabanda Thimmappa, PhD

APPLE
ACADEMIC
PRESS

First edition published 2021

Apple Academic Press Inc.
1265 Goldenrod Circle, NE,
Palm Bay, FL 32905 USA
4164 Lakeshore Road, Burlington,
ON, L7L 1A4 Canada

CRC Press
6000 Broken Sound Parkway NW,
Suite 300, Boca Raton, FL 33487-2742 USA
2 Park Square, Milton Park,
Abingdon, Oxon, OX14 4RN UK

First issued in paperback 2021

Library and Archives Canada Cataloguing in Publication

Title: Phycobiotechnology : biodiversity and biotechnology of algae and algal products for food, feed, and fuel / edited by Jeyabalan Sangeetha, PhD, Devarajan Thangadurai, PhD, Sanniyasi Elumalai, PhD, Shivasharana Chandrabanda Thimmappa, PhD.

Names: Sangeetha, Jeyabalan, editor. | Thangadurai, D., editor. | Elumalai, Sanniyasi, editor. | Thimmappa, Shivasharana Chandrabanda, editor.

Description: Series statement: Innovations in biotechnology ; volume 3 | Includes bibliographical references and index.

Identifiers: Canadiana (print) 20200331272 | Canadiana (ebook) 20200331329 | ISBN 9781771888967 (hardcover) | ISBN 9781003019510 (ebook)

Subjects: LCSH: Algae—Biotechnology. | LCSH: Algae products.

Classification: LCC TP248.27.A46 P59 2021 | DDC 579.8—dc23

Library of Congress Cataloging-in-Publication Data

....

CIP data on file with US Library of Congress

....

ISBN: 978-1-77188-896-7 (hbk)
ISBN: 978-1-77463-760-9 (pbk)
ISBN: 978-1-00301-951-0 (ebk)

INNOVATIONS IN BIOTECHNOLOGY BOOK SERIES

SERIES EDITOR

Devarajan Thangadurai, PhD
Assistant Professor, Karnatak University, Dharwad, South India

BOOKS IN THE SERIES

Volume 1: Fundamentals of Molecular Mycology
Devarajan Thangadurai, PhD, Jeyabalan Sangeetha, PhD, and
Muniswamy David, PhD

**Volume 2: Biotechnology of Microorganisms: Diversity, Improvement,
and Application of Microbes for Food Processing, Healthcare,
Environmental Safety, and Agriculture**
Editors: Jeyabalan Sangeetha, PhD, Devarajan Thangadurai, PhD,
Somboon Tanasupawat, PhD, and Pradnya Pralhad Kanekar, PhD

**Volume 3: Phycobiotechnology: Biodiversity and Biotechnology of
Algae and Algal Products for Food, Feed, and Fuel**
Editors: Jeyabalan Sangeetha, PhD, Devarajan Thangadurai, PhD,
Sanniyasi Elumalai, PhD, and Shivasharana Chandrabanda Thimmappa, PhD

ABOUT THE EDITORS

Jeyabalan Sangeetha, PhD
Assistant Professor, Central University of Kerala at Kasaragod, South India

Jeyabalan Sangeetha, PhD, is Assistant Professor at Central University of Kerala at Kasaragod, South India. She earned her BSc in Microbiology and PhD in Environmental Science from Bharathidasan University, Tiruchirappalli, Tamil Nadu, India. She holds an MSc in Environmental Science from Bharathiar University, Coimbatore, Tamil Nadu, India. She is the recipient of a Tamil Nadu Government Scholarship and a Rajiv Gandhi National Fellowship from the University Grants Commission, Government of India, for her doctoral studies. She served as the Dr. D.S. Kothari Postdoctoral Fellow and a UGC Postdoctoral Fellow at Karnatak University, Dharwad, South India, during 2012–2016 with funding from the University Grants Commission, Government of India, New Delhi. Her research interests are in environmental toxicology, environmental microbiology, and environmental biotechnology, and her scientific/community leadership has included serving as editor of an international journal, *Acta Biololgica Indica.*

Devarajan Thangadurai, PhD
Assistant Professor, Karnatak University, Dharwad, South India

Devarajan Thangadurai, PhD, is Assistant Professor at Karnatak University in South India. In addition, Dr. Thangadurai is Editor-in-Chief of two international journals, *Biotechnology, Bioinformatics and Bioengineering and Acta Biologica Indica.* He received his PhD in Botany from Sri Krishnadevaraya University in South India. During 2002–2004, he worked as a CSIR Senior Research Fellow with funding from the Ministry of Science and Technology, Government of India. He served as a Postdoctoral Fellow at the University of Madeira, Portugal, University of Delhi, India, and ICAR National Research Centre for Banana, India. He is the recipient of the Best Young Scientist Award with a Gold Medal from Acharya Nagarjuna University, India, and the VGST-SMYSR Young Scientist Award of the Government of Karnataka, Republic of India. He has authored and edited nineteen books, including *Genetic Resources and Biotechnology (3 vols.); Genes, Genomes and Genomics (2 vols.); Mycorrhizal Biotechnology;*

viii   About the Editors

Genomics and Proteomics; Industrial Biotechnology; and *Environmental Biotechnology,* with publishers of national and international reputation. He has also visited Portugal, Indonesia, Ukraine, Nepal, Sri Lanka, Bangladesh, Thailand, Oman, Vietnam, Malaysia, China, Georgia, United Arab Emirates, Myanmar, Egypt, Italy, and Russia for academic work, scientific meetings, and international collaborations.

Sanniyasi Elumalai, PhD

Professor, Department of Biotechnology, University of Madras, Chennai, India

Sanniyasi Elumalai, PhD, is Professor in the Department of Biotechnology, University of Madras, Chennai, India, where he also earned his BSc, MSc, MPhil, and PhD degrees. He was a Visiting Scientist at the University of Helsinki, Finland; Durban University of Technology, South Africa; Technical University of Denmark; Stockholm University, Sweden; M.V. Lomonosov State University, Russia; Etovas University, Hungary; and Mexican National Polytechnico Institute. He has published over 120 publications and received several extramural grants for research on algae and plants with reference to biotechnology, genetic engineering, and sustainable biofuel production. For his research contributions and academic achievements in the field of biotechnology and genetic engineering of algae and plants, he has been honored with several awards and fellowships. His expertise is in algal biotechnology, marine biotechnology, genomics, and proteomics of plants. He has done postdoctoral research at the University of Central Florida, Orlando, USA; Murdoch University, Perth, Western Australia; and the University of Madras, Chennai, India.

Shivasharana Chandrabanda Thimmappa, PhD

Assistant Professor of Biotechnology and Microbiology, Karnatak University, Dharwad, India

Shivasharana Chandrabanda Thimmappa, PhD, is Assistant Professor of Biotechnology and Microbiology at Karnatak University, Dharwad, India. He has published more than 30 research articles and delivered over 20 invited lectures in the field of microbiology and biotechnology. His major area of research is algal, environmental, and sustainable biotechnology. He earned his MSc, MPhil, and PhD degrees in Biotechnology from Gulbarga University, Kalaburagi, India.

CONTENTS

CONTRIBUTORS

Mohd. Azmuddin Abdullah
Institute of Marine Biotechnology, University Malaysia Terengganu, 21030 Kuala Nerus, Terengganu, Malaysia

Nazia Ahmad
Department of Biosciences, Jamia Millia Islamia, New Delhi – 110025, India

Anofi Ashafa
Department of Plant Sciences, Faculty of Natural and Agricultural Sciences, University of the Free State-QwaQwa Campus, Phuthaditjhaba, South Africa

M. Jeya Bharathi
Plant Pathology Unit, Tamil Nadu Rice Research Institute, Aduthurai, Tamil Nadu, India

Manjunath Chavadi
Hassan Institute of Medical Science, Sri Chamarjendra Hospital Campus, Department of Microbiology, Hassan, Karnataka, India

Sanniyasi Elumalai
Department of Biotechnology, University of Madras, Guindy Campus, Chennai – 600025, Tamil Nadu, India

Tasneem Fatma
Department of Biosciences, Jamia Millia Islamia, New Delhi – 110025, India

Gustavo Graciano Fonseca
Laboratory of Bioengineering, Faculty of Biological and Environmental Sciences, Federal University of Grande Dourados, Dourados, MS, CEP – 79804-970, Brazil; Center of Biotechnology, Postgraduate Program in Science and Biotechnology, University of West of Santa Catarina (UNOESC), Videira-SC, CEP – 89600-000, Brazil

Jane Mary Lafayette Neves Gelinski
Center of Biotechnology, Postgraduate Program in Science and Biotechnology, University of West of Santa Catarina (UNOESC), Videira-SC, Brazil

Annamalai Jayshree
Center for Environmental Studies, Anna University, Guindy Campus, Chennai – 600025, Tamil Nadu, India

Gopal Rajesh Kanna
Department of Plant Biology and Plant Biotechnology, Presidency College (Autonomous), Chennai – 600005, Tamil Nadu, India

Roshan Kumar
Department of Human Genetics and Molecular Medicine, Central University of Punjab, Bathinda – 151001, Punjab, India

Pek Chin Loh
Department of Biomedical Science, Faculty of Science, Universiti Tunku Abdul Rahman, Jalan Universiti, Bandar Barat, 31900 Kampar, Perak, Malaysia

Ramu Manjula
National Institute of Mental Health and Neurosciences, Department of Biophysics,
Bangalore – 560029, Karnataka, India

Mariana Lara Menegazzo
Laboratory of Bioengineering, Faculty of Biological and Environmental Sciences,
Faculty of Engineering, Federal University of Grande Dourados, Dourados, MS,
CEP – 79804-970, Brazil

Nathanya Nayla De Oliveira
Laboratory of Bioengineering, Faculty of Biological and Environmental Sciences,
Federal University of Grande Dourados, Dourados, MS, CEP – 79804-970, Brazil

Cecilia Oluseyi Osunmakinde
Nanotechnology and Water Sustainability Research Unit, College of Science,
Engineering, and Technology, University of South Africa, Florida, South Africa

Sunil Pabbi
Center for Conservation and Utilization of Blue Green Algae, Division of Microbiology,
ICAR-Indian Agricultural Research Institute, New Delhi – 110012, India

Thinanoor Venugopal Poonguzhali
Department of Botany, Queen Mary's College, Chennai – 600004, Tamil Nadu, India

Moshahid Alam Rizvi
Department of Biosciences, Jamia Millia Islamia, New Delhi – 110025, India

Muthu Sakthivel
Department of Biotechnology, University of Madras, Chennai – 600025, Tamil Nadu, India

Neha Sami
Department of Biosciences, Jamia Millia Islamia, New Delhi, India

Thimarayan Sangeetha
Department of Chemistry, Presidency College (Autonomous), Chennai – 600005,
Tamil Nadu, India

Ramganesh Selvarajan
Department of Plant Sciences, Faculty of Natural and Agricultural Sciences,
University of the Free State-QwaQwa Campus, Phuthaditjhaba, South Africa

Syed Muhammad Usman Shah
Department of Biosciences, COMSATS University Islamabad, Park Road, 44000,
Islamabad, Pakistan

Timothy Sibanda
Department of Biological Sciences, University of Namibia, Windhoek, Namibia

Thangaraj Vimala
Department of Plant Biology and Plant Biotechnology, Quaid-e-Millath Government College (W),
Chennai – 600002, Tamil Nadu, India

Waghmode Ahilya Vitthal
Department of Botany, Shivaji University, Kolhapur – 416003, Maharashtra, India

Hann Ling Wong
Department of Biological Science, Faculty of Science, Universiti Tunku Abdul Rahman,
Jalan Universiti, Bandar Barat, 31900 Kampar, Perak, Malaysia

Anurag Yadav
Department of Microbiology, College of Basic Science and Humanities, S.D. Agricultural University, S.K. Nagar, Gujarat – 385506, India

Kusum Yadav
Department of Biochemistry, University of Lucknow, Lucknow, Uttar Pradesh – 226007, India

Durdana Yasin
Department of Biosciences, Jamia Millia Islamia, New Delhi – 110025, India

ABBREVIATIONS

2-OG	2-oxoglutarate
ACC	acetyl-CoA carboxylase
ACP	acyl carrier protein
AD	Alzheimer's disease
AD	atopic dermatitis
AGO	argonaute
$Al_2(SO_4)_3$	aluminum sulfate
ALA	α-linolenic acid
ALT	alanine transaminase
AMA	apical major antigen
AMD	age-related macular degeneration
APC	allophycocyanin
AST	aspartate transaminase
ASTM	American Society for Testing and Materials
AtDGAT	*Arabidopsis thaliana* acyl-CoA:diacylglycerol acyltransferase
ATP	adenosine triphosphate
BBS	Bardet-Beidl syndrome
BGA	blue-green algae
BHA	butylated hydoxyanisole
BHT	butylated hydoxytoulene
BOD	biological oxygen demand
CAF	chlorophyll autofluorescence
CAP	catabolite activator protein
CB	cyanobacteria
CcbP	calcium-binding protein
CCl_4	carbon tetra chloride
Cdk	cyclin-dependent kinase
CH_4	methane
CHD	coronary heart disease
CikA	circadian input kinase
CikR	cognate-response regulator
CKD	chronic kidney disease

CO	carbon monoxide
CO_2	carbon dioxide
ConA	concanavalin A
COX-2	cyclooxygenase-2
CPK	creatinine phosphokinase
CRISPR	clustered regularly interspaced short palindromic repeats
CrtISO	*cis*-carotene isomerase
CSFV	classical swine fever virus
CTAB	cetyltrimethylammonium bromide
CTB	cholera toxin B
CVD	cardiovascular disease
DAG	diacylglycerol
DCM	diabetic cardiomyopathy
DCs	dendritic cells
DCY1	duplicated carbonic anhydrase 1
DEN	diethyl nitrosamine
DENV-2	dengue virus type 2
DGAT	diacylglycerol acyltransferase
DGCC	1,2-diacylglyceryl-3-O-carboxy-(hydroxymethyl)-choline
DGDG	digalactosyldiacylglycerol
DGTA	1,2-diacylglyceryl-3-*O*-2'-(hydroxymethyl)-(*N,N,N*-trimethyl)-β-alanine
DGTS	1,2-diacylglyceryl-3-*O*-4'-(*N,N,N*-trimethyl)-homoserine
DHA	docosahexaenoic acid
DN	diabetic nephropathy
DNA	deoxyribonucleic acid
DOXP	1-deoxy-D-xylulose-5-phosphate
DR	death receptors
DR	diabetic retinopathy
DT	doubling time
EDTA	ethylenediaminetetraacetic acid
EEA	European Environmental Agency
EEZ	exclusive economic zone
EFA	essential fatty acids
EGP	envelope glycoprotein
EGP	exponential growth phase
EMS	ethyl methanesulfonate
EPA	eicosapentaenoic acid
ER	endoplasmic reticulum

ERK	extracellular signal-regulated kinase
ESRD	end-stage renal disease
EST	expressed sequence tag
FAME	fatty acid methyl ester
FAS	fatty acid synthase
fdxN	ferredoxin gene
Fe/Mo	iron and molybdenum
Fe/V	iron and vanadium
$Fe_2(SO_4)_3$	iron sulfate
$FeCl_3$	ferric chloride
Fe-Mo	iron-molybdenum
FeMo-Co	Fe-Mo cofactors
FGF	fibroblast growth factor
FMDV	foot-and-mouth disease virus
FPP	farnesyl diphosphate
FTIR	Fourier transform infrared
G3P	glycerol-3-phosphate
G6P	glucose-6-phosphate
GAD65	glutamic acid decarboxylase-65
GAF	green autofluorescence
GAGs	glycosaminoglycans
GBSS	granule-bound starch synthase
GE	genetically engineered
GFP	green fluorescent protein
GGPP	geranylgeranyl pyrophosphate
GGPS	geranylgeranyl pyrophosphate synthase
GHGs	greenhouse gases
GLA	gamma-linolenic acid
GMSMs	genetically modified superior microorgansims
GPAT	glycerol 3-phosphate acyltransferase
GRAS	generally recognized as safe
GS	glutamine synthetase
GT	generation time
GW	global warming
H_2SO_4	sulfuric acid
H5N1	avian influenza A
HBsAg	hepatitis B virus surface antigen
HIV	human immunodeficiency virus
HIV-1	human immunodeficiency virus type 1

HIV-AIDS	human immunodeficiency virus-acquired immunodeficiency syndrome
HNG-CoA	3-hydroxy-3-methyl-glutaryl-coenzyme A
HPH	high pressure homogenization
HR	homologous recombination
HSV	herpes simplex virus
hupL	hydrogen uptake gene
HUVEC	human umbilical vein endothelial cells
I/R	ischemia-reperfusion
ICAS	intracranial atherosclerotic stenosis
IFN-γ	interferon
IL-1β	interleukin-1β
iNOS	inducible nitric oxide synthase
IPP	isopentenyl pyrophosphate
ITS	internal transcribed spacer
JAK2	Janus Kinase 2
JNK	Jun N-terminal kinase
KAS III	β-ketoacyl-ACP synthase III
LAB	lactic acid bacteria
LC-PUFA	long chain-PUFA
LDH	lactate dehydrogenase
LDL	low-density lipoprotein
LHC	light-harvesting complex
LMWF	low molecular weight fucoidan
LMWHs	low-molecular-weight heparins
LPAAT	lysophosphatidic acid acyltransferase
LPS	lipopolysaccharide
mAb	monoclonal antibody
MAPK	mitogen-activated protein kinase
Mch	multiple-contiguous-heterocyst
MCI	mild cognitive impairment
MCP-1	monocyte chemoattractant protein-1
MDA	malondialdehyde
ME	malic enzyme
MEP	2-C-methyl-D-erythritol 4-phosphate
MGDG	monogalactosyldiacylglycerol
MI	myocardial infarction
min.	minute
MMA	monomethylalkanes

MMME	multivariate modular metabolic engineering
MMPs	matrix metalloproteinases
MNNG	methylnitronitrosoguanidine
MPO	myeloperoxidase
MSP	major surface protein
MVA	mevalonate
N	nitrogen
N_2O	nitrous oxide
NaCl	sodium chloride
NADPH	nicotinamide adenine dinucleotide phosphate
NaOH	sodium hydroxide
NCAS	non-cerebral atherosclerotic stenosis
nd	not determined
NDV	Newcastle disease virus
NF	natural fucoidan
NF-κB	nuclear factor-kappa B
NGF	nerve growth factor
NK	natural killer
NO	nitric oxide
NOD	non-obese diabetic
NOx	nitrogen oxides
NRPS	non-ribosomal peptide synthetase
NSAIDs	nonsteroidal anti-inflammatory drugs
NSP	neurotoxic shellfish poisoning
OB	oil bodies
OD	optical density
OPP	oxidative pentose phosphate
OSF	oversulfated fucoidan
OtDGAT	*Osteococcus tauri* diacylglycerol acyltransferase
P	phosphorous
PA	phosphatidic acid
PBR	photobioreactors
PC	phosphatidylcholine
PC	phycocyanin
PDAT	phospholipid:diacylglycerol acyltransferase
PDC	pyruvate dehydrogenase complex
PDCT	phosphatidylcholine:diacylglycerol cholinephosphotransferase
PDS	phytoene desaturase

PE	phosphatidylethanolamine
PE	phycoerythrin
PEP	phosphoenolpyruvate
PEPC	PEP carboxylase
PG	phosphatidylglycerol
PGE2	prostaglandin E2
PKS	polyketide synthetase
Plk	polo-like kinase
Pp	product productivity
PS-DHA	phosphatidylserine-containing DHA
PSII	photosystem II
PSPP	presqualene diphosphate
PtDGAT	*Phaeodactylum tricornutum* diacylglycerol acyltransferase
PTO	post-translational oscillator
PUFA	polyunsaturated fatty acid
Pys	phytoene synthase
RA	rheumatoid arthritis
RNAi	RNA interference
ROS	reactive oxygen species
RPE	retinal pigment epithelium
Rubisco	ribulose-1,5-bisphosphate carboxylase/oxygenase
RuBP	ribulose 1,5-bisphosphate
S	substrate concentration
SasA	*Synechococcus* adaptive sensor
SC-CO$_2$	supercritical carbon dioxide
SC-PUFA	short chain-PUFA
SCR	serum creatinine
SD	shine-dalgarno
SDA	stearidonic acid
SDF-1	stromal-derived factor 1
SFE	supercritical fluid extraction
SiC	silicon carbon
SLIC	sequence and ligation-independent cloning
SNP	single nucleotide polymorphism
SPE	solid-phase extraction
SQDG	sulfoquinovosyldiacylglycerol
SSL	squalene synthase-like
STAT1	signal transducer and activator of transcription 1
SUN	serum urea nitrogen

TAG	triacylglyceride
TD-NMR	time-domain nuclear magnetic resonance
TE	thioesterase
TGF-β1	transforming growth factor beta 1
TLRs	toll-like receptors
TNF	tumor necrosis factor
TPP	thiamine pyrophosphate
TRADD	TNFR1-associated death domain
TRAF	TNF receptor associated factor
TRAF2	TNF receptor associated factor-2
TSP	total soluble protein
UDP	uridine diphosphate
UFH	unfractionated heparin
UTRs	untranslated regions
UV	ultraviolet
VCP	violaxanthin/chlorophyll α-binding protein
VEGF	vascular endothelial growth factor
VLDL	very-low-density lipoprotein
VP28	viral protein 28
WHO	World Health Organization
WSSV	white spot syndrome virus
ZDS	ζ-carotene desaturase

SYMBOLS

μ	specific growth rate
μ_{max}	maximum specific growth rate
μ_S	specific rate of substrate consumption
Ks	substrate affinity constant
ln	Napierian logarithm
pH	hydrogenic potential
pI	isoelectric point
pRB	retinoblastoma protein
Px	biomass productivity
Sf	final substrate concentration
Si	initial substrate concentration
t	culture time
t	tonne
tf	final culture time
ti	initial culture time
x_f	final biomass concentration
x_i	initial biomass concentration
XisC	excisase C
XisF	excisase F
$Y_{X/S}$	biomass yield

PREFACE

Algae are photosynthetic organisms that occur in most habitats and vary from small single-celled form to complex multicellular forms such as giant kelps of the eastern Pacific. An estimated 1–10 million species of algae majority being the microalgae used as a source of valuable natural biologically active molecules such as enzymes, carbohydrates, lipids, proteins, phycobiliproteins, carotenoids, chlorophylls, long-chain polyunsaturated fatty acids (PUFAs), starch, cellulose, pigments, antioxidants, and pharmaceuticals. Different algal species of freshwater and marine are compelled to survive in a competitive environment with defense strategies that result in eliciting a significant level of bioactive compound diversity from different metabolite pathways. In this scenario, biotechnology plays its role to assist metabolomic profiling of different growth stages of microalgae in order to maximize its production. Metabolomic approaches offer an advantage in effective de-replication strategies to eliminate the same natural compounds that are already isolated and studied by using powerful databases. Proteomics of microalgae is another biotechnological approach, which helps in understanding many cellular processes and network functions by providing information on post-translational modification including subcellular localization of gene products, thus making it an important study following genomics and transcriptomics.

Algae, Earth's first forms of life, and the first forms of food for subsequent species, now hold the potential to become the planet's next major source of energy and a vital part of the solutions for climate change and dependence on fossil fuels. They are a genetically diverse group of organisms with a wide range of physiological and biochemical characteristics; thus they naturally produce many unusual fats, sugars, and bioactive compounds. Functional components extracted from algal biomass are widely used as dietary and health supplements with a variety of applications in food science and technology. In contrast, the applications of algae in dermal-related products have received much less attention, despite that algae also possess high potential use in anti-infection, anti-aging, skin-whitening, and skin tumor treatments. Algae have a diverse variety of uses and applications and can be found in products ranging from antacids, dentistry molds, energy sources (including biodiesel and ethanol), fertilizers, plastics, prosthetics, pollution control, and pigments. Algae can capture fertilizer runoff from farms and then be re-used as fertilizer themselves produced by fermentation using certain microbes. These are very effective natural products, which can be completely absorbed by plants, capable of increasing seed germination rates, enhancing the resistance of plants to stresses, such as drought, cold, and diseases, and increasing crop yields. An effective biofilter, algae are used to remove pollutants from wastewaters. Also, as an effective bioreactor, algae are used to capture carbon dioxide emissions from industrial plants. An abundant source of vitamins, minerals, and other nutrients, many varieties of algae are known to boost the human immune system.

Unlike most plant species, algae contain a complete protein with essential amino acids. Algae have the potential to produce a volume of biomass and biofuel many times greater than that of our most productive crops. Algae can grow in a wide variety of climates in a multitude of production methods, from ponds to photobioreactors (PBR) to fermenters, and thus will create a wide variety of jobs throughout the world from research to engineering, construction to farming and from marketing to financial services. It's a no denying fact that currently algal industry is a booming sector and will continue to grow manifolds in the coming days and looking at the applications and economic value of algae it would not be wrong to call it, the most potent organism of the century.

Chapters 1 and 2 discuss the diversity and distribution of common seaweeds and microalgal biomarkers for hydrocarbon exploration. Chapters 3 to 6 review microalgal bioactive compounds, biomolecules, and biofuels for sustainable development. The health benefits of fucoidan and the role of cyanobacteria (CB) against various human diseases has well explained in Chapters 7 and 8. Evaluation of protocols for biomass recovery and lipid extraction by microalgae was well illustrated in Chapter 9. In Chapter 10, DNA (deoxyribonucleic acid) rearrangements in cyanobacterial nitrogen fixation were comprehensively defined. Chapters 11 and 12 gives systematic insights into the biotechnological applications of microalgae and exploitation of omega-3 fatty acids. Genetic engineering tools and metabolic engineering aspects of algae constitute Chapters 13 and 14.

As a basis for understanding algal biodiversity and its biotechnological potentiality, chapters in this reference book were extensively reviewed by many well-known subject experts and professionals. This novel book received great support from experts in the field of algal biotechnology. We are immensely indebted to all enthusiastic content contributors for their valuable insights and suggestions to complete this book successfully.

We are also grateful to thank Sandy Jones Sickels, Vice President, and Ashish Kumar, Publisher, and President, Apple Academic Press, Inc., USA for quality production and humble communications to publish this book. Finally, we wish to express our gratitude to individuals who directly or indirectly cooperated and contributed in preparing this book.

Jeyabalan Sangeetha, PhD
Devarajan Thangadurai, PhD
Sanniyasi Elumalai, PhD
Shivasharana Chandrabanda Thimmappa, PhD

CHAPTER 1

DIVERSITY AND DISTRIBUTION OF COMMON SEAWEEDS FROM THE WEST COAST OF MAHARASHTRA IN INDIA

WAGHMODE AHILYA VITTHAL

Department of Botany, Shivaji University, Kolhapur – 416003, Maharashtra, India

1.1 INTRODUCTION

Marine habitats play a very significant role in the ecological, productive habitat, biodiversity, and economical stability of the country. It has an exclusive economic zone (EEZ) of around 2.5 million sq. km., and accounts for about 8% of the global biodiversity (Oza, 2005). Generally, seaweeds are macroscopic and multicellular. Based on their pigmentation (chemical) present, macroalgae are classified into three main groups, viz., Rhodophyta, Phaeophyta, and Chlorophyta (Noel and Sekwon, 2013; Sahayaraj, 2015). Seaweeds have been used as human food from 600 to 800 BC in China and other countries in Asia (Plouguerne et al., 2014; Anand et al., 2016). China and Japan are the two major seaweed harvesting countries, where more than 70 species of seaweeds are consumed as salads directly or after cooking. Seaweeds play a vital role in marine food chains, food sources to humans, especially in East Asia (Japan), the Indo-Pacific (China, Indonesia), and the Pacific (Hawaii). It is recommended that seaweeds could become a foremost food and vigor resource in the 21st century. Presently, there are 42 countries in the world with reports of commercial exploitation of seaweeds. Among them, China holds the first rank, followed by North Korea, South Korea, Japan, Philippines, Chile, Norway, Indonesia, USA, and India. These top 10 countries contribute up to 95% of the world's commercial seaweed utilization (Khan and Satam, 2003).

Nowadays, varieties of health care products like lotions, bath gels, shampoos, soaps, creams, and pharmaceutical components are being produced from the seaweeds. Braune and Guiry (2011) reported that a total of about 400,000 tonnes of seaweeds such as *Porphyra* (Nori), *Laminaria* (Kombu), and *Undaria* (Wakame) are harvested annually throughout the world. About 20,000 species of seaweeds are distributed throughout the world, of which 221 are commercially utilized, which includes 145 species for food and 110 species for phycocolloid production (Chennubhotla et al., 2013a,b). In India, the marine macroalgae include 217 genera, a total of 844 species of seaweeds (Saxena, 2012). The economic utility of seaweeds exposes that there are only very scanty and sporadic works have been carried out in various maritime states of India: Andhra Pradesh (Umamaheswara Rao, 1970, 2011; Anonymous, 1995); Andaman and Nicobar Islands (Jagtap, 1985, 1992; Gopinathan and Panigrahi, 1983; Rao and Rao, 1999; Palanisamy, 2012; Karthick et al., 2013); Gujarat (Thivy, 1958, 1982; Jha et al., 2009); Goa (Dhargalkar, 1981), Lakshadweep (Kaliaperumal et al., 1989; Koya et al., 1999; Koya, 2000; Kaladharan, 2001; Kumar and Kaladharan, 2007); Maharashtra (Dhargalkar et al., 1980); Odisha (Sahoo et al., 2001, 2003) and Tamil Nadu (Chennubhotla, 1977, 1996; Chennubhotla et al., 1981, 1987, 1988; Kaliaperumal et al., 1990, 1992, 1995, 2004). Seaweeds along the southeast coast of Tamil Nadu from Mandapam to Kanyakumari, Gujarat coast, Lakshadweep Island, and Andaman and Nicobar Islands having a luxuriant growth. Whereas abundant seaweeds are present in the locality of Bombay, Raighad, Ratnagiri, Sindhudurgh, Goa, Karwar, Varkala, Kovalam, Vizhinjam, Visakhapatnam, and few other places such as Chilka and Pulicat lakes.

1.2 TAXONOMY OF SEAWEEDS IN INDIA

India, the subcontinent (08°4' to 37°6' N and 68°7' to 97°25' E) is having a coastline of more than 7516 km and 1256 islands including Andaman, Nicobar, and Lakshadweep. The coastline of India is composed of littoral and sublittoral rocky areas of different seaweeds (agarophytes, alginophytes, and other edible seaweeds). The Indian coast is divided into the east coast and the West Coast. The West Coast is usually exposed with heavy surf, rocky shores, and headlands while the east coast is generally shelving with beaches, lagoons, delta, and marshes. Important Gulfs on the West

Coast of India are the Arabian Sea, the Gulf of Kutch, and Gulf of Cambay, while the Arabian Sea washes the shores of Gujarat, Maharashtra, Goa, Karnataka, and Kerala maritime states. Besides the mainland, the Andaman and Nicobar Islands in the Bay of Bengal and Lakshadweep islands in the Arabian Sea are rich in marine macrophytic algal resources. As per the available literature, the *Amphiroa* was first collected by Hermann in 1672 from the Indian Ocean. However, Horenel (1918) gave the first estimates of seaweed resources for the drift *Sargassum* on Okha mandal coast, of the Baroda state. Studies made on Indian marine algae have been reviewed from time to time by various authors particularly Agharkar (1923); Biswas (1932, 1934); Joshi (1949); Iyengar (1957); Randhawa (1960); and Sreenivasan (1965). Seaweed collections on the Indian coast were done by several Europeans in the 20th century.

M. O. P. Iyengar was the first Indian algologist who gave a detailed descriptive account of Indian marine algae occurring on the southeast coast, especially Krusadai Island. Since then, challenges have been made on different traits of seaweed research in India. Borgesen has published a series of paper on the green, brown, and red algae of the northern part of the West Coast (Borgesen, 1930, 1931, 1932a, b, 1933a, b, 1934 a, b, 1935) and brown and red alga of south India (Borgesen, 1937a, b, 1938). After the valuable contribution of Boergesen, much work has been done on the morphology and taxonomy of Indian marine algae during the last four decades. Subsequently, Krishnamurthy (1967b, 1991), Srinivasan Rao (1973), Umamaheswara Rao (1974a) and several other workers published extensively on the Indian marine algae, efforts have been made since 1940s by Biswas (1945), Biswas and Sarma (1950), Dixit (1966, 1968, 1980). Dixit (1966, 1968) recorded 411 species which includes algae from Pakistan and Sri Lanka. A checklist of 520 species of Indian marine algae has been published by Krishnamurthy and Joshi (1970).

Seaweeds are beneficial to mankind in many ways, only 49 species are presently useful as directly edible or as industrial raw materials. Rich vegetation for the seaweeds was found along the east coast, viz. Mahabalipuram (Srinivasan, 1946) Visakhapatnam (Umamaheshwara Rao and Seeramulu, 1970) and along the West Coast, viz. Okha (Misra, 1966). A general review of the marine algae was published by Biswas (1945). Srinivasan (1946) studied on the taxonomy of Indian marine algal flora of Mahabalipuram. According to the estimates given by him, 162 genera and 413 species of marine algae were known from India. Misra (1966) has primed a monograph of the brown algae occurring along the Indian coast. The species of *Ulva* from the Indian

Ocean is published by Krishnamurthy and Joshi (1970). Srinivasan (1966) has published an account of the Indian species of *Sargassum* and their geographical distribution. Taylor (1964) has described the Indian species of *Turbinaria.*

A list of algae collected from Mandapam area has been published by Umamaheswara Rao (1969a, 1969b, 1972c) and gave accounts on the four most important agar and algin-yielding seaweeds, *Gracilaria, Gelidiella, Sargassum, Turbinaria, Gracilaria edulis.* A systematic account of 10 taxa of Indian Gelidiales has been specified by Sreenivasa Rao (1970). An annotated list of 80 alga species growing along the Visakhapatnam coast has been given by Umamaheswara Rao and Sree Ramulu (1970). Joshi and Krishnamurthy (1971) have listed 13 species of *Enteromorpha* from India. The description of 17 species, 2 varieties of *Gracilaria*, 2 species of *Gracilaria*, their habitats and distribution in India are given in detail by Umamaheswara Rao (1972b). Agadi and Untawale (1978) have reported 50 algal species from the different localities of intertidal areas along the Goa coast. An account of 46 species of marine algae occurring at Tiruchendur on the Tamil Nadu coast has been given by Krishnamurthy et al. (1980). Gopalakrishnan (1980) has reported 64 species of algae from the collection grounds of Dwarka, Okha, Adatra, Hanumandandi, and Balapur from Okha coast. Sharma and Khan (1980) have published a checklist of Indian freshwater and marine algae. A taxonomic account of Indian Ectocarpales and Ralfsiaies has been given by Balakrishnan and Kinkar (1981). A systematic list of 44 species of algae collected from 10 localities along the southern Kerala coast was published by Balakrishnan et al. (1982). A total of 74 seaweeds have been reported by Dhargalkar (1981) and Agadi (1986) from the Goa coast. Chennubhotia et al. (1987) reported 35 species of seaweeds occurring along the Kerala Coast. Biodiversity of marine algae along the east and West Coast regions of India was studied by several authors (Kalimuthu, 1995; Sahoo et al., 2003; James, 2004; Venkataraman, 2005; Rath and Adikary, 2006). Subba Rao and Mantri (2006) published a review on the Indian seaweed resources and its sustainable utilization.

The latest systematic account of seaweeds is of 844 species from the Indian coast having been recorded, at the same time 159 species from Maharashtra (Oza and Zaidi, 2001). Sobha et al. (2008) studied six species of seaweeds (*Ulva fasciata, U. reticulata, Caulerpa racemosa, Padina tetrastromatica, Sargassum wightii,* and *Gracilaria corticata*) collected from the southern Kerala coast, as food products in the form of ulva toffy, ulva squash, mixed algae pickle, algae cutlet, algae biriyani and algae thoran. The southern parts of the Kerala coast are extended with rocks, bedrocks, and cliffs which are suitable for luxuriant growth of seaweeds while the central parts of the Kerala coast are fully protected with artificially laid stones, shows less occurrence of seaweeds. However, the northern parts of the Kerala coast are mainly sandy, scattered, and bedrocks which support the maximum numbers of seaweed. The rich diversity and luxuriant growth of seaweeds were recorded at Somatheeram, Mullur Kadalapuram, Vizhinjam, Kovalam, Valliathura, Varkala, Edava, Thangassery, Thirumullavaram, Neendakara, Azheekkal, Kochi fort (Muttenchery), Vypeen island, Munambam, Pudu Ponnani, Baypore, Kappad, Thikkodi, Mahe, Thalassery, Ezhimala, Kappil beach, Hosabettu, and Manjeshwar coasts.

The coastline of Kerala shows a rich diversity, distribution pattern, and exhibits a wide range of variation in different areas of seaweeds. The species of *Gelidium, Gelidiella, Gracilaria,* and *Gelidiopsis* are used as raw materials for the production of agar-agar whereas species of *Hypnea, Spyridia, Acanthophora, Sargassum,* and *Turbinaria* are used for the production of agaroid. In the recent years, species such as *Kappaphycus alvarezii* has been used for the production of biofuels as an alternative source for fossil fuels (Chennubhotla et al., 2011; Khambaty et al., 2012). The observation of the present study shows that the quantitative value of the economically important seaweeds in the Kerala coast is comparatively lesser than other maritime states of India (Chennubhotla et al., 2013a,b) to support or establish the seaweed-based industries. Therefore, it is suggested here that this can be overcome by encouraging the coastal communities to cultivate the economically valuable seaweeds such as *U. fasciata, U. reticulata, C. peltata, C. racemosa, C. taxifolia, P. gymnospora, P. tetrastromatica, Sargassum tenerrimum, S. wightii, Turbinaria conoides, T. ornata, Kappaphycus alvarezii, Gelidium micropterum, G. pusillum, Gelidiella acerosa, Gracilaria corticata, G. edulis, foliifera, Hypnea musciformis, H. valentiae,* and *Acanthophora spicifera.* In addition, this practice would support the livelihood activities of the coastal communities. The importance uses of seaweeds as food in the form of recipes, salads, soups, jellies, and vinegar dishes is well known in many Indo-Pacific countries since long ago (Chennubhotla et al., 2013a,b). However, in the Indian context, the uses of seaweeds in the form of food are still very limited. These seaweeds are very much popular in South Asian countries like China, Japan, Korea, and even in the USA (Chapman and Chapman, 1980). Among all the edible seaweeds, Nori (*Porphyra*) has got maximum popularity in

Japanese cuisine (Sahoo, 2000). As far as the industrial value of seaweeds is concerned, the annual production of the marine algae along the Indian coast has been estimated to 260,876 tonnes (Chennubhotla, 1992) and now it has been revised to 301,646 tonnes (Chennubhotla et al., 2011). Among the various maritime Indian states, the bulk production comes from Tamil Nadu and Andaman and Nicobar Islands amounting to nearly 60% while the rest from the coast of Maharashtra, Goa, and Kerala. Nearly 5% of the raw material is contributed by just single species *Kappaphycus alvarezii*, which is cultivated artificially at a larger scale especially in the Gulf of Mannar region of Tamil Nadu (Chennubhotla et al., 2013a,b).

1.3   SEAWEEDS OF MAHARASHTRA

Marine algal studies from the Kokan region of Maharashtra was initiated in the late 19th century. The first report from Maharashtra goes to Shri Kirtikar (1886) who read out a paper on marine algae collected by Hon. Justice Birdwood from Ratnagiri coast (Deodhar, 1987). Qasim and Wafar (1979) have recorded a total of 72 seaweeds from different regions of Ratnagiri, Malwan, and Redi along the West Coast. Dhargalkar et al. (2001) reported 91 species of macroalgae from the entire West Coast of Maharashtra. According to Untawale et al. (1983), there are 624 species of marine algae belonging to 215 genera and 64 families. Out of these, 60 species are commercially important. *Ulva fasciata*, *U. lactuca*, *Caulerpa* sp., *Chaetomorpha* sp., *Sargassum ilicifolium*, *Padina tetrastromatica*, *Gracilaria corticata*, *Hypnea*, *Acanthophora*, *Jania*, *Corallina*, *Ceramium*, *Amphiroa*, and *Corallina* were present throughout the study period.

Intertidal regions support rich growth of algae belonging to the genera *Ulva*, *Enteromorpha*, and *Chaetomorpha*, blue-green algae like *Microcoleus* and *Lynbya*. Ditches and rocks are observed at Redi, Malvan, Kunakeshwar, Vijaydurg, Ratnagiri near Harnai fort, Kurli-Kasop beach, Harihareshwar, Near Alibag Fort, Murud-Janjira, Rocks between Shrivardhan and Divagar, Colaba, Arnala, Dhanu, Hagi-Ali, Worli C-Link, Bandra (Kartar road), Palli hills (Mud island), Edwan, Kelva, and Tokepada, Bordi (Narpad). The diversity of marine algae in the West Coast showed that the members of Rhodophyta were dominant followed by Chlorophyta. Seaweeds constitute an important part of global marine biodiversity. With few additions, a total of 83 species of seaweeds were reported from the coast of Maharashtra, of these, species of *Ulva*, *Chaetomorpha*, *Enteromorpha*, *Sargassum*, *Padina*, *Amphiroa*, *Jania*, and *Gracilaria* are common in this region. Species of

Porphyra, Gracilaria, Gelidium, Ulva, and *Sargassum* spp. are the economically important seaweeds occurring along this coast (Waghmode, 2017) (Figures 1.1 and 1.2).

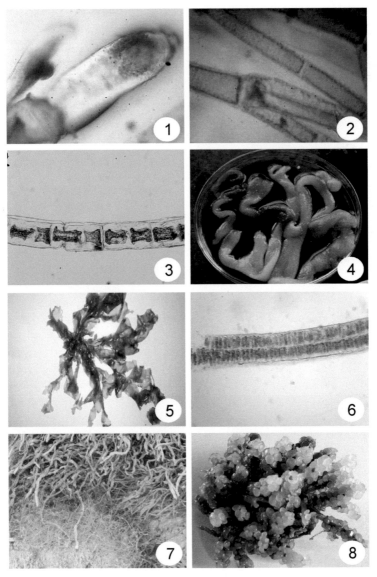

FIGURE 1.1A Common seaweeds of Maharashtra in India: (1) *Cladophora glomerata*; (2) *Cladophora vagabunda*; (3) *Chaetomorpha linum*; (4) *Codium dwarkense*; (5) *Ulva fasciata*; (6) *Ulva lactuca*; (7) *Enteromorpha intestinalis*; (8) *Caulerpa peltata*.

8   Phycobiotechnology

FIGURE 1.1B Common seaweeds of Maharashtra in India: (9) *Ectocarpus simpliciusculus*;
(10) *Padina tetrastromatica*; (11) *Sargassum ilicifolium*; (12) *Sargassum tenerrimum*; (13)
Spatoglossuma sperum; (14) *Dictyopteris woodwardia*; (15) *Stoechospermum marginatum*;
(16) *Rosenvingea intricata*.

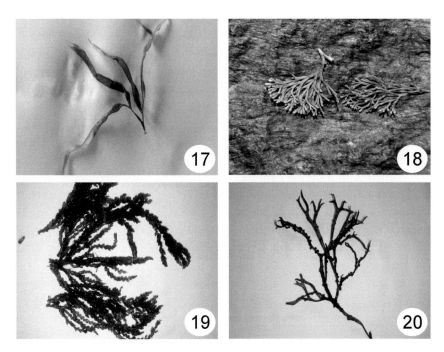

FIGURE 1.1C   Common seaweeds of Maharashtra in India: (17) *Grateloupia lithophila*; (18) *Amphiroa anceps*; (19) *Acanthophora spicifera*; (20) *Gracilaria corticata* var. *corticata*.

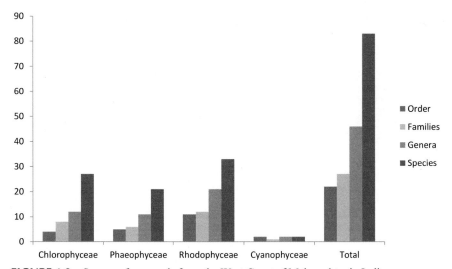

FIGURE 1.2   Survey of seaweeds from the West Coast of Maharashtra in India.

1.4   COLLECTION OF SEAWEED SAMPLES

The seaweed samples were collected randomly from the intertidal regions. Collected samples were thoroughly washed in seawater and subsequently in freshwater without damaging the specimens. A set of herbarium specimens were prepared and the live samples were preserved in 4% formalin. All the collected specimens are deposited at Botany Department, Shivaji University, Maharashtra (India). All the wet and dry specimens were examined carefully under the light microscope (NIKON SMZ 1500 and NIKON ECLIPSE 50*i*) coupled with a digital camera. The taxonomic descriptions were made based on the morphological and anatomical characters of the specimens for each species. Identity of the collected samples was confirmed following the standard available literatures (Srinivasan, 1969, 1973; Desikachary et al., 1990, 1998; Silva et al., 1996; Krishnamurthy, 2000; Jha et al., 2009; Krishnamurthy and Baluswamy, 2010; Braune and Guiry, 2011) (Tables 1.1–1.3).

TABLE 1.1   List of Seaweeds

Name of Marine Algae	Order	Family
Blue-Green Algae/Cyanophyta		
Microcoleus chthonoplastes Thuret ex Gomont	Nostocales	Oscillatoriaceae
Lyngbya majuscula Harvey ex Gomont	Oscillatoriales	Oscillatoriaceae
Chlorophyta/Green Algae		
Monostroma oxyspermum (Kutzing) Doty	Ulvales	Monostromataceae
Enteromorpha compressa (Linnaeus) Nees	Ulvales	Ulvaceae
Enteromorpha flexuosa (Wulfen) J. Agardh	Ulvales	Ulvaceae
Enteromorpha prolifera (Muller) J. Agardh	Ulvales	Ulvaceae
Enteromorpha intestinalis (Linnaeus) Nees	Ulvales	Ulvaceae
Ulva fasciata Delile	Ulvales	Ulvaceae
Ulva lactuca Linnaeus	Ulvales	Ulvaceae
Ulva rigida C. Agardh	Ulvales	Ulvaceae
Chaetomorpha media (Bory de Saint-Vincent)	Cladophorales	Cladophoraceae
Chaetomorpha antennina (Bory) Kutzing	Cladophorales	Cladophoraceae
Chaetomorpha linum (Muller) Kutzing	Cladophorales	Cladophoraceae
Chaetomorpha crassa (C. Agardh) Kutzing	Cladophorales	Cladophoraceae
Cladophora glomerata (Linnaeus) Kutzing	Cladophorales	Cladophoraceae
Cladophora gracilis Grif (ex Harvey)	Cladophorales	Cladophoraceae
Cladophora fascicularis (Mert.) Kutzing	Cladophorales	Cladophoraceae
Cladophora vagabunda (Linnaeus) vanden Hoek	Cladophorales	Cladophoraceae

TABLE 1.1 *(Continued)*

Name of Marine Algae	Order	Family
Cladophoropsis zollingeri (Kutz) Silva	Cladophorales	Cladophoraceae
Rhizoclonium grande Boergesen	Cladophorales	Cladophoraceae
Boodlea composita (Harvey) Brand	Cladophorales	Siphonocladaceae
Ulothrix flacca (Dillwyn) Thuret	Ulotrichales	Ulvophyceae
Caulerpa racemosa (Forsskal) J. Agardh	Bryopsidales	Caulerpaceae
Caulerpa peltata Lamouroux	Bryopsidales	Caulerpaceae
Caulerpa sertularoides (S. Gmelin) Howe	Bryopsidales	Caulerpaceae
Caulerpa taxifolia (M. Vahl) C. Agardh	Bryopsidales	Caulerpaceae
Bryopsis pennata Lamouroux	Bryopsidales	Bryopsidaceae
Bryopsis plumosa	Bryopsidales	Bryopsidaceae
Codium dwarkense	Bryopsidales	Codiaceae
Phaeophyta/Brown Algae		
Ectocarpus simpliciusculus C. Agardh	Ectocarpales	Ectocarpaceae
Rosenvingea intricate (J. Ag.) Boergesen	Ectocarpales	Encoeliaceae
Sphacelaria furcigera Kutzing	Sphacelariales	Sphacelariaceae
Dictyota bartayresiana Lamouroux	Dictyotales	Dictyotaceae
Dictyota maxima Zanardini	Dictyotales	Dictyotaceae
Dictyota dichotoma (Hudson) Lamouroux	Dictyotales	Dictyotaceae
Dictyota divaricata Lamouroux	Dictyotales	Dictyotaceae
Lobophora variegate (Lamouroux) Womersley ex Oliveira	Dictyotales	Dictyotaceae
Padina tetrastromatica Hauck	Dictyotales	Dictyotaceae
Padina gymnospora (Kutzing) Vickers	Dictyotales	Dictyotaceae
Spatoglossum asperum J. Agardh	Dictyotales	Dictyotaceae
Stoechospermum marginatum (C. Agardh) Kutzing	Dictyotales	Dictyotaceae
Sargassum cinereum (J. G. Agardh)	Dictyotales	Dictyotaceae
Sargassum tenerrimum (J. G. Agardh)	Dictyotales	Dictyotaceae
Sargassum wightii (Greville mscr.) J. G. Agardh	Dictyotales	Dictyotaceae
Sargassum swartzii C. Agardh	Fucales	Sargassaceae
Sargassum ilicifolium (Turner) C. Agardh	Fucales	Sargassaceae
Sargassum spp.	Fucales	Sargassaceae
Sargassum spp.	Fucales	Sargassaceae
Dictyopteris woodwardia (R. Brown ex Turner) C. Agardh	Fucales	Sargassaceae
Colpomenia sinuosa (Roth) Derbes et Solier	Scytosiphonales	Scytosiphonaceae
Rhodophyta/Red Algae		
Porphyra vietnamensis T. Tanaka & Pham-Hoang Ho	Bangiales	Bangiaceae

TABLE 1.1 *(Continued)*

Name of Marine Algae	Order	Family
Gelidiella acerosa (Forsskal) J. Feldmann & G. Hamel	Gelidiales	Gelidiellaceae
Gracilaria corticata (J. Agardh) J. Agardh	Gracilariales	Gracilariaceae
Gracilaria eucheumatoides Harvey	Gracilariales	Gracilariaceae
Gracilaria edulis (S. Gmelin) P. Silva	Gracilariales	Gracilariaceae
Gracilaria verrucosa (Hudson) Papefuss	Gracilariales	Gracilariaceae
Gracilaria corticata (J. Agardh) J. Agardh var. *cylindrica* Umamaheshwara Rao	Gracilariales	Gracilariaceae
Acanthophora spicifera (Vahl) Borgesen	Ceramiales	Rhodomelaceae
Acanthophora muscoides (Linnaeus) Bory de Saint-Vincent	Ceramiales	Rhodomelaceae
Acanthophora delilei Lamouroux	Ceramiales	Rhodomelaceae
Polysiphonia platycarpa Borgesen	Ceramiales	Rhodomelaceae
Centroceras clavulatum (C. Ag.) Mont.	Ceramiales	Ceramiaceae
Ceramium diaphanum (Lightfoot) Roth	Ceramiales	Ceramiaceae
Spyridia filamentosa (Wulfen) Harvey	Ceramiales	Ceramiaceae
Amphiroa anceps (Lamarck) Decaisne	Corallinales	Corallinaceae
Amphiroa fragilissima (Linnaeus) Lamouroux	Corallinales	Corallinaceae
Corallina berteroi Montagne ex Kutzing	Corallinales	Corallinaceae
Corallina officinalis Linnaeus	Corallinales	Corallinaceae
Jania rubens (Linnaeus) Lamouroux	Corallinales	Corallinaceae
Jania adhaerens Lamouroux	Corallinales	Corallinaceae
Hypnea musciformis (Wulfen) Lamouroux	Gigartinales	Hypneaceae
Hypnea valentiae (Turner) Montage	Gigartinales	Hypneaceae
Hypnea spinalla (C. Agardh) Kutzing	Gigartinales	Hypneaceae
Sarconema filiforme (Sonder) Kylin	Gigartinales	Solieriaceae
Solieria robusta (Greville) Kylin	Gigartinales	Solieriaceae
Catenella repens (Lightfoot) Batters	Gigartinales	Caulacanthaceae
Champia compressa Harvey	Rhodymeniales	Champiaceae
Portieria hornemannii (Lyngbye) P. Silva	Cryptonemiales	Rhizophyllidaceae
Asparagopsis taxiformis (Delile) Trevisan	Bonnemaisoniales	Bonnemaisoniaceae
Grateloupia filicina (Lamouroux) C. Agardh	Halymeniales	Halymeniaceae
Grateloupia lithophila Borgesen	Halymeniales	Halymeniaceae
Ahnfeltia plicata (Hudson) Fries	Ahnfeltiales	Ahnfeltiaceae

(Source: Reprinted with permission from Waghmode, 2017.)

TABLE 1.2 Distribution of Seaweeds

Species	Place of Collection
Blue-Green Algae	
Microcoleus chthonoplastes Thuret ex Gomont	Sindhudurg, Ratnagiri, Raighad, Bombay, Palghar
Lyngbya majuscula Harvey ex Gomont	Sindhudurg, Ratnagiri, Raighad, Bombay, Palghar
Green Algae	
Monostroma oxyspermum (Kutzing) Doty	Sindhudurg, Raighad
Enteromorpha compressa (Linnaeus) Nees	Sindhudurg, Ratnagiri, Raighad, Bombay, Palghar
Enteromorpha flexuosa (Wulfen) J. Agardh	Sindhudurg, Ratnagiri, Raighad, Bombay, Palghar
Enteromorpha prolifera (Muller) J. Agardh	Sindhudurg, Ratnagiri, Raighad, Bombay
Enteromorpha intestinalis (Linnaeus) Nees	Sindhudurg, Palghar
Ulva fasciata Delile	Sindhudurg, Raighad, Palghar
Ulva lactuca Linnaeus	Sindhudurg, Ratnagiri, Raighad, Bombay
Ulva rigida C. Agardh	Palghar, Bombay
Chaetomorpha media (Bory de Saint-Vincent)	Sindhudurg, Raighad
Chaetomorpha antennina (Bory) Kutzing	Sindhudurg, Ratnagiri, Raighad, Palghar
Chaetomorpha linum (Muller) Kutzing	Sindhudurg, Ratnagiri, Raighad
Chaetomorpha crassa (C. Agardh) Kutzing	Sindhudurg, Ratnagiri, Raighad
Cladophora glomerata (Linnaeus) Kutzing	Sindhudurg, Ratnagiri, Palghar
Cladophora gracilis Grif (ex Harvey)	Sindhudurg, Ratnagiri
Cladophora fascicularis (Mert.) Kutzing	Sindhudurg, Ratnagiri, Palghar
Cladophora vagabunda (Linnaeus) van den Hoek	Ratnagiri, Bombay
Cladophoropsis zollingeri (Kutz) Silva	Ratnagiri, Bombay
Rhizoclonium grande Boergesen	Bombay, Raighad
Boodlea composita (Harvey) Brand	Sindhudurg, Raighad
Ulothrix flacca (Dillwyn)Thuret	Sindhudurg, Raighad
Caulerpa racemosa (Forsskal) J. Agardh	Sindhudurg, Ratnagiri
Caulerpa peltata Lamouroux	Sindhudurg, Ratnagiri
Caulerpa sertularoides (S. Gmelin) Howe	Sindhudurg, Ratnagiri
Caulerpa taxifolia ((M. Vahl) C. Agardh	Sindhudurg, Ratnagiri
Bryopsis pennata Lamouroux	Sindhudurg, Ratnagiri
Bryopsis plumosa	Sindhudurg, Ratnagiri

TABLE 1.2 *(Continued)*

Species	Place of Collection
Brown Algae/Phaeophyta	
Ectocarpus simpliciusculus C. Agardh	Sindhudurg, Ratnagiri
Rosenvingea intricata (J. Ag.) Boergesen	Raighad
Sphacelaria furcigera Kutzing	Sindhudurg, Ratnagiri
Dictyota bartayresiana Lamouroux	Sindhudurg, Ratnagiri
Dictyota maxima Zanardini	Sindhudurg, Ratnagiri
Dictyota dichotoma (Hudson) Lamouroux	Sindhudurg, Ratnagiri
Dictyota divaricata Lamouroux	Sindhudurg, Ratnagiri
Lobophora variegata (Lamouroux) Womersley ex Oliveira	Sindhudurg, Ratnagiri
Padina tetrastromatica Hauck	Sindhudurg, Ratnagiri
Padina gymnospora (Kutzing) Vickers	Sindhudurg, Ratnagiri
Spatoglossum asperum J. Agardh	Sindhudurg, Ratnagiri
Stoechospermum marginatum (C. Agardh) Kutzing	Sindhudurg, Ratnagiri
Sargassum cinereum (J. G. Agardh)	Sindhudurg, Bombay, Palghar
Sargassum tenerrimum (J. G. Agardh)	Sindhudurg, Bombay, Palghar
Sargassum wightii (Greville mscr.) J. G. Agardh	Sindhudurg, Bombay, Palghar
Sargassum swartzii C. Agardh	Ratnagiri
Sargassum ilicifolium (Turner) C. Agardh	Sindhudurg, Bombay, Palghar
Sargassum sp.	Sindhudurg, Ratnagiri
Sargassum sp.	Sindhudurg, Ratnagiri
Dictyopteris woodwardia (R. Brown ex Turner) C. Agardh	Bombay
Colpomenia sinuosa (Roth) Derbes et Solier	Sindhudurg
Red Algae/Rhodophyta	
Porphyra vietnamensis T. Tanaka & Pham-Hoang Ho	Sindhudurg
Gelidiella acerosa (Forsskal) J. Feldmann & G. Hamel)	Sindhudurg, Ratnagiri
Gelidium pusillum (Stackhouse) Le Jolis	Sindhudurg, Ratnagiri
Gracilaria corticata (J. Agardh) J. Agardh	Sindhudurg, Ratnagiri, Palghar
Gracilaria eucheumatoides Harvey	Bombay, Ratnagiri, Palghar
Gracilaria edulis (S. Gmelin) P. Silva	Bombay, Palghar, Ratnagiri

TABLE 1.2 *(Continued)*

Species	Place of Collection
Gracilaria verrucosa (Hudson) Papefuss	Bombay, Ratnagiri, Palghar
Gracilaria corticata (J. Agardh) J. Agardh var. *cylindrica* Umamaheshwara Rao	Bombay, Palghar
Acanthophora spicifera (Vahl) Borgesen	Sindhudurg, Ratnagiri
Acanthophora muscoides (Linnaeus) Bory de Saint-Vincent	Sindhudurg, Ratnagiri
Acanthophora delilei Lamouroux	Sindhudurg, Ratnagiri
Polysiphonia platycarpa Borgesen	Sindhudurg, Ratnagiri, Raighad
Centroceras clavulatum (C. Ag.) Mont.	Sindhudurg, Ratnagiri, Raighad
Ceramium diaphanum (Lightfoot) Roth	Sindhudurg, Ratnagiri, Raighad
Spyridia filamentosa (Wulfen) Harvey	Ratnagiri
Amphiroa anceps (Lamarck) Decaisne	Sindhudurg, Raighad
Amphiroa fragilissima (Linnaeus) Lamouroux	Sindhudurg, Raighad
Corallina berteroi Montagne ex Kutzing	Sindhudurg, Raighad
Corallina officinalis Linnaeus	Sindhudurg, Raighad
Jania rubens (Linnaeus) Lamouroux	Sindhudurg, Raighad
Jania adhaerens Lamouroux	Sindhudurg, Raighad
Hypnea musciformis (Wulfen) Lamouroux	Sindhudurg
Hypnea valentiae (Turner) Montage	Sindhudurg
Hypnea spinalla (C. Agardh) Kutzing	Sindhudurg
Sarconema filiforme (Sonder) Kylin	Bombay
Solieria robusta (Greville) Kylin	Bombay
Catenella repens	Bombay, Palghar
Champia compressa Harvey	Bombay
Portieria hornemannii (Lyngbye) P. Silva	Palghar
Asparagopsis taxiformis (Delile) Trevisan	Sindhudurg
Grateloupia filicina (Lamouroux) C. Agardh	Sindhudurg, Raighad
Grateloupia lithophila Borgesen	Sindhudurg, Raighad
Ahnfeltia plicata (Hudson) Fries	Sindhudurg, Bombay

(Source: Reprinted with permission from Waghmode, 2017.)

TABLE 1.3 Total List of Seaweeds Collected from West Coast of Maharashtra in India

Taxonomic Groups	Chlorophyceae	Phaeophyceae	Rhodophyceae	Cyanophyceae	Total
Order	04	05	11	02	22
Families	08	06	12	01	27
Genera	12	11	21	02	46
Species	27	21	33	02	83

1.5 CONCLUSION

Marine biodiversity needs systematic study. Although, the West Coast is subjected to increasing industrial pollution and habitat destruction, there are some pockets wherein high macroalgal diversity occurs. Thus the adverse effect of pollution is observed more on the marine algal community of Bombay > Thane > Raighad > Ratanagiri > Sindhudurgh. Some economically important seaweed like *Sargassum, Hypnea, Caulerpa, Graciliaria, Acanthophora, Corollina, Enteromorpha, Chaetomorpha,* and *Ulva* are still common along the West Coast of Maharashtra. This may be because of the increased tolerance to the environmental changes. Some factors like industrial and commercial centers are responsible for dwindling down the algal flora. Thus the algal biodiversity is observed more in Sindhudurgh > Ratanagiri > Raighad > Thane > Bombay. However, corrective measures like conservation, research, education, and public awareness are very necessary.

KEYWORDS

- biodiversity
- Chlorophyta
- Cyanophyta
- Flora of West Coast of India
- Pheophyta
- Rhodophyta
- seaweeds distribution
- taxonomic study of seaweeds

REFERENCES

Agadi, V. V., & Untawale, A. G., (1978). Marine algal flora of Goa coast. *Seaweed Res. Utilin., 2*(3), 56–70.

Agadi, V. V., (1986). *Marine Algal Studies of the Central West Coast of India*. PhD Thesis, Karnatak University, India.

Agharkar, S. P., (1923). The present position of our knowledge of the aquatic flora of India. *J. Indian Bot. Soc., 3*, 252–260.

Anand, N., Rachel, D., Thangaraju, N., & Anantharaman, P., (2016). Potential of marine algae (seaweeds) as source of medicinally important compounds. *Plant Genet. Resour., 14*, 303–313.

Anonymous, (1995). *Economically Important Seaweeds* (p. 35). CMFRI Special Publication No. 62.

Balakrishnan, M. S., & Kinker, V. N., (1981). A taxonomic account of Indian *Ectocarpales* and *Ralfsiales*. *Seaweed Res. Utiln., 4*(2), 1–57.

Balakrishnan, N. N., Shoba, N., & Arunachalam, M., (1982). Algae from southern Kerala coast. *Indian J. Mar. Sci., 11*(3), 266–269.

Biswas, K., & Sharma, A. K., (1950). Sargassams of Indian Sea. *J. Roy. Asiat. Soc., Bengal (Science), 16*(1), 79–97.

Biswas, K., (1932). Census of Indian algae. Scope of algological studies in India. *Rev. Algol., 6*, 197–219.

Biswas, K., (1934). Progress of algological studies in India. *Cur. Sci., 3*, 237–241.

Biswas, K., (1945). A general review of the marine algae of the western coast of India. *J. Bombay Nat. Hist. Soc., 45*, 515–530.

Boergesen, F., (1930). Some Indian green and brown algae especially from the shores of the presidency of Bombay. *J. Indian Bot. Soc., 9,* 151–174.

Boergesen, F., (1931). Some Indian *Rhodophyceae* especially from the shores of the presidency of Bombay I. *Kew Bull.,* 1–24.

Boergesen, F., (1932a). Some Indian *Rhodophyceae* especially from the presidency of Bombay: II. *Kew Bull., 3*, 113–134.

Boergesen, F., (1932b). Some Indian green and brown algae especially from the presidency of Bombay. *J. Indian Bot. Soc., 11,* 51–70.

Boergesen, F., (1933a). Some Indian *Rhodophyceae* especially from the presidency of Bombay: III. *Kew Bull., 3*, 113–142.

Boergesen, F., (1933b). Some Indian green and brown algae from the presidency of Bombay. *J. Indian Bot. Soc., 12,* 1–16.

Boergesen, F., (1934a). Some Indian *Rhodophyceae* especially from the presidency of Bombay-IV, *Kew Bull., 4,* 1–30.

Boergesen, F., (1934b). Some marine algae from the northern part of the Arabian Sea with remarks on their geographical distribution. *Kgl Dansk Vidensk Selskab Biol Meddel, 11*(6), 1–72.

Boergesen, F., (1935). A list of marine algae from Bombay. *Kgl Dansk Vidansk Sslskab Biol Meddel, 2*(2), 1–64.

Boergesen, F., (1937a). Contributions to a South Indian marine algal flora-I. *J. Ind. Bot. Soc., 16*, 1–56.

Boergesen, F., (1937b). Contributions to a South Indian marine algal flora-II. *J. Indian. Bot. Soc., 16*, 311–357.

Boergesen, F., (1938). Contributions to a South Indian marine algal flora-III. *J. Indian Bot. Soc., 17,* 205–242.

Braune, W., & Guiry, M. D., (2011). *Seaweeds: A Color Guide to Common Benthic Green, Brown and Red Algae of the World's Oceans* (p. 601). A.R.G. Gantner Verlag K.G., Ruggell, Liechtenstein, Germany.

Chapman, V. J., & Chapman, D. J., (1980). *Seaweeds and Their Uses* (3rd edn., p. 334). Chapman and Hall, London.

Chennubhotla, V. S. K., (1977). Food from the sea: Food from the seaweeds. *Sea Food Export Jour., 9*(3), 1–4.

Chennubhotla, V. S. K., (1981). *Status of Seaweed Industry in India* (pp. 139–145.). UNDP report on the training course on *Gracilaria* algae of the South China Sea fisheries Development and Coordinating Programme, manila, Philippines.

Chennubhotla, V. S. K., (1992). *A Survey of Seaweed Resources of Andaman-Nicobar Islands, Visakhapatnam-Chilka Lake Region and Lakshadweep Group of Islands* (p. 104). Final report on the ICAR Ad-hoc Scheme.

Chennubhotla, V. S. K., (1996). Seaweeds and their importance. *Bull Cent Mar Fishery Res Inst., 48,* 108–109.

Chennubhotla, V. S. K., Kaliaperumal, N., & Kalimuthu, S., (1981). Seaweed recipes and other practical uses of seaweeds. *Seafood Export J., 13*(10), 9–16.

Chennubhotla, V. S. K., Kaliaperumal, N., & Kalimuthu, S., (1987). Economically important Seaweeds. *CMFRI Bulletin, 41,* 3–19.

Chennubhotla, V. S. K., Kalimuthu, S., Najmuddin, M., & Selvaraj, M., (1977c). Field culture of *Gelidiella acerosa* in the inshore waters of Gulf of Mannar. *Jour. Phycol., 13,* S4, 54.

Chennubhotla, V. S. K., Ramachandrudu, B. S., Kaladharan, P., & Dharmaraj, S. K., (1987). *Seaweed Resources of Kerala Coast.* Seminar on Fisheries Research and Development in Kerala. Trivandrum.

Chennubhotla, V. S. K., Ramachandrudu, B. S., Kaladharan, P., & Dharmaraja, S. K., (1988). Seaweed resources of Kerala coast. *Bulletin Aqua Biol., 7,* 69–74.

Chennubhotla, V. S. K., Umamaheswara, R. M., & Rao, K. S., (2011). *Marine Algal Resources of India-Their Role in Biotechnology, Industry, and Medicine.* 9th Indian Fisheries Forum, Chennai.

Chennubhotla, V. S. K., Umamaheswara, R. M., & Rao, K. S., (2013a). Commercial importance of marine macro algae. *Seaweed Res. Utilin., 35*(1/2), 118–128.

Chennubhotla, V. S. K., Umamaheswara, R. M., & Rao, K. S., (2013b). Exploitation of marine algae in Indo-Pacific region. *Seaweed Res. Utilin., 35*(1/2), 1–7.

Chennubhotla, V. S. K., Najmuedin, M., Ramalingam, J. R., & Kaliaperumal, N., (1987). Biochemical composition of some marine algae of Mandapam coast (South India). *Symposium on Research and Development in Marine Fisheries.* Mandapam Camp.

Deodhar, H. D., (1987). *The Biology of Marine Algae of Bombay* (p. 200). PhD Thesis, University of Pune, Pune.

Desikachary, T. V., Krishnamurthy, V., & Balakrishnan, M. S., (1990). *Rhodophyta* (Vol. 1., p. 279). Madras Science Foundation, Madras.

Desikachary, T. V., Krishnamurthy, V., & Balakrishnan, M. S., (1998). *Rhodophyta* (Vol. II., p. 359) Madras Science Foundation, Madras.

Dhargalkar, V. K., (1981). *Studies on Marine Algae of the Goa Coast* (p. 168). PhD Thesis, Bombay University.

Dhargalkar, V. K., Agadi, V. V., & Untawale, A. G., (1981). Occurrence of *Porphyra vietnamensis* (Bangiales, Rhodophyta) along the Goa coast. *Mahasagar, 14*(1), 75–77.

Dhargalkar, V. K., Jagtap, T. G., & Untawale, A. G., (1980). Biochemical constituents of seaweeds along the Maharashtra coast. *Indian J. Mar. Sci., 2, 9*(4), 297–299.

Dhargalkar, V. K., Untawale, A. G., & Jagtap, T. G., (2001). Marine macroalgal diversity along the Maharashtra coast; past and present status. *Indian J. Mar. Sci., 30*, 18–24.

Dixit, S. C., (1966). Species list of Indian marine algae determined by Borgesen. *Journal of University of Bombay (Sci.), 32*(54, 55), 12.

Dixit, S. C., (1968). Species list of Indian marine algae-II. *Journal University of Bombay, 36*(3–5), 9–24.

Dixit, S. C., (1980). Species list of Indian marine algae IV. *Journal of University of Bombay, (Sci.), 48, 49*, 56–80.

Fritsch, F. E., (1935). *The Structure and Reproduction of Algae* (Vol. I., pp. 1–791). Cambridge University Press.

Gopalakrishnan, P., (1980). Some observation on the shore ecology of the Okha coast. *J. Mar. Biol. Asso. India, 12*(1/2), 15–34.

Gopinathan, C. P., & Panigrahy, R., (1983). Mariculture potential of Andaman and Nicobar islands-an indicative survey: Seaweeds resources. *Bulletin of the Central Marine Fisheries Research Institute* (Vol. 34, pp. 47–51). Cochin, India.

Hornell, J., (1918). *Report on the Further Development of Fishery Resources of Baroda State.* Baroda.

Iyengar, M. O. P., (1957). Algology: Progress of science in India: Section 6-Botany. *Proceeding of the National Institute of Science of India*, pp. 229–251.

Jagtap, T. G., (1985). Studies on littoral flora of the Andaman Islands. In: Krishnamurthy, V., (ed.), *Marine Plants* (pp. 43–50).

Jagtap, T. G., (1992). Marine flora of Nicobar group of Islands in the Andaman Sea. *Indian J. Mar. Sci., 21*, 56–58.

James, J. E., Kumar, R. A. S., & Raj, A. D. S., (2004). Marine algal flora from some localities of southeast coast of Tamil Nadu. *Seaweed Res. Utilin., 26*(1/2), 3–39.

Jha, B., Reddy, C. R. K., Thakur, M. K., & Umamaheswara, R. M., (2009). *Seaweeds of India: The Diversity and Distribution of Seaweeds in Gujarat Coast* (p. 215). CSMCRI, Bhavnagar.

Joshi, A. C., (1949). Indian botany; present position and prospects. *4ᵗʰ Ind. Bot. Soc., 28*, 1–15.

Kaladharan, P., (2001). *Seaweed Resource Potential of Lakshadweep* (Vol. 56, pp. 121–124). Geol. Surv. India Spl. Pub.

Kaliaperumal, N., Kaladharan, P., & Kalimuthu, S., (1989). Seaweeds and Sea grass Resources. In: Marine living resources of the union territory of Lakshadweep-an indicative survey with suggestions for development. *Bulletin of the Central Marine Fisheries Res Institute, 43*, 162–175.

Kaliaperumal, N., Kaliamuthu, S., & Ramalingam, J. R., (1990). Studies on phycocolloid contents from seaweeds of south Tamil Nadu coast. *Seaweed Res. Utilin., 12*, 115–119.

Kaliaperumal, N., Kaliamuthu, S., & Ramalingam, J. R., (1992). Studies on agar content of *Gracilaria corticata* var. *corticata* and *G. corticata* var. *cylindrica*. *Seaweed Res. Utilin., 15*, 191–195.

Kaliaperumal, N., Kaliamuthu, S., & Ramalingam, J. R., (1995). *Economically Important Seaweeds* (Vol. 62, pp. 1–35). CMFRI Special Publication.

Kaliaperumal, N., Kaliamuthu, S., & Ramalingam, J. R., (2004). Present scenario of seaweed exploitation and industry in India. *Seaweed Res. Utilin.*, *26*(1/2), 47–53.

Kalimuthu, S., Kaliperumal, N., & Ramalingam, J. R., (1995). Distribution of algae and sea grasses in the estuaries and backwaters of Tamil Nadu. *Seaweed Res Utilin.*, *17*(1/2), 79–86.

Karthick, P., Mohanraju, R., Murthy, K. N., Ramesh, C. H., & Narayana, S., (2013). Seaweed potential of little Andaman, India. *Seaweed Res. Utilin.*, 35(1/2), 17–21.

Karthick, P., Mohanraju, R., Ramesh, C., Murthy, K. N., & Narayana, S., (2013). Distribution and diversity of seaweeds in north and south Andaman Island. *Seaweed Res. Utilin.*, *35*, 8–16.

Khambhaty, Y., Mody, K., Gandhi, M. R., Thampy, S., Maiti, P., Brahmbhatt, H., Eswaran, K., & Ghosh, P. K., (2012). *Kappaphycus alvarezii* as a source of bioethanol. *Bioresour. Technol.*, *103*(1), 180–185.

Khan, S. I., & Satam, S. B., (2003). Seaweed mariculture: Scope and potential in India. *Aquaculture Asia, 8*(4), 26–29.

Kirtikar, (1886). A new species of algae *Conferva thermalis* bird wood. *J. Bombay Natural History Society, 1*, 135–138.

Koya, C. N. H., (2000). *Studies on Ecology, Chemical Constituents and Culture of Marine Macro Algae of Minicoy Island, Lakshadweep.* PhD Thesis, CIFE, Mumbai, India.

Koya, C. N. H., Naseer, A. K. V., & Mohammed, G., (1999). *Productivity of Coral Reef Algae, Halimeda gracilis Harv. Ex. J. Ag. at Minicoy Island, Lakshadweep.* Seaweed Res Development.

Krishnamurthy, V. R., Venugopal, J. G., & Thiagaraj, S. H. N., (1957). Estimating drift seaweeds on the Indian coasts. *Proc. Semi. Sea Salt and Plants* (pp. 315–320). CSMCRI, Bhavnagar.

Krishnamurthy, V., & Baluswami, M., (2010). *Phaeophyceae of India and Neighborhood* (Vol. I, p. 192). Krishnamurthy Institute of Algology, Chennai.

Krishnamurthy, V., & Joshi, H. V., (1970). *A Checklist of the Indian Marine Algae* (pp. 1–36). CSMCRI, Bhavnagar.

Krishnamurthy, V., (1967). Marine algal cultivation necessity, principles, and problems. *Proc. Semi. Sea Salt and Plants* (pp. 327–333). CSMCRI, Bhavnagar.

Krishnamurthy, V., (1991). *Gracilaria* resources of India with particular reference to Tamil Nadu coast. *Seaweed Res. Utilin.*, *14*, 1–7.

Krishnamurthy, V., Raju, V. P., & Thomas, P. C., (1980). On augmenting the seaweed resources of India. *J. Mar. Biol. Assoc. India, 17*, 181–185.

Kumar, V., & Kaladharan, P., (2007). Amino acids in the seaweeds as an alternate source of protein for animal feed. *J. Marine Biol. Ass. India, 49*(1), 35–40.

Misra, J. N., (1966). *Phaeophyceae in India* (p. 203). Indian Council Agricultural Research, New Delhi.

Noel, V. T., & Sekwon, K., (2013). Beneficial effects of marine algal compounds in cosmeceuticals. *Mar. Drugs, 11*, 146–164.

Oza, A., & Rohit, M., (2005). Biodiversity of benthic marine algae along the Indian coast. In: *Handbook of Biotechnology* (p. 48).

Oza, R. M., & Zaidi, S. H., (2001). *A Revised Checklist of Indian Marine Algae* (p. 296). CSMCRI, Publication, Bhavanagar, India.

Palanisamy, M., (2012). *Seaweeds of South Andaman: Chidiyatapu, North Bay, and Viper Island* (pp. 49–58). Uttar Pradesh State Biodiversity Board.

Plouguerné, E., Da Gama, B. A. P., Pereira, R. C., & Barreto-bergter, E., (2014). Glycolipids from seaweeds and their potential biotechnological applications. *Front Cell Infect. Microbiol., 4,* 174.

Qasim, S. Z. M., & Wafar, V. M., (1979). Occurrence of living corals at several places along the West Coast of India. *Mahasagar, 72*(1), 53–58.

Rao, P. S. N., & Umamaheswara, R. M., (1999). On a species of *Kappaphycus* (Solieriaceae, Gigartinales) from Andaman and Nicobar Islands, India. *Phykos, 38*(182), 93–96.

Rao, S. P. V., & Mantri, A. V., (2006). Indian seaweed resources and sustainable utilization: Scenario at the dawn of a new century. *Curr. Sci., 91,* 164–174.

Rath, J., & Adhikary, S. P., (2006). Marine macro algae of Orissa, east coast of India. *Algae, 21*(1), 49–59.

Sahayaraj, K., (2015). Biological values and conservation of marine algae: An overview. In: *Proceedings of the Conservation and Sustainable Utilization of Marine Resources* (pp. 21–24). Tamil Nadu, India.

Sahoo, D. B., (2000). *Farming the Ocean: Seaweeds Cultivation and Utilization.* Aravali Publication Corporation, New Delhi.

Sahoo, D., Nivedita, & Debasish, (2001). *Seaweeds of Indian Coast* (p. 283). A.P.H. Publishing Corporation, New Delhi.

Sahoo, D., Sahu, N., & Sahoo, D., (2003). A critical survey of seaweed diversity of Chilika Lake, India. *Algae, 18*(1), 1–12.

Saxena, A., (2012). Marine biodiversity in India: Status and issues. *National Conference on Marine Biological Diversity: Souvenir,* pp. 127–134.

Sharma, Y. S. R. K., & Khan, M., (1980). *Algal Taxonomy in India* (p. 153). Today and Tomorrow's Printers and Publishers, New Delhi.

Silva, P. C., Basson, P. W., & Moe, R. L., (1996). *Catalog of the Benthic Marine Algae of the Indian Ocean* (p. 1259). University of California Press, London.

Sobha, V., Santhosh, S., Ghita, G., & Valsalakumar, E., (2008). Food products from seaweeds of south Kerala coast. *Seaweed Res. Utilin., 30*(1/2), 199–203.

Sreenivasa, R. P., & Kale, S. R., (1969). Marina algae from a little known place of Gujarat coast. I. Algae from Gopinath. *Phykos, 8,* 71–82.

Sreenivasa, R. P., & Shelat, Y. A., (1979). Antifungal activity of Indian seaweed extracts. *Proc. Int. Symp. Marine Algae of the Indian Ocean Region* (p. 47). CSMCRI, Bhavnagar, India.

Sreenivasa, R. P., (1969). Systematics, ecology and life history of Indian *Gelidiales* with special reference to agarophyte *Gelidiella acerosa* (Forsskal) Feldman et Hamel. *Salt Res. Ind., 6,* 46–47.

Sreenivasa, R. P., (1970). Systematics of Indian *Gelidiales. Phykos, 9,* 63–78.

Sreenivasa, R. P., Iyengar, E. R. R., & Thivy, F., (1964). Survey of algin bearing seaweeds at Adatra reef, Okha. *Curr. Sci., 33,* 464–465.

Srinivasan, K. S., (1946). Ecology and seasonal succession of the marine algae Mahabalipuram (seven pagodas) near madras. In: Sahni, B., (ed.), *Indian Botanical Society* (pp. 267–278).

Srinivasan, K. S., (1960). Distribution patterns of marine algae in Indian seas. *Proc. Symp. Algology.* (pp. 219–242). New Delhi.

Srinivasan, K. S., (1965). Indian botany in retrospect with particular reference to algal systematics. *J. Asiatic. Soc., Bengal, 7,* 49–78.

Srinivasan, K. S., (1966). Conspectus of *Sargassum* species from Indian territorial waters. *Phykos, 5,* 127–129.

Srinivasan, K. S., (1973). *Phycologia Indica (Icones of Indian Marine Algae)* (Vol. II, p. 60). Botanical Survey of India, Calcutta.

Taylor, W. R., (1964). The genus *Turbinaria* in Eastern Seas. *Jour. Linn. Soc. London (Botany), 58*, 475–490.

Thivy, F., (1958). Economic seaweeds. In: Jones, S., (ed.), *Fisheries of West Coast of India* (pp. 74–80). CMFRI, Mandapam Camp.

Thivy, F., (1982). On the importance and prospects of seaweed utilization in India. *Seaweed Res. Utilin., 5*, 53–59.

Umamaheswara, R. M., & Sree, R. T., (1970). An annotated list of the marine algae of Visakhapatnam (India). *Bot. J. Linn. Soc., 63*, 23–45.

Umamaheswara, R. M., (1969a). Catalog of marine algae in the reference collection of the central marine fisheries research institute. *Bull. Cent. Mar. Fish Res. Inst., 9*, 37–48.

Umamaheswara, R. M., (1969b). Agar and algin yielding seaweeds. *Proc. 6th Int. Seaweed Symp.*, pp. 715–721.

Umamaheswara, R. M., (1970). *The Economic Seaweeds of India. Bull. Cent. Mar. Fish Res. Inst., 20*, 1–68.

Umamaheswara, R. M., (1972a). Coral reef flora of Gulf of Mannar and Palk Bay. *Proc. Symp. Corals and Coral Reefs (1969)*, pp. 217–230.

Umamaheswara, R. M., (1972b). On the *Gracilariaceae* of the seas around India. *J. Mar. Biol. Ass. India, 14*(2), 671–696.

Umamaheswara, R. M., (1972c). Ecological observations on some intertidal algae of Mandapam coast. *Proc. Indian Natl. Sci. Acad., 5S*(B3, 4), 298–307.

Umamaheswara, R. M., (1974a). On the cultivation of *Gracilaria edulis* in the near shore areas around Mandapam. *Current Science, 43*(20), 660–661.

Umamaheswara, R. M., (2011). Diversity and commercial feasibility of marine macro algae of India. *Seaweed Res. Utilin., 33*(1/2), 1–13.

Untawale, A. G., Dhargalkar, V. K., & Agadi, V. V., (1983). List of marine algae from India. *Seaweed Res. Utilin., 21*(1/2), 79–84.

Venkataraman, K., & Wafar, M., (2005). Coastal and marine biodiversity of India. *Ind. J. Mar. Sci., 34*, 57–75.

Vinoj, K. V., & Kaladharan, P., (2007). *Amino acids in the seaweeds as an alternate source of protein for animal feed. J. Mar. Biol. Ass. India, 49*(1), 35–40.

Waghmode, A. V., (2017). Diversity and distribution of seaweeds from the West Coast of Maharashtra. *J. Algal. Biomass Utilin., 8*(3), 29–39.

Zingde, M. D., Singbal, S. Y. S., Moraes, C. F., & Reddy, C. V. G., (1976). Arsenic, copper, zinc, and manganese in the marine flora and fauna of coastal and estuarine waters around Goa. *Indian J. Mar. Sci., 5*, 212–217.

CHAPTER 2

MICROALGAL BIOMARKERS FOR HYDROCARBON EXPLORATION IN PENINSULAR INDIA

THIMARAYAN SANGEETHA,[1] SANNIYASI ELUMALAI,[2] and GOPAL RAJESH KANNA[3]

[1]Department of Chemistry, Presidency College (Autonomous), Chennai – 600005, Tamil Nadu, India

[2]Department of Biotechnology, University of Madras, Guindy Campus, Chennai – 600025, Tamil Nadu, India

[3]Department of Plant Biology and Plant Biotechnology, Presidency College (Autonomous), Chennai – 600005, Tamil Nadu, India

2.1 INTRODUCTION

Fuel is a power pack of energy conserved from geological exploration or produced from other sources through conversion. Fossil fuels are explored geologically from the underground earth. Coal, crude oil, and oil shale are the rich sources of fossilized fuel. Crude oil and oil shale are the major sources of petroleum products employed majorly for the process of transportation of goods like food, medicines, and raw materials. Petroleum fuel alone comprises about 97% of the energy utilized for transportation. In another point of view, the rapid depletion of fossil fuel due to unsustainability may cause a huge demand for fuel in the upcoming future (Khan et al., 2009).

2.2 HYDROCARBONS IN SEDIMENTS AND PETROLEUM FUEL

The hydrocarbons are long-chain complex organic compounds deposited long ago as geographical sediments and are strongly reported to be synthesized biologically and modified geologically. Structural changes occurred in the sediments from biologically synthesized compounds include dehydration, hydrogenation, aromatization, isomerization at chiral centers, molecular arrangement, and ring-opening. The biosynthetic chemical reactions occur because of high pressure and temperature. Hence, the biochemical reaction involves the conversion of biological compounds into stable saturated and aromatic hydrocarbons thermodynamically. It takes over 10 to 100 millions of years for the conversion of compounds into a more stable manner. The occurrence of complex lipid compounds from organisms in the geological sediments has recently studied to identify novel compounds and unraveling the biological function of major lipid molecules. The sedimentary lipids can be identified using advanced instrumentation including GC-MS with capillary columns, quick scanning, and more precise spectrometers and processing of data. These applications of instruments are important in the exploration of oil from the underground (Mackenzie, 1983).

The specific class of compounds in sediments includes coal and petroleum, before the isolation of their precursor a product of living organisms were hopanoids (Ourisson et al., 1979). The hopanoids are widespread in geological sediments up to C_{35} with side chains (Van Dorsselaer et al., 1977) and are reported along the geosphere as a precursor distinct from C_{30} to C_{29}. The occurrence of high quantity of sedimentary lipids from marine environments is due to contribution of phytoplankton. More than 90% of the annual photosynthetic products have been reported to sediment as organic matter which are reused or withstand in the oceanic water column. The hydrocarbons of petroleum like triterpanes (18α(H)-oleanane) are the organic compounds belong to marine or terrestrial plants. The geochemical studies on lipids from the ancient sediments and petroleum can provide the information about the biochemistry of the contemporary organisms.

2.3   BIOCHEMICAL MARKERS

The hydrocarbon sediments that occur in the underground soil have been focused in this study as a biochemical marker for the exploration of hydrocarbons from the soil for fuel. The biochemical markers are the organic chemical compounds synthesized biologically due to the occurrence of biological organisms. The range of terpenoids reported in the geological sediments (petroleum and coal) includes monoterpenoids, sesquiterpenoids, diterpenoids, sesterterpenoids, triterpenoids, and tetraterpenoids

(Brassell et al., 1983). The sedimentary environment at low temperature is much possible to sustain the terpenoid lipids for even more than 100 million years.

The acyclic isoprenoids not only include mono-, sesqui-, di-, sester-, tri-, and tetra-terpenoids but also related compounds with series of C_9 to C_{30} as regular isoprenoid alkanes. The biochemicals transferred from organisms have configurations of acyclic isoprenoids which were preserved in the sediments but isomerizes with high thermal pressure into sediments and petroleum. Along with acyclic isoprenoids, free lipids occurred in the solvent extraction of petroleum and sediments and various glycerol tetraethers with isoprenoid structures have been reported among the polar lipids (Chappe et al., 1982).

The monoterpenoids are C_{15+} compounds reported to be obtained from geological materials by fractionation processes used in geological analyses. Some of the examples of monoterpenoids are 2, 6-Dimethyloctane and 2-methyl-3-ethyl-heptane especially as petroleum products of limonene (Maxwell et al., 1971). Sesquiterpenoids occur as high percentage in petroleum which is drimane and cadalene from geological sediments. The structures are less studied among other higher terpenoids (Alexander et al., 1983). The alkanes, alkenes, and aromatic hydrocarbons are bagged under cyclic diterpenoids. The overall biological sources of these diterpenoids are the lipids of higher plant resins and reported to occur in lignite (Shaw et al., 1980). The dispersed occurrence of long-chain diterpenoids in geological sediments is due to tricyclohexaprenol as a membrane constituent of primitive bacteria.

The long-chain steroid compounds recorded in geological sediments are alkanes, alkenes, aromatic hydrocarbons, alcohols, ketones, esters, acids, and ethers (Mackenzie et al., 1982). Some among these components are reported to be synthesized biologically. More than 90% of various kinds of steroids have been reported from marine deposits in addition to unsaturation and alkylation. The sedimentary deposits are due to the deposition of organic matter from biological debris. In addition to this, the major class of sterols in marine sediments derived from the biological debris of algae; gorgosterol depositions from the diatom sediments originated by jellyfish which are not frequently occur in that region. Dinosterone are sterol one of the sterol ketones can be derived directly from organisms usually originated from dinoflagellates.

The hopanoids are the plethora among the triterpenoids found in the sediments of petroleum. The triterpenoids include alkanes, alkenes, aromatic hydrocarbons, alcohols, poly-alcohols, ketones, saturated and unsaturated acids, and aldehydes. The biolipid precursors such as hop-21-ene and hop-17(21)-ene are the sedimentary hopanoids identified from cyanobacteria (CB). The occurrence of these compounds strongly witnessing the hopanoids other than tetraol and pentaol derivatives (Ourisson et al., 1979). The degraded forms of triterpenoids are derived from higher plants and may be formed by photochemical or photomimetic processes (Corbet et al., 1980). The most abundant tetraterpenoids are carotenoids found in the sediments. Zullig in 1982 exemplified the huge deposition of carotenoids during the period of post-glacial lacustrine. Perhydro-β-carotane with saturated alkyl chains were reported to be witnessed in the ancient sediments of the Jurassic period might have occurred due to the accumulation of carotenoids. The phytoplankton origin of α- and γ-tocopherols in marine depositions was reported to obtain from thousands to 140 million years.

2.4 BIOLOGICAL MARKERS

The coorongite is an organic substance obtained as sheet-like deposits occur around the reservoirs in dried basins ephemeral lakes (Darwin River Reservoir, Australia) (Wake and Hillen, 1980; Dubreuil, 1989). The coorongite yield relatively plethora of highly aliphatic oils and these were obtained due to bloom of colonial microalgae *Botryococcus braunii* (Cane, 1977; Wake and Hillen, 1980; McKirdy, 1985; Dubreuil et al., 1989). A coorongite sample was reported to obtain in Lake Balkash, Kazakhstan, CIS, and was confirmed by ^{13}C-NMR, FTIR, Curie point pyrolysis-gas chromatography-mass spectrometry (Gatellier et al., 1993). The coorongite is composed of organic matter of the green microalga *Botryococcus braunii* in which homologous series of n-alkanes and n-alk-1-enes were reported; which are evidencing the accumulation of algaenan from *Botryococcus braunii* race A.

Saturated hydrocarbon fractions like pristine and phytane were obtained in the petroleum and oil shales of the sediments. The major precursor molecule for the saturated hydrocarbon fractions was phytol derived from chlorophyll a (Dean and Whitehead, 1961). The algal-derived liptinite with least amount of vitrinite and inertinite have been reported from the oil shales of Ethiopia and they contain long-chain aliphatic hydrocarbons with low sulfur content. Thus, the sediments can be further utilized to produce bio-fuel and gas (Wolela, 2006). Geologically, rich sources of hydrocarbons in the sediments obtain as dark brown and grey clays were reported by Demetrescu in 1998 during the Pliocene period as a process of South Carpathian's depression. The torbanites and crude oils of Mianas and Duri regions of Indonesia are the major sources of macrocyclic alkanes where witnesses to be produced by the race A and B of green microalgae *Botryococcus braunii*. These alkanes may be used as a biochemical marker for a distinct exploration of hydrocarbons (Audino et al., 2002). The ancient deep-seated marine facies depositions result as the largest reserve in China, a giant Anyue gas field in Sichuan Basin which has been opened for oil exploration (Du et al., 2014).

2.5   BIOFUELS

The shortage of fossil fuels, crude oil price hike, energy security, and accelerated global warming (GW) have led the pathway to search for renewable energy sources. In recent decades, the word 'Biofuel' has become the key factor to the alternative form of energy next to petroleum fuels which include biodiesel, bioethanol, and biomethane. Renewable carbon-neutral transport fuels are necessary for environmental and economic sustainability (Chisti, 2007). The biofuel is attractive towards the developed as well as developing nations to reduce reliance on other nations for energy and to save the globe against GHG (Koh and Ghazoul, 2008). The unprocessed raw biomasses implicated for energy such as burning to heat, and electricity is the primary biofuel. The processed biofuel like bioethanol and biodiesel are the secondary biofuel processed from biomasses. The secondary biofuel can be used for transportation and industrial purposes. The secondary biofuel is categorized into three generations; First generation biofuel, second-generation biofuel, and third-generation biofuel based on type feedstock or their level of development (Nigam and Singh, 2010). In addition, two additional traps were proposed in the North and South regions of Gaoshiti-Moxi region (Wei, 2018).

2.5.1   FIRST GENERATION BIOFUELS

The feedstock for first-generation biofuel includes soybean, rapeseed, oil palm, and mustard. But there are some disadvantages such as loss of biodiversity, excess utilization of water, and increasing GHG. In addition to this, the combustion of rapeseed exhibits nitrous oxide emission to atmospheric air (US EPA), biofuel from food crops like soybean and mustard can increase GHG along with deforestation and loss of biodiversity. The first-generation biofuel can cause serious problems to global food supply and can encourage farmers to farm for biofuel not for food. The World Bank had reported the hike in food prices up to 83% within 2005 and 2008. The food crop cultivation for biofuel requires specific types of farming which may serious issues to the environment. Some of the environmental issues are soil erosion, occurrence of pesticides, insecticides, and chemical fertilizers, eutrophica-tion, and shortage of water for food crop irrigation. However, the important negative issues are the low yield of biomass and loss of fertility of soil due to mono-crop cultivation. The first-generation biofuel yields very low energy which makes impractical for large scale cultivation (Lang et al., 2001).

Microalgae are a very good source for biodiesel production due to high yield and excessive accumulation of lipid molecules. Unlike other oil crop plants, the microalgae can double their biomass in a short period of time. The total lipid content constitutes more than 80% of the dry mass (Metting, 1996; Spolaore et al., 2006; Alam et al., 2012).

2.5.2   SECOND GENERATION BIOFUELS

The main goal of second-generation biofuel is to obtain biofuel from lignocel-luloses. It is the sustainable method of biofuel production by using non-food crop plants like Jatropha, Switchgrass, and residual part of crops including stems, leaves, and husks. The bioethanol production from lignocelluloses can indirectly reduce GHG up to 90% (Lang et al., 2001). Hence, the major problem in second-generation biofuel is utilizing important complex carbohydrates that are locked in plants. The enzymatic hydrolysis of lignocelluloses to bioethanol needs expensive processes and poses technical knowledge.

Second-generation fuels are not produced commercially, but there is a lot of research development going on in nations like the United States and some emerging economies like India, China, and Brazil in order to produce fuel in a more sustainable manner. More research and development of second-generation fuels can prove to be an eco-friendly alternative to conventional fuel production.

2.5.3   THIRD GENERATION BIOFUEL

Biodiesel from the biomass of microalgae is the third-generation biofuel production. It is a renewable, sustainable, and alternative method of biofuel production when compared other modes of biofuel and petroleum fuel production. The feedstock production involves water, sunlight, and CO_2 to grow the microalgae. The algae-for-fuel concept has gained renewed interest in recent decades with wide fluctuation in energy prices (Hu et al., 2007). Algae have many advantages over other conventional crops for biofuel production due to their simpler structure and high rate of biomass production followed by a high percentage of lipid accumulation (Becker, 1994). Approximately, 5000 to 20,000 US gallons of oil have been estimated to be synthesized by algae per acre per year (Maryking, 2007).

Large cultivation of algae also does not require large area as in the case of conventional crop plant cultivation and thereby ensures no competition with food crops (Tsukahara and Sawayama, 2005). The algae can tolerate saline or brackish water for mass cultivation and paved the way to safeguard the freshwater much needful for irrigating food crops. The mass cultivation of algae engulfs enormous amount of CO_2 from atmospheric air thereby reducing CO_2 (GHG) in air (Wang et al., 2009). Some of the biofuels synthesized from cultivating algae are biomethane, biohydrogen, bioethanol, and biodiesel. Biomethane is produced by anaerobic digestion of biomass (Spolaore et al., 2006), biohydrogen, and bioethanol are produced by photobiologically active algae (Kapdan and Kargi, 2006) and biodiesel from microalgal oil (Banerjee et al., 2002).

The microalgae are well known for producing various kinds of lipids, hydrocarbons, and other oils based on different species (Metzger and Largeau, 2005). But the great deal of attention was given to neutral lipids (triglycerides) due to low degree of unsaturation and viable diesel substitute (McGinnis et al., 1997). Some microalgal species can accumulate about 50–60% of neutral lipids of dry mass among triglycerides. Saturated and monounsaturated fatty acids are predominantly reported from microalgae oil (Borowitzka, 1988). The long-chain fatty acids from C_{16} to C_{18} are identical to conventional oleaginous crops (Ohlrogge and Browse, 1995). Long-chain fatty acids can be employed in the production of jet fuels. The Department of Defense, US-sponsored research on the production of jet fuel (JP-8) from microalgae. The microalgae have their own tribute as a very good source for energy production in the biofuel market (Burlew, 1953; Sheehan et al., 1998; Jung and Choe, 2002). The microalgae are photoautotrophic in nature and are the promising and auspicious method for renewable sources of energy production (Sawayama et al., 1999; Tsukahara et al., 2001).

2.6 MICROALGAE

The microalgae are the very old green photoautotrophic microorganisms still exist on the Earth both as prokaryotes and eukaryotes (Song et al., 2008).

They are well known for their fast growth more than a hundred times with high doubling time (DT) when compared with higher plants (Tredici, 2010). Due to the large quantity of lipid accumulation in the cells of microalgae, they are a sustainable and renewable source for biodiesel production (Chisti, 2007). The biomass production of microalgae was found to be in a range from 4.5 to 7.5 ton/ha/year and comprised of approximately 30% of lipid (Tsuka-hara and Sawayama, 2005). This was comparably high when compared to the higher plants (Soybean: 0.4 ton/ha/year; rapeseed: 0.68 ton/ha/year; oil palm: 3.62 ton/ha/year; Jatropha: 4.14 ton/ha/year) (Chisti, 2007; Lam and Lee, 2012). Therefore, biofuel from microalgae is considered an eco-friendly and economically viable sources other than biofuel from food crops which may hike food crop expense (Mitchell, 2008). Among the entire photosyn-thetic biological source, algae alone produce a large amount of lipid in their cell content. Photoautotrophic mode of lipid accumulation from the aquatic ecosystem is a major concept in the microalga biodiesel production. They are the primary producers of the aquatic ecosystems and thus provide energy to zooplanktons and fishes. Due to the accumulation of a large amount of neutral lipids, they are the best alternate for fossil fuel (Casadevall et al., 1985; McGinnis et al., 1997).

At the same time, the biomass production requires very limited land area for cultivation of microalgae than the higher plants. In adding to this, the microalga biomass can also be useful as a premier feedstock for bioethanol production (Harun et al., 2010). The microalgae are the auspicious, sustain-able, and renewable source for biofuel production; because, the microalgae can sequester atmospheric CO_2 from flue gas and can utilize solar energy higher (10–15 times) than terrestrial plants (Li et al., 2008; Khan et al., 2009). The microalgae are the promising sustainable energy source for biofuel production in the near future and need to be studied extensively.

2.7 BOTRYOCOCCUS BRAUNII

The *Botryococcus braunii* is a green, colonial, unicellular, photosynthetic microalga grouped under the member Trebouxiophyceae (Senousy et al., 2004). The microalga *B. braunii* was segregated closely to its neighbor *Characium vaculatum* and *Dunaliella parva* based on the 16S rRNA (small subunit) sequencing (Sawayama et al., 1995). The three chemical races of *B. branii* have formed a monophyletic group by analyzing the 18S rRNA gene sequences of four different *B. braunii* strains (Senousy et al., 2004). This microalga is widespread, viz. cosmopolitan and lives in both fresh and brackish waters of all continents (Chisti, 1980). The universal occurrence of *B. braunii* was witnessed by the strains from the USA (Wolf et al., 1985), Portugal, Bolivia, France, Ivory Coast, Morocco, Philippines, Thailand, and West Indies (Metzger et al., 1985), India (Metzger et al., 1997; Dayananda et al., 2007), Japan (Okada et al., 1995). Thus, supporting the microalga can adapt themselves in temperate, tropical, and alpine zones of the Universe.

The hydrocarbons yielded from the *Botryococcus mahabali* served as a biochemical marker for the Indian isolate to sort out the genus *Botryococcus* (Dayananda et al., 2007). Naturally occurring blooms by the proliferation of the microalga *B. braunii* was reported in lakes and shore deposits which yield highly aliphatic oils (Wake and Hillen, 1981). The accumulation of lipid differs among various strains of *B. braunii* (Singh and Kumar, 1994; Pal et al., 1998) and about thirteen new strains were reported by Komarek and Marvan (1992). Generally, *B. braunii* exists as three various races which are A, B, and L and are segregated based on the different kinds of hydrocarbon accumulation. *B. braunii* race A can synthesize C_{25}–C_{31} odd-numbered n-alkadienes and alkatrienes. Whereas the race B accumulates polymethylated unsaturated triterpenes known as botryococcenes ($C_nH_{2n-10 n=30-37}$). The L race is responsible for the production of a single hydrocarbon $C_{40}H_{78}$ a tetraterpene called lycopadiene (Metzger et al., 1990). A free fatty acid mixture was reported from the bloom of *B. braunii* which are α-linolenic acid (ALA), oleic acid, linolic acid, and palmitic acid. Specifically, oleic acid and ALA were the prominent fatty acid recorded from *B. braunii* (Chiang et al., 2004).

Metzger et al. (1988) discovered the morphological differences between the three races of *B. braunii* based on the cell size and color. Thus, it was described that the race L was 8 to 9 μm × 5μm in size which was comparatively smaller than the other two races A and B which was 13 μm × 7–9 μm in size. In addition to this, the race L consists of low content of pyrenoid. The *B. braunii* races B and L are reddish-orange and orange-brownish in the active state from green color, while the race A was pale yellow in the active state. These changes occur due to the accumulation of keto-carotenoids including canthaxanthin, echinenone, adonixanthin during the stationary phase. The carotenoid production during red pigmentation was directly proportional to the quantitative accumulation of lipid in *B. braunii* (Aaronson et al., 1983). The *B. braunii* races A and B accumulate long-chain aliphatic compounds cross-linked with ether bridges whereas, the race L produces hydrocarbon based on lycopadiene cross-linked with ether bridges. Therefore, the high content of hydrocarbon producing races A and B are much suitable for commercial production of hydrocarbon-based fuel.

Basically, *Botryococcus braunii* is a photosynthetic microalga reported to minimize CO_2 emissions by 1.5×10^5 tons/yr and 8.4×10^3 ha (Sawayama et al., 1999). As a result, it is considered as a promising candidate for sustainable, renewable, and eco-friendly production of lipid including other biological compounds by mitigation of carbon dioxide.

In recent decades, especially the *Botryococcus braunii* race B has more attention due to the occurrence of this ancient algal sediments dating back up to 500 million years ago and are considered as major contributor of coal and oil shale deposits on Earth (Derenne et al., 1997; Glikson et al., 1989; Mastalerz and Hower, 1996).

2.8   *BOTRYOCOCCUS BRAUNII* AS A BIOLOGICAL MARKER

The green microalga *Botryococcus braunii* are the very indicators of hydro-carbon accumulation from the sediments, because, the hydrocarbons were once biologically synthesized by its own. Generally, three different types of hydrocarbons have been reported to be obtained as depositions which are *cis* and *trans* C_{25}–C_{31} n-alkanes and a C_{29} triene; a series of $C_{34}H_{58}$ botryococ-cene and $C_{40}H_{78}$ *trans, trans*-lycopadiene recorded in different races of the green microalga *Botryococcus braunii* A, B, and L respectively (Zhang et al., 2007). Among the three kinds of hydrocarbons, botryococcene, and lycopa-diene are most the abundant hydrocarbons obtained from dry sediments from the freshwater lake El Junco, Galapagos (Zhang et al., 2007).

This was the first-ever report on the occupancy of three races of *B. braunii* along with the three hydrocarbons which were co-existed during the last 460 years. These highly reactive hydrocarbon compounds were evidenced from the water column and sediments under well-oxygenated water for many centuries of years. The lycopadienes were first time reported from the lacus-trine sediments including lycopadiene ($C_{40}H_{78}$) and lycopatriene ($C_{40}H_{76}$) isomers. The plethora of botryococcene was seen in the upper sedimentary regions illustrating the latest depositions of the last 460 years and the occur-rence of *B. braunii* race B. Cyclic or acyclic botryococcenes along with botryococcanes and methyl squalenes were discovered from the lacustrine sediments and crude oils are the very good example of the existence of the green microalga *B. braunii* race B (Moldowan and Seifert, 1980; Brasell et al., 1985; Huang et al., 1999; Summons et al., 2002).

Gatellier et al. (1993) discovered the presence of alkadienes and trienes (C_{27}, C_{29}, and C_{31}) in the flexible organic matter from the Lake Balkash co-existed with *B. braunii* race A. Similarly, monoaromaticlycopane (C_{40}) derivatives are due to the occurrence of *B. braunii* race L. (Derenne et al., 1994; Adam et al., 2006). In addition to this, macrocyclic alkanes and their methylated forms were revealed from Torbanites and crude oils which are originated by *B. braunii* race A (Audino et al., 2002). The *B. braunii* races A and B along with their respective hydrocarbons were discovered from the Lake Overjuyo, Bolivia (Metzger et al., 1989) and also in Devilbend Reservoir, Australia (Wake and Hillen, 1981). Simultaneously, *B. braunii* races B and L with their hydrocarbons were reported from the Lake Yamoussoukro, Ivory Coast (Metzger et al., 1990), and in Khao Kho Hong, Thailand (Metzger and Casadevall, 1987). The sulfur-containing hydrocarbons alkadiene, botryococcene, and lycopadiene were first time reported in the Miocene lacustrine formation sediments co-existing with the three *B. braunii* races (Grice et al., 1998). The persistence of *B. braunii* in the geological past was the well-figured witness for the co-occurrence of botryococcene in sediments and petroleum (Moldowan and Seifert, 1980; Brassell et al., 1985; Grice et al., 1998; Smittenberg et al., 2005).

The hydrogenated product of botryococcene, botryococcane ($C_{34}H_{70}$) was reported to occur in the two crude oil shales of Sumatra (Moldowan and Seifert, 1980). Comparatively, botryococcane was also reported in the range along the Coastal bitumen formations in Australia (McKirdy et al., 1986). About 90% of the microalga *B. braunii* remains were found in the Glen Davis torbanite of Sydney Basin, Australia (Derenne et al., 1988). Lichitfouse et al. (1994) discovered the C_{27}, C_{29}, C_{31} alkanes from the Pula sediment of Pliocene oil shale of Hungary and revealed that the microalga *B. braunii* as their significant source. A C_{30} botryococcene alike compound 7,11-cyclobotryococca 5,12,26-triene was discovered in the organic sediments of the Lake Cadagno, Switzerland, and illustrate the cosmopolitan nature of the microalga *B. braunii* from the past environments (Behrens et al., 2000). The Miocene sediments from the Western Woodlark Basin, Southwest Pacific revealed the presence of the depositions of the microalga *B. braunii* (Testa et al., 2001).

The green microalga *B. braunii* has attained a mass of attraction towards scientists due to the paleobotanical evidences on hydrocarbon accumulation. The fossil studies on the ancient sediments exemplify that the microalga *B. braunii* is an important microorganism meant for lipid-rich deposition of hydrocarbons dating from Ordovician period and still exists (Cane, 1977). The different stages of life cycle with respect the microalga *B. braunii* were revealed from the sediments of Pliocene-Miocene succession in Romania. Therefore, the associated presence of both the microalga *B. braunii* and their respective hydrocarbons accentuate the major task of this important micro-alga for hydrocarbon exploration (Demetrescu, 1998). Here in this chapter, the green microalga *Botryococcus braunii* was used as both biological marker and biochemical marker for the hydrocarbon exploration from fresh and marine, water, and soil samples.

2.9   HYDROCARBONS FROM *BOTRYOCOCCUS BRAUNII*

2.9.1   *RACE A*

Generally, the race A produces odd-numbered compounds from monoenes to tetraenes with a terminal unsaturation. About 50 different hydrocarbons have been witnessed from the wild and cultivating strains of *B. braunii* race A among them 30 hydrocarbons were determined structurally (Metzger and Largeau, 1999). The various kinds of hydrocarbon production are controlled by genetic factors (Metzger et al., 1986). The dienes and trienes are the most common hydrocarbons along with a mid-chain unsaturation in *cis* configuration (Metzger and Largeau, 1999) and two conjugated mid-chain unsaturations with two conjugated double bonds at terminal position (Metzger, 1993), respectively. The tetraenes constitute three conjugated double bonds at the terminal position and consist of only one type (Metzger, 1993). Radio-labeled analysis on the hydrocarbons confirmed that the oleic acid is the major precursor to produce dienes and trienes (Templier et al., 1987).

2.9.2   *RACE B*

The botryococcenes are the only and specific type of hydrocarbon produced *Botryococcus braunii* race B which comprised of cyclic and acyclic compounds. Only 15 botryococcenes were structurally identified among the 50 different kinds of botryococcenes because of the problem in purification process (Metzger and Largeau, 1999). Cox et al. (1973) first time determined a C_{34} compound of botryococcene from a wild *B. braunii* sample followed by five different C_{34} isomers were elucidated till date. The different types of botryococcene distribution in cells are due to strain-dependent (Metzger et al., 1985). The race B of *B. braunii* has a special characteristic feature in producing squalene (C_{31}–C_{34} methylated squalenes) (Huang and Poulter, 1989; Metzger and Largeau, 1999). Very low amount of squalene was reported but recorded high in Bolivian strain (4.5% of dry biomass) (Achitouv et al., 2004).

2.9.3 RACE L

The lycopadiene is the only hydrocarbon synthesized by the race L reported from the strains of Thailand and Ivory Coast (Metzger and Largeau, 1999). An isomer of undiscovered structure was witnessed from the two strains of India (Metzger et al., 1997). A new type of isoprenoid hydrocarbon compound known as Brauniiceae was the first time discovered from *B. braunii* strain of Berkeley (Huang et al., 1987). The three hydrocarbons and its conversion of fuel can be much suitable for combustion engines as an important constituent of petroleum and coal deposits (Audino et al., 2002; Banerjee et al., 2002; Summons et al., 2002). But especially the botryococcene from *B. braunii* race B can be converted into high octane gasoline, kerosene, and diesel fuels (Banerjee et al., 2002).

2.10 HYDROCARBON AND FATTY ACID ANALYSIS

About 50% of hydrocarbons were reported from the *B. braunii* by analyzing the hydrocarbon in GC-MS. In which carbon chain length from C_8 (octane) to C_{40} (lycopadiene) were obtained. Alkanes or alkenes and branched hydrocarbons were eluted with significant retention time intervals around 80 s in comparison with the authentic standards. The *Botryococcus braunii* race A accumulates 45.89% of hydrocarbons among the total compounds extracted. Same way, *B. braunii* race B accumulates 52.70% of hydrocarbons among the extracted compounds (Barupal et al., 2010).

A significant distribution of monomethylalkanes (MMA) ranged from C_{23} to C_{31+} was reported from kerogen an organic matter of Australian and African torbanites. Molecular carbon isotope study revealed that the MMA series was due to the occurrence of *B. braunii* race A based on the structural similarities of botryals (Audino et al., 2001). The *B. braunii* race A was resulted in straight chain hydrocarbons which are dominated by alkadienes and a triene series of odd-numbered C_{25}–C_{31} doublets *cis*/*trans* isomers with a terminal double bond and a mid-chain unsaturation. In addition, two main hydrocarbons were found in race A either a mixture of *cis* dienes and trienes or a mixture of *cis*/*trans* but without triene (Metzger et al., 1997). The high content of lycopadiene was resulted and witnessed from hydrocarbons as *trans*, *trans*-lycopadiene based on authentic standard (Metzger et al., 1997).

Four different monoepoxides and one diepoxide were reported and identified from *B. braunii* (Metzger and Casadevall, 1989; Delahais and Metzger, 1997; Metzger, 1999). *B. braunii* race L showed C_{40} monoepoxide derived from *trans, trans*-lycopadiene along with low quantities of diepoxy-lycopane. The three mid-chain epoxide groups derived from specific n-alkanes was reported from *B. braunii* A. The C_{29} epoxide was much frequent among the three followed by C_{27} and C_{31} (Zhang et al., 2007). The significant amount of botryococcenes was obtained with C_{34} isomers. Two different isomers identified as botryococcenes III and IV obtained as a product of *B. braunii* race B when compared with authentic standards. The C_{34} botryococcene III was reported in large amount from the wild strain of *B. braunii* in Paquemar, Martinique (Metzger et al., 1985) whereas C_{34} botryococcene IV was reported from a *B. braunii* race B of Japan (Okada et al., 1997).

The fatty acid and hydrocarbon composition of *Botryococcus braunii* isolated from the Shira Lake agreed well with published known strains of *Botryococcus* race A. This *Botryococcus* race synthesizes straight-chain C_{23}–C_{33} dienoic and trienoic hydrocarbons with an odd number of carbon atoms. This fact made it possible to assign the organism isolated for the first time from the Shira Lake to the *B. braunii* A race (Volova et al., 2003). The organic-rich compounds extracted from the coorongite of Balkashite sample seems to be rich in high atomic H/C ratio evidenced that the sample is rich in aliphatic compounds. The low contributions of aromatic compounds were represented by hydroxyl, carbonyl, ether, and carboxylic groups. The FTIR, GC-MS, and ^{13}C NMR results were comparatively similar to the algaenans derived from the *B. braunii* race A (Berkaloff et al., 1984; Kadouri et al., 1988). The occurrence of a series of n-alkanes, n-alkenes along with n-alkones represent the fraction compounds. In addition, a series of n-α, ω-alkadienes were also reported. The high molecular weight compounds are due to the result of accumulation of algaenan compounds from *B. braunii* race A (Metzger et al., 1991; Templier et al., 1992). The same way coorongite of Balkashite sample also rich in algaenan type of compounds and assumed to be originated by the Race A of *B. braunii* (Gatellier et al., 1993). A recent study over hydrocarbon exploration reveals that the soil samples from coastal sites are highly rich in hydrocarbon sediments than river bed sites, resulted due to high biochemical depositions of algae (Sangeetha et al., 2017).

2.11   SQUALENE SYNTHASE

The squalene synthase otherwise called farnesyl diphosphate (FPP) farnesyl transferase (EC: 2.5.1.21) which first initiates the catalyzation reaction of isoprenoid pathway towards the production of phytosterols, brassino steroids, and triterpenoids (Abe et al., 1993). As a membrane-bound enzyme, the squalene synthase condensates the two molecules of FPP to produce a linear 30 carbon compound called squalene. The squalene synthase located with the smooth endoplasmic reticulum (ER), with which the carboxyl-terminal fixed to the membrane of ER and the amino-terminal (catalytic site) found at the cytoplasm (Robinson et al., 1993).

The triterpenoids are the secondary metabolites biologically synthesized by the squalene synthase during mevalonate (MVA) pathway and the enzyme squalene synthase is considered to play a significant role in isoprenoid synthesis in Eukaryotes (Hanley et al., 1996). The squalene synthase plays a dual role by condensing two molecules of FPP to form presqualene diphosphate (PSPP) and then it converts to squalene by uptaking NADPH (nicotinamide adenine dinucleotide phosphate) and Mg^{2+} (Lee and Poulter, 2008). The squalene synthase genes have been overexpressed in *Panax ginseng* (Lee et al., 2004) and *Eleutherococcus senticosus* (Seoet al., 2005) and have led to the induced production of phytosterols and triterpenes. Some of the plant squalene synthase genes have been characterized and studied from *Nicotiana tabacum* (Devarenne et al., 1998) and *Glycyrrhiza glabra* (Hayashi et al., 1999).

The B race of *Botryococcus braunii* synthesizes triterpenoids like squalene derivative tetramethyl squalene (Huang and Poulter, 1989), squalene epoxide (Delahais and Metzger, 1997), botryoxanthins (Okada et al., 1996) and braunixanthins (Casadevall et al., 1984) as their secondary metabolites in the extracellular matrix. The C_{30} botryococcene is not the final or the dominant product present in the extracellular matrix of the B race of *Botryococcus braunii*. The radio-labeled methionine or carbon dioxide supplementation to the B race of *Botryococcus braunii* revealed that the C_{30} botryococcene is the initial raw material for the production of all the botryococcenes and which can be converted into homologous a maximum of 34 methylation with S-adenosyl methionine (Casadevall et al., 1984; Wolf et al., 1985; Metzger et al., 1987).

The Botryococcene is the precursor or the raw material for the production of hydrocarbons, the hydrolysis of the linear chain branched botryococcene provide hydrocarbon fuel much suitable for the combustion of engines (Hillen et al., 1982). This innovative and potential implementation of botryococcenes as a renewable energy source paves a new route on the studies of the physico-chemical, molecular, and biosynthesis of botryococcenes. The molecular structural similarities between the squalene and C_{30} botryococcene have resulted in the comparison of both the squalene and C_{30} botryococcene biosynthesis. Both the botryococcene and the squalene are C_{30} compounds along with a backbone of two C_{15} farnesyl residues. Henceforth, there are some differences between both of them which are 10-1 linkage of squalene and 10-3 linkage of botryococcene between the farnesyl residues.

There are some variations in the mechanism of both the enzymes responsible for the biosynthesis of squalene and botryococcene. Squalene can be synthesized by the enzyme squalene synthase which involves two-step reactions; first, it condenses two molecules of FPP to form PSPP (Robinson et al., 1993; Gu et al., 1998) and in the second step, it cleaves the cyclopropane ring of PSPP to form 10-1 linkage by the reduction of NADPH. Similarly, C_{30} botryococcene also synthesized by a two-step reaction, in which the second step; the cleavage of the cyclopropane ring of PSPP alone differs to form 10-3 linkage. Many studies still do not justify whether there are two different enzymes exists for the biosynthesis of squalene and botryococcene or the same enzyme may produce two different by-products due to cellular signals. Some studies on the synthesis of squalene reported that the recombinant squalene synthase enzyme from yeast was incubated in the absence of NADPH have synthesized 10-3 linked (10S, 13S)-10-hydroxybotryococcene in addition to the two 10-1 linked squalene derivatives (Z)-dehydrosqualene and (R)-12-hydroxysqualene (Jarstfer et al., 2002). This resulted that the squalene synthase enzyme under various conditions can synthesize botryococcene. At the same time, there is a separate enzyme found in the microalgae *Botryococcus braunii* for the C_{30} botryococcene biosynthesis.

The microalga *Botryococcus braunii* also possesses three unique squalene synthase-like (SSL) proteins in addition to squalene synthase; they are SSL-1, SSL-2, and SSL-3 (Niehaus et al., 2011). The SSL-1 and SSL-2 are involved in the biosynthesis of squalene whereas the SSL-3 is involved in the biosynthesis of C_{30} botryococcene. The SSL-1 catalyzes the first step of the reaction carried out by squalene synthase by converting FPP into PSPP; the SSL-2 converts the PSPP into squalene and the SSL-3 converts the PSPP into C_{30} botryococcene. At this stage, the SSL-2 and SSL-3 are the key enzymes in order to determine the synthesis of squalene or C_{30} botryococcene biosynthesis respectively. The C_{30} botryococcene can be further methylated from C_{31} to C_{37} botryococcenes (Niehaus et al., 2012) to extract free hydrocarbons as biofuels. Therefore, there are two different ways to synthesize squalene by either the squalene synthase or SSL-1 and SSL-2 genes in the microalga *Botryococcus braunii*.

2.12   CONCLUSION AND FUTURE PERSPECTIVES

Excess amount of capital is being spent on oil and hydrocarbon exploration. Therefore, versatile and feasible exploration methods are needed to cut-short the expenditure. Developing a biological and biochemical marker will be very effective and reported in various parts of the world. By highlighting this aspect, this review constitutes about the hydrocarbons of marine alga as a biochemical marker and the occurrence of marine alga *Botryococcus braunii* itself as a biological marker for oil and hydrocarbon exploration. Developing a PCR based low cost and least time-consuming technique may also be highly effective for hydrocarbon exploration based on the marker genes described above. However, prior to hydrocarbon and oil exploration, testing for these features may also be effective for the successful building of an oil exploration plant.

KEYWORDS

- alkadiene
- biomarker
- botryococcene
- *Botryococcus braunii*
- coorongite
- hydrocarbons
- kerogen
- lycopadiene
- microalgae
- oil exploration
- oil shale
- torbanite

REFERENCES

Aaronson, S., Berner, T., Gold, K., Kushner, L., Patni, N. J., Repak, A., & Rubin, D., (1983). Some observations on the green planktonic alga, *Botryococcus braunii* and its bloom form. *J. Plankton Res., 5*, 693–700.

Abe, I., Rohmer, M., & Preswich, G. D., (1993). Enzymatic cyclization of squalene and oxidosqualene to sterols and triterpenes. *Chem. Rev., 93*, 2189–2206.

Achitouv, E., Metzger, P., Rager, M. N., & Largeau, C., (2004). C_{31}–C_{34} methylated squalenes from a Bolivian strain of *Botryococcus braunii*. *Phytochemistry, 65*, 3159–3165.

Adam, P., Schaeffer, P., & Albrecht, P., (2006). C40 monoaromaticlycopane derivatives as indicators of the contribution of the alga *Botryococcus braunii* race L to the organic matter of Messel oil shale (Eocene, Germany). *Organic Geochem., 37*, 585–596.

Alam, F., Date, A., Rasjidin, R., Mobin, S., Moria, H., & Baqui, A., (2012). Biofuel from algae-is it a viable alternative? *Procedia Engineering, 49*, 221–227.

Alexander, R., Kagi, R. I., & Noble, R. A., (1983). Identification of the bicyclic sesquiterpenes drimane and eudesmane in Petroleum. *J. C. S. Chem. Commun., 0*, 226–228.

Audino, B., Rohit, S., Yusuf, C., & Banerjee, U. C., (2002). *Botryococcus braunii.* A renewable source of hydrocarbons and other chemicals. *Crit. Rev. Biotechnol., 22*, 245–279.

Audino, M., Grice, K., Alexander, R., Boreham, C. J., & Kagi, R. I., (2001). Unusual distribution of monomethylalkanes in *Botryococcus braunii*-rich samples: Origin and significance. *Geochimica et Cosmochimica Acta, 65*(12), 1995–2006.

Banerjee, A., Sharma, R., Chisti, Y., & Banerjee, U. C., (2002). *Botryococcus braunii*: A renewable source of hydrocarbons and other chemicals. *Crit. Rev. Biotechnol., 22*(3), 245–279.

Barupal, D. K., Kind, T., Kothari, S. L., & Fiehn, O., (2010). Hydrocarbon phenotyping of algal species using pyrolysis-gas chromatography mass spectrometry. *BMC Biotechnology, 10*, 40. https://doi.org/10.1186/1472-6750-10-40 (accessed on 26 May 2020).

Behrens, A., Schaeffer, P., Bernasconi, S., & Alrecht, P., (2000). 7,11-cyclobotryococca-5,12,26-triene a novel botryococcene related hydrocarbon occurring in environment. *Organic Lett., 2*, 1271–1274.

Berkaloff, C., Rousseau, B., Coute, A., Casadevall, E., Metzger, P., & Chirac, C., (1984). Variability of cell wall structure and hydrocarbon type in different strains of *Botryococcus braunii. J. Phycol., 20*, 377–389.

Borowitzka, M. A., (1988). Fats, oils, and hydrocarbons. In: Borowitzka, M. A., & Borowitzka, L. J., (eds.), *Micro-Algal Biotechnology* (pp. 257–287). Cambridge University Press, Cambridge, UK.

Brassell, S. C., Eglinton, G., & Fu, J. M., (1985). Biological marker compounds as indicators of the depositional history of the Maoming oil shale. *Org. Geochem., 10*, 927–941.

Brassell, S. C., Eglinton, G., & Maxwell, J. R., (1983). *The Geochemistry of Terpenoids and Steroids* (pp. 575–586). 603rd Meeting, Liverpool.

Burlew, J., (1953). *Algae Culture: From Laboratory to Pilot Plant.* Carnegie Institute Washington.

Cane, R. F., (1977). Coorongite, balkashite and related substances-an annotated bibliography. *Trans. R. Soc. S. Aust., 101*, 153–164.

Casadevall, E., Dif, D., Largeau, C., Gudin, C., Chaument, D., & Desumit, O., (1985). Studies on batch and continuous cultures of *Botryococcus braunii* hydrocarbon production in relation to physiological state, cell ultra-structure, and phosphate nutrition. *Biotechnol. Bioeng., 27*, 286–295.

Casadevall, E., Kadouri, A., & Metzger, P., (1984). In: Schenk, P. A., De Leeuw, J. W., & Lijmbach, G. W. M., (1983), (eds.), *Advances in Geochemistry* (pp. 372–332), Pergamon Press, Oxford, UK.

Casadevall, E., Metzger, P., & Puech, M. P., (1984). Biosynthesis of triterpenoid hydrocarbons in the alga *Botryococcus braunii. Tetrahedron Lett., 25*, 4123–4126.

Chappe, B., Albrecht, P., & Michaelis, W., (1982). Polar lipids of archaebacteria in sediments and petroleums. *Science, 217*, 65–66.

Chiang, I. Z., Huang, W. Y., & Wu, J. T., (2004). Allelochemicals of *Botryococcus braunii* (Chlorophyceae). *J. Phycol., 40*, 474–480.

Chisti, Y., (1980). An unusual hydrocarbon. *J. Ramsay Soc., 27, 28*, 24–26.

Chisti, Y., (2007). Biodiesel from microalgae. *Biotechnol. Adv., 25*, 294–306.

Corbet, B., Albrecht, P., & Ourisson, G., (1980). Photochemical or photomimetic fossil triterpenoids in sediments and petroleum. *J. Am. Chem. Soc., 102*, 1171–1173.

Dayananda, C., Sarada, R., Usha, R. M., Shamala, T. R., & Ravishankar, G. A., (2007). Autotrophic cultivation of *Botryococcus braunii* for the production of hydrocarbons and exopolysaccharides in various media. *Biomass Bioenergy, 31*, 87–93.

Dean, R. A., & Whitehead, E. V., (1961). The occurrence of phytane in petroleum. *Tetrahedron Letters, 21*, 768–770.

Delahais, V., & Metzger, P., (1997). Four polymethylsqualene epoxides and one acyclic tetraterpene epoxide from *Botryococcus braunii*. *Phytochemistry, 44*, 671–678.

Demetrescu, E., (1998). The chloroccalean alga *Botryococcus* and its significance in hydrocarbon exploration. *Proc. Intern. Workshop on "Modern and Ancient Sedimentary Environments and Processes"* (pp. 155–160). Moeciu, Romania.

Derenne, S., Largeau, C., & Behar, F., (1994). Low polarity pyrolysis products of Permian to recent *Botryococcus*-rich sediments: First evidence for the contribution of an isoprenoid algaenan to kerogen formation. *Geochimica et Cosmochimica Acta, 58*, 3703–3711.

Derenne, S., Largeau, C., Casadevall, E., & Connan, J., (1988). Comparison torbanites of various origins and evolutionary stages. Bacterial contribution to their formation. Cause of the lack of botryococcane in bitumens. *Org. Geochem., 12*, 43–59.

Derenne, S., Largeau, C., Hetenyi, M., Brukner-Wein, A., Connan, J., & Lugardon, B., (1997). Chemical structure of the organic matter in a pliocenemaartype shale: Implicated *Botryococcus* race strains and formation pathways. *Geochim. Cosmochim. Acta, 61*, 1879–1889.

Devarenne, T. P., Shin, D. H., Back, K., Yin, S., & Chappell, J., (1998). Molecular characterization of tobacco squalene synthase and regulation in response to fungal elicitor. *Arch. Biochem. Biophys., 349*, 205–215.

Du, J., Zou, C., Xu, C., He, H., Shen, P., Yang, Y., Li, Y., et al., (2014). Theoretical and technical innovations in strategic discovery of a giant gas field in Cambrian Longwangmiao formation of central Sichuan paleo-uplift, Sichuan Basin. *Petroleum Exploration and Development, 41*(3), 268–277.

Dubreuil, C., Derenne, S., Largeau, C., Berkaloff, C., & Rousseau, B., (1989). Mechanism of formation and chemical structure of Coorongite-1. Role of the resistant biopolymer and of the hydrocarbons of *Botryococcus braunii*. Ultra structure of Coorongite and its relationship with Torbanite. *Org. Geochem., 14*, 543–553.

Gatellier, J. P. L. A., De Leeuw, J. W., Sinninghe, D. J. S., Derenne, S., Largeau, C., & Metzger, P., (1993). A comparative study of macromolecular substances of a Coorongite and cell walls of the extant alga *Botryococcus braunii*. *Geochim. Cosmochim. Acta, 57*, 2053–2068.

Glikson, M., Lindsay, K., & Snxav, J., (1989). *Botryococcus*-a planktonic green alga, the source of petroleum through ages: TEM studies of oil shales and petroleum source rocks. *Org. Geochem., 14*, 505–608.

Grice, K., Schouten, S., Nissenbaum, A., Charrach, J., & Sinninghe, D. J. S., (1998). A remarkable paradox: Sulfurised freshwater algal (*Botryococcus braunii*) lipids in a ancient hypersaline euxinic ecosystem. *Org. Geochem., 24*, 195–216.

Gu, P., Ishii, Y., Spencer, T. A., & Shechter, I., (1998). Function-structure studies and identification of three enzyme domains involved in the catalytic activity in rat hepatic squalene synthase. *J. Biol. Chem., 273*, 12515–12525.

Hanley, K. M., Nicolas, O., Donaldson, T. B., Smith-Monroy, C., Robinson, G. W., & Hellmann, G. M., (1996). Molecular cloning, *in vitro* expression and characterization of a plant squalene synthetase cDNA. *Plant Mol. Biol., 30*, 1139–1151.

Harun, R., Danquah, M. K., & Forde, G. M., (2010). Microalgal biomass as a fermentation feedstock for bioethanol production. *J. Chem. Technol. Biotechnol., 85*, 199–203.

Hayashi, H., Hirota, A., Hiraoka, N., & Ikeshiro, Y., (1999). Molecular cloning and characterization of two cDNAs for *Glycyrrhiza glabra* squalene synthase. *Biol. Pharm. Bull., 22*, 947–950.

Hillen, L. W., Pollard, G., Wake, L. V., & White, N., (1982). Hydrocracking of the oils of *Botryococcus braunii* to transport fuels. *Biotechnol. Bioeng., 24*, 193–205.

Hu, Q., Kurano, N., Kawachi, M., Iwasaki, I., & Miyachi, S., (2007). Ultrahigh-cell-density culture of a marine green alga *Chlorococcum littorale* in a flat-plate photobioreactor. *Appl. Microb. Biotechnol., 49*, 655–662.

Huang, Y., Street-Perrott, F. A., Perrott, R. A., Metzger, P., & Eglinton, G., (1999). Glacial-interglacial environmental changes inferred from molecular and compound-specific $\delta^{13}C$ analyses of sediments from Sacred Lake, Mt. Kenya. *Geochim. Cosmochim., 63*, 1383–1404.

Huang, Z., & Poulter, C. D., (1989). Tetramethylsqualene, a triterpene from *Botryococcus braunii* var. Showa. *Phytochemistry, 28*, 1467–1470.

Jarstfer, M. B., Zhang, D. L., & Poulter, C. D., (2002). Recombinant squalene synthase. Synthesis of non-head-to-tail isoprenoids in the absence of NADPH. *J. Am. Chem. Soc., 124*, 8834–8845.

Jung, I. H., & Choe, S. H., (2002). Growth inhibition of freshwater algae by ester compounds released from rotted plants. *Journal of Industrial and Engineering Chemistry, 8*(4), 297–304.

Kadouri, A., Derenne, S., Largeau, C., Casadevall, E., & Berkaloff, C., (1998). Resistant biopolymer in the outer walls of *Botryococcus braunii*, B race. *Phytochem., 27*, 551–557.

Kapdan, I. K., & Kargi, F., (2006). Bio-hydrogen production from waste materials. *Enzyme Microb. Technol., 38*, 569–582.

Khan, S. A., Rashmi, Hussain, M. Z., Prasad, S., & Banerjee, U. C., (2009). Prospects of biodiesel production from microalgae in India. *Renew. Sustain. Energy Rev., 13*, 2361–2372.

Koh, L. P., & Ghazoul, J., (2008). Biofuels, biodiversity and people understanding the conflicts and finding opportunities. *Biol. Conserv., 141*, 2450–2460.

Komarek, J., & Marvan, P., (1992). Morphological differences in natural populations of the genus *Botryococcus* (Chlorophyceae). *Arch. Protisem., 14*, 65–100.

Lam, M. K., & Lee, K. T., (2012). Microalgae biofuels: A critical review of issues, problems and the way forward. *Biotechnology Advances, 30*, 673–690.

Lang, X., Dalai, A. K., & Bakhshi, N. N., (2001). Preparation and characterization of bio-diesels from various bio-oils. *Bio Resour. Technol., 80*, 53–62.

Lee, M. H., Jeong, J. H., Seo, J. W., Shin, C. G., Kim, Y. S., & In, J. G., (2004). Enhanced triterpene and phytosterol biosynthesis in *Panax ginseng* over expressing squalene synthase gene. *Plant Cell Physiol., 45*, 976–984.

Lee, S., & Poulter, C. D., (2008). Cloning, solubilization, and characterization of squalene synthase from *Thermosynechococcus elongatus* BP-1. *J. Bacteriol., 190*, 3808–3816.

Li, Y., Horsman, M., Wu, N., Lan, C. Q., & Dubois-Calero, N., (2008). Biofuels from microalgae. *Biotechnology Progress, 24*(4), 815–820.

Lichtfouse, E., Derenne, S., Mariotti, A., & Largeau, C., (1994). Possible algal origin of long-chain odd *n*-alkanes in immature sediments as revealed by distributions and carbon isotope ratios. *Org. Geochem., 22*, 1023–1027.

Mackenzie, A. S., Lamb, N. A., & Maxwell, J. R., (1982). Steroid hydrocarbons and thermal history of sediments. *Nature, 295*, 223–226.

Mackenzie, K. R., (1983). The clinical application of group measure. In: Dies, R. R., & MacKenzie, K. R., (eds.), *Advances in Group Psychotherapy: Integrating Research and Practice* (pp. 159 –170). International Universities Press, New York.

Maryking, (2007). Will algae beat its competitors to become the king source of biofuels? *Environmental Graffiti*. https://scribol.com/environment/oil-and-gas/will-algae-beat-its-competitors-to-become-the-king-source-of-biofuels/ (accessed on 09 June 2020).

Mastalerz, M., & Hower, J. C., (1996). Elemental composition and molecular structure of *Botryococcus* alginite in Westphalian cannel coals from Kentucky. *Org. Geochem., 24*, 301–308.

Maxwell, J. R., Pillinger, C. T., & Eglinton, G., (1971). Organic geochemistry. *Q. Rev., 25*, 571–628.

McGinnis, K. M., Dempster, T. A., & Sommerfield, M. R., (1997). Characterization of the growth and lipid content of the diatom *Chaetoceros muelleri*. *J. Appl. Phycol., 9*, 19–24.

McKirdy, D. M., (1985). Coorongite, coastal bitumens and their origins from the lacustrine alga *Botryococcus braunii* in the Western Otway Basin. In: *Otway 85, Earth Resources of the Otway Basin.*, pp. 34–49.

McKirdy, D. M., Cox, R. L., Volkman, J. K., & Howell, V. J., (1986). Botryococcane in a new class of Australian non-marine crude oils. *Nature, 320*, 57–59.

Metting, Jr. F. B., (1996). Biodiversity and application of microalgae. *J. Ind. Microbiol. Biot., 17*, 477–489.

Metzger, P., & Casadevall, E., (1987). Lycopadiene, a tetraterpenoid hydrocarbon from new strains of the green alga *Botryococcus braunii*. *Tetrahedron Lett., 28*, 3931–3934.

Metzger, P., & Casadevall, E., (1989). Aldehydes, very long-chain alkenyl phenols, epoxides and other lipids from an alkadiene producing strain of *Botryococcus braunii*. *Phytochemistry, 28*, 2097–2104.

Metzger, P., & Largeau, C., (1999). Chemicals of *Botryococcus braunii*. In: Cohen, Z., (eds.), *Chemicals from Microalgae* (pp. 205–260). Taylor and Francis, London.

Metzger, P., & Largeau, C., (2005). *Botryococcus braunii:* A rich source for hydrocarbons and related ether lipids. *Appl. Microbiol. Biotechnol., 66*, 486–796.

Metzger, P., (1993). n-Heptacosatrienes and tetraenes from a Bolivian strain of *Botryococcus braunii*. *Phytochemistry, 33*, 1125–1128.

Metzger, P., (1999). Two terpenoiddiepoxides from the green microalga *Botryococcus braunii*: Their biomimetic conversion to tetrahydrofurans and tetrahydropyrans. *Tetrahedron, 55*, 167–176.

Metzger, P., Allard, B., Casadevall, E., Berkaloff, C., & Coute, A., (1990). Structure and chemistry of a new chemical race of *Botryococcus braunii* that produces lycopadiene, a tetraterpenoid hydrocarbon. *J. Phycol., 26*, 258–266.

Metzger, P., Berkaloff, C., Casadevall, E., & Coute, A., (1985). Alkadiene-producing and botryococcene-producing races of wild strains of *Botryococcus braunii*. *Phytochemistry, 24*, 2305–2312.

Metzger, P., Largeau, C., & Casadevall, E., (1991). Lipids and macromolecular lipids of the hydrocarbon-rich microalga *Botryococcus braunii*. *Prog. Chem. Org. Nat. Prod., 57*, 1–70.

Metzger, P., Pouet, Y., & Summons, R., (1997). Chemotaxonomic evidence for the similarity between *Botryococcus braunii* L. race and *Botryococcus neglectus*. *Phytochemistry, 44*, 1071–1075.

Metzger, P., Templier, J., Largeau, C., & Casadevall, E., (1986). An n-alkatriene and some n-alkadienes from the race of the green alga *Botryococcus braunii*. *Phytochemistry, 25*, 1869–1872.

Metzger, P., Villarreal-Rosalles, E., Casadevall, E., & Coute, A., (1989). Hydrocarbons, aldehydes and triacylglycerols in some strains of the A race of the green alga *Botryococcus braunii*. *Phytochemistry, 28*, 2349–2353.

Mitchell, D. O., (2008). *A Note on Rising Food Prices*. Research working paper 4682, Development prospects Group. World Bank, Washington, DC.

Moldowan, J. M., & Seifert, W. K., (1980). First discovery of botryococcane in petroleum. *JCS Chem. Commun., 19*, 912–914.

Niehaus, T. D., Kinison, S., Okada, S., Yeo, Y. S., Bell, S. A., Cui, P., et al., (2012). Functional identification of triterpene methyltransferases from *Botryococcus braunii* race B. *J. Biol. Chem., 287*, 8163–8173.

Niehaus, T. D., Okada, S., Devarenne, T. P., Watt, D. S., Sviripa, V., & Chappell, J., (2011). Identification of unique mechanisms for triterpene biosynthesis in *Botryococcus braunii*. *Proc. Natl. Acad. Sci. USA, 108*, 12260–12265.

Nigam, P. S., & Singh, A., (2010). Production of liquid biofuels from renewable resources. *Prog. Energy Combust. Sci., 37*, 52–68.

Ohlrogge, J., & Browse, J., (1995). Lipid biosynthesis. *Plant Cell, 7*, 957–970.

Okada, S., Matsuda, H., Murakami, M., & Yamaguchi, K., (1996). Botryoxanthin A: A member of a new class of carotenoids from the green microalga *Botryococcus braunii* Berkeley. *Tetrahedron Letters, 37*, 1065–1068.

Okada, S., Murakami, M., & Yamaguchi, K., (1995). Hydrocarbon composition of newly isolated strains of the green microalga *Botryococcus braunii*. *J. Appl. Phycol., 7*, 555–559.

Okada, S., Tonegawa, I., Matsuda, H., Murakami, M., & Yamaguchi, K., (1997). Braunixanthins 1 and 2, new carotenoids from the green microalga *Botryococcus braunii*. *Tetrahedron, 53*, 11307–11316.

Ourisson, G., Albrecht, P., & Rohmer, M., (1979). The hopanoids: Palaeochemistry and biochemistry of a group of natural products. *Pure and Applied Chemistry, 51*, 709–729.

Pal, D., Prakash, D., & Amla, D. V., (1998). Chemical composition of the green alga *Botryococcus braunii*. *Cryptogamie. Algologie., 19*(4), 311–317.

Robinson, G. W., Tsay, Y. H., Kienzle, B. K., Smithmonrov, C. A., & Bishop, R. W., (1993). Conservation between human and fungal squalene synthetases: Similarities in structure, function, and regulation. *Mol. Cell. Biol., 13*, 2706–2717.

Sangeetha, T., Elumalai, S., Roop, S. D., & Rajesh, K. G., (2017). Hydrocarbon exploration in Peninsular India by advanced biochemical marker method. *Journal of Petroleum and Environmental Biotechnology, 08*, 353. doi: 10.4172/2157-7463.1000353.

Sawayama, S., Inoue, S., Dote, Y., & Yokoyama, S. Y., (1995). CO_2 fixation and oil production through microalga. *Energy Conversion and Management, 36*, 729–731.

Sawayama, S., Minowa, T., & Yokoyama, S., (1999). Possibility of renewable energy production and carbon dioxide (CO_2) mitigation by thermo chemical liquefaction of microalgae. *Biomass and Bioenergy, 17*, 33–39.

Senousy, H. H., Beakes, G. W., & Hack, E., (2004). Phylogenetic placement of *Botryococcus braunii* (Trebouxiophyceae) and *Botryococcus sudeticus* isolate UTEX 2629 (Chlorophyceae). *J. Phycol., 40*, 412–423.

Seo, J. W., Jeong, J. H., Shin, C. G., Lo, S. C., Hans, S. S., & Yu, K. W., (2005). Over-expression of squalene synthase in *Eleutherococcuss enticosus* increases phytosterol and triterpene accumulation. *Phytochemistry, 66*, 869–877.

Shaw, G. J., Franich, R. A., Eglinton, G., Allan, J., & Douglas, A. G., (1980). Diterpenoid acids in Yallourn lignite. In: Douglas, A. G., & Maxwell, J. R., (eds.), *Advances in Organic Geochemistry* (pp. 281–286). Pergamon Press.

Sheehan, J., Dunahay, T., Benemann, J., & Roessler, P., (1998). *Look Back at the U.S. Department of Energy's Aquatic Species Program: Biodiesel from Algae; Close-Out Report.* The United States. Web. doi: 10.2172/15003040.

Singh, Y., & Kumar, H. D., (1994). Growth of *Botryococcus* sp. in improved culture medium. *Phykos Algiers, 33*(1–2), 77–87.

Smittenberg, R. H., Baas, M., Schouten, S., & Sinninghe, D. J. S., (2005). The demise of the alga *Botryococcus braunii* from a Norwegian fjord was due to early eutrophication. *Holocene, 15*, 133–140.

Song, D., Hou, L., & Shi, D., (2008). Exploitation and utilization of rich lipids-microalgae, as new lipids feedstock for biodiesel production: A review. *Sheng Wu Gong Cheng Xue Bao., 24*, 341–348.

Spolaore, P., Joannis-Cassan, C., Duran, E., & Isambert, A., (2006). Commercial applications of microalgae. *J. Biosci. Bioeng., 101*, 87–96.

Summons, R. E., Metzger, P., Largeau, C., Murray, A. P., & Hope, J. M., (2002). Polymethylsqualanes from *Botryococcus braunii* in lacustrine sediments and crude oils. *Org. Geochem., 33*, 99–109.

Templier, J., Disendorf, C., Largeau, C., & Casadevall, E., (1992). Metabolism of nalkadienes in the A race of *Botryococcus braunii. Phytochemistry, 31*, 113.

Templier, J., Largeau, C., & Casadevall, E., (1987). Effect of various inhibitors on biosynthesis of non-isoprenoid hydrocarbon biosynthesis in *Botryococcus braunii. Phytochemistry, 26*, 377–383.

Teresa, M. M., Antonio, A. M., & Nidia, S. C., (2010). Microalgae for biodiesel production and other applications: A review. *Renewable and Sustainable Energy Reviews, 14*, 217–232.

Testa, M., Gerbaudo, S., & Andri, E., (2001). Data report: *Botryococcus* colonies in Miocene sediments in the western Woodlark Basin, southwest Pacific (ODP Leg 180). In: Huchon, P., Taylor, B., & Klaus, A., (eds.), *Proc. ODP, Sci. Results* (p. 180). [Online]. Available from: World Wide Web: http://www-odp.tamu.edu/publications/180_SR/172/172.htm (accessed on 26 May 2020).

Tredici, M. R., (2010). Photobiology of microalgae mass cultures: Understanding the tools for the next green revolution. *Biofuels, 1*, 143–162.

Tsukahara, H., Ikeda, R., & Yamamoto, K., (2001). *In situ*-stress measurements in a borehole close to the Nojima Fault. *Island Arc., 10*(3/4), 261–265.

Tsukahara, K., & Sawayama, S., (2005). Liquid fuel production using microalgae. *J. Jpn. Petrol Inst., 48*, 251–259.

Van, D. A., Albrecht, P., & Ourisson, G., (1977). Identification of novel (17αH)-hopanes in shales, coals, lignites, sediments and petroleum. *Bulletin De La Societe Chimique De France, 5*, 165–170.

Volova, T. G., Kalacheva, G. S., & Zhila, N. O., (2003). Specificity of lipid composition in two *Botryococcus* strains, the producers of liquid hydrocarbons. *Russ. J. Plant Physiol., 50*(5), 627–633.

Wake, L. V., & Hillen, L. W., (1980). Study of a "bloom" of the oil-rich alga *Botryococcus braunii* in the Darwin river reservoir. *Biotechnol. Bioeng., 22*, 1637–1656.

Wang, B., Li, Y., Wu, N., & Lan, C. O., (2009). CO_2 biomitigation using microalgae. *Appl. Microbiol. Biotechnol., 79*, 707–718.

Wei, G., Yang, W., Zhang, J., Xie, W., Zeng, F., Su, N., & Jin, H., (2018). The pre-Sinian rift in central Sichuan basin and its control on hydrocarbon accumulation in the overlying strata. *Petroleum Exploration and Development, 45*(2), 193–203.

Wolela, A., (2006). Fossil fuel energy resources of Ethiopia: Oil shale deposits. *J. African Earth Sci., 46*, 263–280.

Wolf, F. R., Nonomura, A. M., & Bassham, J. A., (1985). Growth and branched hydrocarbon production in a strain of *Botryococcus braunii* (Chlorophyta). *J. Phycol., 21*, 388–396.

Zhang, Z., Metzger, P., & Sachs, J. P., (2007). Biomarker evidence for the co-occurrence of three races (A, B and L) of *Botryococcus braunii* in El Junco Lake, Galapagos. *Org. Geochem., 38*(9), 1459–1478.

Zullig, H., (1982). Untersuchungenüber die stratigraphie von carotinoid enimgeschichteten sediment von 10 Schweizer Seen zur erkundungfrüherer phytoplankton entfaltungen. *Schweiz. Z. Hydrol., 44*, 1–98.

CHAPTER 3

BIOACTIVE COMPOUNDS FROM ALGAE

M. JEYA BHARATHI

Plant Pathology Unit, Tamil Nadu Rice Research Institute, Aduthurai, Tamil Nadu, India

3.1 INTRODUCTION

Marine group of algae has large divisions and contains huge no of the known and unknown genus. Each alga has its own advantage, *viz.*, pharmaceutical, antifungal, antibacterial, antiviral, human, and animal feed. Microalgae are a huge group of photoautotrophic microorganisms, including species from different phyla such as Cyanophyta (blue-green algae, cyanoprokaryotes, cyanobacteria (CB)), Chlorophyta (green algae), Rhodophyta (red algae), Cryptophyta, Haptophyta, Pyrrophyta, Streptophyta, and Heterokontophyta. Microalgae display a significant ecological plasticity, by the ability to adapt to changing extreme environmental conditions such as temperature, light, pH, salinity, and moisture, which describes their worldwide distribution. Microalgae are a diverse group of photosynthetic microorganisms that convert carbon dioxide into valuable compounds including biologically active compounds such as biofuels, foods, feed, and pharmaceuticals. Algae are considered as a sustainable source of biodiesel which have the ability to synthesize and accumulate significant quantities of lipids as they are sunlight driven oil factories (Chisti, 2007).

Earlier researches have shown that compared to oil crops microalgae have 20–40 times more productivity whereas some of them can accumulate up to 80% of dry lipid biomass weight. Hence, microalgae have greater potential to be a major source for renewable energy production in terms of biofuel. Many algal strains are able to accumulate lipids when they are subjected to stress these strains have the potential to grow in mass culture (Rodolfi et al., 2009).

3.2 CAROTENOIDS

In photosynthetic organisms, including algae and plants, during the light phase of photosynthesis carotenoids act as accessory pigments in light-harvesting and are also able to photoprotect the photosynthetic machinery from surplus light by scavenging reactive oxygen species (ROS) with singlet oxygen and other free radicals. Almost all carotenoids are involved in quenching singlet oxygen and trapping peroxyl radicals.

The bioactivities of astaxanthin including UV-light protection and anti-inflammatory activity have been reported to effect human health conditions because of their stronger antioxidant activity (Guerin et al., 2003). The wide use of carotenoids as colorants has been found in natural foods including egg yolk, chicken, and fish. More than 750 carotenoids have been identified; still, only a few have been used commercially available such as astaxanthin, canthaxanthin, lutein, lycopene, and β carotene are the most common.

As many other antioxidant compounds are present in algal cells, the main advantages of the using microalgae as a carrier of carotenoids are their positive impact on human health. Astaxanthin is synthesized by Chlorophyceae family namely *Chlorella*, *Chlamydomonas*, *Dunaliella*, and *Haematococcus* spp. Though, a few green microalgae such as *Haematococcus* sp. can accumulate xanthophylls in oil bodies (OB) outside the plastids in the cytoplasm (Lemoine and Schoefs, 2010). Formation of xanthophylls in algal cells can be influenced by nitrogen-limitation, temperature, oxidation, light intensity, metal ions, and salts.

β-Carotene is a provitamin A carotenoid and therefore is able to be converted into retinol. Vitamin A has been revealed to reduce the risk of macular degeneration. Proposed mechanisms of action for β-carotene in cancer prevention include inhibiting the growth of cancer, induction of differentiation by modulation of cell cycle regulatory proteins, modifications in insulin-like growth factor-1, hindrance of oxidative DNA damage, and possible augmentation of carcinogen-metabolizing enzymes. Explicitly, β-carotene has been shown to condense cell growth and induce apoptosis in a variety of cancer cell lines, may be through caveolin-1 expression. Lycopene referred as a non-provitamin A carotenoid, lycopene possesses a variety of biological activities. These include antioxidant activity via singlet oxygen quenching and peroxyl radical scavenging, cancer prevention through inhibition of cancer growth, and induction of differentiation by modulation of cell cycle regulatory proteins. Additional reported properties of lycopene include alterations in insulin-like growth factor-1 or vascular endothelial growth factor (VEGF) levels, preventing the oxidative DNA damage, and possible development of carcinogen-metabolizing enzymes, reduced risk for some types of cancers, and some cardiovascular events.

3.3 ASTAXANTHIN

Astaxanthin, another non-provitamin A carotenoid, has recently gained attention due to its antioxidant activity and is also traditionally used to pigment the flesh of salmon and trout through dietary supplementation. Astaxanthin scavenges free radicals, provides protection against cancer, and an implicated role in inflammatory processes, ocular health, and diabetes. The latest studies on astaxanthin have shown its effectiveness against colon cancer cell proliferation, possibly through inhibition of the cell growth and increased apoptosis of human colon cancer cells.

3.4 β-CAROTENE

The value of the worldwide carotenoid market was found to be nearly $1.2 billion in 2010 and is planned to reach over $1.4 billion by 2018. β-Carotene has the leading share of the market. Valued at $250 million in 2007, this part is expected to be worth $309 million by 2018. Red algae and CB and show valuable fluorescent properties and accessory pigments. Some of the common phycobiliproteins include phycoerythrin (PE), phycocyanin (PC), and allophycocyanin (APC). These pigments are hetero-oligomers consisting of a grouping of subunits within producing cells (Cyanophyta) or chloroplasts (Rhodophyta) that are organized into complexes called "phycobilisomes." Phycobiliproteins have been used as natural dyes; moreover, they are extensively being used as nutraceuticals or in other biotechnological applications.

3.5 SECONDARY METABOLITES

The ecological function of secondary metabolites in microalgae or CB is not well understood; however, several possible roles have been intended with recent researches carried out. For example, the brevetoxins responsible for neurotoxic shellfish poisoning (NSP) are generated by marine dinoflagellates which are considered to have an ecological role as a feeding deterrent. Swimmer's itch caused by the molecule Lyngbyatoxin is thought to have an ecological role as a defense against grazing. Some secondary metabolites are said to play roles in sexual communication or symbiotic signals (Maschek and Baker, 2009). Secondary metabolites can be represented in all classes of molecules including isoprenoids, polyketides, peptides, and macromolecules such as nucleic acids, carbohydrates, proteins, and lipids. This level of chemical diversity associates with the huge number of environments inhabited by algae and CB. Secondary metabolites signify unique adaptations to these diverse environments.

Filamentous CB growing on a shallow tropical reef, and would expect unique chemical structures with exclusive biological properties to be produced. With more richness in chemical and biological diversity, the algae and CB are gaining more value as commercial products. For example, the unique structures produced by algal species possess rich value as components in human food, cosmetics, and various pharmaceuticals.

3.6   POLYUNSATURATED FATTY ACIDS (PUFAs)

It is very well known that some of the ω-3 and ω-6 fatty acids are required for humans needs but cannot synthesize this is the reason why some PUFA are called essential fatty acids (EFA). The illness, smoking, or alcohol intake can cause inability to synthesize some fatty acids. EFA, particularly ω-3 and ω-6, are important for the reliability of tissues where they are incorporated. Linoleic acid is used as formulations for treatment of skin hyperplasias. Arachidonic acid can be obtained from linoleic acid. DHA and EPA showed an ability to reduce problems associated to cardiovascular strokes and arthritis, and also to lower hypertension. Furthermore, DHA, and EPA play essential roles in lowering lipid content by reducing triglycerides, augmenting HDL levels and as anti-inflammatory agents. Although it is poorly synthesized, the breast milk contains high amount thus, newborns, fed with artificial milk, should be given ω-3 DHA as an additive.

3.7   PROTEINS AND ENZYMES

Proteins are biopolymers of amino acids, are essential for human beings, as they cannot be obtained without feeding, because of some deficiency in synthesizing them in enough amount. In addition, besides nutritional benefits, some proteins, smaller peptides, and amino acids have functions that contribute to a few health benefits. As microalgal species *Arthrospira* and *Chlorella* are rich in protein and amino acid they may be used as nutra-ceuticals or be included in functional foods to prevent tissue damage and diseases.

3.8 VITAMINS

Marennine, a blue pigment responsible for greening of oysters, diatom *Haslea* (*Navicula*) *ostrearia* is particularly rich in vitamin E. Another microalgae named *Porphyridium cruentum* is rich in vitamins C, E (tocopherols) and provitamin A (β-carotene). In addition, *Dunaliella salina* besides producing β-carotene (provitamin A), it also produces thiamine, pyridoxine, riboflavin, nicotinic acid, biotin, and tocopherol.

3.9 BIOFUELS

The natural source of renewable energy such as microalgae can be formed from natural resources such as sunlight, water, and O_2/CO_2. The production of biodiesel and bioethanol from microalgae that have potential to replace fossil fuels being in an economic sustainable way and leading reduction of GHG emissions. Microalgae are considered to be an outstanding candidate for biomass production (nearly 77% of dry cell mass), photosynthesis process for lipid fabrication, and production of biofuel.

3.10 BIOCONTROL ACTIVITY OF ALGAE

Approximately 30,000 freshwater and marine algal species have been studied and identified (Mata et al., 2010). They are able to produce a wide range of bioactive substances with antibacterial, antiviral, antifungal, immunos-timulant, and antiplasmodial activities (Borowitzka, 1995; Angulo-Preckler et al., 2015). Most of the isolated bioactive substances belong to groups of polyketides, amides, alkaloids, peptides, and fatty acids (Ghasemi et al., 2007; Feng et al., 2013). Research to identify bioactive compounds produced by microalgae has recently received considerable attention as a new source of novel antifouling substances. Marine dinoflagellates could produce diverse bioactive compounds, exemplified by okadaic acid, ciguatoxin, and maitotoxin (Yasumoto and Murata, 1993). The genus Amphidinium was also a rich source of bioactive compounds which showed certain biological activity, such as anti-fungal, anti-tumor, and hemolytic properties (Gopal et al., 1997). For marine dinoflagellate *Amphidinium carterae*, the toxic and red-tide microalgae, numerous efforts had been made to isolate pharmaceuticals and other biologically active compounds. However, the reports about its antifouling potential have not been found by now. Hence, the present study was undertaken to explore the antifouling potential of *Amphidinium carterae* at a low temperature of 35. Secondly, the concentrate was purified by a C18 solid-phase extraction (SPE) column and eluted with 10 mL MeOH. Then, the methanol eluent was dried by nitrogen blowing, solubilized in 20 mL $CHCl_3$, and partitioned with NaOH (0.1 mol L^{-1}) three times. The aqueous partition is equivalent to a saponification, as described in lipidic extraction procedures, and the non-polar compounds were discarded (Bursali et al., 2006). Thirdly, the aqueous phases were combined, acidized by H_2SO_4 (1.0 mol L^{-1}), and extracted three times with $CHCl_3$/MeOH (2:1). Finally, the organic phases were combined, concentrated under vacuum at the low temperature of 30, and dried by nitrogen blowing (the organic extract C). The organic extract C contained lipophilic compounds that were moderately polar and saponifiable.

3.11   FATTY ACIDS

A fatty acid with two or more methylene interrupted double bonds is good for normal cell function, and now has been entered in different areas like biomedical and nutraceutical. As a result of understanding the biological use of fatty acid, it is most commonly used in western society against obesity and cardiovascular problem. Moreover, PUFAs play an important role in cellular and tissue metabolism, PUFAs stand for polyunsaturated fatty acids (PUFAs), it also plays an important role in the regulation of membrane fluidity, oxygen, and electron transport, as well as thermal adaptation (Funk, 2011). In addition, people show more attention towards healthy food in their busy lifestyle (Napier, 1999).

In particular, people show increasing interest towards PUFAs family, namely EPA (5,8,11,14,17-eicosapentaenoic acid). EPA is a fatty acid of 20 carbon chain with having five double bonds located at the carboxy terminus or with the last double bond located at the third carbon from methyl terminus. EPA is normally esterified to form a complex lipid molecule inside the cell and play an important role in higher plants and animals as a precursor of a group of eicosanoids, hormone-like substance such as prostaglandins, thromboxanes are crucial in regulating developmental and regulating physiology. Fish oil seems to be conventional source of EPA. EPA is passed up by food chain *via* consumption by omnivores fish and then by carnivores fish and then finally by human (Wen and Chen, 2003). The biosynthesis of fatty acid occurs in two steps, the first step is the *de novo* synthesis of oleic acid from acetate followed by converting to linolenic acid and α-linolenic acid (ALA). The stepwise saturation and elongation form a ω-3 PUFA. Fish oil is a conventional source of EPA but fish do not synthesize EPA *de novo*, they derived this compound from the marine microorganisms they consumed, thus EPA passed through the food chain and ultimately reaches to the human being. EPA from the marine fish oil is refined for pharmaceuticals, this oil having a very complex fatty acid varying with chain length and unsaturation degrees (Gill and Valivety, 1997). EPA has been found in wide variety of marine algae class but only some of them have the potential to demonstrated industrial production, mainly due to the fact that majority of marine algae have low specific growth rates and low cell densities when grow in autotrophic condition (Wen and Chen, 2003).

3.12 STEROLS

Some species of microalgae have been used for promoting growth of juveniles, especially oysters for their content in sterols. Cholesterol is rarely being found in phytoplankton species such as *Chaetoceros* and *Skeletonema* that were reported to produce up to 27.7 and 2.0 μg sterols/g dry weight cholesterol being the major sterol. Other microalgae, such as *Thalassiosira* and *Pavlova* also show high amounts of sterols. *Pavlova lutheri* contains other uncommon sterols such as brassicasterol, campesterol, stigmasterol, and sitosterol apart from cholesterol. It is well familiar that high levels of cholesterol and cholesterol-lowering of plant sterols are of high risk for heart and coronary diseases.

Sterols are the major nutritional component found in seaweed, they are the most important chemical constituent of microalgae. Mainly sterols are present in plant, animal, and fungi, with the most famous animal sterol known as "cholesterol." Cholesterol is vital to cellular function and affects the fluidity of animal cell membrane. Cholesterol is a precursor to fat-soluble vitamins and steroid hormones. Various species of seaweeds have different sterols, like green seaweeds contain 28 is of mucocholesterol, cholesterol, 24 methylene cholesterol and β-sitosterol while brown seaweeds contain fucosterol, cholesterol, and brassicasterol. Red seaweeds contain sterols like desmosterol, cholesterol, sitosterol, fucosterol, and chalinasterol (Sanchez et al., 2004). Brown seaweeds (*Laminaria* and *Undaria*) contain 83–97% of fucosterol in total sterol content (662–2320 µg/g dry weight) and desmosterol of red seaweeds (*Palmaria* and *Porphyra*) contain 87–93% of total sterol content (87–337 µg/g dry weight). However, red seaweeds of *C. crispus* have cholesterol as major sterol. It is reported that plant sterol like β-sitosterol and fucosterol leads to the decrease in the concentration of cholesterol in the serum experimentation in animals and human (Whittaker et al., 2000).

3.13 POLYSACCHARIDES

Marine algae contain a huge amount of polysaccharides mainly cell wall structure and also mucopolysaccharides and storage polysaccharides (Kumar et al., 2008). Polysaccharides are the polymer of simple chain monosaccharide or simple sugar that is linked together by glycosidic bond. They have many applications like they are use in food, beverages, stabilizers, emulsifiers, thickeners, feed. The green seaweed species contain high content of polysaccharides like *Ulva* contain 65% of dry weight. The other seaweeds species contain high amount of polysaccharide are *Ascophyllum*, *Porphyra*, and *Palmaria*. Mainly the seaweeds contain polysaccharide concentration in the range from 4–76%. According to the nutritional perspective, seaweeds are low lipid content and having high carbohydrate most of this is dietary fibers even though they are not taken up by the human body.

However, dietary fibers are good for human body. The polysaccharide cell wall mainly consists of cellulose, hemicelluloses, and neutral polysaccharides. The content of cellulose and hemicellulose of seaweeds species of interest is 2–10% and 9% of dry weight. *Chlorophyceae* or green algae contain sulfated galactans, sulfuric acid polysaccharide whereas the *Pheophyceae* also known brown algae contain alginic acid, fucoidan or sulfated fucose, laminarian or β-1,3-glucan. *Sargassam* and *Rhodophyceae* or red algae contain carrageenans, amylopectin like sugar also known as floridean starch, water-soluble sulfated galactan, as well as porhyran as mucopolysaccharide that is present in the intracellular spaces (Kumar et al., 2008).

3.14 AGAR

Agar is a mixture of polysaccharide mainly composed of agarose and agro pectin, which having structure and function properties as similar as carrageenan. Agar is the sulfated polysaccharides mainly extracted from

Phaeophyceae, it is also extracted from red seaweeds such as *Gelidium* sp. and *Gracilaria* sp. The use of this compound is mainly in commercial and scientific areas because of its gelling emulsify and viscosity property. Agar is a generic name of seaweeds galactans containing α-(1-4)-3, 6-anhydro-L-galactose and β9-(1-3)-D-galactose residues with having esterification of sulfate in small amount which is up to 6% (w/w) (Hemmingson, 1996). The low quality agar is used in food products like candies, fruit juice, frozen foods, bakery icing, meringues, and desert gels. Agar also has an industrial application include paper coating, adhesives, textile printing dyeing, impressions, casting. The medium quality agar is used in biological culture media as the gel substrate. They are also important and used in the field of medical and pharmaceutical as a bulking agents, anticoagulant agent, laxatives, capsules, and tablets. The high quality of agar or highly purified agar (agarose) is used in molecular biology for separation techniques like electrophoresis, immune diffusion, and gel chromatography. Agar-agar is used for cooking and as a food source in Japan. It is known for the manufacturing of capsules for industrial application and also used as medium for cell culture. Agar structural and functional property is same as carrageenan.

Agar affects the absorption of ultraviolet (UV) rays (Murata and Nakazoe, 2001); it can decrease the blood glucose level and also exerts an anti-aggregation effect on red blood cell. Agar-oligosaccharide has also been shown that it suppresses the production of a pro-inflammatory cytokine and also suppresses the enzyme associated with production of nitric oxide. Agar type polysaccharide which is extracted from cold water extraction of another *Gracilaria* species shows anti-tumor activity. If agar is hydrosylates then it results to form agaro-oligosaccharides with activity against glycosidase and antioxidant ability (Fernandez, 1989). The agar quality and its content are totally depend upon the physiochemical property as well as closely related to environmental parameters, growth, and reproductive cycle. Although agar is also used in commercial level outside the hydrocolloid industry, recently used in medicinal, pharmaceutical areas for the treatment against the cancer cell, since it can induce the apoptosis of these cell *in vitro* (Chen, 2005).

3.15   ALGINATE

Alginate is first discovered by British pharmacist, ECC Stanford in 1880, its first industrial production began in California in 1929. Alginate is a common name for a family of linear polysaccharide containing 1, 4 linked

β-D-mannuronic and α-L-guluronic acid residues which is arrange in a non-regular block wise order (Andrade et al., 2004). Alginate is produced from brown algae and is mainly used in food and pharmaceutical industry, because of its ability to chelate metal ion and to form high viscous solution. It is also used in the textile industry in sizing cotton yarn and also used as a gelling agent. Alginate is available in both forms that is acid and salt, the acid form of alginate is a linear polyuronic acid and it is called alginic acid, whereas the salt form is an important cell wall component of brown seaweeds, consisting of 40–47% dry weight of algal biomass. It has been reported that alginate play an important role as a dietary fiber in human and animal both. It is used for decrease the concentration of cholesterol, exerts hypertension effect, can prevent absorption of toxic chemical substance (Kim and Lee, 2008). This dietary fiber is not present in any land plant, they help protect against the potential carcinogen, and they protect the surface membranes of the stomach and intestine, and they also clear the digestive system. They also have a property to absorb substance such as cholesterol and then eliminated from intestine (Burtin, 2003) and they result in hypocholesterolemic and hypolipidimic responses (Panlasigui et al., 2003).

3.16   CARRAGEENAN

It is the generic name for the family of natural water-soluble sulfated galactans, having alternative backbone consisting of α (1-4)-3,6-anhydro-D-galactose and β (1-3)-D-galactose (Goncalves et al., 2002). Carrageenan is widely used then agar as emulsifiers and stabilizer in numerous foods especially milk-based product. κ- and ι-carrageenan are very importantly used in milk products like chocolate, pudding, deserts gels, ice creams, jellies, jams, evaporated milk, because of its thick and suspension property. Carrageenan commercially divided into five of the following λ-carrageenan, ι-carrageenan, κ-carrageenan, μ-carrageenan, ν-carrageenan. In sulfate level, a large difference that is 20% (w/w) in κ-carrageenan and 40% (w/w) in λ-carrageenan is due to the difference in the seaweeds species and their extraction condition. The μ-carrageenan and ν-carrageenan biologically are the precursor of κ-and ι-carrageenan, can be transferred into sulfotransferase and sulfohydrolase (Van de Velde et al., 2005). Carrageenan can also be used as a potential pharmaceutical as anti-tumor, anti-viral, anti-coagulant, and immunomodulation activity (Zhou et al., 2004). Carrageenan can dissolve in water because of the biomolecules group that composed of linear polysaccharide chain with sulfate half esters attached to the sugar unit. Other medical use of carrageenans is as an anti-coagulant in blood products and also for the treatment of bowel problems, constipation, and dysentery. They also used for making internal poultices to control stomach ulcers. New research from biocide properties shows that carrageenan gel from *Chondrus crispus* may block the transmission of HIV virus as well as other STD viruses such as gonorrhea, genital warts, and the herpes simplex viruses (HSV) (Luescher-Mattli, 2003).

3.17   ANTIVIRAL ACTIVITY

It has been reported that some sulfated polysaccharides from red algae show antiviral activity against viruses that are responsible for human infection. Most notable are *Aghardhiella tenera* and *Nothogenia fastigiata* (Kolender et al., 1995). It was tested that galactan sulfate (from *Aghardhiella tenera*) and xylomannan sulfate (from *Nothogenia fastigiata*) show antiviral activity against the most infectious viruses like human immune deficiency virus or HIV, HSVs types 1, 2, respiratory syncytial virus or RSV. The polysaccharide present in these seaweeds is active during the first stage of RNA replication, when it adsorbs onto the surface of the cells. The most important requirement of antiviral polysaccharide is that it has to be very low cytotoxic effect towards the mammalian cell and most of the seaweeds fulfill this requirement especially *Aghardhiella tenera* and *Nothogenia fastigiata*. Carrageenan is a potential that shows *in vitro* antiviral activity. There are many types of carrageenan like μ-carrageenan, λ-carrageenan, κ-carrageenan, and ι-carrageenan.

It was also reported that some carrageenan also shows potent antiviral activity against different strain HSV 1 and 2. Carragaurd is a carrageenan based on microbicide, is an undergoing phase 3 trials, it is used for blocking of HIV and other sexually transmitted diseases. A sulfated polysaccharide from *Schizymeni apacifica* inhibits the HIV reverse transcriptase *in vitro*, which is a later stage in HIV replication (Nakashima, 1987); they do not have any effect or minimal effect on DNA and RNA polymerase activity. Some high molecular weight galactan sulfate also is known agaroids from *Gracilaria corticata* show antiviral property against HSV type 1 and 2, this is because of the inhibition of initial virus attachment to the host cell (Wu et al., 2012). Fucoidan has potent antiviral activity against HSV types 1 and 2, human cytomegalovirus (Malhotra, 2003), and HIV. Fucoidan shows antiviral property by inhibiting the binding of viral particles to the host cell (Baba, 1988). It also has the property to inhibit the binding of sperm to the zona pellucid (Oehninger, 1991).

3.18   ANTIBIOTIC ACTIVITY

Macroalgae have many compounds that show antibiotic activity. The interesting list of compounds present in macroalgae are halogenated compounds such as halogenated alkanes, haloforms, alkenes, alcohol, aldehyde, hydroquinone, and ketone (Lincoln et al., 1991). Compounds such as sterols, heterocyclic, and phenolic compounds show the antibiotic property. Many of these compounds show antiseptics as well as cleansing property, but their antibiotic activity *in vivo* is often only achieved at toxic concentration (Lincoln et al., 1991). A halogenated furanone also known as fimbrolide which is a promising antibacterial agent, belong to class of lactones from *Delisea pulchra.* It has been examined effective as bacterial anti-fouling (Kjelleberg and Steinberg, 2001) and also used as treatment for chronic *Pseudomonas aeruginosa* infection. *Pseudomonas aeruginosa* infection mainly causes by the production of mucoid alginate and formation of biofilm in the lungs of patients suffering from cystic fibrosis (Hoiby, 2002). Bacterial inhibition mainly occurs by inhibiting the furanone on the quorum sensing mechanism by functioning as an intracellular signal antagonist, as a result disruption of intra and intercellular cell-cell communication occurs (Rasmussen, 2000). This effect mainly occurs in gram-negative bacteria. Compounds like sterols heterocyclic and phenolic compounds sometimes show antibiotic property. These properties may be developed into antiseptics and cleansing agent but the antibiotic property *in vivo* is only achieved at toxic concentration.

3.19   ANTI-CANCER ACTIVITY

Fucoidans obtained from brown algae *Eclonia cava, Sargassum hornery* and *Costaria costalla*, widely spread in the sea of the South Korea, play an inhibitory role in colony formation in human melanoma and colon cancer cells. Hence, these fucoidans may be effective anti-tumor agents (Ermakova et al., 2011). Hydrolyzed fucoidan from sporophyll of *Undaria pinnnatifida* were used to determine the molecular weight and hydrolysis condition on cancer cell growth. Native fucoidan showed anti-cancer effect. A test showed that anti-cancer activity of fucoidan could be significantly enhanced by lowering the molecular weight, only when they are depolymerized by mild condition (Yang et al., 2008).

3.20   CONCLUSION

Algae has multifunctional compounds, *viz.*, carotenoids, astaxanthin, β carotene, secondary metabolites, PUFA, proteins, and enzymes, vitamins, biofuel, biocontrol activity of algae, fatty acids, polysaccharide, agar, alginate, carrageenan, antiviral activity, antibiotic activity, and anticancer activity. All the compounds are very essential for human health and well-being. Hence, for the sustainability of algae availability and its growth, we have to preserve the marine ecosystem.

KEYWORDS

- agar-agar
- allophycocyanin
- biofuel
- carotenoids
- carrageenan
- herpes simplex viruses
- neurotoxic shellfish poisoning
- sterols

REFERENCES

Andrade, L. R., Salgado, L. T., Farina, M., Pereira, M. S., Mourão, P. A., & Amado, F. G. M., (2004). Ultra structure of acidic polysaccharides from the cell walls of brown algae. *Journal of Structural Biology, 145,* 216–225.

Angulo-Preckler, C., Cid, C., Oliva, F., & Avila, C., (2015). Antifouling activity in some benthic Antarctic invertebrates by '*in situ*' experiments at deception Island, Antarctica. *Marine Environmental Research, 105,* 30–38.

Baba, M., Snoeck, R., Pauwels, R., & De Clercq, E., (1988). Sulfated polysaccharides are potent and selective inhibitors of various enveloped viruses, including herpes simplex virus, cytomegalovirus, vesicular stromatitis virus, and human immunodeficiency virus. *Antimicrobial Agents and Chemotherapy, 32,* 1742–1145.

Borowitzka, M. A., (1995). Microalgae as sources of pharmaceuticals and other biologically active compounds. *Journal of Applied Phycology, 7,* 3–15.

Bursali, N., Ertunc, S., & Akay, B., (2006). Process improvement approach to the saponification reaction by using statistical experimental design. *Chemical Engineering and Processing, 45,* 980–989.

Burtin, P., (2003). Nutritional value of sea weeds. *Electronic Journal of Environmental, Agricultural and Food Chemistry, 2,* 498–503.

Chen, H. M., Zheng, L., & Yan, X. L., (2005). The preparation and bioactivity research of agaro-oligosaccharides. *Food Technology and Biotechnology, 43,* 29–36.

Chisti, Y., (2007). Biodiesel from microalgae. *Biotechnology Advances, 25,* 294–306.

De Clercq, E., (2000). Current lead natural products for the chemotherapy of human immunodeficiency virus (HIV) infection. *Medicinal Research Reviews, 20,* 323–349.

Ermakova, S., Sokolova, R., Kim, S. M., Um, B. H., Isakov, V., & Zvyagintseva, T., (2011). Fucoidans from brown seaweeds *Sargassum hornery, Eclonia cava, Costaria costata*: Structural characteristics and anticancer activity. *Applied Biochemistry and Biotechnology, 164,* 841–850.

Feng, D. Q., Qiu, Y., Wang, W., Wang, X., Ouyang, P. G., & Huanke, C., (2013). Antifouling activities of hymenialdisine and debromohymenialdisine from the sponge *Axinella* sp. *International Biodeterioration and Biodegradation, 85,* 359–364.

Fernandez, L. E., Valiente, O. G., Mainardi, V., Bello, J. L., Velez, H., & Rosado, A., (1989). Isolation and characterization of an antitumor active agar-type polysaccharide of *Gracilaria dominguensis. Carbohydrate Research, 190,* 77–83.

Funk, C. D., (2001). Prostaglandins and leukotrienes: Advances in eicosanoid biology. *Science, 294,* 1871–1875.

Ghasemi, Y., Moradian, Mohagheghzadeh, A., Shokravi, S., & Morowvat, M. H., (2007). Antifungal and antibacterial activity of the microalgae collected from paddy fields of Iran: Characterization of antimicrobial activity of *Chroococcus dispersus. Journal of Biological Sciences, 7,* 904–910.

Goncalves, A. G., Ducatti, D. R., Duarte, M. E., & Noseada, M. D., (2002). Sulfated and pyruvylated disaccharide alditols obtained from a red seaweed galactan: ESIMS and NMR approaches. *Carbohydrate Research, 337,* 2443–2453.

Gopal, K. P., Matsumori, N., & Konoki, K., (1997). Chemical structures of amphidionols 5 and 6 isolated from marine dinoflagellate *Arnphidinium klebsii* and their cholesterol dependentment membrane disrtuption. *Journal of Marine Biotechnology, 5,* 124–128.

Guerin, M., Huntley, M. E., & Olaizola, M., (2003). *Trends in Biotechnology, 21,* 210–216.

Hemmingson, J. A., Furneaux, R. H., & Murray-Brown, V. H., (1996). Biosynthesis of agar polysaccharides in *Gracilaria chilensis. Carbohydrate Research, 287,* 101–115.

Høiby, N., (2002). Understanding bacterial biofilms in patients with cystic fibrosis: Current and innovative approaches to potential therapies. *Journal of Cystic Fibrosis, 1,* 249–254.

Kim, I. H., & Lee, J. H., (2008). Antimicrobial activities against methicillin-resistant *Staphylococcus aureus* from macroalgae. *Journal of Industrial and Engineering Chemistry, 14,* 568–572.

Kjelleberg, S., & Steinberg, P., (2001). Surface warfare in the Sea. *Microbiology Today, 28,* 134–135.

Kolender, A. A., Matulewicz, M. C., & Cerezo, A. S., (1995). Structural analysis of antiviral sulfated-α-d-(1→3)-linked mannans. *Carbohydrate Research, 273,* 179–185.

Kumar, C. S., Ganesan, P., Suresh, P. V., & Bhaskar, N., (2008). Seaweeds as a source of nutritionally beneficial compounds: A review. *Journal of Food Science and Technology,* *45,* 1–13.

Lemoine, Y., & Schoefs, B., (2010). *Photosynthesis Research, 106,* 155–177.

Lincoln, R. A., Strupinski, K., & Walker, J. M., (1991). Bioactive compounds from algae. *Life Chemistry Reports, 8,* 97–183.

Luescher-Mattli, M., (2003). Algae, a possible source for new drugs in the treatment of HIV and other viral diseases. *Current Medicinal Chemistry, 2,* 219–225.

Majczak, G. A. H., Brichartz, R. R. T. B., Duarte, M. E. R., & Noseda, M. D., (2003). Antiherpetic activity of heterofucans isolated from *Sargassum stenophyllum* (Fucales, Phaeophyta). In: Chapman, A. R. O., Anderson, R. J., Vreeland, V. J., & Davison, I. R., (eds.), *Proceedings of the 17th International Seaweed Symposium* (pp. 169–174). Oxford University Press, Oxford.

Malhotra, R., Ward, M., Bright, H., Priest, R., Foster, M. R., Hurle, M., Blair, E., & Bird, M., (2003). Isolation and characterization of potential respiratory syncytial virus receptor(s) on epithelial cells. *Microbes and Infection, 5,* 123–133.

Mata, T. M., Martins, A. A., & Caetano, N. S., (2010). Microalgae for biodiesel production and other applications: A review. *Renew Sustainable Energy Review, 14,* 217–232.

Murata, M., & Nakazoe, J., (2001). Production and use of marine algae in Japan. *Japan Agricultural Research Quarterly, 35,* 281–290.

Nakashima, H., Kido, Y., Kobayashi, N., Motoki, Y., Neushul, M., & Yamamoto, N., (1987). Antiretroviral activity in a marine red alga; reverse transcriptase inhibition by an aqueous extract of *Schizymenia pacifica*. *Journal of Cancer Research and Clinical Oncology, 113,* 413–416.

Napier, J. A., Michaelson, L. V., & Stobart, A. K., (1999). Plant desaturases: Harvesting the fat of the land. *Current Opinion in Plant Biology, 2,* 123–127.

Oehninger, S., Clark, G. F., Acosta, A. A., & Hodgen, G. D., (1991). Nature of the inhibitory effect of complex saccharide moieties on the tight binding of human Spermatozoa to the Human Zona Pellucida. *Fertility and Sterility, 55,* 165–169.

Panlasigui, L. N., Baello, O. Q., Dimatangal, J. M., & Dumelod, B. D., (2003). Blood cholesterol and lipid-lowering effects of carrageenan on human volunteers. *Asia Pacific Journal of Clinical Nutrition, 12,* 209–214.

Rodolfi, L., Zittelli, G. C., Bassi, N., Padovani, G., Biondi, N., & Bonini, G., (2009). *Traditional Biotechnology Bioengineering, 102,* 100–112.

Sanchez-Machado, D. I., Lopez-Hernandez, J., Paseiro-Losada, P., & Lopez-Cervantes, J., (2004). An HPLC method for the quantification of sterols in edible seaweeds. *Biomedical Chromatography, 18,* 183–190.

Van, D. V. F., Antipova, A. S., Rollema, H. S., Burova, T. V., Grinberg, N. V., Pereira, L., Gilsenan, P. M., Tromp, R. H., Rudolph, B., & Grinberg, V. Y., (2005). The structure of κ/ι-hybrid carrageenans: II. Coil-helix transition as a function of chain composition. *Carbohydrate Research, 340,* 1113–1129.

Wen, Z. Y., & Chen, F., (2003). Heterothrophic production of eicosapentaenoic acid by microalgae. *Biotechnology Advances, 21,* 273–294.

Whittaker, M. H., Frankos, V. H., Wolterbeek, A. M. P., & Waalkens-Berendsen, D. H., (2000). Effects of dietary phytosterols on cholesterol metabolism and atherosclerosis: Clinical and experimental evidence. *American Journal of Medicine, 109,* 600–601.

Wu, S. C., Kang, S. K., Kazlowski, B., Wu, C. J., & Pan, C. L., (2012). Antivirus and prebiotic properties of seaweed-oligosaccharide-lysates derived from agarase AS-II. *Journal of the Fisheries Society of Taiwan, 39,* 11–21.

Yang, C., Chung, D., Shin, I. S., Lee, H., Kim, J., Lee, Y., & Young, S. G., (2008). Effects of molecular weight and hydrolysis conditions on anticancer activity of fucoidans from sporophyll of *Undaria pinnatifida. International Journal of Biological Macromolecules, 43,* 433–437.

Yasumoto, T., & Murata, M., (1993). Marine toxins. *Chemical Reviews, 93,* 1897–1909.

CHAPTER 4

PROSPECTING MICROALGAE FOR BIOMOLECULES AND BIOFUELS

ROSHAN KUMAR[1], SANNIYASI ELUMALAI[2], and SUNIL PABBI[3]

[1]Department of Human Genetics and Molecular Medicine, Central University of Punjab, Bathinda – 151001, Punjab, India

[2]Department of Biotechnology, University of Madras, Guindy Campus, Chennai – 600025, Tamil Nadu, India

[3]Center for Conservation and Utilisation of Blue-Green Algae, Division of Microbiology, ICAR-Indian Agricultural Research Institute, New Delhi – 110012, India

4.1 INTRODUCTION

Microalgae are microscopic photosynthetic cell factories having a high surface area to volume ratios. These have been reported to be the primary synthesizers of organic matter in aquatic environments. They are capable of rapid nutrient and CO_2 uptake and possess a much faster cell growth rate than most of the land-based plants. Algae have been used in human food and health food additive products since long back (Ciferri, 1983; Khan et al., 2005; Kim, 2010). It is used as feed for fish and livestock (Kim, 1990; Duerr et al., 1998; Iken, 2007), good source of high value-added oils (Molina et al., 1999, 2003; Belarbi et al., 2000; Wen and Chen, 2003; Pulz and Gross, 2004; Spolaore et al., 2006), chemicals, pharmaceutical products (Guerin et al., 2003; Pulz and Gross, 2004; Spolaore et al., 2006) and pigments (Lee and Zhang, 1999; Pulz and Gross, 2004; Spolaore et al., 2006). Microalgae can provide various types of renewable biofuels including biomethane produced by anaerobic digestion of the algal biomass (Spolaore et al., 2006), biodiesel derived from microalgal oil (Roessler et al., 1994; Sawayama et al., 1995; Dunahay et al., 1996; Sheehan et al., 1998; Banerjee et al., 2002;

Gavrilescu and Chisti, 2005), bioethanol produced by fermentation of the microalgal carbohydrates from the biomass (Figure 4.1) (Ragauskas et al., 2006; Dismukes et al., 2007; Huntley and Redalje, 2007) and photobiologically produced biohydrogen (Akkerman et al., 2002; Melis, 2002; Kapdan and Kargi, 2006). The concept of using microalgae as a source of fuel is not new (Nagle and Lemke, 1990; Sawayama et al., 1995; Chisti, 2007), but it is now being taken seriously because of the escalating petroleum prices and more significantly, the emerging concern about global warming (GW) and climate change that is associated to a great extent with burning fossil fuels (Gavrilescu and Chisti, 2005).

Algae are considered as a lower plant, which lacks leaves, stems, and roots. According to the modern classification which based on the molecular phylogenetic tree, algae are not classified under the plantae kingdom although they show maximum resemblance with plants. The eukaryotic algae come under the protista Kingdom where cyanobacteria (CB) or blue-green algae (BGA) are prokaryotes comes under Kingdom Monera (Bacteria). Microalgae are microscopic photosynthetic cell having high surface area to volume ratios. These have been reported to be the primary source of organic matter in aquatic environments. Microalgae are larger in size and it also is known as macroalgae where microscopic algae are smaller in size and known as microalgae. Algal size varies from picoplankton of only 0.2–2.0 mm in diameter to giant kelps with fronds up to 60 m in length. The number of algal species has been estimated to be 1 to 10 million where most of them are microalgae. The potential advantages of microalgae as feedstock are as follows:

- It is an excellent source of natural lipids/oil up to 20–50% DCW.
- Their growth rate is very high (e.g., 1–3 doublings per day).
- Capable of withstanding in saline, brackish, costal sea water for which there are few competing demands.
- Ability to tolerate in marginal lands (e.g., desert, arid-, and semi-arid lands) which are not suitable for conventional agriculture.
- Utilize nitrogen and phosphorus as growth nutrients from wastewater (e.g., industrial, and municipal wastewaters, agricultural run-off, and concentrated animal feed operations) and helping bio-remediation of wastewater.
- Sequester carbon dioxide from flue gases emitted from fossil fuel-fired power plants and other sources, thereby reducing emissions of a major greenhouse gas, and, produce a variety of beneficial bio-products such as proteins, pigments, enzymes, feed, fertilizers, polysaccharides, and biopolymers.

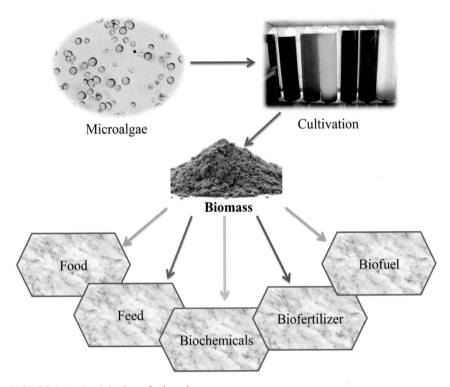

FIGURE 4.1 Exploitation of microalgae.

The products from microalgae can be divided into three categories:

1. Biomass
2. Food and bulk chemicals include protein, carbohydrates, alcohol, oils, and fatty acids
3. Special chemicals such as pigments, vitamins, volatile compounds.

4.2 CHEMICAL COMPOSITION OF MICROALGAE

The major chemical components of microalgae are carbohydrates, proteins, and lipids. Table 4.1 provides an overview of their chemical composition (Becker, 2007). Starch, sugars, glucose, and other polysaccharides are forms of carbohydrate which are found up to 60% of dry weight in the microalgae. Microalgae may contain a high level of protein ranging up to 70% of their dry weight; this is the main reason for the application of microalgae as a protein source (Becker, 2007; Tabassum et al., 2012). Microalgae lipids are composed of saturated or unsaturated fatty acids and the average lipid content varies between 2 and 21% of dry weight, but some species are capable of producing up to 90% of their dry weight as under different stress conditions (Metting, 1996). Microalgae also represent a valuable source of polyunsaturated fatty acids (PUFAs), vitamins, antioxidants, pigments, as well as special products such as toxins and isotopes (Pulz, 2004).

TABLE 4.1   Chemical Composition of Different Algae Biomass (% of Dry Weight)

Algae	Carbohydrate	Lipid	Protein
Anabaena spp.	20–25	10–15	55–56
Nostoc spp.	18.7	30.66–32.85	18–21
Oscillatoria spp.	11–13	12	29.54
Phormedium spp.	11	29	14
Clamydomonas spp.	13–17	20–22	40–46
Chlorella spp.	15–25	5–20	51–68
Dunaliella spp.	30–33	5–8	50–57
Euglena spp.	12–20	12–20	40–60
Porphyridium spp.	35–55	8–16	30–40
Spirogyra spp.	33–64	11–21	6–20
Spirulina spp.	12–18	6–10	55–75
Synechococcus sp.	11–15	10–15	60–65

Source: Adapted from Becker (2007).

4.3   HIGH-VALUE PRODUCTS FROM MICROALGAE

4.3.1   UTILIZATION OF ALGAE IN THE RECENT PAST

In the 1940s, microalgae were used as live feeds in aquaculture (shellfish or fish farming). After 1948, applied algology developed at rapid pace, it was started in Germany and extended to USA, Israel, Japan, and Italy. The aim was to use algal biomass for mass production of protein fat as a nutrition source (Burlew, 1953) the push for this development came from statistical data about population development and predictions of an insufficient protein supply in the future (Spolaore et al., 2006). At that time, the idea of using microalgae for wastewater treatment was launched and the systematic examination of algae for biologically active substances, particularly antibiotics, began (Borowitzka, 1995). In the 1960s, the commercial production of *Chlorella* as a novel health food was started and was success in Japan and Taiwan and in the USA; interest grew in developing algae as photosynthetic gas exchangers for long term space travel (Borowitzka, 1999). The energy crisis in the 1970s triggered considerations about using microalgal biomasses as renewable fuels and fertilizers.

An environmental technology from the USA started worked on the improving of wastewater with the help of microalgae and the subsequent fermentation of the resulting biomass to methane (Pulz and Scheibenbogen, 1998; Spolaore et al., 2006). In addition, in the 1970s, the first large-scale *Spirulina* production plant was established in Mexico (Borowitzka, 1999).

In the 1980s, 46 pilot-scale algae production plants in Asia mainly concentrated on mass production of *Chlorella*. Large scale production of microalgae began in India, and large commercial production facilities in the USA and mass production of halophilic green alga *Dunaliella salina* as a source of β-carotene were also started by Israel (Spolaore et al., 2006). Use of microalgae as a source of food, feed, pigments, and fine chemicals was the beginning of a new trend in 1980s (de la Noue and de Pauw, 1988).

In the 1990s in USA and India, started on pilot-scale production of *Haematococcus pluvialis* which are the best source of carotenoid astaxanthin, which is used in pharmaceuticals, nutraceuticals, agriculture, and animal nutrition (Olaizola, 2000; Spolaore et al., 2006). Particularly during the past two or three decades, algal biotechnology grew steadily into an important global industry with a diversified field of applications, and more and more new entrepreneurs began to realize the potential of algae.

4.3.2   UTILIZATION OF ALGAE: THE PRESENT SITUATION

Nowadays, the algal biotechnology industry harvests 10^7 tons of algae annually for different purposes. Today's commercial algal biotechnology is still a non-transgenic industry that basically produces food, feed, food and feed additives, cosmetics, and pigments. Due to the marine and aquatic applications, algal biotechnology is sometimes also called blue biotechnology. The description of the present situation also outlines the immediate application areas of future genetically optimized transgenic algae. The main objective to grow microalgae commercially is harvesting metabolic products, terrestrial organisms, and feed products for marine and terrestrial organisms, food supplements for humans, or to use the microalgae for an environmental-related process like fertilization of soils, wastewater treatment, biofuels production, and phytoremediation of toxic waste. Algae are the source of various metabolites with different bioactivities that are yet to be fully exploited. *Spirulina* and *Chlorella* are the microalgae have been consumed as food supplements by humans and also used as animal feeds. Microalgae are also considered as a potential source of high-value products for biotechnological exploitation, which include polysaccharides, phycobiliproteins, PUFA, carotenoids, and phycotoxins. Diverse ranges of compounds are produced by microalgae as defense against stress conditions (Figure 4.2). An advantage of exploiting microalgae for bioactive molecules is that they can be cultured on a large scale for production of the desired chemicals. The advent of molecular biology has led to a better understanding of the biosynthesis and physiological functions of the bioactive molecules in microalgae.

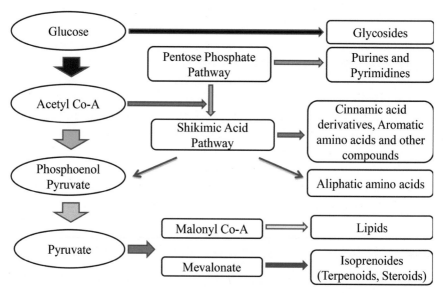

FIGURE 4.2   Lipids, secondary metabolites, and other products from microalgae.

4.4   SOURCE OF NUTRACEUTICALS

Micro-green algae represent an exceptionally diverse and highly specialized group of microorganisms adapted to various ecological habitats, which exhibit oxygenic photosynthesis. They represent one of the most promising natural sources for the mankind as a valuable source of food, feed, primary, and secondary metabolites, natural pigments, biofuel, biofertilizer, enzymes, pharmaceuticals, biopolymers, and other fine chemicals (Kumar et al., 2015b; Prasanna et al., 2013). CB such as *Spirulina, Anabaena, Nostoc*, and *Oscillatoria* produce a diverse variety of secondary metabolites (Burja et al., 2001).

The only comparable group is actinomycetes, which are the tremendous source of various metabolites and it is time now to turn to microalgae and exploit their potential. Because microalgae are largely unexplored, they represent a rich opportunity for discovery; the expected rate of rediscovery is far lower than for other better-studied groups of organisms. Microalgae produce a wide variety of bioactive compounds, which include 5.6% amino acids, 40% lipopeptides, 4.2% macrolides, 4.2% fatty acids, and 9% amides. Microalgal lipopeptides include different compounds being cytotoxic (41%), antitumor (13%), antibiotics (12%), antiviral (4%), and the remaining 18% activities include multi-drug resistance reversers, antimycotics, antimalarial, antifeedant, herbicides, and immune-suppressive agents.

CB are also known as BGA have a cholesterol-lowering ability in animals and humans. In the study, it was found that the level of low-density lipoprotein (LDL) and very-low-density lipoprotein (VLDL) cholesterol and total cholesterol in rat serum was reduced when high cholesterol diet was supplemented with BGA. It was concluded that a high cholesterol diet supplemented with microalgae will cause adipohepatosis. This was due to the activity of lipoprotein lipase, an enzyme for metabolism of triglyceride-rich lipoproteins.

Microalgae are also known to produce antiviral, antitumor, and anti-fungal compounds. Many of the pharmaceutically interesting compounds in microalgae are peptides, including one toxin which is important candidates for anti-cancer drugs. Peptide synthetases are common in microalgae and responsible for the production of hepatotoxins and other peptides. In microalgae, polyketide synthetases (PKS) take part in biosynthesis of various bioactive compounds (e.g., microcystins).

Microalgae are also having good potential as a food. Dried form of *Arthrospira* is sold in the market as a healthy food with $40 million in sales annually. The health benefits of this organism are also attributes to the presence of antioxidant activity of their metabolites. The use of antioxidants to prolong the shelf life of foodstuff is ubiquitous. Nowadays mostly synthetic-based antioxidants compounds such as butylated hydoxytoulene (BHT) and butylated hydoxyanisole (BHA) are used. As these compounds are carcinogens in nature (Namiki, 1990; Pokorny, 1991), the use of natural compound in recent years is promoted to replace these synthetic antioxidants with natural antioxidants.

Over the past two decades, a number of novel antioxidant formulations for use as a food supplement or as food ingredients have been introduced and in many cases, these novel antioxidants were derived from natural origin. Most, if not all, commercially available antioxidants are derived from medicinal and aromatic plant extracts (sage, oregano, rosemerry, cocoa). It is however believed that microalgae are a promising alternative source of antioxidants. The absence of structural damage in microalgae leads to the consideration that these organisms are able to generate the necessary compounds to protect themselves against oxidation. In this respect, microalgae can be considered as an important source of antioxidant compounds that could be suitable for protection against the reactive oxygen species (ROS) formed by our metabolism or induced by external factors (such as pollution, stress, UV radiation).

Microalgae also a good producer of antioxidant substances of very different nature, carotenoids can be highlighted within the fat-soluble fraction, and vitamin E (or α-tocopherol) where most beneficial water-soluble antioxidants found in microalgae are phenolics compounds, phyco-biliproteins, polyphenols, vitamins, and terpenoids (Plaza et al., 2008). A number of studies going on the microalgal bioactive compounds have oriented to the antimicrobial, antihelmintic, cytotoxic, antiviral, anti-inflammatory, and enzyme inhibition properties (Noue and Pauw, 1988; Dufosse et al., 2005; Singh et al., 2005). Moreover, because of phototropic life of microalgae and their permanent exposure to high oxygen and radical stresses, they have a tendency to produce numerous efficient protective chemicals against oxidative and radical stressors (Tsao and Deng, 2004). This scavenger capacity of microalgal contents bring them up as the potential alternative substances against oxidation-associated conditions like inflammation, chronic diseases, aging, or skin UV-exposure. Microalgal phenolic compounds were reported to be potential antioxidants to combat free radicals which are harmful to our health and food systems.

4.5   ANTIOXIDANTS AND PHENOLICS

Microalgae represent an almost untapped resource of natural antioxidants, due to their tremendous biodiversity and considered much more diverse than higher plants. Since, not all groups of microalgae can be used as natural sources of antioxidants, due to their widely varied contents of target products, growth rate, or yields, ease of cultivation, and/or other factors, reports on the antioxidant activity of microalgae are limited. Microalgae have been identified as one of the most promising group of organisms from which novel and biochemically active natural products are isolated.

Nowadays, lot of research work is going on the microalgae as a source of novel, nutraceuticals, and biologically active compounds such as antioxidants, phenolics compounds, phycobiline, phenols, terpenoids, proteins, steroids, and fine chemicals (Abd El-Baky et al., 2008; Li et al., 2008). As compared to higher plants, very less phenolics are reported in microalgae (Colla et al., 2007). In plant the occurrence of phenolics compound is well documented and these compounds are known to possess antioxidant activity in biological systems but the antioxidants and phenolics property are less reported although decreased cholesterol levels have been reported in hypercholesterolemic patients fed with *Spirulina* (Ramamoorthy and Premakumari, 1996) and the antioxidant activity of phycobiliproteins extracted from *S. platensis* has also been demonstrated (Estrada et al., 2001). The effect of growth parameters and conditions on the chemical composition of *Spirulina* has been well studied by many researchers with the main purpose to optimize the production of economic and nutritionally interesting compounds, especially phycocyanin (PC), phycoerytherine, allophycocyanin (APC), and gamma-linolenic acid (GLA) (Cohen et al., 1993; Tanticharoen et al., 1994). Microalgal phenolic compounds were well documented as potential candidate to combat free radicals, which are harmful to our health and body and also food systems (Estrada et al., 2001). *In vitro* studies had reported that the *Spirulina* and *Nostoc* species have several therapeutic and neutraceutical properties due to their ability to scavenge superoxide and hydroxyl radicals and inhibit lipid peroxidation.

Phytophenolic compounds have been associated with antioxidative activity in biological systems, acting as scavengers of singlet oxygen and free radicals (Rice-Evans et al., 1996; Jorgensen et al., 1999). Much of this work has emphasized the role of phenolics in higher plants in relation to human health and cancer (Setchell et al., 1981; Paganga et al., 1999). While Pratt (1992) reported microalgae as a best source of natural antioxidants and a few studies have investigated the role of phenolics and phenolic-antioxidant activity in algae (Foti et al., 1994), especially microalgae. Likewise, compared to higher plants, the pharmaceutical potential of algal secondary metabolites has only received relatively limited attention (Konig and Wright, 1990), even though pharmacologically active microalgal biometabolites such as antioxidants, phenolics, carotenoids, terpenes, and vitamins have been attributed to chemotherapeutic, antimicrobial, and anticancer activity (Hoppe, 1982; Blunden, 1993). Phenylpropanoids function as effective antioxidants due to their ability to donate hydrogen from hydroxyl groups positioned along the aromatic ring to terminate free radical oxidation of lipids and other biomolecules (Foti, 1994). Phenolic antioxidants, therefore, short-circuit a destructive chain reaction that ultimately degrades cellular membranes. Phenolic compounds are also reported to block ultraviolet (UV) irradiation-induced carcinogenesis (Huang and Ferraro, 1992).

4.6   ALGAE AND BIOFUELS

Bio-energy production has recently become a topic of intense interest due to increased concern regarding limited petroleum-based fuel supplies and the contribution of these fuels to atmospheric CO_2 levels. According to European Environmental Agency (EEA), the transportation and energy sectors are the major anthropogenic sources, responsible in the European Union for more than 20% and 60% of greenhouse gas (GHG) emissions, respectively. Agriculture is the third largest anthropogenic source, representing about 9% of GHG emissions, where the most important gases are nitrous oxide (N_2O), and methane (CH_4). It is expected that with the development of new growing economies, such as India and China, the global consumption of energy will rise and lead to more environmental damage. GHG contributes not only to GW but also to other impacts on the environment and human life. Besides these limited supplies, continued use of petroleum-based fuels and accelerating price has raised a question on its sustainability in the near future. Thus, one of the most challenging problems facing mankind in the medium to long term is to find clean and renewable energy sources. At present, the available sources of energy such as solar, wind, geothermal, wave, and nuclear energy have many benefits but they are generally limited to producing electricity rather than liquid and gaseous energy storage forms and they do not provide a solution for other products derived from petroleum. Extensive research has also been carried out to use edible sources such as soybean, sunflower, rapeseed saffola oils (Lang et al., 2001) and non-edible products as frying oil, greases, tallow, jatropha, and mahua oils (Alcantara et al., 2000; Becker and Francis, 2002; Ghadge and Raheman, 2005; Montefri et al., 2010) as biodiesel, however, high feed cost of vegetable oils is a major obstacle in its large-scale commercial exploitation.

Biological systems offer potential of producing liquid or gaseous fuels and biomass has the potential to function as a feedstock for many products that are currently derived from petroleum. Biomass-based fuels are ideal as they are renewable and have the potential to achieve high efficiencies and are generally well distributed geographically. Biofuel production is expected to offer new opportunities to diversify income and fuel supply sources, to promote employment in rural areas, to develop long term replacement of fossil fuels, to reduce GHG emissions, boosting the decarbonization of transportation fuels and increasing the security of energy supply. Among biological systems, many microalgae have the ability to produce substantial amounts (e.g., 20–50% dry cell weight) of triacylglyceride (TAG) as a storage lipid under photo-oxidative stress or other adverse environmental stress condition and depending on the species; microalgae also produce different kinds of lipids.

Currently, our world facing two major issues such as increased energy demand and GW due to over-exploitation of fossil fuel. Finding the better alternate renewable resource to replace the fossil fuel is mandatory in all nations. Biodiesel is quiet interesting and believed to be alternative renewable fuels which have a lot of environmental benefits and merits. Biodiesel are biomass-derived fuel mainly obtained from plants, animal fats and microbes (Williams and Laurens, 2010). The oil composed of triacylglycerides (TAG) are converted into fatty acid methyl esters (FAME) by transesterification with alcohol (e.g., methanol) and catalyst (acid or base). In this process, the first step is to extract lipids using organic solvents (e.g., hexanes, chloroform, and methanol). The solvents are removed by distillation and the TAGs are then reacted with acid or base and an alcohol (typically methanol) to make the FAME (Xu et al., 2006; Chen et al., 2009; Bhakar et al., 2013, 2014; Ngangkham et al., 2013; Kumar et al., 2015a, b, 2017).

The BGA (cyanobacteria) are prokaryotes with eukaryotic photosynthesis (Fogg et al., 1973) found in a different habitat. Substantial information has been gathered on different aspects of these organisms (Singh, 1961; Roger and Kulasooriya, 1980; Venkataraman, 1981). These are ubiquitous in distribution and occur in diverse habitat. Photosynthesis process and the organization of their photosynthetic apparatus are similar to eukaryotic algae and plants (Lang, 1968). But CB contains chlorophyll-a and other photosynthetic pigments such as phycobiliproteins (c-phycoerythrin, c-phycocyanin, and APC) and carotenoids (β-carotene, echinenone, and myxoxanthophyll) (Fay, 1983). Blue-Green Algae contain significant quantities of lipids. Especially have PUFAs whose degree of unsaturation is higher than 3 hence it very much useful for biofuel production compared to saturated fatty acids.

Many microalgal species can be induced to accumulate substantial quantities of lipids (Sheehan et al., 1988) thus contributing to a high oil yield. The average lipid content varies between 1 and 70% but under certain conditions, some species can reach 90% of dry weight (Spolaore et al., 2006; Chisti et al., 2007; Li et al., 2008). At present lipid content and biomass productivities of different microalgae species, showing significant differences between the various species (Renaud et al., 1999; Richmond et al., 2004; Spolaore et al., 2006; Chisti et al., 2007; Li et al., 2008; Gouveia et al., 2009; Rodolfi et al., 2009).

Murphy (2001) studied some of the unicellular algae that are able to produce up to 86% of their total cell dry weight as neutral lipids in environmental and nutritional stress like nitrogen limitation, salinity, or high temperature. Microalgal oil has been studied extensively as a source of methyl ester diesel fuel (Nagle and Lemke, 1990). Many species having the potentiality for rapid growth and high productivity, and many microalgal species can be induced to accumulate substantial quantities of lipids, often greater than 60% of their biomass (Michele et al., 2005).

4.7 CONCLUSION AND FUTURE PROSPECTS

Micro-algae represents an exceptionally diverse and highly specialized group of microorganisms adapted to various ecological habitats. It has been exploited by humankind as a promising source of food, feed, biofuel, and important biochemicals. These have the potential to revolutionize biotechnology in areas like nutrition, aquaculture, pharmaceutical, analytics, cosmetics, biofuel, and fine chemicals. Compared with the past millennium, the biotechnological research in microalgae has increased during the past 50 years considering its vast diversity, but despite the developments, the number of commercially available products is still fairly limited. The commercial exploitation of microalgae for food, feed, pigments, bioactive compounds, and nutraceuticals are in progress but only a few hundreds of microalgal species have been studied. There is a need to further boost various research and developmental activities to overcome the technological barriers, as the potential of these algae has been proven beyond doubt and can serve the mankind with its vast basket of resources.

KEYWORDS

- antioxidants
- biofuel
- microalgae
- nutraceuticals

- **phenolics**
- **pigments**
- **secondary metabolites**
- **value added biomolecules**

REFERENCES

Abd El-Baky, H. H., El-Baz, F. K., & Ek-Baroty, G. S., (2008). Evaluation of marine alga *Ulva lactuca* L. as a source of natural preservatives ingredient. *Am. Eurasian J. Agr. Environ. Sci., 3*(3), 434–444.

Akkerman, I., Janssen, M., Rocha, J., & Wijffels, R. H., (2002). Photobiological hydrogen production: Photochemical efficiency and bioreactor design. *Int. J. Hydrogen Energy, 27,* 1195–1208.

Alcantara, R., Amores, J., Canoira, L., Fidalgo, E., Franco, M. J., & Navarro, A., (2000). Catalytic production of biodiesel from soy-bean oil used frying oil and tallow. *Biomass Bioener., 18,* 515–527.

Banerjee, A., Sharma, R., Chisti, Y., & Banerjee, U. C., (2002). *Botryococcus braunii* a renewable source of hydrocarbons and other chemicals. *Crit. Rev. Biotech., 22*(3), 245–279.

Becker, E. W., (2007). Micro-algae as a source of protein. *Biotechnology Advances, 25,* 207–210.

Becker, K., & Francis, G., (2002). Biodiesel from Jatropha plantations on degraded land. *Natural Resources Forum, 29*(1), 12–24.

Belarbi, H., Molina, E., & Chisti, Y., (2000). A process for high and scaleable recovery of high purity eicosapentaenoic acid esters from microalgae and fish oil. *Enzyme Microb. Technol., 26,* 516–529.

Bhakar, R., Roshan, K., & Pabbi, S., (2013). Total lipids and fatty acid profile of different *Spirulina* strains as affected by salinity and incubation time. *Vegetos, 26,* 148–154.

Bhakar, R., Roshan, K., Elumalai, S., & Pabbi, S., (2014). Lipid biosynthesis and fatty acid characterization in selected *Spirulina* strains under different cultivation systems. *Journal of Pure and Applied Microbiology, 8*(6), 4881–4888.

Blunden, G., (1993). Marine algae as sources of biologically active compounds. *Interdiscip. Sci. Rev., 18,* 73–80.

Borowitzka, M. A., (1995). Microalgae as sources of pharmaceuticals and other biologically active compounds. *Journal of Applied Phycology, 7*(1), 3–15.

Borowitzka, M. A., (1999). Commercial production of microalgae: Ponds, tanks, tubes and fermenters. *Journal of Biotechnology, 70,* 313–321.

Burja, A. M., Banaigs, B., Abou-Mansour, E., Burgess, J. G., & Wright, P. C., (2001). Marine cyanobacteria-a prolific source of natural products. *Tetrahedron, 57*(46), 9347–9377.

Burlew, J. S., (1953). *Algal Culture: From Laboratory to Pilot Plant.* Washington DC: Carnegie Institution of Washington.

Chen, W., Zhang, C., Song, L., Sommerfeld, M., & Hu, Q., (2009). A high throughput Nile red method for quantitative measurement of neutral lipids in microalgae. *J. Microbiol. Meth., 77,* 41–47.

Chisti, Y., (2007). Biodiesel from microalgae. *Biotechnol. Adv., 25*, 294–306.

Ciferri, O., (1983). *Spirulina*, the edible microorganism. *Microbiological Reviews, 47*(4), 551–578.

Cohen, Z., Reungjitchachawali, M., Angdung, W., & Tanticharoen, M., (1993). Production and partial purification of gamma-linolenic acid and some pigments from *Spirulina platensis. J. Appl. Phycol., 5*, 109–115.

Colla, L. M., Reinehr, C. O., Reichert, C. J., & Costa, A. V., (2007). Production of biomass and nutraceutical compounds by *Spirulina platensis* under different temperature and nitrogen regimes. *Biores. Technol., 98*(1), 489–493.

De La Noue, J., & De Pauw, N., (1988). The potential of microalgal biotechnology: A review of production and uses of microalgae. *Biotechnology Advances, 6*(4), 725–770.

Dismukes, R. K., & Nowinski, J., (2007). Prospective memory, concurrent task management, and pilot error. *Attention: From Theory to Practice*, 225–236.

Duerr, E. O., (1998). Cultured microalgae as aquaculture feeds. *Journal of Marine Biotechnology, 6*(2), 65–70.

Dufosse, L., Galaup, P., Yaron, A., Arad, S. M., Blanc, P., Murthy, N. C., & Ravishankar, G. A., (2005). Microorganisms and microalgae as sources of pigments for food use: A scientific oddity or oddity or an industrial reality. *Trends Food Sci. Technol., 16*, 389–406.

Dunahay, T. G., Jarvis, E. E., Dais, S. S., & Roessler, P. G., (1996). Manipulation of microalgal lipid production using genetic engineering. *Appl. Biochem. Biotechnol., 96*, 273–228.

Estrada, J. E. P., Bermejo, B. P., Villar, D., & Fresno, A. M., (2001). Antioxidant activity of different fractions of *Spirulina platensis* protein extract. *Farmaco., 56*, 497–500.

Fay, P., (1983). *The Blue-Greens* (p. 88). London: Edward Arnold.

Fogg, G. E., (1973). Physiology and ecology of marine blue-green algae. In: Carr, N. G., & Whitton, B. A., (eds.), *The Blue-Green Algae*. Academic Press, London.

Foti, M., Piattelli, M., Amico, V., & Ruberto, G., (1994). Antioxidant activity of phenolic meroditerpenoids from marine algae. *J. Photochem. Photobiol., B26*, 159–164.

Gavrilescu, M., & Chisti, Y., (2005). Biotechnology-a sustainable alternative for chemical industry. *Biotechnology Advances, 23*(7, 8), 471–499.

Ghadge, S. V., & Raheman, H., (2005). Biodiesel production from mahua (*Madhuca indica*) oil having high free fatty acids. *Biomass and Bioenergy, 28*, 601–605.

Gouveia, L., Marques, A. E., Silva, D. T. L., & Reis, A., (2009). *Neochlorisole abundans* UTEX 1185, a suitable renewable lipid source for biofuel production. *Industrial Microbiology and Biotechnology, 36*, 821–826.

Guerin, M., Huntley, M. E., & Olaizola, M., (2003). *Haematococcus* astaxanthin: Applications for human health and nutrition. *Trends Biotechnol., 21*(5), 210–216.

Hoppe, R. T., Coleman, C. N., Cox, R. S., Rosenberg, S. A., & Kaplan, H. S., (1982). The management of stage I-II Hodgkins disease with irradiation alone or combined modality therapy, The Stanford experience. *Blood, 59*, 455–465.

Huang, M. T., & Ferraro, T., (1992). Phenolic compounds in food and cancer. In: Huang, M. T., Ho, C. T., & Lee, C. Y., (eds.), *Phenolic Compounds in Food and Their Effects on Health, II* (pp. 8–34). American Chemical Society, Washington, DC.

Huntley, M. E., Niiler, P., Redalje, D., & Leonard, A., (1999). *Method of Control of Haematococcus spp. Growth Process*. US Patent No. 5,882,849.

Iken, K., Amsler, C. D., Hubbard, J. M., McClintock, J. B., & Baker, B. J., (2007). Allocation patterns of phlorotannins in Antarctic brown algae. *Phycologia., 46*(4), 386–395.

Jorgensen, L. V., Madsen, H. L., Thomsen, M. K., Dragsted, L. O., & Skibsted, L. H., (1999). Regulation of phenolic antioxidants from phenoxyl radicals, An ESR, and electrochemical study of antioxidant hierarchy. *Free Radical Research, 30*, 207–220.

Kapdan, I. K., & Kargi, F., (2006). Bio-hydrogen production from waste materials. *Enzyme Microb. Technol., 38*, 569–582.

Khan, M., Shobha, C. J., Rao, U. M., Sundaram, C. M., Singh, S., Mohan, J. I., Kuppusamy, P., & Kutala, K. V., (2005). Protective effect of *Spirulina* against doxorubicin-induced cardiotoxicity. *Phytother. Res., 19*, 1030–1037.

Kim, H. C., Chong, S. T., Nunn, P. V., & Klein, T. A., (2009). Seasonal prevalence of mosquitoes collected from light traps in the Republic of Korea. *Entomol. Res., 40*, 136–144.

Konig, G. M., Wright, A. D., & Sticher, O., (1990). A new polyhalogenated monoterpene from the red alga *Plocamium cartilagineum*. *Journal of Natural Products, 53*, 1615–1618.

Kumar, R., Biswas, K., Singh, P. K., Singh, P. K., Elumalai, S., Shukla, K., & Pabbi, S., (2017). Lipid production and molecular dynamics simulation for regulation of accD gene in cyanobacteria under different N and P regimes. *Biotechnol. Biofuels, 10*, 94.

Kumar, R., Elumalai, S., & Pabbi, S., (2015a). Evaluation of antioxidant activity and total phenolic content in cyanobacteria isolated from different regions of India. *International Journal of Scientific Research, 4*(1), 2277–8179.

Kumar, R., Bhowmick, A., Chakdar, H., Elumalai, S., & Pabbi, S., (2015b). Biochemical characterization and diversity analysis of cyanobacteria isolated from different locations. *Vegetos, 28*(1), 38–48.

Lang, N. J., (1968). The fine structure of blue-green algae. *Annual Reviews in Microbiology, 22*(1), 15–46.

Lang, X., Dalai, A. K., Bakhshi, N. N., Reaney, M. J., & Hertz, P. B., (2001). Preparation and characterization of bio-diesels from various bio-oils. *Biores. Technol., 8*, 53–62.

Lee, Y. K., & Zhang, D. H., (1999). Production of astaxanthin by *Haematococcus*. In: Cohen, Z., (ed.), *Chemicals from Microalgae* (pp. 173–195). Taylor and Francis: London, UK.

Li, H. B., Wong, C. C., Cheng, K. W., & Chen, F., (2008). Antioxidant properties *in vitro* and total phenolic contents in methanol extracts from medicinal plants. *LWT Food Sci. Technol., 41*, 385–390.

Melis, A., (2002). Green alga hydrogen production: Progress, challenges and prospects. *Int. J. Hydrogen Energy, 27*, 1217–1228.

Metting, F. B., (1996). Biodiversity and application of microalgae. *J. Ind. Microbiol. Biotechnol., 17*(5/6), 477–489.

Michele, A., Dibenedetto, A., Carone, M., Colonna, T., & Carlo, T., (2005). Fragile production of biodiesel from macroalgae by supercritical CO_2 extraction and thermochemical liquefaction. *Environmental Chemistry Letters, 3*(3), 136–139.

Molina, G. E., (1999). Microalgae, mass culture methods. In: Flickinger, M. C., & Drew, S. W., (eds.), *Encyclopedia of Bioprocess Technology: Fermentation, Biocatalysis and Bioseparation* (Vol. 3., pp. 1753–1769) New York: Wiley.

Montefrio, M. J., Xinwen, T., & Philip, O. J., (2010). Recovery and pre-treatment of fats, oil, and grease from grease interceptors for biodiesel production. *Applied Energy, 87*, 3155–3161.

Murphy, J., (2001). The biogenesis and functions of lipid bodies in animals, plants, and microorganisms. *Prog. Lipid Res., 40*, 325–438.

Nagle, N., & Lemke, P., (1990). Production of methyl-ester fuel from microalgae. *Appl. Biochem. Biotechnol., 24*(5), 355–361.

Namiki, M., (1990). Antioxidants/antimutagens in food. *Crit. Rev. Food Sci.*, *29*, 273–300.

Ngangkham, M., Ratha, S., Kumar, R., Babu, S., Dhar, D. W., Sarika, C., Prasad, R., & Parasanna, P., (2013). Substrate amendment mediated enhancement of the valorization potential of microalgal lipids. *Biocatalysis and Agricultural Biotechnology*, *2*, 240–246.

Noue, D. L. J., & Pauw, D. N., (1988). The potential of microalgal biotechnology: A review of production and uses of microalgae. *Biotechnol. Adv.*, *6*(4), 725–770.

Olaizola, M., (2003). Commercial development of microalgal biotechnology: From the test tube to the market place. *Biomol. Eng.*, *20*, 459–466.

Paganga, G., Miller, N., & Rice-Evans, C. A., (1999). The polyphenolic content of fruit and vegetables and their antioxidant activities. What does a serving constitute? *Free Rad. Res.*, *30*, 153–162.

Plaza, M., Cifuentes, A., & Ibanez, E., (2008). In the search of new functional food ingredients from algae. *Food Sci. Tech.*, *19*, 31–39.

Prasanna, R., Sharma, E., Sharma, P., Kumar, A., Roshan, K., Gupta, V., Pal, R. K., Shivay, Y. V., & Nain, L., (2013). Soil fertility and establishment potential of inoculated cyanobacteria in rice crop grown under non flooded conditions, paddy and water environment. *Paddy Water Environent.*, *11*, 175–183.

Pulz, O., & Gross, W., (2004). Valuable products from biotechnology of microalgae. *Applied Microbiology and Biotechnology*, *65*, 635–648.

Pulz, O., & Scheibenbogen, K., (1998). Photobioreactors: Design and performance with respect to light energy input. In: *Bioprocess and Algae Reactor Technology, Apoptosis* (pp. 123–152). Springer, Berlin, Heidelberg.

Ragauskas, A. J., Williams, C. K., Davison, B. H., Britovsek, G., Cairney, J., Eckert, C. A., & Mielenz, J. R., (2006). The path forward for biofuels and biomaterials. *Science*, *311*(5760), 484–489.

Ramamoorthy, A., & Premakumari, S., (1996). Effect of supplementation of *Spirulina* on hyper cholesterolemic patients. *J. Food Sci. Technol.*, *33*, 124–128.

Renaud, S. M., Thinh, L. V., & Parry, D. L., (1999). The gross chemical composition and fatty acid composition of 18 species of tropical Australian microalgae for possible use in mariculture. *Aquaculture*, *170*(2), 147–159.

Rice-Evans, C. A., Miller, N. J., & Paganga, G., (1996). Structure-antioxidant activity relationships of flavonoids and phenolic acids. *Free Radical Biology and Medicine*, *20*(7), 933–956.

Richmond, A., (2004). Biological principles of mass cultivation. In: Richmond, A., (ed.), *Handbook of Microalgal Culture, Biotechnology and Applied Phycology* (pp. 125–177). Blackwell.

Richmond, A., (2008). *Handbook of Microalgal Culture: Biotechnology and Applied Phycology*. John Wiley & Sons, London, UK.

Rodolfi, L., Chini-Zittelli, G., Bassi, N., Padovani, G., Biondi, N., Bonini, G., & Tredici, M. R., (2009). Microalgae for oil: Strain selection, induction of lipid synthesis and outdoor mass cultivation in a low-cost photobioreactor. *Biotechnol. Bioeng.*, *102*(1), 100–112.

Roessler, P. G., Brown, L. M., Dunahay, T. G., Heacox, D. A., Jarvis, E. E., & Schneider, J. C., (1994). Genetic-engineering approaches for enhanced production of biodiesel fuel from microalgae. *ACS Symp. Ser.*, pp. 255–270.

Roger, P. A., & Kulasooriya, S. A., (1980). *Blue-green Algae and Rice* (p. 112). IRRI, Los Banos, Laguna.

Sawayama, S., Inoue, S., Dote, Y., & Yokoyama, S. Y., (1995). CO_2 fixation and oil production through microalga. *Energy Convers Manag.*, *130*, 652–658.

Setchell, K. D., Lawson, A. M., Borriello, S. P., Harkness, R., Gordon, H., Morgan, D. M., Kirk, D. N., et al., (1981). Lignan formation in man microbial involvement and possible roles in relation to cancer. *Lancet*, *2*, 4–7.

Sheehan, J., Dunahay, T., Benemann, J., & Roessler, P., (1998). A look back at the U.S Department of energy's aquatic species program, Biodiesel from algae. *National Renewable Energy Laboratory* (pp. 758–767). Golden, CO.

Singh, R. N., (1961). *Role of Blue-Green Algae in Nitrogen Economy of Indian Agriculture* (pp. 83–98). Indian Council of Agricultural Research, New Delhi.

Singh, S., Kate, B., & Banerjee, U. C., (2005). Bioactive compounds from cyanobacteria and microalgae: An overview. *Crit. Rev. Biotechnol.*, *125*, 73–95.

Spolaore, P., Joannis-Cassan, C., Duran, E., & Isambert, A., (2006). Commercial applications of microalgae. *Journal of Bioscience and Bioengineering*, *101*(2), 87–96.

Tabassum, R., Roshan, K., Yadav, R., Dhar, D. W., & Bhatnagar, S. K., (2012). Selected growth attributes in cyanobacterial isolates from Rohilkhand region of Uttar Pradesh. *Vegetos*, *25*(1), 174–177.

Tanticharoen, M., Reungjitchachawali, M., Boonag, B., Vondtaveesuk, P., Vonshak, A., & Cohen, Z., (1994). Optimization of gamma linolenic acid (GLA) production in *Spirulina platensis*. *J. Appl. Phycol.*, *6*, 295–300.

Tsao, R., & Deng, Z., (2004). Separation procedures for naturally occurring antioxidant phytochemicals. *J. Chromatogr.*, *812*, 85–99.

Venkataraman, G. S., (1981). *Blue-Green Algae for Rice Production, a Manual for its Promotion* (Vol. 46, p. 102). Food and Agricultural Organization, Italy.

Wen, Z., & Chen, F., (2003). Heterotrophic production of eicosapentaenoic acid by microalgae. *Biotechnol. Adv.*, *21*, 273–294.

While-Pratt, D. E., (1992). In: Huang, M. T., Ho, C. T., & Lee, C. Y., (eds.), *Phenolic Compounds in Food and Their Effects on Health* (Vol. II, pp. 2–7.). Washington, DC, American Chemical Society.

Williams, P. J. L. B., & Laurens, L. M., (2010). Microalgae as biodiesel and biomass feed stocks, review, and analysis of the biochemistry, energetics and economics. *Energy and Environmental Science*, *3*(5), 554–590.

Xu, H., Faber, C., Uchiki, T., Racca, J., & Dealwis, C., (2006). Structures of eukaryotic ribonucleotide reductase I define gemcitabine diphosphate binding and subunit assembly. *Proc. Natn. Acad. Sci. USA.*, *103*(11), 4028–4033.

CHAPTER 5

MICROALGAL BIOFUELS: CURRENT STATUS AND OPPORTUNITIES

ANURAG YADAV[1] and KUSUM YADAV[2]

[1]*Department of Microbiology, College of Basic Science and Humanities, S.D. Agricultural University, S.K. Nagar, Gujarat – 385506, India*

[2]*Department of Biochemistry, University of Lucknow, Lucknow, Uttar Pradesh – 226007, India*

5.1 INTRODUCTION

Currently, approximately 90% of the energy is driven by burning fossil fuels, and the rest of the 10% from green energy sources (Demirbas, 2010). Due to the increasing load of the growing population and anticipated huge future land demands, the biofuels could not be efficiently generated on land. The first generation biofuels failed to achieve commercial viability due to the involvement of arable land and crop-based feedstock. It has been reported that the use of arable land and clearing of forest for biofuel production actually enhances greenhouse gas emission compared to fossil fuel combustion (Clearing, 2008). Moreover, such practices negatively impact the food production chain (Johansson and Azar, 2007).

Exploring the potential of marine algae for biofuel generation turned out to be a strong driver for making biofuel generation more sustainable. Notably, marine biomass has a higher energy potential of 100 EJ/year compared to land biomass (22 EJ/year) (Chynoweth et al., 2001). Algae are aquatic plants that lack roots, stems, and leaves. They are enriched with active compounds like vitamins, carotenoids, proteins, and lipids, and are the fastest-growing organism with a growth rate faster than plants per unit area. They can yield 7–31 times higher oil content compared to oil palms (Demirbas and Demirbas, 2011). Algae are efficient photosynthesizers (3–8%) compared to plants (0.5%) (Lardon et al., 2009). Algal biomass has the advantage of high productivity, broader working temperature range, high nutrient content, salinity survival, and no strives with land crops (Lüning and Freshwater, 1988; Jones and Mayfield, 2012). Optimal algal biomass production requires appropriate temperature, suitable nutrients, effective gas exchange, and provision for light with a desirable wavelength range for photosynthesis. Algae chiefly require the sources of C, N, and P along with traces of Si, Ca, Mg, K, Fe, Mn, S, Zn, Cu, and Co (Christenson and Sims, 2011) for growth.

Algae possess greater than 50% of sugar content which can be used in fermentation to obtain bioethanol (Wi et al., 2009). Algae are generally referred to as microalgae, cyanobacteria (CB), and macroalgae (seaweed). Microalgae could be divided into three major groups: brown seaweed (Phaeophyceae), red seaweed (Rhodophyceae) and green seaweed (Chlorophyceae). Among all the groups, the algae from class Chlorophyceae and Bacillariophyceae are potent members for biofuel production. They are highly efficient compared to terrestrial plants in fixing CO_2. Sometimes CB are also reported under microalgae due to their small size. Macroalgae are larger in size and are not fully tapped for their potential in biofuel production. However, most of these microalgae have not been commercialized due to the higher cost of cultivation medium and inferior biomass production. Among all algal groups, microalgae seem economically reliable feedstock for biodiesel production. They can be grown throughout the year. In addition, the amount of oil produced by microalgae fairly exceeds the oil yield of the best feedstock crop.

Microalgae could be efficiently produced in photobioreactors (PBR) and raceway ponds. During their growth, microalgae can treat wastewater and remove harmful gases from power plants. They have a doubling time (DT) of 24 h which generally reduces to 3.5 h during exponential growth. Microalgae are enriched with oil content as high as 80% of biomass dry weight. The oil contents between 20–50% are not uncommon in microalgae (Chisti, 2007). Microalgal farming does not require the application of herbicides and pesticides. In addition, the residual biomass left after oil extraction can be used as a protein source for animal feed, fertilizer, or can be fermented to obtain biomethane or bioethanol.

During the previous few decades, the cost of biodiesel production stayed higher due to our over-dependence on food crop feedstocks (first-generation biofuels). There are four generations of biofuels known. The first generation biofuel system employs food crops like sugarcane, beets, and maize for generating biofuel. The second-generation system deals with waste materials and other non-food crops enriched with lignocellulose like wood chips, pellets, animal waste, forest, and crop residues, landfill gas as fuel.

Whereas microalgal derived biofuel is considered under the third generation. The fourth-generation biofuel involves genetic modification of microalgae for higher CO_2 fixation and lipid production.

As the global price of food crops are continuously increasing, the cost of first-generation biofuels always stayed higher than petroleum fuels. The second-generation biofuel system was introduced as an alternate of first-generation. The feedstock for second-generation biofuels is obtained from the processing of primary types and includes bioethanol, butanol, biodiesel, and biohydrogen (Dragone et al., 2010; Russo et al., 2012). The third-generation biofuel system uses microalgae for biofuel, which has reduced the overall cost of operation. However, even with this system, the current cost of biodiesel production is higher than the market natural fuel price. Efforts are underway to reduce the cost of biofuel production from third-generation biofuel systems. Such modifications fall under the fourth generation biofuel system. The measures involve screening best microalgal strains, strain improvement of useful microalgae, optimal nutrient sourcing and utilization, harvesting, biofuel extraction, product refining, and reuse of residual biomass. Table 5.1 lists some of the advantages and disadvantages of using marine microalgal biofuel.

The efforts of utilizing microalgal feedstocks were initiated quite early and the initial microalgal biomass production units were started at USA, Germany, and Japan (Burlew, 1953) which was followed by countries like Israel, Czech Republic, and Taiwan (Shelef and Soeder, 1980). Now the technology has spread to China (Li et al., 2011b), Iran (Najafi et al., 2011), Malaysia (Lim and Teong, 2010), and several other countries. Table 5.2 lists country-specific screening of new microalgal species with bioenergy potentials.

TABLE 5.1   Advantages and Disadvantages of Using Microalgae for Biofuel Production

Advantages	Disadvantages
High growth rate	Lower biomass yield
Minimal land requirement	The initial cost of investment is higher
Higher photosynthetic efficiency	Only useful for countries with longer coastlines
Higher microalgal diversity improves the chances of screening ideal species	Land bioremediation is not possible
Less land required	Environmental specificity of microalgae restricts large-scale usage
Negligible effect on freshwater resources	Can deplete non-renewable resources needed for oil extraction (e.g., phosphorous)

TABLE 5.1 *(Continued)*

Advantages	Disadvantages
No impact on global food prices and supply	Limited information of microalgal versatility
Environmentally beneficial	Poses a risk to animals
The versatile fuel source for automobiles, machines, and jets	Produces unstable biodiesel with many polyunsaturated compounds
Oil processing requires less energy	Oil extraction from microalgal cell wall needs additional steps
Produce numerous byproducts of commercial importance	Costlier than fossil fuels
Higher combustion efficiencies	–
Higher lipid content compared to plant-based biofuels	–
Algal oil extracts can be used as livestock feed	–
Unlike plant biomass feedstocks of the microalgae could work efficiently in colder climates	–
Nontoxic	–

TABLE 5.2   Country-Wise Algae Screened for Biodiesel Production

Country	Algae Screened	References
China	*Scenedesmus* sp. Y5, *Scenedesmus* sp. Y7, *Chlorella* sp. Y9	Chen et al., 2014
Brazil	Unknown LEM-IM 11, *Botryococcus braunii*, *Chlorella vulgaris*	Sydney et al., 2011
Iran	*Amphora* sp., *Dunaliella* sp., *Chlorella vulgaris*	Talebi et al., 2013
Sweden	*Skeletonema marinoi* GF-0410-J	Johansson, 2012
Singapore	*Nannochloropsis*	Doan et al., 2011
Germany	*Scenedesmus obliquus*	Abomohra et al., 2013

5.2   MICROALGAL BIOFUEL TYPES

5.2.1   BIOHYDROGEN

Biohydrogen is biological hydrogen produced by anaerobic digestion of several types of biomass. Hydrogen as a fuel has the advantage of zero greenhouse gas emission during combustion (Nasr et al., 2013a) with greater energy per unit weight compared to other fuels (Nasr et al., 2013b). It is also used as a fuel cell and could be produced from several types of feedstocks, mostly through dark fermentation using sugars. However, the use of renewable feedstock could improve the economic viability and sustainability of the process. Adopting biohydrogen as fuel has several advantages. It is a renewable source that consumes CO_2 during growth and has a negligible environmental impact compared to fossil fuels. However, the intensity of hydrogen produced by natural microalgal strains is inefficient, economically non-viable, and requires process optimization. Optimization of H_2 photoproduction by microalgae requires O_2 tolerant hydrogenases since the active site metal cluster of hydrogenase A is O_2 sensitive (Dubini and Ghirardi, 2015). Through gene shuffling a diverse hydrogenase recombinant library could be made to screen O_2 tolerance ability and stability of microalgae (Nagy et al., 2007). With a more sophisticated approach, the natural diversity of hydA could be screened using degenerate PCR primers (Boyd et al., 2009).

5.2.2   BIOMETHANE

Biomethane is a natural gas produced during anaerobic digestion of various kinds of biomass. Biomethanation involves microorganisms that degrade and stabilize organic wastes in the absence of oxygen. The process generates methane and biomass, which can be used as fertilizer in arable lands (Rajagopal et al., 2011). The advantages of biomethanation involve low energy input, low waste production, and environment-friendly energy recovery (Xia et al., 2012). The factor that demonstrates the economic viability of biomethane generation is usability of sludge left after oil extraction from microalgal cells, which could recover additional energy from oil extracted microalgal cells (Sialve et al., 2009). Biogas (biomethane) formation from microalgal biomass is a neglected approach owing to the associated limitations like anaerobic degradation resistant cell wall, the release of toxic metabolites by microalgal cells that restricts the growth of anaerobic bacteria, and fermentation unfriendly C:N ratio of biomass (Chynoweth et al., 1993; Wu et al., 2010). However, a study confirmed the commercial viability of marine microalgal biomass for bioenergy production (Harun et al., 2011). In this study, multi-variant calculations were performed to estimate the energetics and profitability of marine microalgae.

For efficient biogas production the microalgal technology has to over-come difficulties like the resistance of microalgal cell wall to degrade under anoxic conditions, antagonism of anaerobic bacteria during fermentation, and non-useful C:N ratio of biomass (Mata-Alvarez et al., 2000; Wu et al., 2010). Fortunately, most of the marine microalgae possess a high content of polysaccharides and lipid, are free from the non-degradable type of lignocel-luloses (Vergara-Fernández et al., 2008), and hold a higher growth rate. The possibility of harvest from aquifers presents them as the preferred choice for methane generation (Stephens et al., 2010).

5.2.3 BIOETHANOL

Bioethanol or ethyl alcohol is a volatile liquid which is made from the fermentation of sugary, starchy, and cellulosic biomass. Bioethanol is prepared from yeast fermentation containing glucose derivatives in the medium. Bioethanol could be one of the alternatives of petroleum due to similar physical and chemical properties with gasoline (Naik et al., 2010). The first generation bioethanol used sugar cane, sugar beet, corn, and wheat, making the process highly debatable due to food security and cost issues. The second-generation bioethanol production used lignocellulosic biomass feedstock with no food security issues. Such kind of biomass is present in huge quantities on earth. But due to the additional steps involved in pre-treatment for breaking apart lignin and cellulose (Cardona and Sánchez, 2007), the cost of processing cannot be reduced beyond certain limits. Third generation bioethanol production refers to the use of microalgae as they are known to store carbohydrates in their cells, which is used as a substrate in bioethanol production. Microalgae specifically store carbohydrate in outer (e.g., pectin, agar, alginate), or inner (e.g., cellulose, hemicellulose) cell wall layers and within the cell (e.g., starch) (Chen et al., 2013).

Few microalgal species are efficient in producing high percentages of carbohydrates instead of oils. The screening and use of such species is required for extracting fermentable sugars. An estimate shows that microalgae can roughly produce 5000–15,000 gal of ethanol/acre/year (46,760–140,290 L/ha) (Nguyen, 2012), which is way higher than yields obtained from crop feedstock. The fourth-generation bioethanol setup involves genetic modifi-cations of microalgae for optimized output. In seaweeds like red algae, the carbohydrate fraction is altered by the formation of agar, a linear polymer of galactose. The research is underway for developing methods to break off galactose units from agar and glucose from cellulose for higher ethanol yield (Yoon et al., 2010). Bioethanol production from microalgae has the additional advantage of lower lignin and hemicellulose components compared to plant cells (Harun et al., 2010). Bioalcohol production from microalgae is simpler than plant wastes as their cells are devoid of lignin and hemicellulose. Hence, negligible enzymatic pre-treatment steps are needed for extracting sugars from cells. A simple mechanical shear releases fermentable sugars from microalgal cells. However, microalgal feedstock still faces drawbacks due to the lower concentration of fermentable sugars compared to feedstock crops like maize.

5.2.4   BIODIESEL

The biodiesel production from microalgae is an area of core interest in research. Biodiesel is a renewable form of diesel composed of long-chain alkyl esters. Biodiesel is fatty acid methyl ester (FAME) produced by transesterification of vegetable oils, microalgal-derived oils, and animal fats by adding methanol. The biodiesel has an equivalent composition like petroleum diesel but the former is biodegradable and non-toxic and has a low emission rate (Krawczyk, 1996). Biodiesel is prepared through blending, microemulsions, thermal cracking (pyrolysis), and transesterification (Ma and Hanna, 1999). Transesterification is the most preferred method of biodiesel generation process as it reduces the viscosity of oil or fat. Although the viscosity of oil can be moderated using solvents and microemulsions, nevertheless due to the problem of carbon deposition over engines and lubricating oil contamination, such methods are rarely used. The biodiesel after blending with petroleum-derived diesel can be used in diesel engines without any structural modification of engine. Biodiesel production and microalgal life cycle analysis are crucial for calculating the net energy consumed and carbon generated until the last step of fuel generation. By quantifying the inputs and outputs, measures can be adopted for making the process more sustainable. Figure 5.1 describes some major step in biodiesel production from microalgae.

FIGURE 5.1   Major steps in converting microalgal biomass to biodiesel.

5.2.5   BIOJET FUEL

Biojet fuel is the newest form of biofuel which has tremendous potential in the aviation industry as fuel is one of the biggest operating costs of aviation (Adhikari, 2018). The global aviation consumes approximately $17.88–20.27\times10^{10}$ L fuel annually (Stratton, 2010). Although bioethanol and biodiesels are used in automobiles but due to lower octane values, are unsuitable for aviation. Moreover, there are stringent specifications of the American Society for Testing and Materials (ASTM D7566-09) for maintaining the quality of jet fuel. As per the guidelines, the aviation fuel must be safe, usable in a broad range of operational conditions and should have a high performance rate. In addition, the produced biojet fuel has to be sustainable, cost-effective, and greener. Meeting such specifications is a challenge for the biofuel industry and requires core research in optimized microalgal biomass production and biojet fuel production. The main driver for pushing the research in this area is the uncertainty over petroleum oil prices which significantly affect air carriers. The price of petroleum-derived jet fuel proportionates to crude oil value. The aviation industry has already started testing biofuels for fuelling jets to reduce carbon footprint (Hari, 2015).

5.3   ECONOMICS OF MICROALGAL BIOFUELS

The global shortage of fossil fuel supply during the 1970s raised worldwide interest in alternative fuel sources (Horowitz, 2005). Since then, considerable work has been done in biofuels for designing self-sustainable technology. Such transformation requires structural modification of diesel engine to accommodate highly viscous and less volatile fuel, biodiesel. Several methods like pyrolysis, blending, and microemulsification are available to lower the viscosity of biodiesel.

The biomass production cost of microalgae can be further reduced by utilizing the leftover after oil extraction as animal feed, fertilizer, for obtaining biogas through fermentation or pyrolysis to extract useful chemical compounds (Danielo, 2005; Rana and Spada, 2007). Due to the large-scale production of microalgae and seaweeds in the coastal area, the economy of local fishermen could be uplifted (Lim and Teong, 2010). The family of fishermen could be taught the methods of seaweed cultivation. Similar schemes are already running with Governmental and non-Governmental support in several countries (Buschmann et al., 2001; Neish, 2013; Sievanen et al., 2005).

Such leaps in the technological development of biofuel generation are crucial for uplifting the nation's self-esteem and hold numerous benefits. The direct benefits could be in the form of revenues generated from biodiesel export. Moreover, such moves can fetch more job opportunities for the country. It will also create motivational and competitive spur for other countries. Corresponding growth opportunities could be more significant for developing countries because such technological application and commercialization could transform a nation to become globally competitive.

5.4   STEPS IN ACHIEVING THE SUSTAINABILITY

5.4.1   MICROALGAL SCREENING

The proper selection of microalgae is crucial for optimizing oil output, which requires careful screening of microalgae for high lipid content (Dasgupta and Nayaka, 2017). The type and amounts of lipids produced by microalgal species influence the quality of biofuel. Therefore, screening local microalgal species should be prioritized due to their geographic, climatic, and ecological competitive advantage over other algal types. It is noteworthy to mention that change in microenvironment alters the biomass yield of microalgae; therefore, the screening process must include an in-depth understanding of an entire taxonomic group of microalgae over the extended geographical area (Rismani-Yazdi et al., 2011). By correlating, the data obtained from microalgal taxonomy, oil content, and environmental growth conditions, predictive tools can be developed which could screen microalgae more effectively. The advanced statistical tools can assist in the screening and identification of effective strains. The microalgal phylogenetic study could facilitate the screening of similar strains with bioenergy production ability.

The biomass obtained from one week microalgal growth roughly equals the biomass gained by higher crops in one season which represents its edge over feedstock crops (Hall and Benemann, 2011). As a result, microalgae, and seaweeds are considered ideal feedstocks for bioethanol and biodiesel production. They can grow photosynthetically, assimilate CO_2, possess a high growth rate, can easily grow in oceans, and require less space in indoor bioreactors. In addition, they can grow heterotrophically in absence of light using organic carbon sources, or mixotrophically in presence of light using CO_2 from photosynthesis and organic carbon sources (Pandey et al., 2013), thereby reducing the cost associated with lighting in close bioreactors (Chen and Chen, 2006).

Like other microorganisms, microalgal cells are commonly isolated on enrichment medium. Fatty acid profiling is also a reliable method for screening potent biodiesel producing strains (Talebi et al., 2013). The composition of FAME is a direct marker of the fuel characteristics. Hence, FAME profiling can be used to predict the fuel value over physicochemical methods of determination. Nowadays automated single-cell isolation systems are available to screen microalgal cells through flow cytometry-based on chlorophyll autofluorescence (CAF) and through green autofluorescence (GAF) by distinguishing microalgal species and strains. In one study microalgal cells were screened using flow cytometry through 2D distribution of microalgal cells for red fluorescence (for CAF) against forward light scattering (to represent cell size) (Doan et al., 2011).

Microalgae are screened on the basis of growth kinetics, biomass concentration, lipid content, and profiling of fatty acids (Doan et al., 2011). Due to metabolic diversity and capacity to mitigate CO_2, large-scale screening of microalgae is required. Microalgal groups have diverse growth requirements. As a result, they can grow in varied places. Moreover, environmental conditions influence microalgal populations and growth. Therefore, it is wise to screen microalgal strains from vivid sites for traits linked with higher lipid content, higher growth rate, and scalability for large-scale cultivation.

5.4.2 MICROALGAL STRAIN IMPROVEMENT

Microalgal strain improvement is a crucial step in the commercial development of fermentation processes for biofuel generation. Like CB, microalgae hold advantages like high rate of triacylglycerol accumulation, synthesis of starch, and efficient coupling of photosynthetic electron transfer to H production. By improving microalgal strains, useful attributes of microalgae could be augmented. Obtaining a higher conversion of solar energy into biomass requires molecular intervention to bioengineer the microalgal photosynthetic apparatus like modification of light-harvesting pigments, and reduction in light saturation effects. In addition, the microalgal cell can be bioengineered for altering metabolism towards lipid or carbohydrate production instead of growth. Some remarkable advances have been made in improving microalgal strains by manipulating the cellular metabolism of organisms (Radakovits et al., 2010). In the near future, these advances could be extended to industry for optimizing biofuel yield. The methods of microalgal strain improvement are as follows.

5.4.2.1 MUTATIONS

Although, several newer methods are available for microalgal strain improvement, however, mutation, and selection (random screening) are preferred
due to lower cost, the simplicity of procedure and quicker outcomes. Mutagens are of two types, chemical, and physical. Chemical mutagens include
alkylating agents like methylnitronitrosoguanidine (MNNG) and ethyl
methanesulfonate (EMS). Physical mutagens include UV rays, gamma rays,
and heavy-ion beams. UV based mutations are easy to perform without the
need for specific equipment.

Mostly temperature-sensitive microalgal mutants are developed which
have mutations in cell cycle regulators. The temperature-sensitive mutants,
which are restricted to a specific temperature are in use for lipid biosynthesis. The temperature switch halts the microalgal growth after a specific
temperature, thus funneling the photosynthetic energy for lipid and starch
biosynthesis. This approach has been successfully tested on *Chlamydomonas
reinhardtii* and *Chlorella vulgaris* (Yao et al., 2012). However, the outcomes
of mutations are not always reliable due trial and error approach. Moreover,
extra care and precautions are required for the maintenance and survival of
mutants.

5.4.2.2 ALTERATION OF GENOME

From the past two decades, molecular research has enriched our understanding of eukaryotic microalgal gene expression and is facilitating workers
to manipulate microalgal metabolism through metabolic engineering (Cheng
and Ogden, 2011; Singh et al., 2011). Metabolic engineering is paving the
path for developing strains optimized for biofuel production through carbon
partitioning, manipulation of triacylglycerides (TAG), and by improving
biohydrogen production (Sander and Murthy, 2010; Yacoby et al., 2011).
The core research in microalgal molecular engineering could (i) improve
biomass output by enhancing photosynthetic efficiency, (ii) improve the
oil content of biomass, and (iii) provide greater environmental adaptability
like higher temperature tolerance that can reduce the cost associated with
temperature regulation of microalgae (Chisti, 2008).

Many transformation methods were successfully adopted for introducing
DNA fragment in microalgal cells, which include agitation with the help of
glass beads or silicon carbide whisker, electroporation, biolistic microparticle

bombardment, and *Agrobacterium tumefaciens*-mediated transfer. Microalgal transformants are isolated using selection markers of antibiotic resistance, fluorescens, or the marker with biochemical significance.

Algae from all the groups, including green red and brown algae, have been genetically transformed leading to stable or temporary expression of transgenes. The transgene stability was improved using appropriate codons, endogenous promoters and species-specific 5' and 3' intron sequences (Eichler-Stahlberg et al., 2009).

Microalgal genome sequencing has hugely facilitated genetic manipulation for achieving large-scale industrial outputs for future advantages. Lot of microalgal types including *Cyanidioschyzon merolae, C. reinhardtii, Micromonas pusilla, Ostreococcus lucimarinus, Ostreococcus tauri, Phaeodactylum tricornutum,* and *Thalassiosira pseudonana* have been sequenced (Radakovits et al., 2010). Additionally, efforts were made to sequence chloroplast as well as mitochondrial DNA and transcriptome from several microalgae (Radakovits et al., 2010). Recent attempts for establishing expressed sequence tag (EST) databases and sequencing nuclear, mitochondrial, and chloroplast genomes of several microalgae were successfully accomplished.

Significant breakthrough is required for improving the stability of gene manipulation methods in eukaryotic microalgae with the least number of steps. The major hurdle in microalgal genetic modification is the heterologous expression of enzymes. The enzymes which are laboratory or nature optimized may work with variable efficiencies under bioreactor, especially during scale up. In addition, it is imperative to screen strains that could work under saline conditions due to limited freshwater resources. Thus, enzyme systems optimized for work in the marine environment needs screening.

5.4.2.3 ALTERATION OF LIPID METABOLISM

On the basis of chemical composition, lipids can be classified into two types, polar (phospholipids and glycolipids) and non-polar (tri-, di-, and monoglycerides, waxes, and isoprenoid-type lipids) (Christie, 2003). Most of the microalgae are known to be rich in lipids (tri- and diglycerides, phospho-, and glycolipids) and are used as a nutrient supplement in aquaculture to fulfill the dietary requirement of aquatic organisms. The composition of lipid in feedstock influence the efficiency and yield of biofuel produced. Microalgal lipid composition varies with species, nutrition, environment, and developmental condition of cells.

Microalgal cells synthesize triglycerides when cellular C supply surpasses metabolic demand during nutritional starvation and environmental stresses (Roessler, 1990). During biosynthesis, the cellular growth as well as division cease and the triglycerides accumulated during the day are mobilized and metabolized to funnel energy for cell division in dark. Bioengineering the lipid biosynthesis could help in getting higher triglyceride concentrations in the cell. Successful modification of microalgal genome has been done to alter the biochemistry of microalgal lipids (Radakovits et al., 2011; Shunni et al., 2011). Sometimes, even after metabolic engineering, the desirable results are not obtained due to the delicate balance of core metabolic and energy storage pathways. It has been found that microalgal enhanced lipid production is obtained on the cost of reduction in cell division and growth rate by overexpressing ACCase (Dunahay et al., 1996).

5.4.2.4   ALTERATION OF RUBISCO AND OTHER ENZYMES

The ribulose-1,5-bisphosphate carboxylase/oxygenase (Rubisco) plays a pivotal role in plant biomass generation from carbohydrates by incorporating CO_2 into Ribulose 1,5-bisphosphate (RuBP). However, the lower affinity of Rubisco to CO_2 limits plant biomass generation. Efforts are underway to screen organisms possessing Rubisco with high CO_2 specificity. In two thermophilic red algae, namely, *Galdieria partita* and *Cyanidium caldarium* (Cyanidiophyceae) a strong Rubisco carboxylase specific activity was found, which was 2.4–2.5 times greater than Rubisco of higher plants (Uemura et al., 1997). Certain algae-like *Hydrodictyon africanum*, *Macrocystis integrifolia*, *Nereocystis luetkeana*, *Lessoniopsis littoralis*, *Laminaria saccharina*, and *Fucus serratus* has C_4 type characteristic and they fix CO_2 via β-carboxylation of phosphoenolpyruvate (PEP) (Kremer, 1981; Raven and Glidewell, 1978). Such studies are imperative to understand the functioning of Rubisco. Rubisco is the prime target of genetic engineering for improving photosynthetic efficiency. Recent advances in molecular biology have enriched our understanding of Rubisco functioning and its influence on photosynthesis. The cutting edge technological advances in genetic engineering are paving paths for modifying the catalytic properties of Rubisco.

It is evident that C_4 plants have high photosynthetic rates than C_3 flora due to evolved strategies for concentrating CO_2 in tissues. The C_4 plants can effectively operate under reduced CO_2 levels. Research is underway to incorporate strategies of CO_2 concentration in C_3 plants and microalgae

(Peterhansel et al., 2008). Other approaches include introducing CB derived CO_2/HCO_3^- transporter proteins in chloroplast membranes or bioengineer pathways to bypassing photorespiration (Whitney et al., 2010).

5.5   LARGE-SCALE MICROALGAL BIOMASS PRODUCTION

The large-scale microalgal biomass is produced using raceway ponds, tubular PBR, and microalgal heterotrophic bioreactors (Jacob-Lopes et al., 2018; Maroneze and Queiroz, 2018). Raceway ponds are shallow structures that involve the loop recirculation channel. The nutrients remain in continuous circulation through paddle wheels. They are the most commonly used ponds for mass microalgal production due to the low cost of operation, scalability, and flexibility (Chang et al., 2017). Also, heterotrophic microalgal cultivation uses conventional reactors with stirred tank and bubble column type setups. The conventional design, high scalability, ease of use, and cheaper operation costs make the process preferable for few microalgal species. However, due to a higher contamination rate and no means for light exposure the system is not commonly used. Recently, a newer system called tubular photobioreactor has been introduced for mass microalgal cultivation which consists of an array of transparent tubes constructed in the straight, bent, or spiral pattern. The system allows fair control over the availability of nutrition and sunlight. Due to the compact design, it can be installed indoors with artificial light supply.

5.6   IMPROVING MICROALGAL CULTIVATION SYSTEMS

5.6.1   OXYGEN SUPPLY

Broadly speaking, all organisms use similar metabolic pathway for respiration, the process in which O_2 is consumed and CO_2 is released. As per expectation, the metabolism of microalgae closely resembles that of higher plants. Oxygen supply is a critical factor in microalgal growth. Higher levels of aeration positively influence microalgal biomass (Griffiths et al., 1960). The experiment with *Chlorella* sp. proved that oxygen limitation reduces the microalgal biomass output (Wu and Shi, 2007). Microalgae from several genera show heterotrophism with considerable metabolic versatility and diversity. Heterotrophic algae absorb nutrients from a complex organic substance. Oxygen supply is considerably important in heterotrophic cultivation of microalgae. Heterotrophic growth culture is preferred over autotrophic systems due to the higher growth rate. In an experiment on *Chlorella* sp., the biomass production was found 5.5 times higher with the heterotrophic system, compared to autotrophic microalgae (Yang et al., 2000).

5.6.2   NUTRIENT SUPPLY

Every organism needs nutrition to sustain its livelihood. The microalgae could have an autotrophic or heterotrophic mode of nutrition. Autotrophic microalgae get their energy and nutrition through light by reducing CO_2 and oxidizing substances, mostly water, by releasing O_2. Heterotrophic microalgae obtain their nutrition from organic compounds produced by other organisms. Heterotrophic microalgae are preferred in closed bioreactors for mass cultivation. The concentration of microalgal cells in artificial culture systems is generally higher than those in nature. The microalgal cultures must be enriched with nutrients to make up for the deficiencies of the seawater. Microalgal mass culture requires nitrogenous (especially nitrate, ammonia, and urea) and carbonaceous macronutrients (CO_2 or HCO^{3-}). In addition, micronutrients like Mg^{2+}, Na^+, Ca^{2+}, SO^{4-} and Cl^-; trace elements; some chelating agent such as ethylenediaminetetraacetic acid (EDTA) and vitamins are also required (Richmond, 2008). Enhanced percentages of oil can be obtained from microalgae by regulating the nutrient supply. For example, reports show that nitrogen starvation of microalgal cells triggers the accumulation of lipids and carbohydrates (Illman et al., 2000; Ho et al., 2012).

5.6.3   MIXING AND GROWTH MEDIUM VISCOSITY

The carbon sources used for microalgal nutrition impart viscosity to the medium. The viscosity also increases with higher cell density due to the secretion of primary and secondary metabolites. Mixing is necessary for uniform flow of nutrients and gas exchange in the bioreactor. The commercial-scale microalgal cell mixing is aided by impellers and baffles or airlift bubble column systems (Perez-Garcia and Bashan, 2015).

5.7   MICROALGAL HARVESTING

For making the process, energy sustainable, efficient harvesting of microalgae is important. Microalgal biofuel harvesting is done through mechanical, chemical, electrical, and biological methods (Christenson and Sims, 2011) through centrifugation, flocculation, flotation, sedimentation, filtration or their combinations. Techniques like flocculation using magnetic microparticles (Vergini et al., 2016), magnetic membrane filtration (Bilad et al., 2013), sedimentation with the help of polymers are in use (Zheng et al., 2011). Some other electrochemical methods like electro-coagulation-filtration (Gao et al., 2010) and electrochemical harvesting are in use (Misra et al., 2015).

On the small scale, the oils from microalgae can be separated by sonification of freeze-dried biomass followed by extraction with organic solvents. However, due to the cost restrain, the large-scale fuel extraction is done by liquefaction of biomass through pyrolysis (Sawayama et al., 1999; Miao et al., 2004; Li et al., 2011a; Thangalazhy-Gopakumar et al., 2012). Large-scale biofuel production requires (1) bulk production of microalgal biomass, (2) recovery, (3) extraction of metabolites from biomass, and (4) purification of metabolite (Grima et al., 2003).

Due to the involvement of multiple steps in microalgal harvesting, the capital cost, and energy input increases. The cost of microalgal harvesting ranges between 20–30% (Mata et al., 2010) which can reach up to 50% of the total biomass cost (Greenwell et al., 2009). The microalgal biomass concentration is merely 1–2% in the growth medium. To extract some meaningful amount for fuel from scanty microalgal biomass, a huge sum of energy is required. The bio-oil harvesting technology is newer and holds the great possibility of improvement through innovation in the forthcoming years. The oil extraction from microalgae is costly, energy-intensive, and could be up to 30% of production cost (Alabi et al., 2009). Henceforth, newer methods are required for reducing the energy input.

5.8   FUTURE AND LIMITATIONS

In the forthcoming years, human society has to learn ways for efficient energy usage without altering lifestyles. The commercial production of biofuels holds the key for efficient energy usage. The current microalgal biofuel production process is not economically viable and environmentally stable. Therefore, modifications in the biofuel harvesting process are needed to accommodate the latest research outcomes for achieving economic viability to compete with petroleum-based oils. For this modification, several issues need rectification to be rectified like higher densities of microalgal oils. The microalgal oil holds higher densities than diesel with 10–20 times higher viscosities. As a result, they burn slowly, incompletely, and get deposited on diesel engine fuel injectors (Demirbas, 2008). In actual, we lack the technology to produce biofuel on a commercial scale. Most of the biofuel generation plants produce microalgal biofuel only up to pilot scale. The projections of many studies for commercial microalgal biofuel production are exaggerated and sometimes defy thermodynamic limits (Hall and Benemann, 2011).

Nevertheless, microalgal biofuels have clear potential to aid in environmental and economic sustainability. Microalgae used in biofuel production consume CO_2 for growth when their energy is released in the form of biofuel, the CO_2 is released. Superficially, the biofuel production seems carbon neutral as the amount of CO_2 sequester by microalgal cell equals that from burning biofuel. However, the steps involved in the production from microalgal biomass production to harvesting and processing are carbon positive and energy-intensive. In other words, the process consumes more energy than what is finally obtained along with a huge carbon load. The advanced technological leap is required to make the process carbon neutral or negative with lesser energy input. Apart from research, some other factors that may decide the commercial potential of microalgal biofuels are described below.

5.8.1 FUEL PRICE

The biofuel price relies on the cost of feedstock and production. The precise cost assessment of microalgal oil is difficult due to its dependability on several factors like the country of production, microalgal type, and site of biomass production. From current technologies, the cost of producing biofuels is approximately three times the current petroleum price (Kovacevic and Wesseler, 2010). Therefore, the major technological breakthrough is required to make the production cheaper than petroleum. Nevertheless, the potential yield of biomass is increasing with evolving technologies.

Biofuel prices can be regulated by implementing three major modification in oil extraction process (Robertson et al., 2011), which involves (1) optimization of microalgal growth for maximum alkane production, (2) engineer the pathway for direct alkane production, and (3) optimized alkane extraction method. Through metabolic engineering, microalgal strains with the trait of switching between the growth phase and energy storage step in the form of oil could be developed. The ideal condition would be the evolution of strains that could rapidly transform into biomass and then could easily stop the growth and switch to oil reserve mode.

5.8.2   TRADE BARRIERS

The development of biofuel supply chains for international trade is crucial for the environment and the nation's economy. Setup of such investment opportunities could offer constant, stable, and higher returns. However, these investments have failed to attract major investors and stakeholders, as they are perceived as risky and uncertain. Insufficient finance is the major trade barrier that needs critical discussion with policymakers, investors, and stakeholders. Initiation of such establishment from countries with longer coastlines could prove more beneficial to them. Such countries will have a competitive advantage over others for bulk microalgal biomass production. Developing countries can benefit most from this technology, as the cost of microalgal biomass production in such places would be much lower. However, the contribution of developing countries in global biofuel production is negligible due to technological and economic issues. In addition, most of the countries have implemented conservative policies towards biofuel industry due to food security and environmental issues. The import-related technical regulation for the biofuel characteristics are applicable in almost all the countries to ensure fuel safety and to protect the consumer. Biofuels are derived from several feedstocks, causing variation in chemical composition (variation in chain length and double bonds) which influences fuel quality specifications. Often the fuel quality specification varies with country. Such differences could be moderated by blending biofuels from various sources to meet country-specific needs for fuel quality and emissions.

5.8.3   GOVERNMENT POLICIES

Current biofuel demand far exceeds the supply. The huge demand-supply gap could be fulfilled through the exploration of newer channels of energy and commercial accommodation of newer technologies. The use of microalgae for biofuel generation is one such avenue which has an environmental benefit. Its use can reduce land competition to positively affect food prices and biodiversity. The Governments from almost all the countries, at different levels of support, are promoting biofuels for reducing greenhouse gas (GHG) emissions, security in energy supply, revenue generation and job creation (Lamers et al., 2014). It was difficult to adopt a firm policy for first and second-generation biofuel production due to higher prices of food grains and limited land resources. Nevertheless, the policy interventions were done which included tax rebates, lower fuel excise, and subsidies in infrastructure establishment (De Gorter and Just, 2009). The resurgence of fourth-generation microalgal technology could boost the biofuel industry by providing cost-competitive and natural bio-oil. For attaining the economic competitiveness, the biofuel producing industries should work in synergy with local communities for mutual benefit and future development. Therefore, similar to the first-, second-, and third-generation biofuel systems, the implementation of appropriate policies is desirable for the fourth-generation biofuel system (Lim and Teong, 2010). The policy reform in terms of tax incentives, governmental purchase of produced oil, the establishment of fuel quality standards, and public-private partnerships are the expectable initiatives from policymakers and the government. Such reforms will reduce the overall biofuel price to prove its viability and sustainability against fossil fuels (Lee, 2011). The Governments of several counties has initiated such moves. For example, the Philippines Government has made it mandatory to supplement 1% biodiesel in diesel fuel for governmental use.

5.8.4   FEEDSTOCK AVAILABILITY AND COST

Microalgae and seaweeds are desirable feedstocks for biofuel generation as their growth is season independent, possess a higher growth rate, and require less water and land compared to food crops. At the commercial level, continuous biomass supply for biofuel generation is difficult to maintain as the system has a complex supply chain which is influenced by season, the interest of farmers, landholders, project developers, and transport companies. The long-term contracts for securing a significant amount of feedstocks are rare as there is a lack of integration in supply chains. Moreover, each biofuel type requires specific conversion technologies to power engines and the specifications change with time. Also, the low liquidity of the international biofuel market has not attracted large-scale investments.

5.8.5   PUBLIC SUPPORT

Public awareness for newer biofuel technologies is important for social acceptance and improvement in consumer's energy behavioral patterns (Karytsas and Theodoropoulou, 2014). Before developing, a common conception about general awareness the deeper analysis of public belief system related to biofuel use is needed. Such studies should try to determine the interest of stakeholders in the development of renewable energy sources. From public surveys, it is evident that the public support in this field is negligible due to under-awareness (Balogh et al., 2015). Such studies should involve the support of social phycologists, consumer scientists and behavioral economists to observe the factors that influence the choice of fuel types; sociologists and political scientists to study the role of institutions that regulate the public perception, and, individual researchers to critically analyze the subject and outcomes in the long run (Wegener et al., 2014). Biofuel production has to overcome the issues of price restriction and negligible public support.

5.9 CONCLUSION

Oil plays a leading role in fulfilling our current energy demands and driving a nation's economy. However, our over-dependency on petroleum fuel has funneled several environmental problems. Due to the non-renewability factor and uncertainty over crude oil prices, it is imperative to edge our dependability on petroleum-based oils and endeavor newer, cleaner, and reliable sources. The biological generation of fuel is an exciting conception and is the focus of core research from few decades. However, expectable results did not turn out. There were great expectations with first and second-generation biofuels for generating locally made natural biofuel. Nevertheless, due to issues of fertilizer and pesticide use, food security, dependability on arable land, and water requirements, biofuel production failed to express its commercial viability. However, the idea of using marine microalgae for commercial biofuel generation has infused new aspirations in the biofuel industry. Microalgal biofuel generation prospects as the most sophisticated system that could reduce our dependency on natural oils. Adopting genetically modified microalgae as feedstocks for the fourth generation biofuel system seems efficient, faster, and carbon-neutral way of generating biofuels with the least use of land and water. However, the biofuel production cost needs to be pushed down for commercial viability. This requires the active participation of corporates to commercialize a technology, government to device a suitable policy and citizens to be aware of positive environmental aspects of biofuels.

In forthcoming years, the biofuel industry will sustain only if the economic harvest of microalgae and extraction of intracellular oils is made possible. Nevertheless, microalgal biofuels production could be made technically feasible and economically competitive with petroleum fuels, provided that microalgal biomass production, harvesting, oil extraction process is optimized by adopting newer technologies and methodologies. Mathematical data modeling through computer simulations may be applied for tapping the commercial advantages of microalgae. Data modeling of microalgae for biofuel generation process could optimize microalgal growth, bioreactor design as well as operation and product recovery processes. The biofuel output optimization in the form of the end product could be approached by co-modeling microalgal physiology, bioreactor types, and varied nutrient types to sort effective combinations.

KEYWORDS

- **biodiesel**
- **biomethane**
- **cyanobacteria**
- **marine microalgae**
- **microalgal biofuels**
- **strain improvement**
- **sustainability**

REFERENCES

Abomohra, A. E. F., Wagner, M., El-Sheekh, M., & Hanelt, D., (2013). Lipid and total fatty acid productivity in photoautotrophic fresh water microalgae: Screening studies towards biodiesel production. *Journal of Applied Phycology, 25*(4), 931–936.

Adhikari, D. K., (2018). Bio-jet Fuel. In: Kumar, S., & Sani, R. K., (eds.), *Biorefining of Biomass to Biofuels* (pp. 187–201). Springer, Cham.

Alabi, A. O., Bibeau, E., & Tampier, M., (2009). *Microalgae Technologies and Processes for Biofuels-Bioenergy Production in British Columbia: Current Technology, Suitability and Barriers to Implementation* (p. 88). British Columbia Innovation Council.

ATAG, (2009). *Beginner's Guide to Aviation Biofuels.* Air Transport Action Group.

Balogh, P., Bai, A., Popp, J., Huzsvai, L., & Jobbágy, P., (2015). Internet-orientated Hungarian car drivers' knowledge and attitudes towards biofuels. *Renewable and Sustainable Energy Reviews, 48*, 17–26.

Bilad, M. R., Discart, V., Vandamme, D., Foubert, I., Muylaert, K., & Vankelecom, I. F., (2013). Harvesting microalgal biomass using a magnetically induced membrane vibration (MMV) system: Filtration performance and energy consumption. *Bioresource Technology, 138*, 329–338.

Boyd, E. S., Spear, J. R., & Peters, J. W., (2009). [FeFe] hydrogenase genetic diversity provides insight into molecular adaptation in a saline microbial mat community. *Applied and Environmental Microbiology, 75*(13), 4620–4623.

Burlew, J. S., (1953). *Algal Culture: From Laboratory to Pilot Plant* (p. 357). Carnegie Institution of Washington, Washington, DC.

Buschmann, A. H., Correa, J. A., Westermeier, R., Del Carmen Hernandez-Gonzalez, M., & Norambuena, R., (2001). Red algal farming in Chile: A review. *Aquaculture, 194*(3/4), 203–220.

Cardona, C. A., & Sánchez, Ó. J., (2007). Fuel ethanol production: Process design trends and integration opportunities. *Bioresource Technology, 98*(12), 2415–2457.

Chang, J. S., Show, P. L., Ling, T. C., Chen, C. Y., Ho, S. H., Tan, C. H., Nagarajan, D., & Phong, W. N., (2017). Photobioreactors, In: Pandey, A., Negi, S., & Carlos, R. S., (eds.), *Current Developments in Biotechnology and Bioengineering* (pp. 313–352). Elsevier.

Chen, C. Y., Zhao, X. Q., Yen, H. W., Ho, S. H., Cheng, C. L., Lee, D. J., Bai, F. W., & Chang, J. S., (2013). Microalgae-based carbohydrates for biofuel production. *Biochemical Engineering Journal, 78*, 1–10.

Chen, G. Q., & Chen, F., (2006). Growing phototrophic cells without light. *Biotechnology Letters, 28*(9), 607–616.

Chen, X., He, G., Deng, Z., Wang, N., Jiang, W., & Chen, S., (2014). Screening of microalgae for biodiesel feedstock. *Advances in Microbiology, 4*(7), 365–376.

Cheng, K. C., & Ogden, K. L., (2011). Algal biofuels: The research. *Chemical Engineering Progress, 107*(3), 42–47.

Chisti, Y., (2007). Biodiesel from microalgae. *Biotechnology Advances, 25*(3), 294–306.

Chisti, Y., (2008). Biodiesel from microalgae beats bioethanol. *Trends in Biotechnology, 26*(3), 126–131.

Christenson, L., & Sims, R., (2011). Production and harvesting of microalgae for wastewater treatment, biofuels, and bioproducts. *Biotechnology Advances, 29*(6), 686–702.

Christie, W. W., & Han, X., (2003). *Lipid Analysis: Isolation, Separation, Identification, and Structural Analysis of Lipids* (p. 448). Elsevier Science.

Chynoweth, D. P., Owens, J. M., & Legrand, R., (2001). Renewable methane from anaerobic digestion of biomass. *Renewable Energy, 22*(1–3), 1–8.

Chynoweth, D., Turick, C., Owens, J., Jerger, D., & Peck, M., (1993). Biochemical methane potential of biomass and waste feedstocks. *Biomass and Bioenergy, 5*(1), 95–111.

Clearing, L., (2008). The biofuel carbon debt. *Science, 319*(5867), 1235–1238.

Danielo, O., (2005). An algae-based fuel. *Biofutur., 255*, 1–4.

Dasgupta, C. N., & Nayaka, S., (2017*). Comprehensive Screening of Micro- and Macroalgal Species for Bioenergy* (pp. 39–56). Algal Biofuels. Springer.

De Gorter, H., & Just, D. R., (2009). The economics of a blend mandate for biofuels. *American Journal of Agricultural Economics, 91*(3), 738–750.

Demirbas, A., & Demirbas, M. F., (2011). Importance of algae oil as a source of biodiesel. *Energy Conversion and Management, 52*(1), 163–170.

Demirbaş, A., (2008). Production of biodiesel from algae oils. *Energy Sources, Part A: Recovery, Utilization, and Environmental Effects, 31*(2), 163–168.

Demirbas, A., (2010). Social, economic, environmental and policy aspects of biofuels. *Energy Education Science and Technology Part B-Social and Educational Studies, 2*(1/2), 75–109.

Doan, T. T. Y., Sivaloganathan, B., & Obbard, J. P., (2011). Screening of marine microalgae for biodiesel feedstock. *Biomass and Bioenergy, 35*(7), 2534–2544.

Dragone, G., Fernandes, B. D., Vicente, A. A., & Teixeira, J. A., (2010). Third generation biofuels from microalgae. *Current Research, Technology and Education Topics in Applied Microbiology and Microbial Biotechnology, 2*, 1355–1366.

Dubini, A., & Ghirardi, M. L., (2015). Engineering photosynthetic organisms for the production of biohydrogen. *Photosynthesis Research, 123*(3), 241–253.

Dunahay, T. G., Jarvis, E. E., Dais, S. S., & Roessler, P. G., (1996). Manipulation of microalgal lipid production using genetic engineering. *Applied Biochemistry and Biotechnology, 57*(1), 223.

Eichler-Stahlberg, A., Weisheit, W., Ruecker, O., & Heitzer, M., (2009). Strategies to facilitate transgene expression in *Chlamydomonas reinhardtii*. *Planta, 229*(4), 873–883.

Gao, S., Yang, J., Tian, J., Ma, F., Tu, G., & Du, M., (2010). Electro-coagulation-flotation process for algae removal. *Journal of Hazardous Materials, 177*(1–3), 336–343.

Greenwell, H., Laurens, L., Shields, R., Lovitt, R., & Flynn, K., (2009). Placing microalgae on the biofuels priority list: A review of the technological challenges. *Journal of the Royal Society Interface, 7*(46), 703–726.

Griffiths, D., Thresher, C., & Street, H., (1960). The heterotrophic nutrition of *Chlorella vulgaris* (Brannon No. 1 strain). *Annals of Botany, 24*(1), 1–11.

Grima, E. M., Belarbi, E. H., Fernández, F. A., Medina, A. R., & Chisti, Y., (2003). Recovery of microalgal biomass and metabolites: Process options and economics. *Biotechnology Advances, 20*(7/8), 491–515.

Hall, C. A. S., & Benemann, J. R., (2011). Oil from algae? *Bioscience, 61*(10), 741–742.

Hari, T. K., Yaakob, Z., & Binitha, N. N., (2015). Aviation biofuel from renewable resources: Routes, opportunities and challenges. *Renewable and Sustainable Energy Reviews, 42*, 1234–1244.

Harun, R., Danquah, M. K., & Forde, G. M., (2010). Microalgal biomass as a fermentation feedstock for bioethanol production. *Journal of Chemical Technology and Biotechnology, 85*(2), 199–203.

Harun, R., Davidson, M., Doyle, M., Gopiraj, R., Danquah, M., & Forde, G., (2011). Technoeconomic analysis of an integrated microalgae photobioreactor, biodiesel, and biogas production facility. *Biomass and Bioenergy, 35*(1), 741–747.

Ho, S. H., Chen, C. Y., & Chang, J. S., (2012). Effect of light intensity and nitrogen starvation on CO_2 fixation and lipid/carbohydrate production of an indigenous microalga *Scenedesmus obliquus* CNW-N. *Bioresource Technology, 113*, 244–252.

Horowitz, D., (2005). *Jimmy Carter and the Energy Crisis of the 1970s: The 'Crisis of Confidence' Speech of July 15, 1979: A Brief History with Documents.* Boston: Bedford, St. Martin's.

IATA, (2013). *Report on Alternative Fuels.* International Air Transport Association.

Illman, A., Scragg, A., & Shales, S., (2000). Increase in *Chlorella* strains calorific values when grown in low nitrogen medium. *Enzyme and Microbial Technology, 27*(8), 631–635.

Jacob-Lopes, E., Zepka, L. Q., & Queiroz, M. I., (2018). *Energy from Microalgae* (p. 306). Springer International Publishing, Switzerland.

Johansson, D. J., & Azar, C., (2007). A scenario based analysis of land competition between food and bioenergy production in the US. *Climatic Change, 82*(3/4), 267–291.

Johansson, J., (2016). *Screening Microalgae Strains for Biodiesel Production on the Swedish West Coast.* https://bioenv.gu.se/digitalAssets/1595/1595192_bio602-vt16-jimmy-johansson-examensarbete-kandidat-15hp-920515-2474.pdf.

Jones, C. S., & Mayfield, S. P., (2012). Algae biofuels: Versatility for the future of bioenergy. *Current Opinion in Biotechnology, 23*(3), 346–351.

Karytsas, S., & Theodoropoulou, H., (2014). Socioeconomic and demographic factors that influence publics' awareness on the different forms of renewable energy sources. *Renewable Energy, 71*, 480–485.

Kovacevic, V., & Wesseler, J., (2010). Cost-effectiveness analysis of algae energy production in the EU. *Energy Policy, 38*(10), 5749–5757.

Krauss, C., (2008). *Taking Flight on Jatropha Fuel.* New York Times Online.

Krawczyk, T., (1996). Biodiesel-alternative fuel makes inroads but hurdles remain. *Inform, 7,* 801–815.

Kremer, B. P., (1981). C4-metabolism in marine brown macrophytic algae. *Zeitschrift für Naturforschung C, 36*(9, 10), 840–847.

Lamers, P., Rosillo-Calle, F., Pelkmans, L., & Hamelinck, C., (2014). *Developments in International Liquid Biofuel Trade, International Bioenergy Trade* (pp. 17–40). Springer, Dordrecht.

Lardon, L., Helias, A., Sialve, B., Steyer, J. P., & Bernard, O., (2009). Life-cycle assessment of biodiesel production from microalgae. *Environmental Science and Technology, 43,* 6475–6481.

Lee, D., (2011). Algal biodiesel economy and competition among bio-fuels. *Bioresource Technology, 102*(1), 43–49.

Li, D., Chen, L., Zhang, X., Ye, N., & Xing, F., (2011a). Pyrolytic characteristics and kinetic studies of three kinds of red algae. *Biomass and Bioenergy, 35*(5), 1765–1772.

Li, Y. G., Xu, L., Huang, Y. M., Wang, F., Guo, C., & Liu, C. Z., (2011b). Microalgal biodiesel in China: Opportunities and challenges. *Applied Energy, 88*(10), 3432–3437.

Lim, S., & Teong, L. K., (2010). Recent trends, opportunities, and challenges of biodiesel in Malaysia: An overview. *Renewable and Sustainable Energy Reviews, 14*(3), 938–954.

Lüning, K., & Freshwater, W., (1988). Temperature tolerance of Northeast Pacific marine algae 1. *Journal of Phycology, 24*(3), 310–315.

Ma, F., & Hanna, M. A., (1999). Biodiesel production: A review. *Bioresource Technology, 70*(1), 1–15.

Maroneze, M. M., & Queiroz, M. I., (2018). *Microalgal Production Systems with Highlights of Bioenergy Production, Energy from Microalgae* (pp. 5–34). Springer, Cham.

Mata, T. M., Martins, A. A., & Caetano, N. S., (2010). Microalgae for biodiesel production and other applications: A review. *Renewable and Sustainable Energy Reviews, 14*(1), 217–232.

Mata-Alvarez, J., Mace, S., & Llabres, P., (2000). Anaerobic digestion of organic solid wastes: An overview of research achievements and perspectives. *Bioresource Technology, 74*(1), 3–16.

Miao, X., Wu, Q., & Yang, C., (2004). Fast pyrolysis of microalgae to produce renewable fuels. *Journal of Analytical and Applied Pyrolysis, 71*(2), 855–863.

Misra, R., Guldhe, A., Singh, P., Rawat, I., Stenström, T. A., & Bux, F., (2015). Evaluation of operating conditions for sustainable harvesting of microalgal biomass applying electrochemical method using non sacrificial electrodes. *Bioresource Technology, 176,* 1–7.

Nagy, L. E., Meuser, J. E., Plummer, S., Seibert, M., Ghirardi, M. L., King, P. W., Ahmann, D., & Posewitz, M. C., (2007). Application of gene-shuffling for the rapid generation of novel [FeFe]-hydrogenase libraries. *Biotechnology Letters, 29*(3), 421–430.

Naik, S. N., Goud, V. V., Rout, P. K., & Dalai, A. K., (2010). Production of first and second generation biofuels: A comprehensive review. *Renewable and Sustainable Energy Reviews, 14*(2), 578–597.

Najafi, G., Ghobadian, B., & Yusaf, T. F., (2011). Algae as a sustainable energy source for biofuel production in Iran: A case study. *Renewable and Sustainable Energy Reviews, 15*(8), 3870–3876.

Nasr, M., Tawfik, A., Ookawara, S., & Suzuki, M., (2013a). Biological hydrogen production from starch wastewater using a novel up-flow anaerobic staged reactor. *Bioresources, 8*(4), 4951–4968.

Nasr, M., Tawfik, A., Ookawara, S., & Suzuki, M., (2013b). Environmental and economic aspects of hydrogen and methane production from starch wastewater industry. *Journal of Water and Environment Technology, 11*(5), 463–475.

Neish, I. C., (2013). Social and economic dimensions of carrageenan seaweed farming in Indonesia. In: Diego, V. D., Cai, J., Hishamunda, N., & Ridler, N., (eds.), *Social and Economic Dimensions of Carrageenan Seaweed Farming* (pp. 61–89). FAO Fisheries and Aquaculture Technical Paper.

Nguyen, T. H. M., (2012). Bioethanol production from marine algae biomass: Prospect and troubles. *Journal of Vietnamese Environment, 3*(1), 25–29.

OECD, (2012). *Green Growth and the Future of Aviation*. 27th round table on sustainable development.

Pandey, A., Lee, D. J., Chisti, Y., & Soccol, C. R., (2013). *Biofuels from Algae* (p. 348). Elsevier, Amsterdam.

Perez-Garcia, O., & Bashan, Y., (2015). Microalgal heterotrophic and mixotrophic culturing for bio-refining: From metabolic routes to techno-economics. *Algal Biorefineries* (pp. 61–131) Springer, Cham.

Peterhansel, C., Niessen, M., & Kebeish, R. M., (2008). Metabolic engineering towards the enhancement of photosynthesis. *Photochemistry and Photobiology, 84*(6), 1317–1323.

Radakovits, R., Eduafo, P. M., & Posewitz, M. C., (2011). Genetic engineering of fatty acid chain length in *Phaeodactylum tricornutum*. *Metabolic Engineering, 13*(1), 89–95.

Radakovits, R., Jinkerson, R. E., Darzins, A., & Posewitz, M. C., (2010). Genetic engineering of algae for enhanced biofuel production. *Eukaryotic Cell,* 9(4), 486–501.

Rajagopal, R., Rousseau, P., Bernet, N., & Béline, F., (2011). Combined anaerobic and activated sludge anoxic/oxic treatment for piggery wastewater. *Bioresource Technology, 102*(3), 2185–2192.

Rana, R., & Spada, V., (2007). Biodiesel production from ocean biomass. *Proceedings of the 15th European Biomass Conference and Exhibition from Research to Market Deployment* (pp. 2050–2053). ETA-Renewable energies and WIP renewable energy, Florence, Italy.

Raven, J. A., & Glidewell., S. M., (1978). C4 characteristics of photosynthesis in the C3 alga *Hydrodictyon africanum*. *Plant, Cell and Environment, 1*(3), 185–197.

Richmond, A., (2008). *Handbook of Microalgal Culture: Biiotechnology and Applied Phycology* (Vol. 577, p. 584). John Wiley and Sons, New York.

Rismani-Yazdi, H., Haznedaroglu, B. Z., Bibby, K., & Peccia, J., (2011). Transcriptome sequencing and annotation of the microalgae *Dunaliella tertiolecta*: Pathway description and gene discovery for production of next-generation biofuels. *BMC Genomics, 12*(1), 148. doi: https://doi.org/10.1186/1471-2164-12-148.

Robertson, D. E., Jacobson, S. A., Morgan, F., Berry, D., Church, G. M., & Afeyan, N. B., (2011). A new dawn for industrial photosynthesis. *Photosynthesis Research, 107*(3), 269–277.

Roessler, P. G., (1990). Environmental control of glycerolipid metabolism in microalgae: Commercial implications and future research directions. *Journal of Phycology, 26*(3), 393–399.

Russo, D., Dassisti, M., Lawlor, V., & Olabi, A., (2012). State of the art of biofuels from pure plant oil. *Renewable and Sustainable Energy Reviews, 16*(6), 4056–4070.

Sander, K., & Murthy, G. S., (2010). Life cycle analysis of algae biodiesel. *The International Journal of Life Cycle Assessment, 15*(7), 704–714.

Sawayama, S., Minowa, T., & Yokoyama, S. Y., (1999). Possibility of renewable energy production and CO_2 mitigation by thermochemical liquefaction of microalgae. *Biomass and Bioenergy, 17*(1), 33–39.

Shelef, G., & Soeder, C. J., (1980). *Algae Biomass: Production and Use, International Symposium on the Production and Use of Micro-Algae Biomass* (pp. 17–22). National Council for Research and Development, Acre, Israel. Elsevier, Amsterdam.

Shunni, Z., Zhongming, W., Changhua, S., Weizheng, Z., Kang, Y., & Zhenhong, Y., (2011). Lipid biosynthesis and metabolic regulation in microalgae. *Progress in Chemistry, 23*(10), 2169–2176.

Sialve, B., Bernet, N., & Bernard, O., (2009). Anaerobic digestion of microalgae as a necessary step to make microalgal biodiesel sustainable. *Biotechnology Advances, 27*(4), 409–416.

Sievanen, L., Crawford, B., Pollnac, R., & Lowe, C., (2005). Weeding through assumptions of livelihood approaches in ICM: Seaweed farming in the Philippines and Indonesia. *Ocean and Coastal Management, 48*(3), 297–313.

Singh, A., Nigam, P. S., & Murphy, J. D., (2011). Renewable fuels from algae: An answer to debatable land based fuels. *Bioresource Technology, 102*(1), 10–16.

Stephens, E., Ross, I. L., King, Z., Mussgnug, J. H., Kruse, O., Posten, C., Borowitzka, M. A., & Hankamer, B., (2010). An economic and technical evaluation of microalgal biofuels. *Nature Biotechnology, 28*(2), 126. doi: http://hdl.handle.net/1721.1/59694 (accessed on 11 June 2020).

Stratton, R. W., (2010). *Life Cycle Assessment of Greenhouse Gas Emissions and Non-CO$_2$ Combustion Effects from Alternative Jet Fuels* (pp. 1–144). PhD Thesis, Massachusetts Institute of Technology, Massachusetts, USA.

Sydney, E. B., Da Silva, T. E., Tokarski, A., Novak, A. C., De Carvalho, J. C., Woiciecohwski, A. L., Larroche, C., & Soccol, C. R., (2011). Screening of microalgae with potential for biodiesel production and nutrient removal from treated domestic sewage. *Applied Energy, 88*(10), 3291–3294.

Talebi, A. F., Mohtashami, S. K., Tabatabaei, M., Tohidfar, M., Bagheri, A., Zeinalabedini, M., Mirzaei, H. H., et al., (2013). Fatty acids profiling: A selective criterion for screening microalgae strains for biodiesel production. *Algal Research, 2*(3), 258–267.

Thangalazhy-Gopakumar, S., Adhikari, S., Chattanathan, S. A., & Gupta, R. B., (2012). Catalytic pyrolysis of green algae for hydrocarbon production using H$^+$ ZSM-5 catalyst. *Bioresource Technology, 118*, 150–157.

Uemura, K., Anwaruzzaman, Miyachi, S., & Yokota, A., (1997). Ribulose-1,5-bisphosphate carboxylase/oxygenase from thermophilic red algae with a strong specificity for CO$_2$ fixation. *Biochemical and Biophysical Research Communications, 233*(2), 568–571.

Vergara-Fernández, A., Vargas, G., Alarcón, N., & Velasco, A., (2008). Evaluation of marine algae as a source of biogas in a two-stage anaerobic reactor system. *Biomass and Bioenergy, 32*(4), 338–344.

Vergini, S., Aravantinou, A. F., & Manariotis, I. D., (2016). Harvesting of freshwater and marine microalgae by common flocculants and magnetic microparticles. *Journal of Applied Phycology, 28*(2), 1041–1049.

Wegener, D. T., Kelly, J. R., Wallace, L. E., & Sawicki, V., (2014). Public opinions of biofuels: Attitude strength and willingness to use biofuels. *Biofuels, 5*(3), 249–259.

Whitney, S. M., Houtz, R. L., & Alonso, H., (2010). Advancing our understanding and capacity to engineer nature's CO$_2$ sequestering enzyme, Rubisco. *Plant Physiology, 1*, 110.

Wi, S. G., Kim, H. J., Mahadevan, S. A., Yang, D. J., & Bae, H. J., (2009). The potential value of the seaweed Ceylon moss (*Gelidium amansii*) as an alternative bioenergy resource. *Bioresource Technology, 100*(24), 6658–6660.

Wu, X., Yao, W., Zhu, J., & Miller, C., (2010). Biogas and CH_4 productivity by co-digesting swine manure with three crop residues as an external carbon source. *Bioresource Technology, 101*, 4042–4047.

Wu, Z., & Shi, X., (2007). Optimization for high-density cultivation of heterotrophic *Chlorella* based on a hybrid neural network model. *Letters in Applied Microbiology, 44*(1), 13–18.

Xia, Y., Massé, D. I., McAllister, T. A., Beaulieu, C., & Ungerfeld, E., (2012). Anaerobic digestion of chicken feather with swine manure or slaughterhouse sludge for biogas production. *Waste Management, 32*(3), 404–409.

Yacoby, I., Pochekailov, S., Toporik, H., Ghirardi, M. L., King, P. W., & Zhang, S., (2011). Photosynthetic electron partitioning between [FeFe]-hydrogenase and ferredoxin: NADP+-oxidoreductase (FNR) enzymes *in vitro*. *Proceedings of the National Academy of Sciences USA, 108*(23), 9396–9401.

Yang, C., Hua, Q., & Shimizu, K., (2000). Energetics and carbon metabolism during growth of microalgal cells under photoautotrophic, mixotrophic and cyclic light-autotrophic/dark-heterotrophic conditions. *Biochemical Engineering Journal, 6*(2), 87–102.

Yao, S., Brandt, A., Egsgaard, H., & Gjermansen, C., (2012). Neutral lipid accumulation at elevated temperature in conditional mutants of two microalgae species. *Plant Physiology and Biochemistry, 61*, 71–79.

Yoon, J. J., Kim, Y. J., Kim, S. H., Ryu, H. J., Choi, J. Y., Kim, G. S., & Shin, M. K., (2010). Production of polysaccharides and corresponding sugars from red seaweed. In: Xiao, Z. H., & Alan, K. T. L., (eds.), *Advanced Materials Research* (pp. 463–466). Trans Tech Publications.

Zheng, H., Yin, J., Gao, Z., Huang, H., Ji, X., & Dou, C., (2011). Disruption of *Chlorella vulgaris* cells for the release of biodiesel-producing lipids: A comparison of grinding, ultrasonication, bead milling, enzymatic lysis, and microwaves. *Applied Biochemistry and Biotechnology, 164*(7), 1215–1224.

CHAPTER 6

BIOFUELS FROM MICROALGAE: FUTURE BIO-ENERGIES FOR SUSTAINABLE DEVELOPMENT

RAMGANESH SELVARAJAN,[1] CECILIA OLUSEYI OSUNMAKINDE,[2] TIMOTHY SIBANDA,[3] ANOFI ASHAFA,[1] and SANNIYASI ELUMALAI[4]

[1]Department of Plant Sciences, Faculty of Natural and Agricultural Sciences, University of the Free State-QwaQwa Campus, Phuthaditjhaba, South Africa

[2]Nanotechnology and Water Sustainability Research Unit, College of Science, Engineering, and Technology, University of South Africa, Florida, South Africa

[3]Department of Biological Sciences, University of Namibia, Windhoek, Namibia

[4]Department of Biotechnology, University of Madras, Chennai, Tamil Nadu, India

6.1 INTRODUCTION

Energy is a crucial factor on which the global economy literally runs (Patil, Tran, and Giselrød, 2008). It is anticipated that by 2030, the world's energy requirements will increase by more than 60% of the current status (Singh et al., 2014). Transportation is one of the fastest-growing sectors accounting for more than 30% of the world's total energy consumption (Atabani et al., 2012), and is expected to increase by an average of 2.2% per year from 2010 to 2040 (U.S. Energy Information Agency, 2013). The potential threat of global climate change is on the increase (Brennan and Owende, 2010), and upto one-fifth of CO_2 emissions are from the transport sector, which is now standing out as a significant contributor to global warming (GW) (Rawat et al., 2013).

Fossil fuels are the largest contributor of greenhouse gases (GHGs) to the biosphere. As of 2010, fossil fuel associated CO_2 emissions were 31.2 billion metric tons, and are projected to continue to rise up to 45.5 billion metric tons by 2040 (U.S. Energy Information Agency, 2013). If the evils caused by GW have to be avoided, alternatives to fossil fuels and compatible mitigation strategies are required to neutralize the excess CO_2 (Bilanovic et al., 2009). From an environmental perspective, depletion of fossil fuel reserves is a welcome "evil" since it has the potential to force a rise in the cost of crude oil, thereby pushing technologists into the search for fuels from alternate sources (Rawat et al., 2013).

Biofuels have the potential to replace existing fossil fuels, reinforce energy security, and reduce the emission of GHGs and other air pollutants (Bharathiraja et al., 2015). Biofuels such as bio-diesel, bio-ethanol, and bio-butanol are produced from different sources are listed in Table 6.1. However, in recent year's research into the use of algae as a feedstock for biofuels has grown rapidly (Mata et al., 2010; Savage and Hestekin, 2013). Biodiesel is an attractive alternate fuel for diesel engines because it can be made from any vegetable oil (edible or non-edible oils), used cooking oils, animal fats as well as microalgae oils (Chisti, 2007) which can reduce 70% of particulate emissions, 40% hydrocarbon emissions, and completely eliminates sulfur foul emissions including carbon monoxide (CO) and nitrogen oxides (NOx) (Blumberg et al., 2003).

The first generation fuels produced from edible plant oils like soybean, rapeseed, palm oil and mustard (Peterson and Auld, 1983; Kalam and Masjuki, 2002; Xie and Huang, 2006) had many disadvantages such as loss of biodiversity, excess utilization of water and increased greenhouse gas emissions (Chisti, 2007). Another major concern with the first generation fuels is their low energy yield which makes them impractical candidates for large scale production (Lang et al., 2001). Sources from non-edible crops such as Jatropha, almond, sesame, and residual parts of crops were used in second generation fuels (Bak et al., 1996; Patil et al., 2009; Ahmad et al., 2010). However, that option has not been exploited commercially and the use of plant oils or non-edible crops requires large areas of cultivable land which may not be available due to population demand and food security (Lee, 2011). Therefore, to meet the sufficient global energy requirement, and to rescue the agricultural land; aquatic biomass, and especially micro-algae is being considered as the most suitable petro-diesel alternative (Chisti, 2007).

TABLE 6.1 Sources of Biofuels

Biofuels	Sources	References
Bio-diesel (vegetable oils)	Canola	Kulkarni et al., 2006; Baroi et al., 2014
	Cotton seed	Nabi et al., 2009
	Groundnut	Bello and Agge, 2012; Oniya and Bamgboye, 2014
	Rapeseed oil	Peterson and Auld, 1983
	Safflower	Ilkiliç et al., 2011
	Soybean	Cao et al., 2005; Xie and Huang, 2006; Jaffar Al-Mulla et al., 2015
Bio-diesel (non-edible oils)	Almond	Abu-Hamdeh and Alnefaie, 2015
	Brassica carinata	Cardone et al., 2003; Bouaid et al., 2009
	Camelina	Fröhlich and Rice, 2005; Patil et al., 2009
	Jatropha	Shah et al., 2004; Lu et al., 2009; Patil et al., 2009
	Palm	Kalam and Masjuki, 2002
	Rice bran	Bak et al., 1996; Kumar, 2007
	Sesame	Ahmad et al., 2010; Dawodu et al., 2014
Bio-diesel (animal fats)	Tallow and Poultry fat	Goodrum et al., 2003; Bhatti et al., 2008; Panneerselvam et al., 2011
	Salmon oïl	Reyes and Sepúlveda, 2006; El-Mashad et al., 2008
	Fish oil	Preto et al., 2008; Lin and Li, 2009
Bio-diesel (microorganisms and seaweeds)	Bacteria	Atsumi and Liao, 2008; Lu et al., 2008; Escobar-Niño et al., 2014
	Fungi	Vicente et al., 2009; Zheng et al., 2012
	Macroalgae (seaweeds)	Nelson et al., 2002; Maceiras et al., 2011; Selvarajan et al., 2011
	Microalgae	Dayananda et al., 2010; Sanniyasi et al., 2011; Tan and Lin, 2011; Song et al., 2013; Selvarajan et al., 2014, 2015; Valdez-Ojeda et al., 2015; Vidyashankar et al., 2015
Bio-ethanol	Sugarcane	Dias et al., 2009; Cardona et al., 2010
	Rice straw	Binod et al., 2010
	Corn	Mojović et al., 2006
	Sweet sorghum	Reddy et al., 2005
	Microalgae	Harun et al., 2011; Hernández et al., 2015

TABLE 6.1 *(Continued)*

Biofuels	Sources	References
Bio-butanol	Barley straw	Qureshi et al., 2014
	Corn Stover	Qureshi et al., 2014
	Bacteria	Wang et al., 2014
	Macroalgae	Potts et al., 2012
	Microalgae residues	Cheng et al., 2015

6.2 MICROALGAE

Microalgae are microscopic hetero-autotrophic photosynthesizing organisms that are able to capture solar energy along with carbon dioxide for biomass production and converted to chemical energy (Fuentes-Grünewald et al., 2009; Patel et al., 2012). Microalgae are present in all existing earth ecosystems (Mata et al., 2010), and they fall under different groups such as *Chlorophyceae, Cyanophyceae, Chrysophyceae,* and *Bacillariophyceae* (diatoms). *Bacillariophyceae* represent the largest group of biomass producers on earth and dominant life form in the phytoplankton system (Bharathiraja et al., 2015). *Chlorophyceae* (green algae) are abundantly found in fresh and brackish water than in marine waters, and are efficient producers of lipids and other metabolites that can be used for biofuel production and other applications (Talebi et al., 2013). *Cyanophyceae* (blue-green algae/cyanobacteria (CB)) are found in a wide range of extreme environments, and are one of the richest sources of known and novel bioactive compounds including toxins with wide pharmaceutical applications (Santhose et al., 2011). *Chrysophyceae* (golden alga) are similar to diatoms and are excellent producers of lipids and carbohydrates, making them ideal for biofuel applications (Bharathiraja et al., 2015). There are many species within these groups as listed in Table 6.2. One common feature among them is that they exhibit rapid growth and high productivity to accumulate substantial quantities of lipids, variety of metabolites such as carbohydrates, proteins, pigments for potential and feasible biotechnological applications including biofuels (Demirbas, 2011).

TABLE 6.2   Potential and Feasible Microalgal Species of Different Groups for Existing and Future Biotechnological Applications

Algal Groups	Species	By-Products	Applications	References
Chlorophyceae	*Chlorella* sp.	Biomass, lipids, ascorbic acid	Food industry, biofuels	Xu et al., 2006; Xu and Hu, 2013; Selvarajan et al., 2015
	Dunaliella salina	Carotenoid, β carotene	Health food, nutritional supplement, feed	Pisal and Lele, 2005
	Haematococcus pluvialis	Carotenoids, astaxanthin	Food colorant, antioxidant, cancer-preventive properties, pigment for pharmaceuticals	Kobayashi et al., 1998
	Isochrysis galbana	Fatty acids	Animal nutrition	Custódio et al., 2014
	Scenedesmus sp.	Lipids and fatty acids	Biofuels	Arias-Peñaranda et al., 2013; Mandotra et al., 2014
	Nannochloropsis	Lipids and fatty acids	Biofuels	Ma et al., 2014
	Botryococcus braunii	Lipids-triglycerides and hydrocarbons	Biofuels	Banerjee et al., 2002
	Tetraselmis sp.	Fatty acids	Pharmaceutical and functional food industries	Custódio et al., 2014
	Chlamydomonas	Lipids and fatty acids	Biofuels	Morowvat et al., 2010
Cyanophyceae	*Spirulina platensis*	Phycocyanins, biomass	Health food, natural dye for health food and cosmetics (lipsticks and eyeliners) antioxidant, pharmaceuticals	Dartsch, 2008; Capelli and Cysewski, 2010
	Nostoc sp. *Lyngbya* sp. *Tolypohrix* sp.	Immune modulators, drug formulations	Pharmaceuticals, nutrition	Singh et al., 2011
	Phormidium	Carotenoids	Pharmaceuticals, nutrition	Santhose et al., 2011

TABLE 6.2 *(Continued)*

Algal Groups	Species	By-Products	Applications	References
Bacillariophyceae	*Phaedactylum tricornutum*	Lipids, fatty acids	Biofuels	Valenzuela et al., 2012
	Odontella aurita	Fatty acids	Pharmaceutical and functional food industries	Keerthi et al., 2013
	Nitzschia sp.	Lipids and fatty acids	Biofuel	Abou-Shanab et al., 2011
Chrysophyceae	*Prymnesium parvum*	Fatty acids	Biofuels	Culver, 2009

6.2.1 ADVANTAGES OF MICROALGAE

The utilization of microalgae for biofuels production has numerous advantages compared to other feedstocks (Raheem et al., 2015). To begin with, microalgae have a high growth rate in a short duration of time compared to land-based crops and it could be harvested continuously throughout the year. They undergo a simple cell division cycle and double up in mass by using photosynthesis to convert light energy into chemical energy (Amin, 2009; Mata et al., 2010; Raheem et al., 2015). Secondly, microalgae can tolerate different types of environments such as desert, arid, and semi-arid lands that are not suitable for conventional agriculture (Khan et al., 2009). They do not require any special nutrients and chemicals such as herbicides or pesticides; and can be cultivated in freshwater, saline/brackish water, and coastal seawater thus reducing costs and environmental impacts (Rawat et al., 2011). Microalgae can even grow in wastewater and in some industrial effluents and is exploited in bioremediation processes where its role is the removal of water contaminants as nutrients (Li et al., 2008). The tolerance of microalgae to high CO_2 levels allows high-efficiency CO_2 mitigation (Wang et al., 2008). Microalgae synthesize and accumulate large quantities of lipids to a significant portion of their biomass along with value-added co-products or by-products such as biopolymers, proteins, polysaccharides, and pigments (Khan et al., 2009). Because of this variety of high-value biological derivatives, microalgae can potentially revolutionize a number of biotechnology sectors including biofuels, bio-pharmaceuticals, aquaculture, nutrition, and food additives, and phycoremediation industries (Raja et al., 2008; Rosenberg et al., 2008; Mata et al., 2010). This chapter reports on recent advancements in upstream and downstream processes of microalgae biofuel production and its potential to replace fossil fuels for future energy needs.

6.3   ALGAL BIOFUEL PRODUCTION TECHNOLOGY

The illustration in Figure 6.1 shows the several steps involved in upstream and downstream process for the production of algal biofuels. The selection of microalgae from specific to certain environments is a key point to the successful biomass cultivation and lipid production of potential microalgae species for biofuel production (Brennan and Owende, 2010; Mutanda et al., 2011).

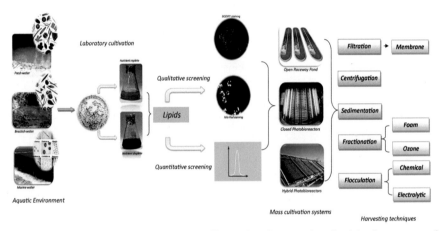

FIGURE 6.1   Simple schematic diagram illustrating the steps involved in the process of microalgae for biofuels production.

6.3.1   CULTIVATION TECHNIQUES

To establish an effective microalgae production the selection of a suitable cultivation method is of vital importance. Growing and collection of microalgae has been done over decades using the technologies of either the indoor growth system (photobioreactors (PBR)) or an outdoor system (ponds, lakes) (Chisti, 2007). The main differences and characteristics between open and closed culturing systems are summarized in Table 6.3.

TABLE 6.3 Characteristics Between the Open and Closed Pond Culturing Systems

Parameters	Open Pond	Closed System (Photobioreactors)
Ease of use of land and construction cost	High	Flexible
Water loss	Very high, may also cause salt precipitation	Low, and maybe high if water spray is used for cooling
Biomass quality	Difficult to control	Easy to control
Hydrodynamic stress on algae	Very low	Low-High
Gas transfer control	Low	High
CO_2 loss	High	Depending on pond depth
Temperature	Highly variable	Cooling often required
Startup time	6–8 weeks	2–4 week
Risk of pollution	High, difficult species control	Low pollution control, easy species control
Weather dependence	High light intensity, temperature, and rainfall dependent	Moderate intensity of light, cooling is required
Maintenance	Easy to clean and maintain Susceptibility to overheating and excessive oxygen is low	Hard to clean, susceptibility to overheating and excessive oxygen level is high
Cell density in culture and biomass concentration	Low between 0.1 and 0.5 g l^{-1}	High between 2 and 8 g l^{-1}
Applicability to variable species	Low	High
Repeatability	Low	High
Duration of culturing	Long	Relatively short

Open ponds are the most widely used systems for large scale outdoor microalgae cultivation. However, there are numerous classifications of open cultivation systems which may vary in size, shape, material used for construction design, type of agitation and inclination (Grobbelaar, 2000; Lee, 2001; Shen et al., 2009). Microalgae have been mass cultured in man-made circular ponds, tanks, shallow ponds, raceway open ponds and open cascades (Lee, 2001; Molina Grima et al., 2003; Chisti, 2007; Harun et al., 2010). The most common forms are shallow raceway ponds with a mixing done using a paddle wheel which allows for the penetration of light penetration, gas diffusion, and nutrient distribution (Christenson and Sims, 2011; Gallagher, 2011). The cultivation of algae is achieved in open pond system through the availability of sunlight and temperature is the major precursors for microalgae grown in open ponds. The open system culture has gained more popularity due to its ease of operation and cost effectiveness for the micro-algae biomass production (Harun et al., 2010; Christenson and Sims, 2011). In these open-culture systems, nutrients can be normally provided through runoff water from nearby land area or by channeling the water from sewage/water treatment plants. Generally, these cultivation systems are less expen-sive to build and operate, more durable than large closed reactors and with a large production capacity when compared with closed systems (Chisti, 2007; Harun et al., 2010). The disadvantages of open system ponds in photoauto-trophic growth is that they are prone to contamination by invading species, air, ground, insects, low biomass productivity due to poor carbon dioxide utilization efficiency (Chisti, 2007; Ashokkumar and Rengasamy, 2012).

Closed culture systems involve the use of PBR to provide a controlled environment which enables high productivity of microalgae (Chisti, 2007). PBR is closed system that is designed mainly for the culturing of micro-algae with a large surface volume ratio. As it is a closed system, all growth requirements are introduced into the system and controlled according to the requirements of the species, thereby enabling a better control of culture surroundings such as carbon dioxide supply, pH, water, optimized tempera-ture, and adequate exposure to light (Chisti, 2007; Christenson and Sims, 2011; Gallagher, 2011). PBR allow for comprehensive varieties of micro-algae cultivation and reduce the risk of contamination. PBR are also clas-sified based according to the dimension and shapes which could be in form of plastic bags, flat panels, tubes or fermenters and mode of operation (Suali and Sarbatly, 2012). Despite the advancement in technology, there is no best reactor system to achieve maximum productivity with minimum operation costs (Carvalho and Meireles, 2006). Selection of the PBR depends on the ability to maximize productivity and photosynthetic efficiency. The design and operation of suitable microalgal biomass production systems have been highlighted and discussed extensively in literature (Carvalho and Meireles, 2006; Christenson and Sims, 2011). Numerous types of enclosed PBRs suitable for large-scale cultivation have been designed in an attempt to best control the growth factors of microalgae (Mata et al., 2010; Christenson and Sims, 2011). There are four main categories which are tubular/horizontal, column/vertical, flat plate or flat panel and the helical/tubular, each with its strengths and limitations (Shen et al., 2009; Christenson and Sims, 2011).

The setbacks associated with photobioreactor systems is that it is more capital intensive than the open pond systems, and also that it depends mainly on temperature and light (Chisti, 2007; Harun et al., 2010; Zhu et al., 2011, 2013; Wang et al., 2014).

The cultivation of microalgae has been enhanced through four different cultivation methods namely, photoautotrophic, heterotrophic, mixotrophic or photoheterotrophic methods (Mata et al., 2010; Wang et al., 2014). Photoautotrophic cultivation occurs when algae specie uses sunlight or an artificial source of light as the energy source and inorganic carbon as the carbon source to form chemical energy through the process of photosynthesis (Huang et al., 2010). This technique is the most widely used approach for cultivating algae because it is economically friendly for large-scale production of algae biomass (Gouveia and Oliveira, 2009; Yoo et al., 2010). Photoautotrophic cultivation can be carried out in open ponds system as well as closed system, under photoautotrophic cultivation there is a large variation in the lipid content of microalgae depending on the type of microalgae species (Kamalanathan et al., 2018). Nevertheless, nutrient-limiting or nitrogen limiting conditions can be used to boost the lipid content in microalgae (Mata et al., 2010; Zhu et al., 2014).

The heterotrophic algae growth is another growth technique whereby the algae are not able to synthesize its own food; hence, its propagation is dependent on the use of complex organic substances for nutrition and source of energy (Kamalanathan et al., 2018). Heterotrophic techniques make use of a bioreactor system where there is usually the introduction of light is by an artificial source, the cultivation of the microalgae are performed in huge vessels to generate a high biomass concentration (Lee, 2001; Medipally et al., 2015). There is usually an increase in the growth patterns of microalgae in heterotrophic systems and a decrease with the cost of harvesting due to the high cell density (Mata et al., 2010; Kamalanathan et al., 2018). Heterotrophy culturing systems have paved way for the use of a wide variety of organic compounds such as glucose, acetate, ethanol, glycerol, sucrose, lactose, galactose, mannose, and fructose as nutrient sources for microalgae species (Zhu et al., 2014). Heterotrophic cultivation offers several advantages over photoautotrophic cultivation such as the removal of complications relating to limited access to light which hamper the high cell density in large scale PBR during phototrophic cultivation (Huang et al., 2010). The main limitation of heterotrophic algae culture is the high cost accompanying the procedure and sourcing of the organic carbon substrates (Huang et al., 2010). Also, there is possibility of contamination by other microorganisms, which may reduce the quality and quantity of the products of interest (Medipally et al., 2015). Species such as *Chlorella protothecoides* have shown a greater amount of lipid content during heterotrophic growth (Xiong et al., 2008; Zhang et al., 2014). Further studies by Kamalanathan et al. (2018) also revealed that the biomass of *Scenedesmus* sp. was greater when grown in a molasses-based growth medium environment. Heterotrophic cultivations have also been effectively used for algal biomass and its metabolites production (Chen et al., 1996; Miao and Wu, 2006).

Mixotrophic cultivation is when microalgae undergo photosynthesis and use both organic and inorganic compounds as a carbon source for growth (Wang et al., 2014). This implies that the microalgae have the potential to be propagated either through phototrophic or heterotrophic culturing conditions, or both (Wang et al., 2014). Microalgae assimilate organic compounds and carbon dioxide as a carbon source. The carbon dioxide is then released by microalgae through respiration which will then be trapped and reused under phototrophic cultivation (Mata et al., 2010). The capability of the mixotrophs to use organic substrates indicates that the cell growth is not strictly dependent on photosynthesis; therefore, light energy is not an absolutely limiting factor for growth (Andrade et al., 2007). A number of microalgae species are able to change between photoautotrophic and heterotrophic growth mode thereby leading to a growth rate which is higher than in the photoautotrophic algae (Wang et al., 2014; Medipally et al., 2015). Examples of microalgae that display mixotrophic metabolism processes for growth include the CB *Spirulina platensis* and the green alga *Chlamydomonas reinhardtii* (Chen et al., 1996).

6.3.2 HARVESTING TECHNIQUES

A wide range of techniques or combinations of methods has been proposed for the harvesting of algae biomass. However, there are some challenges associated with some of these techniques. The choice of the most suitable harvesting technique of algae depends mainly on the type of species, morphology, mode of cultivation, specific size, extracellular organic matter composition, and production cost (Kröger and Müller-Langer, 2012). The significant objective of harvesting is to get slurry with a remarkable percentage of total solid matter. The common technologies employed for the harvesting of microalgae biomass include centrifugation, sedimentation, flocculation, flotation, and filtration or a combination of these techniques (Kröger and Müller-Langer, 2012; Kim et al., 2013; Chutia et al., 2017). These techniques are discussed briefly with the advantages and disadvantages of harvesting methods for microalgae presented in Table 6.4. After harvesting, the next phase is dewatering and drying of the biomass, drying is energy-intensive and thus is the economic tailback of the entire procedure (Mata et al., 2010; Uduman et al., 2010). The most common methods for drying include spray-drying, drum-drying, freeze-drying, and sun-drying. Sun-drying is cheap, but it is geography dependent, and at the mercy of availability of space and significant time.

Flocculation is a harvesting technique that can be used to aggregate particles to increase the particle size and thereby easing other separation methods such as sedimentation, filtration, and centrifugation (Molina Grima et al., 2003). The use of inorganic and organic multivalent cations known as flocculants are added to encourage flocculation (Molina Grima et al., 2003; Knuckey et al., 2006; Harun et al., 2010; Papazi et al., 2010; Vandamme et al., 2010). The principle governing this concept is mainly based on the negative and positive charge associated with microalgae attachment to the microscopic particles' surfaces. This phenomenon makes the microalgae to settle at the bottom of the pond and then stick together thereby increasing their weight (Ugwu et al., 2008). Flocculation is a fast and simple technique, which has be applied for concentration of massive biomass of microalgae by dewatering and reducing energy consumption (Vandamme et al., 2013; Ndikubwimana et al., 2014). While, on the other hand, this harvesting flocculants technique can be very expensive and toxic to ecosystem.

Filtration a separation technique which usually uses a bed of granular media or a porous membrane filter using pressure to harvest microalgal biomass, the algae accumulate on the surface of the filters thereby allowing only the liquid medium to pass through. The filtration systems are categorized based on the mode of filtration as well as the pore size of the membrane that is been used. Generally, small scale microalgae samples are filtrated by this method (Barros et al., 2015). Several types of filtration systems exist for algae collection including dead-end, tangential flow, pressure filtration, and microfiltration (Harun et al., 2010). Filtration and centrifugation processes can be combined for harvesting of microalgae from enclosed PBR (Gong and Jiang, 2011). In spite of filtration's simplicity, there are several drawbacks to using filtration. Because filtration is based on the exclusion of particles based on size via a membrane, it suffers from continuous fouling problems and requires constant maintenance and changing of filters and membranes (Mata et al., 2010). Filtration has been cited as being relatively slow compared to other harvesting methods, which makes it an unfavorable approach for large scale operations (Molina Grima et al., 2003).

In the centrifugation harvesting technique, the centrifuge employs the use of a centrifugal force to isolate, separate, and concentrate the suspended macromolecules from the solutions. The largely dense materials have a tendency to settle faster than they would under normal gravitational force, which is typically based on their physical characteristics such as the shape, size, and density. A wide range of centrifugation techniques have proposed and applied for microalgae separation, such as tubular centrifuge, multi-chamber centrifuges, imperforate basket centrifuge, decanter, solid retaining disc centrifuge, nozzle type centrifuge, solid ejecting-type disc centrifuge, and hydro-cyclone (Bhatt et al., 2014). Centrifugation is a very effective cell harvesting method for microalgae but it is energy-intensive, there is usually a high capital and operational costs associated with this technique (Molina Grima et al., 2003; Shen et al., 2009). Centrifugation has been used for microalgae with high value metabolites and extended shelf-life concentrates for hatcheries and nurseries in aquaculture evaluation (Heasman et al., 2000). Microalgal species cells such as *Spirulina*, *Botryococcus* sp., *C. vulgaris*, and *Scenedesmus* sp. with large size and a high density has been isolated using centrifugation systems (Brennan and Owende, 2010).

Electrical based processes are another harvesting approach for microalgal harvesting but they are not generally spread out (Chutia et al., 2017). When an electrical field is applied to the culture broth, the microalgal cells become negatively charged and the cell tends to separate (Chutia et al., 2017). This is a very resourceful method as they are applicable to a wide variety of microalgal species. Sedimentation is a harvesting technique whereby particles are removed from a liquid under the influence of gravity. Particle removal by sedimentation requires the downward settling velocity of the particle or floc to be greater than the overflow rate of the vessel. This requirement leads to the fundamental design parameter of a vertical settling system. This method uses more energy-intensive than centrifugation; it has less efficiency and lower concentration factors. Another harvesting technique is flotation; flotation is a process in which a pressurized gas is dissolved into the liquid medium. As air is released from solution, small bubbles nucleate on particles in the fluid. As the bubbles rise toward the open atmosphere, they bring particles with them. At this point, the concentrated float algal biomass can then be skimmed from the top of the solution with a higher concentration. Table 6.4 shows the advantages and disadvantages of the different harvesting techniques for microalgae (Christenson and Sims, 2011; Barros et al., 2015; Chutia et al., 2017).

TABLE 6.4 Summary of the Advantages and Disadvantages of Harvesting Techniques

Harvesting Technique	Solid Concentration (%)	Recovery (%)	Advantages	Disadvantages
Chemical coagulation/ flocculation	–	83–92	Simple and fast method, no energy requirements	Expensive and toxic to microalgae biomass; not applicable to small scale production
Auto and bio flocculation	–	80–90	Inexpensive, allow culture medium recycling, non-toxic to microalgae biomass	Changes in cellular composition and contamination
Sedimentation	0.5–3	10–90	Simple and inexpensive approach	Time consuming, deterioration of biomass
Flotation	3–6	50–90	Low cost method, feasible for large scale applications	Requires the use of chemical flocculation
Electrical based methods	–	–	Applicable to a wide variety of microalgal species	High equipment cost and very energetic
Filtration	5–27	70–90	High efficiency and recovery	Cleaning of the membrane should be done frequently- membrane fouling
Centrifugation	12–22	>90	Fast, high recovery efficiency	Expensive and high energy requirements

6.3.3 EXTRACTION OF MICROALGAE LIPIDS

There is usually a huge diversity within the microalgal lipids species, therefore screening of microalgae and the optimization of the appropriate culturing condition helps to increase lipid productivity. General screening procedures for microalgae consist of sampling, isolation, and purification, and the evaluation of the potential microalgal for lipid production (Mutanda et al., 2011). Microalgae can be extracted as either in a dried or wet processes, the wet extraction procedure are fast and energy input is not a perquisite for drying of microalgae. On the other hand, drying procedure needs energy input but is efficient for lipid extraction (Mutanda et al., 2011; Islam et al., 2013). In order to produce biodiesel from algal biomass cellular lipids must be extracted from the cell and then collected. This requires the disruption or ruptures of the algal cell and can be accomplished using a variety of methods including mechanical and chemical disruption, solvent extraction, super-critical fluid extraction (SFE), and combinations of some new distinctive techniques (Harun et al., 2010; Lee et al., 2010; Mercer and Armenta, 2011).

The mechanical disruption techniques use approaches such as mechanical pressing, bead milling, and homogenization, it is achieved by applying high pressures on the cells that need to be extracted, forcing the cell wall to rupture, allowing the intracellular lipids to be extracted and collected (Mercer and Armenta, 2011; Ramesh, 2013; Bharathiraja et al., 2015). The technique is expensive, allows for the introduction of pigments along with the extracted oil in this manner increasing the cost incurred during the produc-tion (Ramesh, 2013; Bharathiraja et al., 2015). Bead beating or milling is a technique that is achieved by agitating the algal biomass in the presence of beads. Agitation permits the beads to batter the algal cells, breaking them apart by mechanical force hence enabling the process to extract the lipids (Mercer and Armenta, 2011). This approach has been significantly used in both laboratory and industrial set up for size reduction of particles and the disruption of cells (Doucha and Lívanský, 2008).

Solvent extraction techniques are regarded as the conventional and effec-tive approach for the extensive extraction of lipids from microalgae (Ramesh, 2013; Bharathiraja et al., 2015). This is because lipids have high solubility potential in non-polar solvents such as chloroform, hexane, petroleum ether (Ahmad et al., 2010; Ramesh, 2013; Bharathiraja et al., 2015). Solvent extraction methods such as the Folch extraction, Bligh, and Dyer, and the Soxhlet or Gold-fisch techniques have been developed and standardized with the appropriate solvent ratio for the extraction of lipids from biomass with specific apparatus (Bligh and Dyer, 1959; Folch et al., 1987; Ramesh, 2013; Bharathiraja et al., 2015). Even though the use of solvents to extract algal lipids is equitably direct, there are a number of drawbacks when applied to microalgae. The Folch approach makes use of chloroform and methanol for extracting the lipids from the cells and the resulting mixture is allowed to separate into layers (Bharathiraja et al., 2015). In the Bligh and Dyer method, lipid extraction and partitioning are performed simultaneously; in which proteins are precipitated in the line of two liquid phases (Bharathiraja et al.,

2015). Standard methods such as the Folch and Bligh and Dyer methods of lipid extraction have proven inefficient when applied to microalgae, because sometimes the solvents used are too costly or difficult to use on large scale production (Campbell et al., 2011; Russin et al., 2011).

SFE of lipids is an alternative for lipid extraction; SFE usually makes use of carbon dioxide that has been pushed beyond its critical point to its supercritical state. The properties or state of the carbon dioxide changes, thereby possessing the properties of both a liquid and a gas, making it more diffusive with low viscosity (Herrero et al., 2006; Gong and Jiang, 2011). Due to these changes, the carbon dioxide is able to quickly penetrate into the solids and extract the target molecules of interest. There are main benefits of using supercritical CO_2 for the extraction of algal lipids is that there is no need for the use of organic solvents for the extraction of lipids in this method (Carrapiso and García, 2000). Another significant advantage of supercritical CO_2 extraction approach is that it has the potential to extract lipids from wet algal biomass, since the use of organic solvents involves the dewatering of algal biomass to greater than 90% solids to avoid reductions in lipid yield (Ferrell and Sarisky-Reed, 2010). SFE is a widely pleasant technique because it does not leave any harmful solvent residues, has a faster extraction time than mechanical disruption and solvent extraction, and is used for thermally sensitive products. However, the high cost of instrumentation and high energy demand of SFE processes makes it less attractive method for extracting lipids from microalgae (Carrapiso and García, 2000; Ferrell and Sarisky-Reed, 2010; Halim et al., 2011).

In addition to the above-mentioned extraction techniques, the ultra-sound and microwave methods has also been developed and studied for oil extraction from microalgae (Mata et al., 2010). The ultrasonic-assisted extraction is an extraction technique whereby there is the disintegration of cell structures by means of ultrasound and it is used for the extraction of intracellular compounds. The sound waves that propagate into the liquid media lead to an alternating high-pressure and low-pressure cycles during sonication of liquids at high intensities (Adam et al., 2012; Bharathiraja et al., 2015). This procedure easy to set up leading to a high purity on the final product (Bharathiraja et al., 2015). It is an economical and friendly and can be completed in a very short time with high reproducibility (Bharathiraja et al., 2015). Added advantage of the ultra-sonication approach is that there is a reduced amount of thermal denaturation of biomolecules due to the low temperature is generated (Bharathiraja et al., 2015). In addition, another advantage ultrasound extraction process are the ability to increase the yield of algae oil and reduce the time of the extraction process with moderate or low cost (Mata et al., 2010). Nevertheless, when ultra-sonication is extended, there is usually the production of free radicals; hence, the quality of the oil is compromised (Bharathiraja et al., 2015).

Other innovative and new generation extraction methods for lipids include the solvent-free extraction and non-mechanical methods such as the osmotic pressure and isotonic solution (Bharathiraja et al., 2015). The solvent-free methods are able to overcome the limitations associated with the tradition organic solvent-based approach (Bharathiraja et al., 2015). These approaches are economically feasible and ecofriendly. Osmotic pressure method has the potential to disrupt the balance of the osmotic pressure in the algal cell walls by the use a rapid increase or decrease in the concentration of salt within the aqueous media (Adam et al., 2012; Bharathiraja et al., 2015). A few studies have revealed that there is a possibility of extracting oil from *Chlamydomonas reinhardtii* (Lee et al., 2010; Bharathiraja et al., 2015), *Botryococcus* sp., *Chlorella vulgaris*, and *Scenedesmus* sp. (Kim et al., 2013; Bharathiraja et al., 2015) using osmotic pressure method. However, this is still an ongoing research that still needs to be validated. Isotonic extraction method is an extraction technique that makes use of an ionic liquid for the algal lipid extraction (Bharathiraja et al., 2015). It is also an upcoming and advanced pre-treatment technology to bypass the organic solvent approach (Bharathiraja et al., 2015). On the other hand, only limited studies have implemented the use of isotonic extraction methods on microalgal species (Bharathiraja et al., 2015). An unusual method that is gaining momentum right now for the extraction of microalgal lipids is the application of enzymes to enhance cell disruption with enzymes such as the cellulose and trypsin (Taher et al., 2014; Bharathiraja et al., 2015). A large amount of energy is required to dissociate the cell wall at a lower temperature (Taher et al., 2014).

6.3.4   DETERMINATION OF MICROALGAE LIPIDS

There are different methods that can be used for quantification and quality detection of lipids from an algae biomass. A significant feature of algae biofuel production is the ability to measure the lipid content of microalgae species. The conventional methods of lipid detection in microalgae include solvent extraction and gravimetric method (Folch et al., 1957; Bligh and Dyer, 1959) and chromatographic-based methods. A major drawback of these conventional methods can be that they are time consuming and expensive, making it difficult to screen large numbers of microalgae. Other quantitative methods used include the Nile red (9-diethylamino-5H-benzo[α]phenoxazine-5-one) red (9-diethylamino-5H-benzo[α]phenoxazine-5-one) method, BODIPY 505/515 (4,4-difluoro-1,3,5,7-tetramethyl-4-bora-3α,4adiaza-s-indacene) method, time-domain nuclear magnetic resonance (TD-NMR) and colorimetric methods to quantify the neutral fraction of the lipids (Akoto et al., 2005).

Nile red is a lipid-soluble fluorescence dye that can be used for *in-situ* lipid determination of microalgae (Bertozzini et al., 2011). This dye is photostable in hydrophilic and hydrophobic solutions and the maximum emission peak is shifted when the polarity of solvent decreases (Pick and Rachutin-Zalogin, 2012). This method is high throughput method for screening of microalgae strains (Huang et al., 2010). The accuracy of the Nile Red methods can be adversely affected by the fact that it cannot stain dead microalgae cells making it difficult to rely on as a sole quantification method for lipid. In addition to the Nile red dye, another dye used for broad detection of intracellular lipid in algae is BODIPY 505/515 (4,4-difluoro-1,3,5,7-tetramethyl-4-bora-3α,4adiaza-s-indacene) method is lipid soluble fluorescence dye as well, which has the ability to dye lipids from their initial color to a green color. This dye is non-destructive and it can also be reused for other experiments after lipid determination (Govender et al., 2012). It is more effective for vital staining of intracellular lipid bodies and single-cell sorting than Nile Red, though the Nile red and BODIPY 505/515 are good for rapid semi-quantitative detection of lipid in algae, however, they are best when combined with other quantitative methods such as mass spectrometry, gravimetry, and Fourier transform infrared (FTIR) spectroscopy. In calorimetric method, fatty acids react with copper and with addition of substrates, it occurs in colored product and it is measured using the spectrophotometer (Wawrik and Harriman, 2010). The advantages and disadvantages of the different methods of lipid determination in microalgae are summarized in Table 6.5.

6.3.5   BIODIESEL FROM MICROALGAE VIA TRANSESTERIFICATION

The oils extracted from the algal biomass cannot be used directly for biodiesel due to the high viscosity in the oil; hence, the oils first need to be trans-formed to fatty acid methyl ester (FAME) via transesterification reactions or through supercritical transesterification (Yaakob et al., 2014). Biodiesel can be produced by three common esterification methods: acid-catalyzed transesterification, base-catalyzed transesterification and chemical-catalyzed transesterification of fatty acids to alkyl esters. Acid-catalyzed transesterification is mostly achieved with the use of homogeneous catalysts; Homogenous acid catalysts used are strong acids such as hydrochloric, sulfuric, sulfonic, and phosphoric acids (Vyas et al., 2010). The strong acid is mixed into the alcohol, dissolved, and then contacted with oil permitting the formation of the alkyl esters (Demirbas, 2009). Base catalyzed transesterification is the established means of processing biodiesel and is the overwhelming option used in the industry because it is faster than acid catalyzed transesterification (Vyas et al., 2010). The higher rate of reaction is due to the strong nucleophilic nature of the alkoxides species formed from the catalyst and alcohol (Lotero et al., 2005). A number of catalysts can be used such as sodium hydroxide, potassium hydroxide, and sodium methoxide (Meher et al., 2006). Parameters, such as temperature, alcohol, and catalyst concentration are also needed in this reaction which makes it quite similar to the acids catalysts reaction (Vyas et al., 2010).

TABLE 6.5 Different Methods of Quantification and Qualification of Microalgae

Methods	Advantages	Disadvantages
Solvent extraction and gravimetric method	High accuracy and reproducibility	Time consuming, labor intensive, large amount of biomass
Chromatographic Techniques	Good reproducibility, single analysis generates data of both quantity and profile of fatty acids, small sample required	Requirement of cell disruption, requirement of expensive analytical equipment
Nile Red staining (NR) I	*In situ* measurement, high throughput, simple, rapid, and efficient	Variable efficiencies in some microalgae, accuracy can be affected by many factors
Time domain nuclear magnetic resonance (TD-NMR)	*In situ* measurement, rapid, and less expensive	Accuracy is dependent on high lipid content
Colorimetric quantification	Rapid, simple, cheap	Not applicable to detect fatty acids with length chain of less than 12-C atoms

For the chemical conversion or enzymatic transesterification procedure of biodiesel from lipid feedstock, the oils are converted from their initial form as triglycerides to alkyl esters, which are comparable to the petroleum-based diesel (Meher et al., 2006; Vyas et al., 2010). This conversion can be performed using a number of methods by reacting lipids with an alcohol with, or without, the presence of a catalyst (Meher et al., 2006; Huang et al., 2010). A purification step which increases the production cost is always required due to the presence of water in the oil feedstock (Vyas et al., 2010). Transesterification of lipids via enzyme catalysts leads to lower concentrations of contaminants in the crude biodiesel, due to enzyme specificity. Hence, the threat of undesirable byproducts is reduced, ensuing a less effort for downstream purification of the crude biodiesel (Vyas et al., 2010). Even though the advantages of using enzymatic catalysts are numerous, large scale use of lipases for biodiesel production has not been practiced (Demirbas, 2009). Production of enzymes is costly and can drive up biodiesel production costs (Meher et al., 2006; Demirbas, 2009). Until the production costs of enzyme catalysts are reduced and the process streamlined, this method will continue to remain too costly to scale up (Vyas et al., 2010).

Supercritical transesterification method is based on the use of supercritical solvents for the simultaneous extraction and conversion of oils from biomass to biodiesel and for the extraction of high value pigments and compounds (Herrero et al., 2006). This method of extraction and conversion makes use of solvents, such as methanol or ethanol, which are beyond their critical point, at this phase the physical properties of the solvent change allowing them to penetrate solids and effectively dissolve compounds not soluble in the solvent at normal conditions. Changes in the solvent's properties allow for supercritical solvents to break down cell matter, dissolve oils or other desired products, and extract the target compounds much more efficiently and quickly (Herrero et al., 2006; Demirbas, 2009; Halim et al., 2011). Supercritical transesterification can be performed without a catalyst due to the catalytic nature of the alcohol at the supercritical state (Vyas et al., 2010). At this state the dielectric constant of methanol, for example, decreases, lowering its polarity and allowing it to become soluble in the oil phase (Vyas et al., 2010). The biodiesel obtained from the supercritical transesterification method is exceedingly pure requiring little purification after the reaction (Vyas et al., 2010).

6.4   OTHER APPLICATIONS OF MICROALGAE

Microalgae are mainly well-thought-out as a prospective biofuel source due to their lipid content. They are a leading source of animal feed particularly in aquaculture. Due to the various capabilities of microalgae these has enabled them to be used in a variety of ways. They are of immense benefit to the environment as they address pollution problem. Microalgae they can be used for a variety of applications such as wastewater remediation, carbon sequestration from flue gases (coal-fired power plants), and removal of heavy metals from industrial wastewaters while generating biomass for biofuel and bioproduct production (Harun et al., 2010; Pokoo-Aikins et al., 2010; Rawat et al., 2011).

Wastewater samples contain domestic, agricultural, and industrial waste which contains some nutrients, suspended solids, and dissolved inorganic and organic materials and pathogens. These waters are toxic and hazardous to human health and the environment, and treating them before discharge to the river or land is a mandatory. Wastewaters contain essential nutrients for microalgae growth such as nitrogen, phosphorus, trace metals, carbon, however, the concentrations of these nutrients vary in wastewater depending on the source (Metcalf et al., 1995). Studies have shown that microalgae have the capacity to remove contaminants such as N, P, and heavy metals from wastewater effluents (Aslan and Kapdan, 2006; Bouaid et al., 2009). The use of microalgae for wastewater treatment offers a prospect for the effective recycling of nutrients, and economic benefits by avoiding high-cost chemical treatment (Wang et al., 2010; Rawat et al., 2011). During remediation, microalgae can be used for disinfection purposes at wastewater systems through increase in pH of the wastewater samples as a result of their photosynthetic activity which helps to reduce the BOD (biological oxygen demand) and coliform bacteria in the wastewater (Abdel-Raouf et al., 2012).

Microalgae can also been used for fertilizer application because algae have the ability to improve soil fertility by enhancing the water retention capacity of the soil and macronutrient composition. The presence of some vital vitamins such as vitamin A, B_1, B_2, B_6, B_{12}, C, E, nicotinate, biotin, and folic acid in microalgae has made them suitable for use in both human and animal nutrition. Microalgae are used in human nutrition in various forms as nutritional supplements, colorants, cosmetics, and medicine (Spolaore et al., 2006), and they can also be integrated into some snacks and beverages in diverse forms like tablets, capsules, and liquids (Rodriguez-Garcia and Guil-Guerrero, 2008). In the cosmetic industry, microalgae species such as *Spirulina* and *Chlorella*, are mostly used as a skincare product for the anti-aging cream, hair care products, and sun protection lotions (Singh et al., 2016; Spolaore et al., 2006).

Microalgae have the potential to remove carbon dioxide, which is an important GHGs. The carbon dioxide captured by algae during cultivation is a feasible technology that can be used for the mitigating the emissions of fossil fuels for algal flue gas sequestration in a power plant. CO_2 fixation by algal cultures is not only a method for greenhouse gas mitigation but can also be used for producing algae biomass, which can then be converted into a biofuel (Jeffrey and Humphrey, 1975). Due to the challenges faces in the agricultural sector, the growth of some algae species has shown tremendous use as a good source of fertilizer application (Singh et al., 2016). The distinctive features of some species generation such as high yield biomass and growth in a different water source have enhanced its soil fertility ability. Cyanobacterial biomass is an effective bio-fertilizer source to improve soil physico-chemical characteristics such as water-holding capacity and mineral nutrient status of the degraded lands (Singh et al., 2016).

6.5   CONCLUSION AND FUTURE PERSPECTIVES

The exploitation of microalgal biomass cultivation is not only beneficial for CO_2 fixation but also results in the bioproduction of triglycerides which after esterification with methanol get converted into biodiesel. Though biofuels produced from microalgae biomass has been touted as one of the best alternatives to petroleum-based transportation fuels, there have been limited technological breakthroughs to make the commercialization of microalgal biodiesel a reality. Therefore, in future, new approaches must be considered as follows: (a) the approaches must be applicable to any microalgal species without significant differences in effectiveness; (b) the technologies must have a strong potential to be integrated into existing upstream and downstream process for biodiesel production; and (c) the technologies must be scalable to meet the challenge of commercial bio-oil production. Researchers should apply their knowledge to improve the both upstream and downstream process to achieve the best solution in the near future.

KEYWORDS

- bio-butanol
- biodiesel
- bio-ethanol
- biofuels
- biomass

- **carbon dioxide**
- **cultivation**
- **downstream processes**
- **fatty acid**
- **lipids**
- **microalgae**
- **photobioreactors**
- **supercritical fluid extraction**
- **upstream processes**

REFERENCES

Abdel-Raouf, N., Al-Homaidan, A. A., & Ibraheem, I. B. M., (2012). Microalgae and wastewater treatment. *Saudi J. Biol. Sci., 19,* 257–275.

Abou-Shanab, R. A. I., Hwang, J., Cho, Y., Min, B., & Jeon, B., (2011). Characterization of microalgal species isolated from fresh water bodies as a potential source for biodiesel production. *Appl. Energy, 88,* 3300–3306.

Abu-Hamdeh, N. H., & Alnefaie, K. A., (2015). A Comparative study of almond biodiesel-diesel blends for diesel engine in terms of performance and emissions. *Biomed Res. Int., 8.*

Adam, F., Abert-Vian, M., Peltier, G., & Chemat, F., (2012). "Solvent-free" ultrasound-assisted extraction of lipids from fresh microalgae cells: A green, clean, and scalable process. *Bioresour. Technol., 114,* 457–465.

Ahmad, M., Khan, M. A., Zafar, M., & Sultana, S., (2009). Environment-friendly renewable energy from sesame biodiesel. *Energy Sources, Part A Recover. Util. Environ. Eff., 32,* 189–196.

Akoto, L., Pel, R., Irth, H., Brinkman, U. A. T., & Vreuls, R. J. J., (2005). Automated GC-MS analysis of raw biological samples: Application to fatty acid profiling of aquatic micro-organisms. *J. Anal. Appl. Pyrolysis, 73,* 69–75.

Amin, S., (2009). Review on biofuel oil and gas production processes from microalgae. *Energy Convers. Manag., 50,* 1834–1840.

Andrade, M. R., & Costa, J. A. V., (2007). Mixotrophic cultivation of microalga *Spirulina platensis* using molasses as organic substrate. *Aquaculture, 264,* 130–134.

Arias-Peñaranda, M. T., Cristiani-Urbina, E., Montes-Horcasitas, C., Esparza-García, F., Torzillo, G., & Cañizares-Villanueva, R. O., (2013). *Scenedesmus incrassatulus* CLHE-Si01: A potential source of renewable lipid for high quality biodiesel production. *Bioresour. Technol., 140,* 158–164.

Ashokkumar, V., & Rengasamy, R., (2012). Mass culture of *Botryococcus braunii* Kutz. under open raceway pond for biofuel production. *Bioresour. Technol., 104,* 394–399.

Aslan, S., & Kapdan, I. K., (2006). Batch kinetics of nitrogen and phosphorus removal from synthetic wastewater by algae. *Ecol. Eng., 28,* 64–70.

Atabani, A. E., Silitonga, A. S., Badruddin, I. A., Mahlia, T. M. I., Masjuki, H. H., & Mekhilef, S., (2012). A comprehensive review on biodiesel as an alternative energy resource and its characteristics. *Renew. Sustain. Energy Rev., 16,* 2070–2093.

Atsumi, S., & Liao, J. C., (2008). Metabolic engineering for advanced biofuels production from *Escherichia coli*. *Curr. Opin. Biotechnol., 19,* 414–419.

Bak, Y., Choi, J., Kim, S., & Kang, D., (1996). Production of Bio-diesel fuels by transesterification of rice bran oil. *Korean J. Chem. Eng., 13*(3), 242–245.

Banerjee, A., Sharma, R., Chisti, Y., & Banerjee, U. C., (2002). *Botryococcus braunii*: A renewable source of hydrocarbons and other chemicals. *Crit. Rev. Biotechnol., 22,* 245–279.

Baroi, C., Mahto, S., Niu, C., & Dalai, A. K., (2014). Biofuel production from green seed canola oil using zeolites. *Appl. Catal. A. Gen., 469,* 18–32.

Barros, A. I., Gonçalves, A. L., Simões, M., & Pires, J. C. M., (2015). Harvesting techniques applied to microalgae: A review. *Renew. Sustain. Energy Rev., 41,* 1489–1500.

Bello, E. I., & Agge, M., (2012). Biodiesel production from ground nut oil. *J. Emerg. Trends Eng. Appl. Sci., 3,* 276–280.

Bertozzini, E., Galluzzi, L., Penna, A., & Magnani, M., (2011). Application of the standard addition method for the absolute quantification of neutral lipids in microalgae using Nile red. *J. Microbiol. Methods, 87,* 17–23.

Bharathiraja, B., Chakravarthy, M., Ranjith, K. R., Yogendran, D., Yuvaraj, D., Jayamuthunagai, J., Praveen, K. R., & Palani, S., (2015). Aquatic biomass (algae) as a future feed stock for bio-refineries: A review on cultivation, processing, and products. *Renew. Sustain. Energy Rev., 47,* 634–653.

Bhatt, N. C., Panwar, A., Bisht, T. S., & Tamta, S., (2014). Coupling of algal biofuel production with wastewater. *Sci. World J.,* 210504. doi:10.1155/2014/210504.

Bhatti, H., Hanif, M., Qasim, M., & Ataurrehman, A., (2008). Biodiesel production from waste tallow. *Fuel, 87,* 2961–2966.

Bilanovic, D., Andargatchew, A., Kroeger, T., & Shelef, G., (2009). Freshwater and marine microalgae sequestering of CO_2 at different C and N concentrations-response surface methodology analysis. *Energy Convers. Manag., 50,* 262–267.

Binod, P., Sindhu, R., Singhania, R. R., Vikram, S., Devi, L., Nagalakshmi, S., Kurien, N., et al., (2010). Bioethanol production from rice straw: An overview. *Bioresour. Technol., 101,* 4767–4774.

Bligh, E., & Dyer, W., (1959). A rapid method of total lipid extraction and purification. *Canadian Journal of Biochemistry and Physiology, 37,* 911–917.

Blumberg, K. O., Walsh, M. P., & Pera, C., (2003). *Low-sulfur Gasoline and Diesel: The Key to Lower Vehicle Emissions.* The International Council on Clean Transportation, Napa, California.

Bouaid, A., Martinez, M., & Aracil, J., (2009). Production of biodiesel from bioethanol and *Brassica carinata* oil: Oxidation stability study. *Bioresour. Technol., 100,* 2234–2239.

Brennan, L., & Owende, P., (2010). Biofuels from microalgae: A review of technologies for production, processing, and extractions of biofuels and co-products. *Renew. Sustain. Energy Rev. 14,* 557–577.

Campbell, K. A., Glatz, C. E., Johnson, L. A., Jung, S., De Moura, J. M. N., Kapchie, V., & Murphy, P., (2011). Advances in aqueous extraction processing of soybeans. *J. Am. Oil Chem. Soc., 88,* 449–465.

Cao, W., Han, H., & Zhang, J., (2005). Preparation of biodiesel from soybean oil using supercritical methanol and co-solvent. *Fuel, 84,* 347–351.

Capelli, B., & Cysewski, G. R., (2010). Potential health benefits of *Spirulina* microalgae. *Nutrafoods, 9,* 19–26.

Cardona, C. A., Quintero, J. A., & Paz, I. C., (2010). Production of bioethanol from sugarcane bagasse: Status and perspectives. *Bioresour. Technol., 101,* 4754–4766.

Cardone, M., Mazzoncini, M., Menini, S., Rocco, V., Senatore, A., Seggiani, M., et al., (2003). *Brassica carinata* as an alternative oil crop for the production of biodiesel in Italy: Agronomic evaluation, fuel production by transesterification and characterization. *Biomass and Bioenergy, 25,* 623–636.

Carrapiso, A. I., & García, C., (2000). Development in lipid analysis: Some new extraction techniques and *in situ* transesterification. *Lipids, 35,* 1167–1177.

Carvalho, A. P., & Meireles, L. A., (2006). Microalgae reactors: A review of enclosed systems and performances. *Biotechnol. Prog., 3,* 1490–1506.

Chen, F., Zhang, Y., & Guo, S., (1996). Growth and phycocyanin formation of *Spirulina platensis* in photoheterotrophic culture. *Biotechnol. Lett., 18,* 603–608.

Cheng, H. H., Whang, L. M., Chan, K. C., Chung, M. C., Wu, S. H., Liu, C. P., Tien, S. Y., Chen, S. Y., Chang, J. S., & Lee, W. J., (2015). Biological butanol production from microalgae-based biodiesel residues by *Clostridium acetobutylicum. Bioresour. Technol., 184,* 379–385.

Chisti, Y., (2007). Biodiesel from microalgae. *Biotechnol. Adv., 25,* 294–306.

Christenson, L., & Sims, R., (2011). Production and harvesting of microalgae for wastewater treatment, biofuels, and bioproducts. *Biotechnol. Adv., 29,* 686–702.

Chutia, S., Gohain, M., Deka, D., & Kakoty, N. M., (2017). A review on the harvesting techniques of algae for algal based biofuel production. *JERET, 4,* 58–62.

Culver, B., (2009). *Total Fatty Acid Production in Golden Alga Prymnesium parvum a Potential Bio-Diesel Feedstock.* Virginia Tech., Blacksburg, Virginia, USA.

Custódio, L., Soares, F., Pereira, H., Barreira, L., Vizetto-Duarte, C., Rodrigues, T., et al., (2014). Fatty acid composition and biological activities of *Isochrysis galbana* T-ISO, *Tetraselmis* sp. and *Scenedesmus* sp.: Possible application in the pharmaceutical and functional food industries. *J. Appl. Phycol., 26,* 151–161.

Dartsch, P. C., (2008). Antioxidant potential of selected *Spirulina platensis* preparations. *Phyther. Res., 22,* 627–633.

Dawodu, F. A., Ayodele, O. O., & Bolanle-Ojo, T., (2014). Biodiesel production from *Sesamum indicum* L. seed oil: An optimization study. *Egypt. J. Pet., 23,* 191–199.

Dayananda, C., Kumudha, A., Sarada, R., & Ravishankar, G. A., (2010). Isolation, characterization, and outdoor cultivation of green microalgae *Botryococcus* sp. *Sci. Res. Essays., 5,* 2497–2505.

Demirbas, A., (2009). Progress and recent trends in biodiesel fuels. *Energy Convers. Manag., 50,* 14–34.

Demirbas, M. F., (2011). Biofuels from algae for sustainable development. *Appl. Energy, 88,* 3473–3480.

Dias, M. O. S., Ensinas, A. V., Nebra, S. A., Maciel, F. R., Rossell, C. E. V., & Maciel, M. R. W., (2009). Production of bioethanol and other bio-based materials from sugarcane bagasse: Integration to conventional bioethanol production process. *Chem. Eng. Res. Des., 87,* 1206–1216.

Doucha, J., & Lívanský, K., (2008). Influence of processing parameters on disintegration of *Chlorella* cells in various types of homogenizers. *Appl. Microbiol. Biotechnol., 81,* 431–440.

El-Mashad, H. M., Zhang, R., & Avena-Bustillos, R. J., (2008). A two-step process for biodiesel production from salmon oil. *Biosyst. Eng., 99*, 220–227.

Escobar-Niño, A., Luna, C., Luna, D., Marcos, A. T., Cánovas, D., & Mellado, E., (2014). Selection and characterization of biofuel-producing environmental bacteria isolated from vegetable oil-rich wastes. *PLoS One, 9.* doi: 10.1371/journal.pone.0104063.

Ferrell, J., & Sarisky-Reed, V., (2010). *National Algal Biofuels Technology Roadmap.* U.S. Department of Energy, USA.

Folch, J., Lees, M., & Sloane, S. G. H., (1987). A simple method for the isolation and purification of total lipids from animal tissues. *J. Biol. Chem., 55*, 999–1033.

Fröhlich, A., & Rice, B., (2005). Evaluation of *Camelina sativa* oil as a feedstock for biodiesel production. *Ind. Crops Prod., 21*, 25–31.

Fuentes-Grünewald, C., Garcés, E., Rossi, S., & Camp, J., (2009). Use of the dinoflagellate *Karlodinium veneficum* as a sustainable source of biodiesel production. *J. Ind. Microbiol. Biotechnol., 36,* 1215–1224.

Gallagher, B. J., (2011). The economics of producing biodiesel from algae. *Renew. Energy, 36,* 158–162.

Gong, Y., & Jiang, M., (2011). Biodiesel production with microalgae as feedstock: From strains to biodiesel. *Biotechnol. Lett., 33,* 1269–1284.

Goodrum, J. W., Geller, D. P., & Adams, T. T., (2003). Rheological characterization of animal fats and their mixtures with #2 fuel oil. *Biomass and Bioenergy, 24*, 249–256.

Gouveia, L., & Oliveira, A. C., (2009). Microalgae as a raw material for biofuels production. *J. Ind. Microbiol. Biotechnol., 36,* 269–274.

Govender, T., Ramanna, L., Rawat, I., & Bux, F., (2012). BODIPY staining, an alternative to the Nile Red fluorescence method for the evaluation of intracellular lipids in microalgae. *Bioresour. Technol., 114*, 507–511.

Grima, E. M., Acie, F. G., Medina, A. R., & Chisti, Y., (2003). Recovery of microalgal biomass and metabolites: Process options and economics. *Biotechnol. Adv., 20*, 491–515.

Grobbelaar, J. U., (2000). Physiological and technological considerations for optimizing mass algal cultures. *J. Appl. Phycol., 12,* 201–206.

Halim, R., Gladman, B., Danquah, M. K., & Webley, P. A., (2011). Oil extraction from microalgae for biodiesel production. *Bioresour. Technol., 102*, 178–185.

Harun, R., Jason, W. S. Y., Cherrington, T., & Danquah, M. K., (2011). Exploring alkaline pre-treatment of microalgal biomass for bioethanol production. *Appl. Energy, 88,* 3464–3467.

Harun, R., Singh, M., Forde, G. M., & Danquah, M. K., (2010). Bioprocess engineering of microalgae to produce a variety of consumer products. *Renew. Sustain. Energy Rev., 14,* 1037–1047.

Heasman, M., Diemar, J., O'Connor, W., Sushames, T., & Foulkes, L., (2000). Development of extended shelf-life microalgae concentrate diets harvested by centrifugation for bivalve mollusks: A summary. *Aquac. Res., 31,* 637–659.

Hernández, D., Riaño, B., Coca, M., & García-González, M. C., (2015). Saccharification of carbohydrates in microalgal biomass by physical, chemical, and enzymatic pre-treatments as a previous step for bioethanol production. *Chem. Eng. J., 262*, 939–945.

Herrero, M., Cifuentes, A., & Ibañez, E., (2006). Sub- and supercritical fluid extraction of functional ingredients from different natural sources: Plants, food-by-products, Algae, and microalgae: A review. *Food Chem., 98,* 136–148.

Huang, G. H., Chen, F., Wei, D., Zhang, X. W., & Chen, G., (2010). Biodiesel production by microalgal biotechnology. *Appl. Energy, 87*, 38–46.

Ilkiliç, C., Aydin, S., Behcet, R., & Aydin, H., (2011). Biodiesel from safflower oil and its application in a diesel engine. *Fuel Process Technol., 92*, 356–362.

Islam, M. A., Magnusson, M., Brown, R. J., Ayoko, G. A., Nabi, M. N., & Heimann, K., (2013). Microalgal species selection for biodiesel production based on fuel properties derived from fatty acid profiles. *Energies, 6*, 5676–5702.

Jaffar Al-Mulla, E. A., Issam, A. M., & Al-Janabi, K. W. S., (2015). A novel method for the synthesis of biodiesel from soybean oil and urea. *Comptes Rendus Chim., 18*, 525–529.

Jeffrey, S. W., & Humphrey, G. F., (1975). New spectrophotometric equations for determining chlorophylls a, b, c1, and c2 in higher plants, algae and natural phytoplankton. *Biochem. Und. Physiol. Der. Pflanz., 167*, 191–194.

Kalam, M., & Masjuki, H., (2002). Biodiesel from palmoil - an analysis of its properties and potential. *Biomass and Bioenergy, 23*, 471–479.

Kamalanathan, M., Chaisutyakorn, P., Gleadow, R., & Beardall, J., (2018). A comparison of photoautotrophic, heterotrophic, and mixotrophic growth for biomass production by the green alga *Scenedesmus* sp. (*Chlorophyceae*). *Phycologia, 57*, 309–317.

Keerthi, S., Sujatha, A., Devi, K. U., & Sarma, N. S., (2013). Bioprospecting for nutraceutically useful marine diatom, *Odontella aurita* in the South-East Coast of India and medium optimization. *Int. J. Curr. Sci., 6*, 22–28.

Khan, S. A., Hussain, M. Z., Prasad, S., & Banerjee, U. C., (2009). Prospects of biodiesel production from microalgae in India. *Renew. Sustain. Energy Rev., 13*, 2361–2372.

Kim, J., Yoo, G., Lee, H., Lim, J., Kim, K., Kim, C. W., et al., (2013). Methods of downstream processing for the production of biodiesel from microalgae. *Biotechnol. Adv., 31*, 862–876.

Knuckey, R. M., Brown, M. R., Robert, R., & Frampton, D. M. F., (2006). Production of microalgal concentrates by flocculation and their assessment as aquaculture feeds. *Aquac. Eng., 35*, 300–313.

Kobayashi, M., Todoroki, Y., & Ohigashi, H., (1998). Biological activities of abscisic acid analogs in the morphological change of the green alga *Haematococcus pluvialis. J. Ferment. Bioeng., 85*, 529–531.

Kröger, M., & Müller-Langer, F., (2012). Review on possible algal-biofuel production processes. *Biofuels, 3*, 333–349.

Kulkarni, M. G., Dalai, A. K., & Bakhshi, N. N., (2006). Utilization of green seed canola oil for biodiesel production. *J. Chem. Technol. Biotechnol., 81*, 1886–1893.

Kumar, N., (2007). Production of biodiesel from high FFA rice bran oil and its utilization in a small capacity diesel engine. *J. Sci. Ind. Res. (India), 66*, 399–402.

Lang, X., Dalai, A. K., Bakhshi, N. N., Reaney, M. J., & Hertz, P. B., (2001). Preparation and characterization of bio-diesels from various bio-oils. *Bioresour. Technol., 80*, 53–62.

Lee, D. H., (2011). Algal biodiesel economy and competition among bio-fuels. *Bioresour. Technol. 102*, 43–49.

Lee, J. Y., Yoo, C., Jun, S. Y., Ahn, C. Y., & Oh, H. M., (2010). Comparison of several methods for effective lipid extraction from microalgae. *Bioresour. Technol., 101*, S75–S77.

Lee, Y. K., (2001). Microalgal mass culture systems and methods: Their limitation and potential. *J. Appl. Phycol., 13*, 307–315.

Li, Y., Horsman, M., Wu, N., Lan, C. Q., & Dubois-Calero, N., (2008). Biofuels from microalgae. *Biotechnol. Prog., 24*, 815–820.

Lin, C. Y., & Li, R. J., (2009). Fuel properties of biodiesel produced from the crude fish oil from the soapstock of marine fish. *Fuel Process Technol., 90,* 130–136.

Lotero, E., Liu, Y., Lopez, D. E., Suwannakarn, K., Bruce, D. A., & Goodwin, J. G., (2005). Synthesis of biodiesel via acid catalysis. *Ind. Eng. Chem. Res., 44,* 5353–5363.

Lu, H., Liu, Y., Zhou, H., Yang, Y., Chen, M., & Liang, B., (2009). Production of biodiesel from *Jatropha curcas* L. oil. *Comput. Chem. Eng., 33,* 1091–1096.

Lu, X., Vora, H., & Khosla, C., (2008). Overproduction of free fatty acids in *E. coli*: Implications for biodiesel production. *Metab. Eng., 10,* 333–339.

Ma, Y., Wang, Z., Yu, C., Yin, Y., & Zhou, G., (2014). Evaluation of the potential of 9 *Nannochloropsis* strains for biodiesel production. *Bioresour. Technol., 167,* 503–509.

Maceiras, R., Rodríguez, M., Cancela, A., Urréjola, S., & Sánchez, A., (2011). Macroalgae: Raw material for biodiesel production. *Appl. Energy, 88,* 3318–3323.

Mandotra, S. K., Kumar, P., Suseela, M. R., & Ramteke, P. W., (2014). Fresh water green microalga *Scenedesmus abundans*: A potential feedstock for high quality biodiesel production. *Bioresour. Technol., 156,* 42–47.

Mata, T. M., Martins, A. A., & Caetano, N. S., (2010). Microalgae for biodiesel production and other applications: A review. *Renew. Sustain. Energy Rev., 14,* 217–232.

Medipally, S. R., Yusoff, F. M., Banerjee, S., & Shariff, M., (2015). Microalgae as sustainable renewable energy feedstock for biofuel production. *Biomed Res. Int., 519513.*

Meher, L. C., Sagar, D. V., & Naik, S. N., (2006). Technical aspects of biodiesel production by transesterification: A review. *Renew Sust. Ener. Rev., 10*(3), 248–268.

Mercer, P., & Armenta, R. E., (2011). Developments in oil extraction from microalgae. *Eur. J. Lipid Sci. Technol., 113,* 539–547.

Metcalf, T. G., Melnick, J. L., & Estes, M. K., (1995). Environmental virology: From detection of virus in sewage and water by isolation to identification by molecular biology-a trip of over 50 years. *Annu. Rev. Microbiol., 49,* 461–487.

Miao, X., & Wu, Q., (2006). Biodiesel production from heterotrophic microalgal oil. *Bioresour. Technol., 97,* 841–846.

Mojović, L., Nikolić, S., Rakin, M., & Vukasinović, M., (2006). Production of bioethanol from corn meal hydrolyzates. *Fuel, 85,* 1750–1755.

Morowvat, M. H., Rasoul-Amini, S., & Ghasemi, Y., (2010). *Chlamydomonas* as a "new" organism for biodiesel production. *Bioresour. Technol., 101,* 2059–2062.

Mutanda, T., Ramesh, D., Karthikeyan, S., Kumari, S., Anandraj, A., & Bux, F., (2011). Bioprospecting for hyper-lipid producing microalgal strains for sustainable biofuel production. *Bioresour. Technol., 102,* 57–70.

Nabi, M. N., Rahman, M. M., & Akhter, M. S., (2009). Biodiesel from cotton seed oil and its effect on engine performance and exhaust emissions. *Appl. Therm. Eng., 29,* 2265–2270.

Ndikubwimana, T., Zeng, X., Liu, Y., Chang, J. S., & Lu, Y., (2014). Harvesting of microalgae *Desmodesmus* sp. F51 by bioflocculation with bacterial bioflocculant. *Algal Res., 6,* 186–193.

Nelson, M. M., Phleger, C. F., & Nichols, P. D., (2002). Seasonal lipid composition in macroalgae of the Northeastern Pacific Ocean. *Bot. Mar., 45,* 58–65.

Oniya, O. O., & Bamgboye, A. I., (2014). Production of biodiesel from groundnut (*Arachis hypogea* L.) oil. *Agric. Eng. Int. CIGR J., 16,* 143–150.

Panneerselvam, S. I., Parthiban, R., & Miranda, L. R., (2011). Poultry fat: a cheap and viable source for biodiesel production. *2nd Int. Conf. Environ. Sci. Technol., 6,* 371–374.

Papazi, A., Makridis, P., & Divanach, P., (2010). Harvesting *Chlorella minutissima* using cell coagulants. *J. Appl. Phycol., 22,* 349–355.

Patel, B., Tamburic, B., Zemichael, F. W., Dechatiwongse, P., & Hellgardt, K., (2012). Algal biofuels: A credible prospective? *ISRN Renew. Energy, 631574,* 1–14.

Patil, P. D., Gude, V. G., & Deng, S., (2009). Biodiesel production from *Jatropha curcas*, waste cooking, and *Camelina sativa* oils. *Ind. Eng. Chem. Res., 48,* 10850–10856.

Patil, V., Tran, K. Q., & Giselrød, H. R., (2008). Towards sustainable production of biofuels from microalgae. *Int. J. Mol. Sci., 9,* 1188–1195.

Peterson, D. L., & Auld, R. K., (1983). Winter rape oil fuel for diesel engines: Recovery and utilization. *J. Am. Oil Chem. Soc., 60,* 1579–1587.

Pick, U., & Rachutin-Zalogin, T., (2012). Kinetic anomalies in the interactions of Nile red with microalgae. *J. Microbiol. Methods, 88,* 189–196.

Pisal, D. S., & Lele, S. S., (2005). Carotenoid production from microalga, *Dunaliella salina*. *Indian J. Biotechnol., 4,* 476–483.

Pokoo-Aikins, G., Nadim, A., El-Halwagi, M. M., & Mahalec, V., (2010). Design and analysis of biodiesel production from algae grown through carbon sequestration. *Clean Technologies and Environmental Policy, 12*(3), 239–254.

Potts, T., Du, J., Paul, M., May, P., Beitle, R., & Hestekin, J., (2012). The production of butanol from Jamaica Bay Macro Algae. *Environ. Prog. Sustain. Energy, 31,* 29–36.

Preto, F., Zhang, F., & Wang, J., (2008). A study on using fish oil as an alternative fuel for conventional combustors. *Fuel, 87,* 2258–2268.

Qureshi, N., Cotta, M. A., & Saha, B. C., (2014). Bioconversion of barley straw and corn stover to butanol (a biofuel) in integrated fermentation and simultaneous product recovery bioreactors. *Food Bioprod. Process, 92,* 298–308.

Raheem, A., Wan, A. W. A. K. G., Taufiq, Y. Y. H., Danquah, M. K., & Harun, R., (2015). Thermochemical conversion of microalgal biomass for biofuel production. *Renew. Sustain. Energy Rev., 49,* 990–999.

Raja, R., Hemaiswarya, S., Kumar, N. A., Sridhar, S., & Rengasamy, R., (2008). A perspective on the biotechnological potential of microalgae. *Crit. Rev. Microbiol., 34,* 77–88.

Ramesh, D., (2013). Lipid identification and extraction techniques. In: Bux, F., (ed.), *Biotechnological Applications of Microalgae: Biodiesel and Value-Added Products* (pp. 89–97). CRC Press: Boca Raton, FL.

Rawat, I., Kumar, R. R., Mutanda, T., & Bux, F., (2011). Dual role of microalgae : Phycoremediation of domestic wastewater and biomass production for sustainable biofuels production. *Appl. Energy, 88,* 3411–3424.

Rawat, I., Ranjith, K. R., Mutanda, T., & Bux, F., (2013). Biodiesel from microalgae: A critical evaluation from laboratory to large scale production. *Appl. Energy, 103,* 444–467.

Reddy, B. V. S., Ramesh, S., & Reddy, P. S., (2005). Sweet Sorghum: A potential alternate raw material for bio-ethanol and bio-energy. *ISMN, 46*(1), 1–8.

Reyes, J. F., & Sepúlveda, M. A., (2006). PM-10 emissions and power of a diesel engine fueled with crude and refined biodiesel from salmon oil. *Fuel, 85,* 1714–1719.

Rodriguez-Garcia, I., & Guil-Guerrero, J. L., (2008). Evaluation of the antioxidant activity of three microalgal species for use as dietary supplements and in the preservation of foods. *Food Chem., 108,* 1023–1026.

Rosenberg, J. N., Oyler, G. A., Wilkinson, L., & Betenbaugh, M. J., (2008). A green light for engineered algae: Redirecting metabolism to fuel a biotechnology revolution. *Curr. Opin. Biotechnol., 19,* 430–436.

Russin, T. A., Boye, J. I., Arcand, Y., & Rajamohamed, S. H., (2011). Alternative techniques for defatting soy: A practical review. *Food Bioprocess Technol., 4,* 200–223.

Sanniyasi, E., Velu, P., Selvarajan, R., Elumalai, S., Prakasam, V., & Selvarajan, R., (2011). Optimization of abiotic conditions suitable for the production of biodiesel from *Chlorella vulgaris*. *Indian J. Sci. Technol., 4,* 91–97.

Santhose, I., Gnanadoss, J., Selvarajan, R., & Sanniyasi, E., (2011). Enhanced carotenoid synthesis of *Phormidium* sp. in stressed conditions. *J. Exp. Sci., 1,* 38–44.

Savage, P. E., & Hestekin, J. A., (2013). A perspective on algae, the environment, and energy. *Environ. Prog. Sustain. Energy, 32,* 877–883.

Selvarajan, R., Felföldi, T., Tauber, T., Sanniyasi, E., Sibanda, T., & Tekere, M., (2015). Screening and evaluation of some green algal strains (*Chlorophyceae*) isolated from freshwater and soda lakes for biofuel production. *Energies 8,* 7502–7521.

Selvarajan, R., Senthil, K. R. M., & Elumalai, S., (2011). Pilot scale production of biodiesel from macro algae collected from coromandal coast, east coast India. *Int. J. Curr. Res. Rev., 3,* 58–76.

Selvarajan, R., Velu, P., & Sanniyasi, E., (2014). Molecular characterization and fatty acid profiling of indigenous microalgae species with potential for biofuel production in Tamil Nadu, India. *Golden Res. Thoughts, 3,* 1–9.

Shah, S., Sharma, S., & Gupta, M. N., (2004). Biodiesel preparation by lipase-catalyzed transesterification of Jatropha oil. *Energy and Fuels, 18,* 154–159.

Shen, W., Yuan, Z. J., Pei, Q., Wu, E., & Mao, A., (2009). Microalgae mass production methods. *Trans. ASABE, 52,* 1275–1287.

Singh, B., Guldhe, A., Rawat, I., & Bux, F., (2014). Towards a sustainable approach for development of biodiesel from plant and microalgae. *Renew. Sustain. Energy Rev., 29,* 216–245.

Singh, J. S., Kumar, A., Rai, A. N., & Singh, D. P., (2016). Cyanobacteria: A precious bio-resource in agriculture, ecosystem, and environmental sustainability. *Front. Microbiol., 7,* 1–19.

Singh, R. K., Tiwari, S. P., Rai, A. K., & Mohapatra, T. M., (2011). Cyanobacteria: An emerging source for drug discovery. *J. Antibiot. (Tokyo), 64,* 401–412.

Song, M., Pei, H., Hu, W., & Ma, G., (2013). Evaluation of the potential of 10 microalgal strains for biodiesel production. *Bioresour. Technol., 141,* 245–251.

Spolaore, P., Joannis-cassan, C., Duran, E., & Isambert, A., (2006). Commercial applications of microalgae. *J. Biosci. Bioeng., 101,* 87–96.

Suali, E., & Sarbatly, R., (2012). Conversion of microalgae to biofuel. *Renew. Sustain. Energy Rev., 16,* 4316–4342.

Taher, H., Al-Zuhair, S., Al-Marzouqi, A. H., Haik, Y., & Farid, M., (2014). Effective extraction of microalgae lipids from wet biomass for biodiesel production. *Biomass and Bioenergy, 66,* 159–167.

Talebi, A. F., Mohtashami, S. K., Tabatabaei, M., Tohidfar, M., Bagheri, A., Zeinalabedini, M., Hadavand, M. H., et al., (2013). Fatty acids profiling: A selective criterion for screening microalgae strains for biodiesel production. *Algal Res., 2,* 258–267.

Tan, Y., & Lin, J., (2011). Biomass production and fatty acid profile of a *Scenedesmus rubescens*-like microalga. *Bioresour. Technol., 102,* 10131–10135.

U.S. Energy Information Agency, (2013). *International Energy Outlook.*

Uduman, N., Qi, Y., Danquah, M. K., Forde, G. M., & Hoadley, A., (2010). Dewatering of microalgal cultures : A major bottleneck to algae-based fuels. *J. Renew. Sustain. Energy, 2*(1), 12701.

Ugwu, C. U., Aoyagi, H., & Uchiyama, H., (2008). Photobioreactors for mass cultivation of algae *Bioresour. Technol.*, *99*(10), 4021–4028.

Valdez-Ojeda, R., González-Muñoz, M., Us-Vázquez, R., Narváez-Zapata, J., Chavarria-Hernandez, J. C., López-Adrián, S., et al., (2015). Characterization of five fresh water microalgae with potential for biodiesel production. *Algal Res., 7,* 33–44.

Valenzuela, J., Mazurie, A., Carlson, R. P., Gerlach, R., Cooksey, K. E., Peyton, B. M., & Fields, M. W., (2012). Potential role of multiple carbon fixation pathways during lipid accumulation in *Phaeodactylum tricornutum. Biotechnol. Biofuels, 5*, 1–17.

Vandamme, D., Foubert, I., & Muylaert, K., (2013). Flocculation as a low-cost method for harvesting microalgae for bulk biomass production. *Trends Biotechnol., 31*, 233–239.

Vandamme, D., Foubert, I., Meesschaert, B., & Muylaert, K., (2010). Flocculation of microalgae using cationic starch. *J. Appl. Phycol., 22*, 525–530.

Vicente, G., Bautista, L. F., Rodríguez, R., Gutiérrez, F. J., Sádaba, I., Ruiz-Vázquez, R. M., et al., (2009). Biodiesel production from biomass of an oleaginous fungus. *Biochem. Eng. J., 48*, 22–27.

Vidyashankar, S., VenuGopal, K. S., Swarnalatha, G. V., Kavitha, M. D., Chauhan, V. S., Ravi, R., et al., (2015). Characterization of fatty acids and hydrocarbons of Chlorophycean microalgae towards their use as biofuel source. *Biomass and Bioenergy, 77*, 75–91.

Vyas, A. P., Verma, J. L., & Subrahmanyam, N., (2010). A review on FAME production processes. *Fuel, 89*, 1–9.

Wang, B., Li, Y., Wu, N., & Lan, C. Q., (2008). CO_2 bio-mitigation using microalgae. *Appl. Microbiol. Biotechnol., 79*(5), 707–718.

Wang, J., Yang, H., & Wang, F., (2014). Mixotrophic cultivation of microalgae for biodiesel production: Status and prospects. *Appl. Biochem. Biotechnol., 172*, 3307–3329.

Wang, L., Li, Y., Chen, P., Min, M., Chen, Y., Zhu, J., & Ruan, R. R., (2010). Anaerobic digested dairy manure as a nutrient supplement for cultivation of oil-rich green microalgae *Chlorella* sp. *Bioresour. Technol., 101*, 2623–2628.

Wang, Y., Guo, W., Lo, Y., Chang, J., & Ren, N., (2014). Characterization and kinetics of bio-butanol production with *Clostridium acetobutylicum* ATCC824 using mixed sugar medium simulating microalgae-based carbohydrates. *Biochem. Eng., J. 91*, 220–230.

Wawrik, B., & Harriman, B. H., (2010). Rapid, colorimetric quantification of lipid from algal cultures. *J. Microbiol. Methods, 80*, 262–266.

Xie, W., & Huang, X., (2006). Synthesis of biodiesel from soybean oil using heterogeneous KF/ZnO catalyst. *Catal. Letters, 107*, 53–59.

Xiong, W., Li, X., Xiang, J., & Wu, Q., (2008). High-density fermentation of microalga *Chlorella protothecoides* in bioreactor for microbio-diesel production. *Appl. Microbiol. Biotechnol., 78*(1), 29–36.

Xu, H., Miao, X., & Wu, Q., (2006). High quality biodiesel production from a microalga *Chlorella protothecoides* by heterotrophic growth in fermenters. *J. Biotechnol., 126*, 499–507.

Xu, J., & Hu, H., (2013). Screening high oleaginous *Chlorella* strains from different climate zones. *Bioresour. Technol., 144*, 637–643.

Yaakob, Z., Narayanan, B. N., Padikkaparambil, S., Unni, K. S., & Akbar, P. M., (2014). A review on the oxidation stability of biodiesel. *Renew. Sustain. Energy Rev., 35*, 136–153.

Yoo, C., Jun, S. Y., Lee, J. Y., Ahn, C. Y., & Oh, H. M., (2010). Selection of microalgae for lipid production under high levels carbon dioxide. *Bioresour. Technol., 101*, S71–S74.

Zhang, Y., Ma, N., Zhou, Y., Fu, T., & Meng, J., (2014). Treatment of nutrient-rich municipal wastewater using mixotrophic strain *Chlorella kessleri* GXLB-9. *Journal of Sustainable Development Studies, 7,* 1–18.

Zheng, Y., Yu, X., Zeng, J., & Chen, S., (2012). Feasibility of filamentous fungi for biofuel production using hydrolysate from dilute sulfuric acid pretreatment of wheat straw. *Biotechnol. Biofuels, 5*(1), 1.

Zhu, J., Rong, J., & Zong, B., (2013). Factors in mass cultivation of microalgae for biodiesel. *Chinese J. Catal., 34,* 80–100.

Zhu, L. D., Hiltunen, E., Antila, E., Zhong, J. J., Yuan, Z. H., & Wang, Z. M., (2014). Microalgal biofuels: Flexible bioenergies for sustainable development. *Renew. Sustain. Energy Rev., 30,* 1035–1046.

Zhu, L. D., Naaranoja, M., & Hiltunen, E., (2011). Environmental sustainability of microalgae production as a biofuel source. *Adv. Mater. Res.*, pp. 378, 379, 433–438.

CHAPTER 7

POTENTIAL HEALTH BENEFITS OF FUCOIDAN: AN UPDATE

THANGARAJ VIMALA,[1] THINANOOR VENUGOPAL POONGUZHALI,[2] and MUTHU SAKTHIVEL[3]

[1]Department of Plant Biology and Plant Biotechnology, Quaid-e-Millath Government College (W), Chennai – 600002, Tamil Nadu, India

[2]Department of Botany, Queen Mary's College, Chennai – 600004, Tamil Nadu, India

[3]Department of Biotechnology, University of Madras, Chennai – 600025, Tamil Nadu, India

7.1 INTRODUCTION

Seaweeds or macroalgae grow in the deep sea areas of depth up to 180 meters, in estuaries and shallow water on the solid substrates like rocks, dead corals, shells, and other plant material. They are classified into three higher taxa, Green (Phylum: *Chlorophyta*), Brown (Phylum: *Phaeophyceae*), and Red (Phylum: *Rhodophyta*) based on their pigmentation. A larger diversity in biochemical composition of seaweeds paves the trail to explore the form of compounds with a good varies of physiological and biochemical characteristics, many of which are rare or absent in other taxonomic groups (Holdt and Kraan, 2011). Seaweeds have emerged as a potential source in the biomedical area, mainly due to their bioactive substances which show great efficacy as anti-inflammatory, antimicrobial, antiviral, and anti-tumoral properties (Smit, 2004). Indeed, several species of algae have been found to be the sources of polysaccharides and glycoproteins with immune-stimulant, anti-coagulant, anti-tumoral, and anti-viral activity (Nishino et al., 1991). Sulfated polysaccharides are among the most abundant and broadly studied seaweed polysaccharides and they have attracted much attention in the fields of pharmacology and biochemistry.

Fucoidan is a sulfated polysaccharide mainly found in the cell-wall matrix of various brown seaweed species (Kim et al., 2010). Algal fucoidans are present in several orders, mainly Fucales and Laminariales but also in *Chordariales, Dictyotales, Dictyosiphonales, Ectocarpales, Scytosiphonales,* and seems to be absent in *Chlorophyceae, Rhodophyceae, Xanthophyceae,* and in freshwater algae and terrestrial plants (Berteau and Mulloy, 2003).

Fucoidans from seaweeds are heterogenic mixtures of structurally related polysaccharides. Two different types of backbone chains are found in fucoidans such as repeating (1→3) linked alpha-L-fucopyranose residues and alternating (1→3) and (1→4) linked alpha-L-fucopyranose residues. Both chains have carbohydrate (L-fucopyranose, alpha-D glucuronic acid) and non-carbohydrate substituent (sulfate and acetyl groups). The chemical composition and structure of fucoidans are very diverse (Figure 7.1) (Rioux et al., 2007) and significantly vary depending on the algae source, place of cultivation and harvesting time. The composition also changes according to algal species, the extraction process, season of harvest, and local climatic conditions (Honya et al., 1999; Ale and Meyer, 2013).

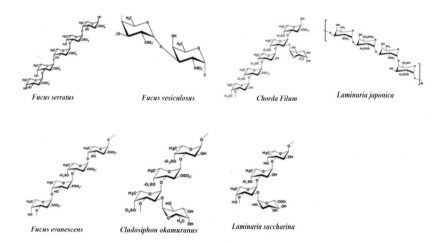

Fucus serratus   *Fucus vesiculosus*   *Chorda Filum*   *Laminaria japonica*

Fucus evanescens   *Cladosiphon okamuranus*   *Laminaria saccharina*

FIGURE 7.1   Different structures of fucoidan vary with different brown algae.

(Source: Reprinted from Menshova et al., 2016. Creative Commons Attribution License (CC BY).

Apart from seaweeds, polysaccharides of low molecular weight (less than 30kDa) that show fucoidan-like properties have also been reported in marine invertebrates such as sea cucumbers and sea urchins (Figure 7.2). However, these polysaccharides are simpler than fucoidans derived from marine brown algae and are referred to as sulfated fucans. Table 7.1 provides an overview of various studies conducted on brown algae and other marine sources to investigate the structure and composition of fucoidans.

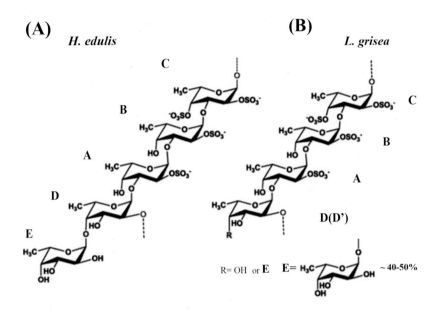

FIGURE 7.2 Sulfated fucan isolated from two sea cucumbers *H. edulis* (A) and *L. grisea* (B).

TABLE 7.1 Fucoidan from Different Marine Sources

Group	Species	References
Dictyotales	*Dictyota menstrualis*	De et al., 2004
Laminariales	*Kjellmaniella crassifolia*	Sakai et al., 2002
	Ecklonia kurome	Nishino et al., 1991
	Chorda filum	Chizhov et al., 1999
Ectocarpales	*Cladosiphon okamuranus*	Nagaoka et al., 1999; Sakai et al., 2003
Fucales	*Pelvetia canaliculata*	Colliec et al., 1994
	Fucus vesiculosus	Takashi Nishino et al., 1994
	F. evanescens	Kuznetsova et al., 2003
	F. serratus	Bilan et al., 2006
	F. distichus	Bilan et al., 2004

TABLE 7.1   *(Continued)*

Group	Species	References
	Sargassum stenophyllum	Duarte et al., 2001
	Ascophyllum nodosum	Medcalf et al., 1978; Chevolot et al., 2001a
	Sargassum fusiforme	Li et al., 2006
Ralfsiales	*Analipus japonicus*	Bilan et al., 2007
Sea cucumber	*Ludwigothurea grisea*	Ana-Cristina Ribeiro et al., 1993
Sea urchins	*Lytechinus variegates*	Mulloy et al., 1994
	Arbacia lixula	Alves et al., 1997
	Strongylocentrotus purpuratus	Alves et al., 1998
	S. franciscanus	Vilela-Silva et al., 1999
	S. droebachiensis	Vilela-Silva et al., 2002
	Acaudina molpadioides	Yu et al., 2014

Biological activity and medicinal importance of fucoidans depend strongly on their structural properties viz., the type of glycosidic bond, composition, species, conformations, molecular weight, and sulfate content. Extraction techniques are crucial for obtaining the relevant structural features required for specific biological activities and for elucidating structure-function relations (Ale and Meyer, 2013). Since fucoidan structures are extremely diverse and isolated under different extraction conditions; preservation of the structural integrity of the fucoidan molecules essentially depends on the extraction methodology.

Fucoidans form specific conformation to enhance its various biological functions. Structural analysis of several fucoidans demonstrates that their biological properties are determined not only by the charge density, but also by fine chemical structure. Therefore, the elucidations of the molecular structure, chemical structure, and conformation could be useful in the application of a particular polysaccharide. However, as the structure and extraction of fucoidan is beyond the scope of this review, some of their bioactive applications have been discussed in this review. Several studies have revealed the biological properties of brown algal fucoidans and this has opened up potential opportunities in pharmaceutical, cosmeceutical, nutraceutical, and functional food industries (Figure 7.3). While fucoidans of different species have broad similarities with regard to its bioactivity, there are sometimes differences in their activity. For example, *C. okamuranus* fucoidan did not inhibit breast cancer cell adhesion to platelets *in vitro*, whereas fucoidans from several other species including *F. vesiculosus*, inhibited adhesion by more than 80% (Cumashi et al., 2007). On the other hand, *Cladosiphon* fractions attenuated tumor growth and increased survival in a mouse colon cancer model (Azuma et al., 2012). Hence, it is apparent that fucoidan fractions need to be standardized, assessed and validated for each particular application and their main mechanism of action has to be understood. This chapter focuses on the various pharmaceutical applications of fucoidan.

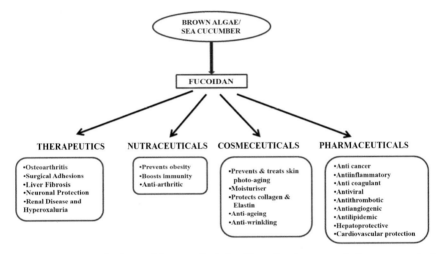

FIGURE 7.3 Applications of brown algal fucoidans in pharmaceutical, cosmeceutical, therapeutical, and nutraceutical industries.

7.2 PHARMACEUTICAL APPLICATIONS OF FUCOIDAN

In recent years, algal fucoidans have been the subject of numerous scientific studies due to their diverse and potential biological functions. Table 7.2 provides a summary of the biological activities of fucoidan isolated from various brown seaweeds. The functions include antitumor (Synytsya et al., 2010; Senthilkumar et al., 2013), immunomodulatory (Byon et al., 2008; Do et al., 2010), antioxidant (Chevolot et al., 2001; Zhang et al., 2003); antiviral (Hemmingson et al., 2006; Synytsya et al., 2014), antithrombotic, and anticoagulant (Kuznetsova et al., 2003; De et al., 2005; Chandía and Matsuhiro, 2008), anti-inflammatory (Park Hye-Young et al., 2011; Lee et al., 2012), anti-angiogenic activity (Matou et al., 2002; Cong et al., 2016); cardioprotection (Thomes et al., 2010) as well as their effects against various renal, hepatic, and uropathic disorders (Wang et al., 2012).

TABLE 7.2   Health Effects of Fucoidans Isolated from Different Marine Sources

Health Effects	Source	References
Antioxidant	*Porphyra haitanesis*	Zhang et al., 2003b
	Lessonia nigrescens	Qu et al., 2014
	L. trabeculata	Chevolot et al., 2001a
	F. vesiculosus	
Antilipidemic	*Laminaria japonica*	Huang et al., 2010
Anticoagulant/ antithrombotic	*Ecklonia cava*	Athukorala et al., 2009
	E. cava	Jung et al., 2007
	Fucus evanescens	Kuznetsova et al., 2003
	E. cava	Wijesinghe et al., 2011
	Padina gymnospora	De et al., 2005
	Ascophyllum nodosum	Chevolot et al., 2001
	Sargassum fulvellum	De Zoysa et al., 2008
	Hizikia fusiforme	Dobashi et al., 1989
	L. cichorioides	Yoon et al., 2007
	L. saccharina	Cumashi et al., 2007
	L. digitata	
	Holothuria edulis	Wu et al., 2015
	Ludwigothurea grisea (sea cucumber)	
	Isostichopus badionotus	Shiguo Chen et al., 2012
Immunomodulation	*F. vesiculosus*	Kim and Joo, 2008
	F. vesiculosus	Do et al., 2010
	F. vesiculosus	Jintang et al., 2010
	Undaria pinnatifida	Yoo et al., 2007
	F. vesiculosus	Yang et al., 2008
Anti-inflammation	*L. japonica*	Li et al., 2011
	E. cava	Kang et al., 2011
Antitumor/anti-proliferation/ anticancer	*U. pinnatifida*	Synytsya et al., 2010
	E. cava	Athukorala et al., 2009
	F. evanescens	Alekseyenko et al., 2007
	L. guryanovae	Lee et al., 2008
	Cladosiphon okamuranus	Kawamoto et al., 2006
		Teruya et al., 2007
	S. plagiophyllum	Suresh et al., 2013

TABLE 7.2 *(Continued)*

Health Effects	Source	References
Angiogenesis	*A. nodosum*	Matou et al., 2002
Cardioprotection	*C. okamuranus*	Thomes et al., 2010
Antiviral	*U. pinnatifida*	Hemmingson et al., 2006
Neuroprotection	*L. japonica*	Luo et al., 2009
Renal protective activity	*L. japonica*	Wang et al., 2012
Anti-obesity	*F. vesiculosus*	Park et al., 2011
Gastric protection	*C. okamuranus*	Kawamoto et al., 2006
	Acaudina molpadioides (sea cucumber)	Yanchao Wang et al., 2012
Anti-allergy	*U. pinnatifida*	Maruyama et al., 2005

7.2.1 FUCOIDAN AND CANCER

Polysaccharides, especially fucoidan from marine sources as anti-cancer drugs recently have caught increasing attention because of their biological activities and lower side effects when compared to synthetic anti-cancer drugs. The extensive research on the fucoidan of various brown algae isolated from different marine environments against various cancer cell lines shows promising anti-cancer potential. Fucoidan shows a high efficiency in the treatment of a variety of cancers, including lung cancer, prostate cancer, breast cancer, hepatoma, and leukemia (Boo et al., 2013; Park Hyun-Soo et al., 2013; Choo et al., 2016). Its anti-tumor activity is exerted by regulating multiple signaling pathways in cancer cells.

The anti-cancer property of fucoidan has been demonstrated *in vivo* and *in vitro* in different types of cancers. Fucoidan, from different brown seaweeds with anti-cancer effects studied is shown in Table 7.3. Investigations have shown that fucoidan has a cytotoxic effect on tumor cells, but not on normal cells (Xue et al., 2013; Wang Yan et al., 2015). A derivative of fucoidan with 3,4-O-sulfated chains which are repeats of (1–3)- and (1–4)-α-l-fucopyranose was found to be non-cytotoxic to human diploid fibroblast cells WI-38, while the other linkage of this derivative of fucoidan was cytotoxic to human diploid fibroblast cells WI-38 (Kasai et al., 2015).

Fucoidan exerts anti-cancer effects directly and also indirectly kills cancer cells by enhancing immunity. The direct action of fucoidan are inducing apoptosis, arresting cell cycle, especially the G1 phase, antiangiogenesis, and inhibiting cellular migration. Studies on the antitumor mechanism of fucoidan reported that varieties of signaling pathways are involved. The apoptosis-inducing effect of fucoidan is by Bcl-2 family proteins, the PI3K/ Akt signaling pathway, MAPK pathways, ER stress, the generation of ROS, the caspase-independent AIF pathway and the Wnt/β-catenin pathway. The effect of fucoidan on inhibiting cellular migration is by MMPs expression, hypoxia condition, suppressing CXCL12, increasing miR-29 and the loss of Mtss1 (Wu Lei et al., 2016).

TABLE 7.3   Effect of Fucoidan on Different Cancer Cells

Species of Brown Algae	Cell Type	Mechanism of Anti-Cancer Activity	References
C. okamuranus	Stomach cancer cell line of MKN45	Growth inhibitory activity	Kawamoto et al., 2006
	U937, human leukemia	Induction of apoptosis MAPK and p38 MAPK inhibitor activated	Park et al., 2013
	MCF-7 cells, human breast cancer	↑ sub-G0/G1 fraction Apoptosis blocked	Yamasaki-Miyamoto et al., 2009
	Human T cell leukemia	Induction of apoptosis	Haneji et al., 2005
F. vesiculosus	EJ human bladder cancer cells	G1 arrest in cell cycle induces apoptosis	Park et al., 2015
	HLF cells-Hepatocellular carcinoma (HCC)	Arrested HLF cells in G1/S phase Fucoidan suppressed proliferation, but not apoptosis	Kawaguchi et al., 2015
	HCT-15, colon carcinoma cells	↑ G1-phase. Apoptosis induced by ERK and p38 kinase activation Inactivation of PI3K/Akt	Hyun et al., 2009
	HT-29 and HCT116, human colon cancer cells	Caspase-8,-9,-7 and -3 increased Bak level increased Apoptosis induced via death receptor and mitochondria pathways	Kim et al., 2010
	Human lymphoma HS-Sultan cell line	No G0/G1 or G2/M arrest Induction of apoptosis downregulation of ERK pathway	Aisa et al., 2005
	4T1 cells	Induction of apoptosis	Xue et al., 2013
	HeLa and MCF-7 cells	Induction of apoptosis	Kasai et al., 2015

TABLE 7.3 *(Continued)*

Species of Brown Algae	Cell Type	Mechanism of Anti-Cancer Activity	References
F. evanescens	Lewis lung adenocarcinoma	Antimetastatic and antitumor activity	Alekseyenko et al., 2007
U. pinnatifida	Human lung cancer A549 cells	↑ Sub-G1 fraction Downregulation of phosphor-P13K/Akt pathway Activation of ERK1/2MAPK pathway	Boo et al., 2013
	SMMC-7721 cells	Inducing apoptosis	Yang et al., 2013
S. hemiphyllum	SK-Hep1 and HepG2 cells	Suppressing invasion and migration	Yan et al., 2015
S. filipendula	HeLa cells	Inhibiting proliferation	Costa et al., 2011
Sargassum plagiophyllum	HepG2 and A549 cells	Induction of apoptosis	Suresh et al., 2013
Saccharina latissima	Human Burkitt's lymphoma cells	Inhibiting proliferation and migration	Schneider et al., 2015

7.2.1.1   EFFECT OF FUCOIDAN ON CELL CYCLE AND APOPTOSIS

Cell cycle deregulation and apoptosis resulting in uncontrolled cell proliferation are the most frequent alterations that occur during the development and progression of cancer. For this reason, a blockade of the cell cycle and apoptosis induction are regarded as effective strategies for eliminating cancer. Wide research on cell cycle deregulations in cancers has promoted the usage of marine bioactive compounds that can either modulate signaling pathways leading to cell cycle regulation or directly alter cell cycle regulatory molecules, in cancer therapy. In mammalian cell cycle regulation, cyclins, and cyclin-dependent kinase (Cdk) inhibitors control regulator proteins, Cdks at specific points of the cell cycle (Lee and Yang, 2001). In particular, the G1 checkpoint is the most significant target for many anti-cancer agents. D-type cyclins, cyclin E and Cdk4/6, Cdk inhibitors, including p21 and p27, and retinoblastoma protein (pRB) are the central players of G1 phase transition to the S phase cell cycle (Paternot et al., 2010). Alteration or modification in the formation of Cyclin/Cdk complexes could lead to increased or decreased cell growth and proliferation followed by differentiation and/or cell death by apoptosis (Canavese et al., 2012).

Several studies have reported the effect of different species of fucoidan on different tumor cells cycle arrest, especially the G1 phase. In a recent study by Park Hyun-Soo et al. (2015), fucoidan was found to have arrested G1 phase in human bladder cancer cell cycle progression by down-regulation of G1 related cyclins and Cdks, without any change in Cdk inhibitors, such as p21 and p27 and by inhibiting pRB phosphorylation (Figure 7.4). Furthermore, dephosphorylation of pRB by fucoidan was associated with enhanced binding of pRB with the transcription factors E2F-1 and E2F-4. Reports also say that fucoidan suppressed proliferation of cancer cell and inhibited the growth of transplanted tumor xenografts by inducing apoptosis and by blocking abnormal cell cycle progression at the G1 or G2/M phase (Riou et al., 1995; Hsu et al., 2013; Park Hyun-Soo et al., 2013; Zhang et al., 2013). It has been found to also induce cell cycle arrest in other phases too. However, studies show that this effect was selective for cancer cells and fucoidan did not affect the cell cycle and apoptosis of normal cell lines. Also, fucoidan inhibited migration and invasion of highly metastatic cancer cells through down-regulation of matrix metalloproteinases (MMPs) and inhibition of the phosphoinositide 3-kinase/Akt and nuclear factor-kB signaling pathways (Lee Seung-Hong et al., 2012; Wang Peisheng et al., 2014).

Fucoidan treatment in a variety of cell types has resulted in sub G0/G1 cell accumulation. Fucoidan of *C. okamuranus* increased the G0/G1-phase population in hepatocarcinoma cell line (Huh7) accompanying by a decrease in the S phase, suggesting that fucoidan may cause the cell cycle arrest at the G0/G1 phase (Fukahori et al., 2008). In addition, fucoidan from *U. pinnatifida* also found to arrest cell cycle arrest at G0/G1 phase of prostate cancer cell line PC-3 through down-regulation of the Wnt/β-catenin signaling pathway to result in down-regulating E2F-1 and up-regulating p21 by using flow cytometry and Western blot methods *in vitro* (Boo et al., 2013). Crude fucoidan from *F. vesiculosus* arrested G1 cell cycle of mouse breast cancer cell line 4T1 by downregulating β-catenin levels in nucleus and cytoplasm and thereby reducing the expression of its target-survivin, c-myc, and cyclin D1 *in vitro* and *in vivo* (Xue et al., 2013). If the researchers investigate more on signaling mechanism in response to a kind of fucoidan, the mechanism of cell cycle arrest may be more clearly understood.

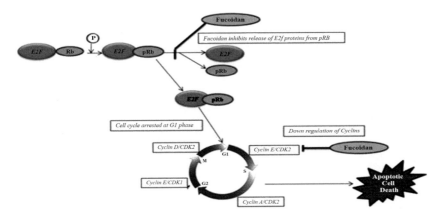

FIGURE 7.4 Fucoidan arrests G1 phase in human bladder cancer cell cycle progression. Downregulation of Cyclin D1/E and Cdks associated.

Apoptosis characterized by cytoplasmic shrinkage and chromatin condensation facilitates the removal of cells without inducing inflammation. The role of fucoidan in the induction of apoptosis is a critical factor in cancer therapy and oncology. Apoptosis occurs through either the extrinsic (cytoplasmic) pathway or by intrinsic (mitochondrial) pathway (Figure 7.5). Extrinsic apoptosis indicates a form of death induced by extracellular signals that result in the binding of ligands to specific trans-membrane receptors, collectively known as death receptors (DR) belonging to the tumor necrosis factor (TNF)/nerve growth factor (NGF) family. The intrinsic pathway is activated in response to variety of stressing conditions such as injury, oxidative stress and many others. Both pathways activate caspases that cleave regulatory and structural molecules. Several studies on cancers such as hematopoietic, lung, breast, and colon cancers have shown that fucoidan-mediated cell death occurs through triggering apoptosis.

The direct effects fucoidan on cancer cells are by inducing cell apoptosis through a nuclear factor kappa-light-chain-enhancer of activated B cells (NF-κB) pathway, mediated by PI3K/Akt and ERK signaling pathways (Burz et al., 2009; Atashrazm et al., 2015) and also through the activation of caspase-cascades, extracellular signal-regulated kinase mitogen-activated protein kinase (ERK1/2 MAPK) and the inactivation of p38 MAPK and phosphatidylinositol 3-kinase (PI3 K)/protein kinase B (Akt) (Nagamine et al., 2009; Zhang et al., 2011; Boo et al., 2013). Fucoidan inhibited proliferation in DU-145 prostate cancer cells and modulated protein expression associated with apoptosis through the PI3K/Akt and MAPK signaling pathway (Choo et al., 2016).

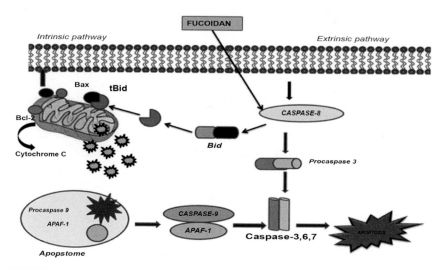

FIGURE 7.5   Schematic representation of apoptosis induced by fucoidan through molecular pathways. In the extrinsic pathway, caspase 8 is activated when ligand binds to specific receptors. In the Intrinsic pathway, apoptosome forms due to Cyt C release from mitochondria, and caspase 9 is activated. Caspase 8 and 9 activate downstream caspase cascade resulting in cell death. The two pathways are connected through BH3, a protein BID.

Fucoidan isolated from *U. pinnatifida* was found to induce apoptosis through ROS-mediated mitochondrial pathway in human hepatocellular carcinoma SMMC-7721 cells (Yang et al., 2013). Apoptotic cells displayed typical features such as chromatin condensation and marginalization, decrease in the number of mitochondria, mitochondrial swelling, and vacuolation. Fucoidan-induced cell death was associated with depletion of reduced glutathione (GSH), accumulation of high intracellular levels of reactive oxygen species (ROS), and accompanied by damage to the mitochondrial ultrastructure, depolarization of the mitochondrial membrane potential and caspase activation.

Fucoidan inhibited proliferation and apoptosis through a mitochondrial pathway in human lymphoma HS-Sultan cell lines (Aisa et al., 2005). However, phosphorylation of p38 and Akt was not altered. More recent research indicates that fucoidan may induce apoptosis in breast and colon cancer cells through modulation of the endoplasmic reticulum (ER) stress cascades (Chen et al., 2014). The fucoidan obtained from *Cladosiphon navae-caledoniae* showed reduction in the expression of Bcl-xL and Mcl-1, up-regulation of Bax and ROS in human breast cancer cell line MCF-7; and reduction in the expression of Bcl-xL, Mcl-1 as well as increased generation of ROS in human breast cancer cell line MDA-MB-231 *in vitro* (Zhang et al., 2013).

7.2.1.2   FUCOIDAN AND ANGIOGENESIS

Angiogenesis is a physiological process which is responsible for tissue organ regeneration. However, uncontrolled, and persistent angiogenesis may lead to several tumor progressions. The growth and spread of cancer are highly dependent on angiogenesis for feeding growing tumors with nutrients and oxygen (Carmeliet and Jain, 2000; Potente et al., 2011). Therefore, suppressing tumor angiogenesis is a critical target for preventing or slowing cancer growth. Many researchers have investigated the effect of different species fucoidan on angiogenesis of different tumor cells. Fucoidan shows both pro- and anti-angiogenic activity.

7.2.1.3   ANTIANGIOGENIC EFFECTS OF FUCOIDANS

Antitumor action of fucoidan may be due to its anti-angiogenic potency and the number of sulfate groups in the fucoidan molecule contributes to the effectiveness of its anti-angiogenic and antitumor activities. Research has shown that high-molecular-weight fucoidans (MW > 30kDa) with a high degree of sulfation demonstrate an antiangiogenic effect, whereas low-molecular-weight fucoidans (MW < 15kDa) promote angiogenesis (Ustyuzhanina et al., 2014). However, due to the fact that the structure of fucoidan has not been elucidated in detail in many cases, it is not possible to conclude that particular structural characteristics of fucoidans determine the mode of action in angiogenesis.

Vascular endothelial growth factor (VEGF) induces angiogenesis and lymphangiogenesis, because it is a highly specific mitogen for endothelial cells. Xue et al. (2012) investigated the antiangiogenic effect of fucoidan purified from F. vesiculosus on breast cancer 4T1 cells and results showed that fucoidan significantly reduced the expression of VEGF. Natural fucoidan (NF) and Oversulfated fucoidan (OSF) significantly suppressed the mitogenic and chemotactic actions of VEGF in tumor cells (Koyanagi et al., 2003). The OSF showed a more suppressive effect than that of NF, suggesting an important role for the numbers of sulfate groups in the fucoidan molecule.

Fucoidan isolated from C. novae-caledoniae demonstrated antiangiogenic activity (Ye et al., 2005) by decreasing the expression and secretion of VEGF in human uterine carcinoma HeLa cells, and suppressing tubulogenesis induced by tumor cells in vitro. Studies have shown that fucoidan inhibited the binding of VEGF to its cell membrane receptor and also inhibited tube formation following migration of human umbilical vein endothelial cells (HUVEC) and its chemical over-sulfation enhanced the inhibitory potency. In a recent study on multiple Myeloma cells, fucoidan interfered with angiogenesis of myeloma cells both *in vitro* and *in vivo* by decreasing VEGF secretion (Liu et al., 2016). Also, fucoidan decreased HUVEC, formed tube structures and inhibited HUVEC migration, and suppressed the angiogenic ability of multiple myeloma cells. The study also showed that fucoidan down-regulated the expression of several kinds of proteins, which may be correlated with the reduction of angiogenesis induced by myeloma cell. Fucoidan from *S. fusiforme* was found to inhibit lung cancer cell growth by disrupting VEGF-induced angiogenesis both *in vitro* and *in vivo* through targeting VEGFR2/VEGF and blocking VEGFR2/Erk/VEGF signaling suggesting that fucoidan could be a potential novel leading compound to inhibit lung cancer cell growth (Chen et al., 2016).

7.2.1.4   PROANGIOGENIC EFFECTS OF FUCOIDANS

Glycosaminoglycans (GAGs) and chemokines are important to regulate angiogenesis. Fucoidans are used as GAG mimetic. Small and highly sulfated fucoidans interact with pro-angiogenic chemokines, such as RANTES and SDF-1 which leads to revascularization and increase endothelial cell migration. However, this pro-angiogenic activity remains unclear and depends on the type of fucoidan. A recent study demonstrated that sulfated fucoidan (5kDa with the ratio sulfate/fucose at 1.87) presented high affinity to biotinylated-SDF-1 and RANTES. Also, the fucoidan significantly increased HUVEC migration and microvascular network formation (Marinval et al., 2015).

Fucoidans can also induce angiogenesis *in vitro* by modulating the proangiogenic properties of heparin-binding growth factors such as basic fibroblast growth factor (FGF). This pro-angiogenic effect is related to their molecular weight and sulfate content (Koyanagi et al., 2003). Fucoidan can also exhibit anti-cancer effect through inhibition of the proliferation of cancer cells without suppressing the angiogenesis induced by the cancer cells (Zhu et al., 2013). Not many literatures are available on the effect of fucoidan on angiogenesis as the effect is weak on angiogenesis. We can conclude from the literature that fucoidan has an effect on VGEF and different sources and molecular weight may be important for its angiogenic activity.

7.2.1.5   FUCOIDAN AND METASTASIS

Tumor metastasis of primary tumor cells to distant organs is a multistep process that follows a typical tumor metastatic cascade which includes uncontrolled cell proliferation, tissue remodeling, angiogenesis, and invasion. Metastasis, a complex process, can be summarized as invasion, adhesion, and angiogenesis (Teng et al., 2015). Cancer metastasis leads to disease recurrence and high mortality in cancer patients and also, it impairs quality of life and results in a poor prognosis. Therefore, inhibiting metastasis process or killing metastatic cancer cells by inducing apoptosis is significant in improving cancer patient survival.

P-Selectin, a protein residing on the platelet surface facilitates the adhesion between tumor cells and platelets, thus, enabling the tumor cells to survive the attack from immune cells. A study conducted by Cumashi et al. (2007) in highly metastatic MDA-MB-231 breast cancer cells, fucoidan inhibited P-Selectin residing on the platelet surface and thus reduced the number of attached tumor cells. The fucoidan from *U. pinnatifida* sporophylls exerted inhibitory effect on hepatocarcinoma Hca-F tumor metastasis *in vivo* and inhibited Hca-F cell growth, migration, invasion, and adhesion capabilities *in vitro* through the mechanism involving inactivation of the NF-κB pathway mediated by PI3K/Akt and ERK signaling pathways (Wang et al., 2014). It also inhibited growth and metastasis by downregulating VEGFC/VEGF receptor 3, hepatocyte growth factor/c-MET, cyclin D1, Cdk-4, phosphorylated (p) phosphoinositide 3-kinase, p-Akt, p-ERK 1/2, and nuclear transcription factor-κB (NF-κB), and suppressed adhesion and invasion by downregulating L-Selectin. Fucoidan, from *S. latissima,* could inhibit the migration of Burkitt's lymphoma cells and MMP-9 level by suppressing upstream CXCL12/CXCR4 *in vitro* (Schneider et al., 2015). Fucoidan from *F. vesiculosus* could inhibit the migration of human lung carcinoma cell line CL1-5 and mouse Lewis lung carcinoma cell line LLC1 by inhibiting TGFβ/TGFR pathway and down-regulating the downstream FAK signaling pathway in tumor tissue (Hsu et al., 2014).

7.2.2   ANTI-INFLAMMATORY ACTIVITY OF FUCOIDAN

Inflammation is a complex biological response of the host tissues to harmful stimuli such as damaged cells, pathogens, UV irradiation, irritants, and is a protective response involving immune cells, blood vessels and molecular mediators that initiate the inflammatory responses. Excessive inflammation can be harmful contributing to the pathogenesis of variety of diseases like chronic asthma, multiple sclerosis, inflammatory bowel disease, psoriasis, cancer. Anti-inflammatory substances like N-steroidal anti-inflammatory drugs (NSAIDs) can cause fatal side effects on long term use. Natural products with anti-inflammatory activity have long been used as a folk remedy for inflammatory conditions such as fevers, pain, migraine, and arthritis. Marine sources are becoming more popular because of its easy availability and potential anti-inflammatory activity. Fucoidan have been demonstrated to inhibit inflammatory response in many recent studies. The beneficial effects of fucoidan on inflammatory diseases are attracting investigators' attention.

Nitric oxide (NO) and prostaglandin E2 (PGE2) are major factors involved in the pathogenesis of many inflammatory diseases and TNF-α-NO and COX-II-PGE2 pathways have been considered as two main streams of inflammatory processes, which are blocked by inhibitors of inducible nitric oxide synthase (iNOS) and COX (NSAIDs), respectively. Kang et al. (2011) evaluated the inhibitory effect of fucoidan isolated from *E. cava* in lipopolysaccharide (LPS)-stimulated macrophage cell line RAW 264.7 cells. The fucoidan inhibited NO and PGE2 production and suppressed iNOS and cyclooxygenase-2 (COX-2) expression in RAW264.7.

Park Hye-Young et al. (2011) explained the mechanism of the inhibitory effects of fucoidan on production of LPS-induced pro-inflammatory mediators in BV2 microglia. They found that fucoidan inhibited excessive production of NO and PGE2 and attenuated expression of iNOS, COX-2, monocyte chemoattractant protein-1 (MCP-1), and pro-inflammatory cytokines, including interleukin-1β (IL-1β) and TNF-α. Moreover, fucoidan exhibited anti-inflammatory properties by suppressing nuclear factor-kappa B (NF-κB) activation and down-regulation of extracellular signal-regulated kinase (ERK), c-Jun N-terminal kinase (JNK), p38 MAPK, and AKT pathways.

Cui et al. (2010) showed that fucoidan from *L. japonica* on exhibited inhibitory effect on NO production and expression by suppressing p38 and ERK phosphorylation. Li et al. (2011) evaluated the inhibitory mechanisms of fucoidan on rat myocardial ischemia-reperfusion (I/R) model and the study revealed that fucoidan could regulate the inflammation response via high-mobility group and NF-jB inactivation in I/R-induced myocardial damage. Fucoidan reduced the myocardial IS, serum levels of pro-inflammatory factors TNF-α and IL-6, and the activity of myeloperoxidase (MPO) and also fucoidan down-regulated the expression of high mobility group, phosphor-IjB-a, and NF-jB. At the site of injury, the over-secretion of inflammatory cytokines such as IL-1β triggers the uncontrolled expression of MMPs by leukocytes and resident mesenchymal, epithelial, or endothelial cells. Fucoidan isolated from *A. nodosum* interfered with factors involved in connective tissue degradation such as MMPs secretion, MMPs/TIMPs association, and leukocyte elastase activity. Fucoidan also minimized human leukocyte elastase activity resulting in the protection of human skin elastic against the enzymatic proteolysis (Senni et al., 2006).

7.2.3   ANTICOAGULANT AND ANTITHROMBOTIC ACTIVITY

Heparin, an anticoagulant, accelerates serine proteinase inhibitor plasma factor-like thrombin (factor IIa) and factor Xa. Heparin is from animal source and hence, can induce diseases in mammals. These reasons necessitate finding a new anticoagulant and antithrombotic agent replacing heparin. Fucoidans have a wide variety of biological activities of which their anti-coagulant and antithrombotic activities are well studied. Many studies showed that the anticoagulant activity of fucoidan was related to sulfate content and position (Haroun-Bouhedja et al., 2000), molecular weight, and sugar composition (Suwan et al., 2009). Few studies have shown that anticoagulant activity of the fucoidans did not depend on the content of fucose, the other neutral sugars, and sulfates in the preparations, and also on the structure of the backbone of molecule. Stronger anticoagulant activity in native fucoidans was found due to higher content of sulfate groups like *Ecklonia kurome, H. fusiforme*. However, the anticoagulant and the antithrombin effects gradually decreased with an increase in the sulfate content of the fucans (Nishino et al., 1991).

It was previously reported that only homofucans induced anticoagulant activity. Chevolot et al. (2001) demonstrated that the anticoagulant activity of a homofucan from *A. nodosum* with a high proportion of (1→4) linkage was related to 2-O-sulfation and 2,3-disulfation. It was also observed that desulfation resulted in loss of anticoagulant activity. Few studies have interpreted the biological activity of fucans in terms of molecular structure. The anticoagulant activity of fucan is unlikely to be merely a charge density effect; rather it depends critically on the distribution pattern of sulfate groups (Nader et al., 2001) and the size of the molecule (Nardella et al., 1996).

Structural features of fucoidan also influence the mechanism of anticoagulant activity (Pereira et al., 1999). Duarte et al. (2001) reported that the anticoagulant properties of fucoidans were mainly determined by the fucose sulfated chains, especially by the disulfated fucosyl units. Shiguo Chen et al. (2011) confirmed that the difference in the anticoagulant activities of sulfated polysaccharides might depend on the different patterns of sulfation of the fucose branch of the chondroitin sulfate, especially 2,4-O-disulfation.

The mechanism by which fucoidan exerts anti-coagulant activity remains unclear probably, due to the structural variations of fucoidan between algal species and different extraction methodologies have produced fucoidans of different composition, structure, and size. This has led to conflicting results in the studies of mechanisms of anticoagulant activity of fucoidan.

Thromboembolic diseases continue to be the leading cause of death throughout the world (Streiff et al., 2011). Most thromboembolic processes require anticoagulant therapy. Unfractionated heparin (UFH) and low-molecular-weight heparins (LMWHs), the only sulfated polysaccharides currently used as anticoagulant drugs, have several side effects such as hemorrhagic effects, development of thrombocytopenia, ineffectiveness in congenital or acquired antithrombin deficiencies, incapacity to inhibit thrombin bound to fibrin (Warkentin et al., 1995). Since heparin is an animal source, anticoagulant from other natural sources are being explored.

Fucoidans isolated from sea cucumbers *Holothuria edulis* and *Ludwigothurea grisea*, are composed of a central core of regular α (1→3)- and α (1→2)-linked novel tetrasaccharide repeating units together with an unsulfated fucose residue as a side chain which may contribute to the anticoagulant action. The anticoagulant and platelet aggregation assays indicated that the sea cucumber fucans strongly inhibited human blood clotting through the intrinsic pathways of the coagulation cascade. The mechanism was selective anti-thrombin activity by heparin cofactor II without causing platelets to aggregate, whose anticoagulant mechanisms are different from those of heparin-like drugs (Wu et al., 2015). These results provided an insight into the structure-function relationships of the well-defined polysaccharides from invertebrate as new types of safer anticoagulants.

Thrombin plays an important role in thrombosis; hence, thrombin inhibitor becomes the main content of studies on antithrombotic drugs. Various researches show that anticoagulant activity of fucoidan is mainly mediated by thrombin inhibition by heparin cofactor II. They form ternary complexes with thrombin and thrombin inhibitor (AT) or HCII. A slight decrease in the molecular size of the sulfated fucans, however, reduced its effect on thrombin inactivation mediated by heparin cofactor II. Fucoidan requires a long sugar-chain and a conformation to bind the thrombin to achieve anticoagulant activity. Studies using a sulfated fucan from the brown alga *F. vesiculosus* suggested that the antithrombin activity is mediated mainly by heparin cofactor II, with a minor contribution of antithrombin (Mourao, 2004). However, Kuznetsova et al. (2003) reported that anticoagulant properties of fucoidan from *F. vesiculosus* were determined by thrombin inhibition mediated via plasma antithrombin-III *in vitro* and *in vivo* experiments, whose anticoagulant activity was similar to that of heparin.

Fucoidans extracted from brown algae have been documented to have excellent antithrombotic activity when administered by either intravenous or subcutaneous route in animal models. However, it is unknown if the fucoidans also have antithrombotic activity when administered orally, a highly desirable feature of oral antithrombotic agents. Qiu et al. (2006) reported that the OSF showed four times higher anticoagulant activity in doubling prothrombin time of normal human plasma in comparison with native fucoidan. Fucoidan also expressed antithrombotic activity when tested on *in vivo* models of venous and arterial thrombosis in experimental animals.

7.2.4   FUCOIDAN AND IMMUNE MODULATION

Immune modulation by fucoidan can be used as a tool to disrupt disease processes, including cancer and pathogen infections (Fitton et al., 2015). One mechanism by which the fucoidan exerts its immunomodulatory effect is provided by the abundant supply of fucose. Many polysaccharides of natural sources are considered as biological response modifiers and have been shown to enhance various immune responses (Li et al., 2008). Fucoidan modulates the function of various immune cells such as neutrophils (Zen et al., 2002), lymphocytes (Oomizu et al., 2006), natural killer (NK) cells (Namkoong et al., 2012), and dendritic cells (DCs) (Kim and Joo, 2008).

DCs direct immune function within tissues and recognize the pathogens and initiate a specific response in immune cells. Fucoidan functions as an adjuvant, enhancing an adaptive immune response and activating cytotoxic T cells. Fucoidan purified from brown seaweed *F. vesiculosus* was found to have immunostimulating and maturing effects on bone marrow-derived DCs through a pathway involving at least NF-κB (Kim and Joo, 2008). The study suggested that fucoidan is an activator of lymphocytes and macrophages. This property may contribute to fucoidan's effectiveness in the immunoprevention of cancer. Zhang et al. (2014) also demonstrated that fucoidan from *A. nodosum* induced dendritic cell maturation and enhanced immune response *in vitro* and *in vivo*. According to Yang et al. (2014), fucoidan may be used on DCs-based vaccines for cancer immunotherapy. Their study demonstrated the effects of fucoidan on maturation process and activation of human monocyte-derived DCs.

Toll-like receptors (TLRs) are a class of proteins that are single, membrane-spanning, non-catalytic receptors that play a key role in innate immunity system. Studies have shown that fucoidans isolated from *L. japonica, L. cichorioides, F. evanescens* activate and bind to TLRs in human embryonic kidney cells (Makarenkova et al., 2012). Fucoidans from these three species activated transcription nuclear factor NF-κB by binding specifically to TLR-2 and TLR-4 but not with TLR-5, and were nontoxic for the cell cultures, thus, indicating that fucoidans could induce *in vivo* defense from pathogenic microorganisms of various classes.

7.2.5   FUCOIDAN AND CARDIOVASCULAR HEALTH

Cardiovascular diseases (CVDs), including heart attacks and strokes, continue to be the leading causes of deaths resulting in more than 17 million deaths each year worldwide. CVD manifests as atherosclerosis, high cholesterol, high blood pressure, arrhythmia, and heart failure and has many more symptoms affecting the heart and blood vessels. Additionally, other metabolic diseases such as obesity and diabetes further increase the risk of CVD. Several experiments with isolated and intact hearts have demonstrated the preventative effect of fucoidan. In isolated rat hearts subject to no-flow ischemia followed by reperfusion, treatment with fucoidan (0.36 mg/mL blood) significantly reduced the leukocyte accumulation in both capillaries and venules. In addition, fucoidan significantly reduced the persistence of leukostasis in both capillaries and venules, indicating that it affected a transient adhesion process (Ritter et al., 1998).

Fucoidan also prevented proliferation of intimal lesions and reduced intimal thickening in a recent study done on grafted atherosclerotic aortas in animals. Furthermore, fucoidan helped stimulate the rebuilding of the endothelial cells in the grafted aorta. Fucoidan may be useful for myocardial ischemic patients as it enhanced tissue repair in myocardial I/R by promoting revascularization (*in situ* VEGF and SDF-1α over expression) and limiting fibrosis (Manzo-Silberman et al., 2011). Fucoidan extracted from *C. okamuranus* was found to provide cardioprotection against isoproterenol-induced myocardial infarction (MI) (Thomes et al., 2010). Fucoidan treatment reduced myocardial damage by improving parameters such as creatinine phosphokinase (CPK), lactate dehydrogenase (LDH), alanine transaminase (ALT) and aspartate transaminase (AST). Fucoidan also improved the antioxidant defense system and considerably reduced the oxidative stress exerted by isoproterenol.

Fucoidan extracted from *L. japonica* was shown to have anti-atherosclerotic activity, which was mediated through inhibition of the inflammation and oxidative stress (Wang et al., 2016). Fucoidan was shown to suppress the ROS related pathway and reduce the expression of LOX-1 and the diminishing the inflammation response. Studies indicate that fucoidan could be exploited as an innovative cardiovascular drug to prevent or retard the pathogenesis of atherosclerotic CVDs. The effect of Low Molecular Weight Fucoidan (LMWF) isolated from *L. japonica* on the development of Diabetic cardiomyopathy (DCM) was studied by Yu et al. (2014). The fucoidan inhibited oxidative stress and resultant cardiomyocyte apoptosis in diabetic heart by potentiating the antioxidant enzymatic activity and inhibiting PKCβ-dependent ROS production. In light of its efficacy and safety, LMWF may serve as a potential therapeutic drug for prevention and treatment of DCM.

7.2.5.1   FUCOIDAN, STEM CELLS, AND CARDIOVASCULAR HEALTH

Mobilization of stem cells and progenitor cells to ischemic sites is an important step in new vessel formation. After a period of ischemia-when blood vessels and cardiac tissue are injured due to lack of oxygen-these progenitor cells migrate from the bone marrow to the site of injury and encourage the growth of new blood vessels to help repair the damage, a process known as neovascularization. Fucoidan is now thought to have a role to play in this stem-cell-induced repair of cardiovascular damage. Studies have begun to emerge indicating fucoidan might influence the mobilization of endothelial progenitor cells and their incorporation in ischemic tissue. Studies have shown that treatment with the fucoidan increases circulating mature white blood cells and progenitor/stem cells in mice and nonhuman primates. Recent studies suggest that fucoidan may work by enhancing the activity of stromal-derived factor 1 (SDF-1), which plays a critical role at several steps of progenitor cell mobilization. *In vitro* and *in vivo* data suggest that fucoidan displaces certain factors that normally trap the SDF-1 in bone marrow, on endothelial cell surfaces or other tissues, helping to release the SDF-1 and allowing it to more easily mobilize the stem cells.

7.2.6   FUCOIDAN AS THERAPEUTIC FOR MAJOR BLINDING DISEASES

Age-related macular degeneration (AMD) and diabetic retinopathy (DR) are major causes for vision loss and blindness. Environmental factors such as smoking and genetic susceptibilities have been shown to be significantly associated with the development of AMD. DR is induced by hyperglycemia characterized by microvascular dysfunction and neovascularization. High glucose causes oxidative damage on cells of the retinal pigment epithelium (RPE) and causes pathological changes, which are responsible for the loss of vision associated with DR.

VEGF is the most important angiogenic factor in development of a disease (Ferrara, 2000). One of the sources of VEGF in the retina is the RPE. The secretion of VEGF can be elevated by many factors, such as oxidative stress or hypoxia (Klettner et al., 2013). The up-regulation of VEGF by the RPE due to age-dependent or pathological alterations is considered an important factor in the development of wet AMD. Fucoidan reduces VEGF secretion in RPE cells and RPE/choroid organ cultures and also it reduces the expression of intracellular VEGF (Dithmer et al., 2014). Anti-VEGF therapies are the recent treatment for AMD and DR. Fucoidan displayed several anti-VEGF functions in a variety of systems and seems to interfere with VEGF-induced signaling (Figure 7.6). Fucoidan has reduced the expression of the VEGF-receptors and even more of the VEGF co-receptors neuropilin (Narazaki et al., 2008). Fucoidan from *L. japonica* was found to reduce retinal damage and retinal neovascularization by inhibiting HIF-1α activation of VEGF.

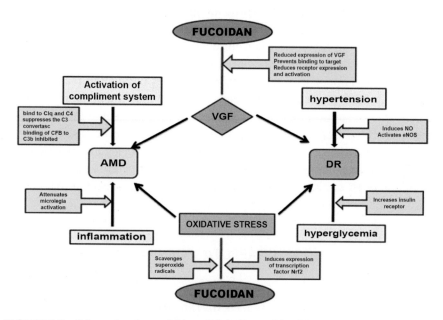

FIGURE 7.6   Schematic of potential beneficial effects of fucoidan on AMD or DR.

Another important feature of DR and AMD is the prevalence of oxidative stress. Overproduction of ROS contributes to oxidative stress-induced cell injury. ROS may lead to hyperglycemic complications in diabetes by mediating RPE cell dysfunction. Therefore, intracellular ROS production in RPE cells must be inhibited to protect the retina from oxidative damage. A recent study by Li Xiaoxia et al. (2015) on the role of fucoidan in DR showed that fucoidan protected ARPE-19 (Human Retinal Pigment Epithelial) cells from high glucose-induced cell death and normalized high glucose-induced generation of ROS. Fucoidan also inhibited high glucose-induced cell apoptosis, as well as the Ca^{2+} influx and ERK1/2 phosphorylation in ARPE-19 cells.

7.2.7   FUCOIDAN FOR LIVER AND KIDNEY HEALTH

Liver fibrosis is a dynamic reversible pathological process in the development of chronic liver disease to cirrhosis in conjunction with the progressive accumulation of fibrillar extracellular matrix proteins (Lieber, 1999). The main causes of liver fibrosis are infection with hepatitis B or C, alcohol abuse, and non-alcohol steatohepatitis. Fucoidan has different properties such as anti-inflammatory, anti-oxidant, and anti-fibrotic activities. Several studies indicate that fucoidan has hepatoprotective property which includes inhibiting liver fibrosis and this can be achieved by oral dosing in the models tested (Table 7.4).

TABLE 7.4   Effect of Fucoidan on Liver Diseases

Fucoidan Source	Model	Effect of Fucoidan	References
F. vesiculosus	Concanavalin induced liver injury in mice	Significantly inhibited raised levels of TNF-alpha and IFN-gamma. Increased endogenous IL-10 production	Saito et al., 2006
F. vesiculosus	CCl_4 induced liver fibrosis in mice	Protection of normal hepatocytes. Inhibition of hepatic stellate cell proliferation	Hayashi et al., 2008
F. vesiculosus	Ex vivo human hepatoma (hepatitis B and C) and Cirrhosis	Fucoidan decreased the disease elevated activity of biotinidase	Nagamine et al., 1993
F. vesiculosus	Alcohol-induced liver cirrhosis-intragastric feeding model	Fucoidan reduced the production of inflammation-promoting cyclooygenase-2 and nitric oxide; increased the expression of the hemeoxygenase-1 enzyme.	Lim et al., 2015

TABLE 7.4 *(Continued)*

Fucoidan Source	Model	Effect of Fucoidan	References
C. okamuranus	*N*-nitrosodiethylamine (DEN) induced liver fibrosis model in Sprague dawley rats	Protection from damage Increased metallothionein Down regulation of TGFbeta 1 and SDF1	Nakazato et al., 2010
F. evanescens	Mouse model of endotoxemia inducted by lipopolysaccharide (LPS)	Fucoidan inhibited the increased levels of TNFα and IL-6, decreasing of the processes of hypercoagulability.	Kuznetsova et al., 2014
F. vesiculosus	Acetaminophen-induced acute liver injury in rats	Increased glutathione, superoxide dismutase, and glutathione peroxidase, decreased the expression of tumor necrosis factoralpha, interleukin 1 beta, and inducible nitric oxide synthase.	Hong et al., 2012

Hayashi et al. (2008) demonstrated that fucoidan reduced acute and chronic liver failure in carbon tetra chloride (CCl_4) induced liver injury. Fucoidan was also found to attenuate Hepatic fibrosis induced by CCl_4. Liver fibrosis includes damage to hepatocytes and activation of hepatic stellate cells, and treatment of hepatocytes with fucoidan prevented CCl_4-induced cell death and inhibited the proliferation hepatic stellate cells. Saito et al. (2006) showed that fucoidan isolated from *F. vesiculosus* prevented Con A-induced liver injury by mediating the endogenous IL-10 production and the inhibition of proinflammatory cytokine. Research with fucoidan derived from *C. okamuranus* (Mozuku) showed anti-fibrogenesis in diethyl nitrosamine (DEN)-induced liver cirrhosis through the down-regulation of transforming factor-beta 1 and chemokine ligand 12 expressions (Nakazato et al., 2010). Fucoidan attenuated increase of hepatic messenger RNA levels and immunochemistry of transforming growth factor beta-1 by DEN. Fucoidan also suppressed Hepatic chemokine ligand 12 expression that was increased by DEN. Fucoidan significantly decreased the DEN-induced malondialdehyde (MDA) levels and increased metallothionein expression in the liver.

Fucoidan isolated from *F. vesiculosus* administered on alcohol-induced murine liver damage prevented the increase of alcohol-induced enzymes like liver enzyme aspartate (AST) and alanine (ALT) transaminases (Lim et al., 2015). Transforming growth factor beta 1 (TGF-β1), a liver fibrosis-inducing factor, was highly expressed in the alcohol-fed group and human hepatoma HepG2 cell. However, fucoidan reduced TGF-β1 expression and the production of inflammation-promoting COX-2 and nitric oxide, whereas markedly augmented the expression of the hepatoprotective enzyme, heme-oxygenase-1, on murine liver and HepG2 cells.

The mechanism of fucoidan from *F. vesiculosus* in inhibiting concanavalin A (ConA)-induced acute liver injury was studied by (Li et al., 2016). Fucoidan attenuated acute liver injury induced by ConA through the suppression of both intrinsic and extrinsic apoptosis. The mechanism may be associated with the anti-inflammatory action of fucoidan identified by the reduction in TNF-α and IFN-γ release which inhibited the TNF receptor-associated factor (TRAF2)/TNFR1-associated death domain (TRADD) and Janus Kinase 2 (JAK2)/signal transducer and activator of transcription 1 (STAT1) pathways, respectively. These findings may offer a potential powerful therapy for T cell-related hepatitis diseases.

Diabetic nephropathy (DN) is one amongst the foremost serious micro-vascular complications of diabetes. DN may lead to end-stage renal disease (ESRD) and chronic renal failure. In a recent study, it was found that LMWF may be developed as drug to prevent or inhibit fibrotic progression in DN (Chen et al., 2015). Treatment with LMWF attenuated renal dysfunction and fibrogenesis in kidneys of type 1 and type 2 diabetic rats. The renal beneficial effect of LMWF might be associated with its blockage of TIF via repressing TGF-β1 and the downstream pathways, which was proved by both *in vivo* and *in vitro* models.

Fucoidan protect against kidney damage through several ways. Various studies have shown that fucoidan had renoprotective effects in chronic renal failure model (Zhang et al., 2003; Wang et al., 2012), ameliorated metabolic dysfunctions in db/db mice (Kim and Lee, 2012) and inhibited α-amylase and α-glucosidase activities (Kim et al., 2014). In recent studies, fucoidan administration increased insulin secretion in overweight or obese adults (Hernández-Corona et al., 2014) and modulated metabolic abnor-malities and reduced blood glucose levels and delayed the progression of diabetic renal complications (Wang et al., 2014). A study using fucoidan isolated from *L. japonica* exhibited renoprotective activity in chronic kidney disease (CKD). It was found that fucoidan decreased the level of serum urea nitrogen (SUN) and serum creatinine (SCR). Histopathological changes of renal tubules and interstitium were markedly alleviated and the mesangial areas were also reduced. Alterations were observed in the activities/levels of serum enzymic (CAT, GSH-PX) and non-enzymic (GSH) antioxidants, along with high level of MDA (Wang et al., 2012). Albuminuria is an aggravating factor for progressive renal damage in

CKD. LMWF could protect renal function and tubular cells from albumin overload caused injury (Jia et al., 2016).

7.2.8 ANTI-VIRAL ACTIVITY

Fucoidan can protect the cells from viral infections besides antiproliferative, antiadhesive effects on cells. Antiviral activity of fucoidan against poliovirus III, adenovirus III, ECHO6 virus, coxsackie B3 virus, and coxsackie A16 have been studied. The ability of fucoidan to inhibit the replication of enveloped viruses including herpes simplex virus (HSV), human immunodeficiency virus (HIV), human cytomegalovirus, dengue virus is also well established (Witvrouw et al., 1994; Hayashi et al., 2008; Hidari et al., 2008; Borst et al., 2013). The chemical structure including the degree of sulfation, constituent sugars, molecular weight, conformation, and dynamic stereochemistry play a significant role in the antiviral activity of fucoidans (Ngo and Kim, 2013). Sulfate is necessary for the antiviral activity.

Ponce et al. (2003) reported the presence of two different types of fucoidans, galactofuran, and uronofucoidan, in the seaweeds *Adenocystis utricularis* and of the two fucoidans the galactofuran showed a high inhibitory activity against HSV with no cytotoxicity whereas uronofucoidans had no anti-viral activity. Since negatively charged sulfated groups can be involved in antiviral efficacy, the size and degree of sulfation present in fucoidan correlates relatively well with their ability to inhibit viral infection of cells (Ghosh et al., 2008). An advantage of sulfated polysaccharides from algae is the high content of polyanions in their extracellular matrix, low cost, and easy production, and they possess species-specific structural variations that appear to affect their antiviral activity (Harden et al., 2009). Studies have proven that fucoidan blocks the entry of viral pathogens-including HSV. By reinforcing cell defenses, they are less susceptible to infection. Neutralizing antibodies from the fucoidan significantly slow the virus' ability to replicate, while reinforcing immune defense functions.

Fucoidans from *A. utricularis* (Ponce et al., 2003), *Stoechospermum marginatum* (Adhikari et al., 2006), *Cystoseira indica* (Mandal et al., 2007), and *U. pinnatifida* (Ee et al., 2004; Hemmingson et al., 2006) was found to show antiviral activities against HSV-1 and HSV-2 without cytotoxicity for Vero cell cultures. Sulfate located at C-4 of (1→3)-linked fucopyranosyl units are significant for the anti-herpetic activity of fucoidan (Mandal et al., 2007). While there is no cure for HSV-1 (oral herpes) or HSV-2 (genital herpes), fucoidan acts as an antiviral, and is capable of suppressing outbreaks of sores when taken preventatively and can decrease the amount of time it takes for sores to heal when taken during an outbreak.

Fucoidan of *L. japonica* showed antiviral activity against avian influenza A (H5N1) virus and no cytotoxic activity (Makarenkova et al., 2012). The fucoidan protected the cell cultures from the cytopathogenic activity of influenza virus and suppressed influenza A/H5N1 virus production within 24 hours of infection when prophylactic and therapeutic-and-prophylactic treatment regimens were used. Fucoidan from *C. okamuranus* composed of glucuronic acid and sulfated fucose units potently inhibited infection of BHK-21 cells with dengue virus type 2 (DENV-2). However, the fucoidan showed little effect on the other three serotypes of the virus (Hidari et al., 2008). The study showed that DEN2 particles bound exclusively to fucoidan, indicating that fucoidan interacted directly with envelope glycoprotein (EGP) on DEN2. Structure-based analysis suggested that Arg323 of DEN2 EGP, which is conformationally proximal to one of the putative heparin-binding residues, Lys310, is critical for the interaction with fucoidan. In conclusion, both the sulfated group and glucuronic acid of fucoidan account for the inhibition of DEN2 infection.

Fucoidan is also a potential therapeutic agent in veterinary and animal health applications. Newcastle disease virus (NDV) causes a serious infectious disease in birds those results in severe losses in the poultry industry. Despite vaccination, NDV outbreaks have increased the necessity of alternative prevention and control methods. Fucoidan from *C. okamuranus* exhibited good antiviral activity against NDV and it did not exhibit significant toxicity at effective concentration (Elizondo-Gonzalez et al., 2012). In a recent study, Fucoidan from *C. okamuranus* is found to inhibit cellular entry of paramyxovirus type virus which causes Canine distemper in dogs and other carnivores (Trejo-Avila et al., 2014).

Treatment of AIDS caused by human immunodeficiency virus type 1 (HIV-1) infection represents a major challenge in antiviral therapeutics. Toxicity, resistance, and high costs are some of the difficulties that are associated with the treatment. In order to overcome these limitations research for novel compounds is the need of the hour. Fucoidan from seaweeds have proven to be a promising compound with antiretroviral activity. Antiviral agents that interfere with HIV at different stages of viral replication have been developed since the first report on the HIV inhibitory property of sulfated polysaccharides from seaweeds was published in 1987 (Nakashima et al., 1987). Fucoidan from seaweeds have great potential for the development of new generation of anti-HIV therapeutics, as reported by several studies.

The presence of sulfate groups seems to be necessary for the anti-HIV activity. However, in a recent study by Thuy et al. (2015) using fucoidan of three brown algae, viz., *S. mcclurei, S. polycystum* and *T. ornata* for antiretroviral activity, it was found that neither sulfate content nor position of sulfate groups was related to the anti-HIV activity suggesting the involvement of other structural parameters like molecular weight, the type of glycosidic linkage or even a unique fucoidan sequence. Hence, the antiviral activity of fucoidans is due to their binding with HIV-1 blocking the early steps of HIV entry. The fucoidan might have exerted their anti-HIV-1 activity by shielding off the positively charged amino acids present in the viral EGP gp120 (Moulard et al., 2000) or by the strong binding with a specific sulfation motif. Queiroz et al. (2008) found that the sulfated fucans from seaweed species, viz., *Dictyota mertensii, Lobophora variegata, F. vesiculosus*, and *Spatoglossum schroederi* could prevent HIV infection via blocking the activity of reverse transcriptase and their results strongly indicated the necessity of sulfate and carboxyl group in the inhibitory activity of these polysaccharides. Trinchero et al. (2009) have shown that galactofucan fractions from the brown algae *A. utricularis* have the capacity to inhibit HIV-1 replication *in vitro* with low cytotoxicity by blocking viral entry.

7.3 COSMECEUTICAL APPLICATIONS OF FUCOIDAN

Cosmeceuticals are gaining increased attention because of their beneficial effects on human health. Cosmetics prepared from natural sources instead of synthetic ingredients have become more popular among the consumers. Marine organisms including seaweeds are rich sources of structurally diverse biologically active compounds with great cosmeceutical potential. Bioactive substances derived from marine algae have diverse functional roles as a secondary metabolite that can be used in the development of cosmeceuticals. Till date, only a few marine organisms have been exploited for the screening of cosmeceutical compounds. Several studies of recent times have provided insight into biological activities of marine algae in promoting skin health and beauty products. Among the marine bioactive compounds, polysaccharides, especially fucoidan play an important role in the field of cosmetics. There has been a significant growth of *in vitro* research on potential of fucoidan as cosmetic or cosmeceutical ingredient for the past few years. The main focus of research has been the inhibitory effects of topically applied fucoidan on skin related issues like aging and photo-damaged skin. Many research studies have provided evidence of efficacy of fucoidan in topical skin cancer treatments, wound healing and preventative skin care health in the form of cosmeceuticals. Fucoidan may enhance dermal fibroblast proliferation and deposition of collagen; protect the elastic fiber network in human skin culture; and have potential for inclusion in lotion as a natural whitening agent. The prospects of using fucoidan as a protective agent in topical cosmetic formulations thus seem very promising. Since fucoidan are water soluble, they can easily be incorporated into lotions, creams, and other beauty products.

7.3.1   SKIN HEALTH PROTECTION: ANTI-AGING AND SKIN WHITENING

Marine algae have been used for skin-related diseases since ancient times. Several metabolites especially fucoidan isolated from marine algae are being used in cosmeceutical formulations for many decades due to their emollient, viscosity controlling, and skin conditioning properties, as well as their inherent stability, bioactive properties, physical, and natural marine source (Kim et al., 2005). Fucoidan has both soothing and restorative effects on skin. Epidemiological and experimental studies have suggested that fucoidan is useful in skin protecting, antioxidant, and antiaging activities, in addition to its anti-viral, anti-inflammatory, anticoagulant, and anti-tumor properties (Kim et al., 2005). Table 7.5 presents the uses of fucoidan in cosmetic products for skin related problems.

TABLE 7.5   Uses of Fucoidan in Cosmetic Products for Skin Related Problems

Seaweed Source	Possible Functions of Fucoidans Used in Cosmetic Products	References
L. japonica	Skin protection	Lee et al., 2009
F. vesiculosus	Soothing, smoothing, emollient, skin conditioning, masking	Senni et al., 2006a; Moon et al., 2009; Fujimura et al., 2016
U. pinnatifida	Skin protection, skin conditioning, smoothing, soothing, masking	Fitton et al., 2015
A. nodosum	Skin protection	Senni et al., 2006b
Costaria costata	Skin protection	Moon et al., 2009

7.3.1.1   FUCOIDAN AND ANTI-AGING

The stimulation of skin matrix enzymes such as collagenase and elastase, that breakdown collagen and elastin of the skin, is the significant aspect of skin aging. Degraded collagen fibrils prevent new tissue formation and causes skin degradation by inducing more enzyme activity. Reducing the enzyme activity may reduce skin degradation and formation of new matrix. Clinical trials of fucoidan isolated from *U. pinnatifida* has shown that it inhibits enzymes such as elastase and collagenase and is therefore highly effective in reducing inflammation, tissue damage and enhancing dermal condition (Fitton et al., 2015).

Fucoidan isolated from *F. vesiculosus* has been proven to have anti-aging activity that includes increase of the number of dermal fibroblasts and deposition of collagen, collagen tightness and facial elasticity (Fujimura et al., 2016). Purified fucoidan extracts of *F. vesiculosus* can be incorporated into creams and lotions, providing cosmetic anti-aging and anti-wrinkle benefits, such as inhibition of matrix enzymes against hyaluronidase, heparanase, phospholipase A2, tyrosine kinase and collagenase expression, and anti-inflammatory activity. Fucoidan helps in the promotion of collagen gel contraction which is caused by an increased expression of cell surface integrins that mediate interactions between fibroblast and extracellular matrix proteins in the dermis. These results therefore suggest that fucoidan may alter the thickness and the mechanical properties of skin, by enhancing integrin expression of skin fibroblasts.

Skin can age in two ways, viz. intrinsic or chronologic aging and extrinsic aging which are due to external factors (Figure 7.7). Photoaging caused by external factors like UV B irradiation increases skin fragility, laxity, blister formation, leathery appearance, and formation of wrinkles. UVB reduces type I procollagen levels and induces the production of MMPs which are responsible for the degradation or synthesis inhibition of collagenous extracellular matrix in connective tissues. Fucoidan is of interest due to their inhibitory effects on aging and photo-damaged skin when applied topically. The effects of fucoidan on MMP-1 expression by various *in vitro* experiments reported that fucoidan inhibited UVB-induced MMP-I expression at the protein and mRNA levels in human skin fibroblasts (Moon et al., 2008).

Fucoidans isolated from *Costaria costata* have been reported to possess the MMP inhibition activities. The fucoidan showed significant decrease in the UVB-induced MMP-1 expression in the human keratinocyte (HaCaT) cell lines. Moreover, fucoidan reduced the expression MMP-1 mRNA and inhibited UVB-induced MMP-1 promoter activity by 37.3%, 53.3%, and

58.5% at 0.01, 0.1, and 1 mg/mL, respectively, compared to UVB irradiation alone (Moon et al., 2009). Studies have demonstrated that fucoidan enhances dermal fibroblast proliferation and the deposition of collagen (Senni et al., 2006). *Ex vivo,* studies have reported that fucoidan was able to minimize human leukocyte elastase activity, to protect human skin elastic fibers against enzymatic proteolysis (Senni et al., 2006). These findings clearly suggest the potential role of seaweed fucoidans in reducing the risk of some inflammatory pathologies involving extracellular matrix degradation by MMPs.

Sirtuins are a family of seven proteins in humans (SIRT1–SIRT7) and among them SIRT1, SIRT3, and SIRT6 are induced by calorie restriction conditions and are considered anti-aging molecules. Sirtuins are involved in cellular pathways related to skin structure and function, including aging, ultraviolet (UV)-induced photoaging, inflammation, epigenetics, cancer, and a variety of cellular functions including cell cycle, DNA repair and proliferation (Serravallo et al., 2013). Exposure to UV irradiation and oxidative damage may depress SIRT1 levels leading to aging of the skin. Increasing the levels of SIRT1 may assist in maintaining skin functions by reversing the external effects on the skin. *In vitro* study on anti-aging applications of fucoidan extracted from *U. pinnatifida* reported that expression of the SIRT1 protein *in vitro* was increased by the fucoidan (Fitton et al., 2015).

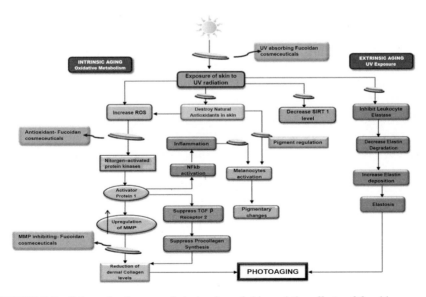

FIGURE 7.7   Schematic diagram of photoaging of skin and the effects of fucoidan as an anti-aging cosmeceutical.

7.3.1.2 SKIN WHITENING

Melanogenesis is a physiological process in which tyrosinase enzyme is involved in melanin synthesis. It leads to hyperpigmentation disorders and abnormal accumulation of melanin pigments, which can be improved by treatment with depigmenting agents. Tyrosinase inhibitors which are melanin-reducing compounds are most promising for preventing and treating pigmentation disorder and are used as skin-whitening agents in the cosmetic industry. Very few tyrosinase inhibitors are capable of inducing effects in the clinical trials. Hence, development of novel tyrosinase inhibitor from natural sources is gaining attention and in recent years, marine sources have been proved to be more promising as natural tyrosinase inhibitor. Melanocyte hormone (a-MSH) stimulates melanin production and tyrosinase activity. In B16 melanoma cells fucoidan isolated from *U. pinnatifida* was shown to decrease melanin synthesis and tyrosinase activity (Jung et al., 2009).

7.3.2 ATOPIC DERMATITIS (AD)

Atopic dermatitis (AD) is a chronic inflammatory skin disease that affects 5–20% of children worldwide (Eric Simpson et al., 2006). While the exact causes of AD are not understood, deregulation of immune systems and environmental allergenic responses are suggested (Cantani, 2001). Many therapeutic approaches including corticoid steroids, UV exposure, and immuno-suppressant are currently applied to improve this skin disease. However, these approaches have a limited duration of effects and exert side-effects after long-term use. Alternative approaches with safer and more effective therapeutic potency are now in great need.

Recent studies reveal that the coupling of the IgE onto the Fcε receptors without the help of antigens could elicit inflammatory responses of AD. Hence, inhibition of IgE production or reduction in the concentration of IgE would be the best therapeutic approach in treating AD. Recently, a recombinant humanized anti-IgE monoclonal antibody (mAb) has shown promising effects on AD. However, mAB is very expensive and enormous amount of mAB is required to remove IgE from AD patients. Fucoidan has shown promising effect in many cases of AD treatment. Fucoidan modulates the functions of immune cells and also suppresses IgE production in peripheral blood mononuclear cells from AD patients suggesting its potential use in allergic reactions (Iwamoto et al., 2011).

AD has been associated with the development of Th2-mediated inflammatory disease and is related to high levels of chemokine. It has been reported that stimulation with TNF-α and interferon (IFN-γ) synergistically increases the production of pro-inflammatory cytokines and chemokine in human keratinocytes. Fucoidan showed anti-inflammatory effects by suppressing the expression of TNF-α/IFN-γ-induced chemokines by blocking NF-κB, STAT1, and ERK1/2 activation, suggestive of as used as a therapeutic application in inflammatory skin diseases, such as AD (Ryu and Chung, 2015).

7.4   NUTRACEUTICAL APPLICATIONS OF FUCOIDAN

Nutraceutical refers to food or food ingredients with medical or health benefits. Through food-based approaches, active compounds with medicinal values are used to prevent or treat certain diseases linked to food. In the nutraceutical industry, marine algae are used as sources of food and food ingredients. In recent years, there has been tremendous development in the fields of pharmaceuticals, nutraceuticals, and cosmeceuticals. Over the years, nutraceuticals have attracted considerable interest due to their potential nutritional, safety, and therapeutic effects. Both producers and public are more interested to explore those areas that may provide health benefits beyond basic nutrition. Increasing awareness among consumers about health-promoting foods has aroused interest in food supplement research and has encouraged researchers to search for new functional ingredients that can contribute to develop new opportunities in the relevant applications. Although terrestrial plants have been studied widely for functional ingredients seaweeds are being currently explored as a nutraceutical.

Seaweeds are rich in polysaccharides, minerals, and certain vitamins and also contain bioactive substances like polysaccharides, proteins, lipids, and polyphenols, with antibacterial, antiviral, and antifungal properties. Even though seaweeds low in lipids carbohydrate content is high, most of it is dietary fiber which is not taken up by human body. However, dietary fibers are good for human health (Holdt and Kraan, 2011). Several active compounds produced by different marine organisms have a wide role in the nutraceutical applications. In recent times, isolation, and characterization of the biologically important compounds that include polysaccharides like fucoidan have gained lot of attention from various research groups across the world. Seaweeds are considered as dietary components and also as alternative medicine in Asian countries like Japan, Korea, and China. Seaweeds containing fucoidan have been used as food such as sea vegetables for centuries in many Asian countries. Some important species of brown algae viz., *A. nodosum, E. cava, E. kurome, L. digitata, Lessonia flavicans, Saccharina japonica* (kombu), *S. horneri, U. pinnatifida* (wakame), *S. fusiforme* (hijiki), *C. okamuranus* (mozuku) are of nutritional interest.

Fucoidans are being used in formulations ranging from beverages to tablets. Like alginates and carageenans, fucoidans are relatively stable. Fucoidans are also being integrated into immune-boosting and anti-inflammatory formulations as well as in nutritional beverages and functional foods. Fucoidan as nutraceuticals might have a positive effect on human health as they can protect human body against damage by ROS which attack macromolecules such as membrane lipids, proteins, and DNA and lead to many health disorders like cancer, diabetes, and neurodegenerative diseases. For centuries the Chinese have, known referred to Fucoidan as "Virgin Mother's Milk" because it contains all the same nutrients and healing properties as breast milk. Fundamental nutrients including vitamins, minerals, antioxidants, amino acids, glyconutrients, polyphenols, and more than seventy necessary nutrients are found in fucoidan.

The oxidants such as superoxide anion, hydrogen peroxide, hydroxyl radicals, and singlet oxygen are well-known to cause various chronic diseases (Waris and Ahsan, 2006). Synthetic commercial antioxidants cause side effects on long term use and hence, natural sources are being considered as safe alternatives which have evoked considerable interest in the food industry as well as pharmaceutical industry. Among the natural antioxidants, fucoidans from seaweeds have produced promising effects through their scavenging effect on harmful oxidants. Fucoidan from the edible seaweed *F. vesiculosus* was shown to prevent the formation of superoxide radicals, hydroxyl radicals, and lipid peroxidation (Rocha de Souza et al., 2007). Potential anti-inflammatory nutraceuticals like fucoidan play a significant role in the prevention of chronic inflammation either by inhibiting the production of nitric oxide (NO), COX-2 or proinflammatory cytokines (Rajapakse and Kim, 2011). Nutricosmetics are specific macro and micronutrients that are consumed to provide positive changes in the quality of skin and general appearance. Seaweeds are one such potential nutricosmetic known for inhibitory effects on skin damage. Fucoidan is being used in wide varieties of topical cosmetic preparations and its role in anti-aging products as a detoxifying ingredient and micronutrient provider is gaining popular.

7.5 CONCLUSION

Fucoidans of marine sources are complex and heterogeneous and their structure is diverse and varies with algal species, location, and extraction technique. Every extraction technique yields different composition within the same species, and hence specific structure of fucoidan remains to elude the researchers till date. Future conformational studies of well-defined structures should lead to better understanding of the biological properties of fucoidans.

Many reports have shown that the biological activities of fucoidan are by various molecular mechanisms and can be affected by many factors. But, the reports have proven that fucoidan has antitumor, anticoagulant, anti-inflammatory, etc., effects on different cells types and has little toxicity even if it is from distinct brown seaweeds species. However, heterogeneous fucoidan is difficult to study structure-activity relationships because of the presence of bioactive impurities (Wu et al., 2016). Hence, considering the medical applications of fucoidan in the future, the fucoidan with lower potential contaminants, homogeneous structure, and higher biological activity can be used as suitable materials. Therefore, it is necessary for researchers to investigate the structure, composition, and bioactivity of purified fucoidan.

Research on bioactivity of fucoidan is still gaining popular in pharma-ceutical, cosmeceutical, and nutraceutical applications. Their vegetal origin, absence of adverse effects and an affordable price due to easy handling in production processes makes them promising for human health. Fucoidan activity in promoting vaccine responses is an example of currently available complementary therapeutic use. There is growing support for the role of fucoidan as an adjunct dietary therapy in cancer and inflammatory diseases.

Fucoidan is also being used in nanomedicine and nanosystems for diag-nostic, drug delivery, and tissue engineering. However, there are few bottle-necks for developments in nanomedicine due to the difficulty in obtaining reproducible chemical structures and molecular weights. It is apparent that each type of fucoidan needs to be screened and validated for a particular thera-peutic activity. Although different species and different preparation processes can produce different research results, at least, we know that fucoidan has active ingredients and hence, can analyze the production structure by certain species and preparation methods. The activity of fucoidan has been proved by many investigators, but whether we can find one or several exactly active structures and whether they can go into the market as chemotherapeutics still needs more researchers, especially multidisciplinary partnerships, to realize the relay race. Assessment of pharmacokinetics, uptake, and distribution are aspects that require more insight.

KEYWORDS

- antiaging activities
- antiangiogenesis
- anti-atherosclerotic activity
- anticoagulant activities
- anti-fibrogenesis
- anti-herpetic activity
- anti-inflammatory properties
- antithrombotic drugs
- anti-VEGF
- atopic dermatitis
- galactofuran
- hepatoprotective property
- immunomodulatory effect
- melanogenesis
- MMP inhibition activities
- nutraceuticals
- nutricosmetics
- pro-angiogenic effect
- renoprotective activity
- sulfated fucosyl units

REFERENCES

Adhikari, U., Mateu, C. G., Chattopadhyay, K., Pujol, C. A., Damonte, E. B., & Ray, B., (2006). Structure and antiviral activity of sulfated fucans from *Stoechospermum marginatum*. *Phytochemistry*, *67*(22), 2474–2482.

Aisa, Y., Miyakawa, Y., Nakazato, T., Shibata, H., Saito, K., Ikeda, Y., & Kizaki, M., (2005). Fucoidan induces apoptosis of human HS-sultan cells accompanied by activation of caspase-3 and down-regulation of ERK pathways. *Am. J. Hematol.*, *78*(1), 7–14.

Ale, M. T., & Meyer, A. S., (2013). Fucoidans from brown seaweeds: An update on structures, extraction techniques, and use of enzymes as tools for structural elucidation. *RSC Adv.*, *3*(22), 8131–8141.

Atashrazm, F., Lowenthal, R. M., Woods, G. M., Holloway, A. F., & Dickinson, J. L., (2015). Fucoidan and cancer: A multifunctional molecule with anti-tumor potential. *Marine Drugs*, *13*(4), 2327–2346.

Azuma, K., Ishihara, T., Nakamoto, H., Amaha, T., Osaki, T., Tsuka, T., et al., (2012). Effects of oral administration of fucoidan extracted from *Cladosiphon okamuranus* on tumor growth and survival time in a tumor-bearing mouse model. *Marine Drugs*, *10*(10), 2337–2348.

Berteau, O., & Mulloy, B., (2003). Sulfated fucans, fresh perspectives: Structures, functions, and biological properties of sulfated fucans and an overview of enzymes active toward this class of polysaccharide. *Glycobiology*, *13*(6), 29R–40R.

Boo, H. J., Hong, J. Y., Kim, S. C., Kang, J. I., Kim, M. K., Kim, E. J., et al., (2013). The anticancer effect of fucoidan in PC-3 prostate cancer cells. *Marine Drugs*, *11*(8), 2982–2999.

Borst, E. M., Ständker, L., Wagner, K., Schulz, T. F., Forssmann, W. G., & Messerle, M., (2013). A peptide inhibitor of cytomegalovirus infection from human hemofiltrate. *Antimicrobial Agents and Chemotherapy*, *57*(10), 4751–4760.

Burz, C., Berindan-Neagoe, I., Balacescu, O., & Irimie, A., (2009). Apoptosis in cancer: Key molecular signaling pathways and therapy targets. *Acta Oncologica*, *48*(6), 811–821.

Byon, Y. Y., Kim, M. H., Yoo, E. S., Hwang, K. K., Jee, Y., Shin, T., et al., (2008). Radioprotective effects of fucoidan on bone marrow cells: Improvement of the cell survival and immunoreactivity. *Journal of Veterinary Science*, *9*(4), 359.

Canavese, M., Santo, L., & Raje, N., (2012). Cyclin dependent kinases in cancer. *Cancer Biology and Therapy*, *13*(7), 451–457.

Cantani, A., (2001). Pathogenesis of atopic dermatitis (AD) and the role of allergic factors. *European Review for Medical and Pharmacological Sciences*, *5*(3), 95–117.

Carmeliet, P., & Jain, R. K., (2000). Angiogenesis in cancer and other diseases. *Nature*, *407*(6801), 249–257.

Chandía, N. P., & Matsuhiro, B., (2008). Characterization of a fucoidan from *Lessonia vadosa* (Phaeophyta) and its anticoagulant and elicitor properties. *International Journal of Biological Macromolecules*, *42*(3), 235–240.

Chen, H., Cong, Q., Du, Z., Liao, W., Zhang, L., Yao, Y., & Ding, K., (2016). Sulfated fucoidan FP08S2 inhibits lung cancer cell growth *in vivo* by disrupting angiogenesis via targeting VEGFR2/VEGF and blocking VEGFR2/Erk/VEGF signaling. *Cancer Letters*, *382*(1), 44–52.

Chen, J., Cui, W., Zhang, Q., Jia, Y., Sun, Y., Weng, L., et al., (2015). Low molecular weight fucoidan ameliorates diabetic nephropathy via inhibiting epithelial-mesenchymal transition and fibrotic processes. *Am. J. Transl. Res.*, *7*(9), 1553–1563.

Chen, S., Xue, C., Yin, L., Tang, Q., Yu, G., & Chai, W., (2011). Comparison of structures and anticoagulant activities of fucosylated chondroitin sulfates from different sea cucumbers. *Carbohydrate Polymers*, *83*(2), 688–696.

Chen, S., Zhao, Y., Zhang, Y., Zhang, D., Bilan, M., Grachev, A., et al., (2014). Fucoidan induces cancer cell apoptosis by modulating the endoplasmic reticulum stress cascades. *PLoS One*, *9*(9), e108157.

Chevolot, L., Mulloy, B., Ratiskol, J., Foucault, A., & Colliec-Jouault, S., (2001). A disaccharide repeat unit is the major structure in fucoidans from two species of brown algae. *Carbohydrate Research*, *330*, 529–535.

Choo, G. S., Lee, H. N., Shin, S. A., Kim, H. J., & Jung, J. Y., (2016). Anticancer effect of Fucoidan on DU-145 prostate cancer cells through inhibition of PI3K/Akt and MAPK pathway expression. *Marine Drugs*, *14*(7), 126.

Cong, Q., Chen, H., Liao, W., Xiao, F., Wang, P., Qin, Y., et al., (2016). Structural characterization and effect on anti-angiogenic activity of a fucoidan from *Sargassum fusiforme*. *Carbohydrate Polymers*, *136*, 899–907.

Cui, Y. Q., Zhang, L. J., Zhang, T., Luo, D. Z., Jia, Y. J., Guo, Z. X., et al., (2010). Inhibitory effect of fucoidan on nitric oxide production in lipopolysaccharide-activated primary microglia. *Clinical and Experimental Pharmacology and Physiology*, *37*(4), 422–428.

Cumashi, A., Ushakova, N. A., Preobrazhenskaya, M. E., D'Incecco, A., Piccoli, A., Totani, L., et al., (2007). A comparative study of the anti-inflammatory, anticoagulant, antiangiogenic, and antiadhesive activities of nine different fucoidans from brown seaweeds. *Glycobiology*, *17*(5), 541–552.

De, L., Silva, T. M. A., Alves, L. G., De Queiroz, K. C. S., Santos, M. G. L., Marques, C. T., et al., (2005). Partial characterization and anticoagulant activity of a heterofucan from the brown seaweed *Padina gymnospora*. *Brazilian Journal of Medical and Biological Research*, *38*(38), 523–533.

Dithmer, M., Fuchs, S., Shi, Y., Schmidt, H., Richert, E., Roider, J., et al., (2014). Fucoidan reduces secretion and expression of vascular endothelial growth factor in the retinal pigment epithelium and reduces angiogenesis *in vitro. PLoS One*, *9*(2), e89150.

Do, H., Pyo, S., & Sohn, E. H., (2010). Suppression of iNOS expression by fucoidan is mediated by regulation of p38 MAPK, JAK/STAT, AP-1 and IRF-1, and depends on up-regulation of scavenger receptor B1 expression in TNF-α-and IFN-γ-stimulated C6 glioma cells. *The Journal of Nutritional Biochemistry*, *21*, 671–679.

Duarte, M. E., Cardoso, M. A., Noseda, M. D., & Cerezo, A. S., (2001). Structural studies on fucoidans from the brown seaweed *Sargassum stenophyllum*. *Carbohydrate Research*, *333*(4), 281–93.

Elizondo-Gonzalez, R., Cruz-Suarez, L. E., Ricque-Marie, D., Mendoza-Gamboa, E., Rodriguez-Padilla, C., & Trejo-Avila, L. M., (2012). *In vitro* characterization of the antiviral activity of fucoidan from *Cladosiphon okamuranus* against Newcastle disease virus. *Virology Journal*, *9*(1), 1.

Ferrara, N., (2000). Vascular endothelial growth factor and the regulation of angiogenesis. *Recent Progress in Hormone Research*, *55*, 15–36.

Fitton, J. H., Stringer, D. N., & Karpiniec, S. S., (2015). Therapies from Fucoidan: An update. *Marine Drugs*, *13*(9), 5920–5946.

Fitton, J., Dell'Acqua, G., Gardiner, V. A., Karpiniec, S., Stringer, D., & Davis, E., (2015). Topical benefits of two fucoidan-rich extracts from marine macroalgae. *Cosmetics*, *2*(2), 66–81.

Fujimura, T., Tsukahara, K., Moriwaki, S., Kitahara, T., & Takema, Y., (2000). Effects of natural product extracts on contraction and mechanical properties of fibroblast populated collagen gel. *Biological and Pharmaceutical Bulletin*, *23*(3), 291–297.

Ghosh, T., Chattopadhyay, K., Marschall, M., Karmakar, P., Mandal, P., & Ray, B., (2008). Focus on antivirally active sulfated polysaccharides: From structure-activity analysis to clinical evaluation. *Glycobiology*, *19*(1), 2–15.

Harden, E. A., Falshaw, R., Carnachan, S. M., Kern, E. R., & Prichard, M. N., (2009). Virucidal activity of polysaccharide extracts from four algal species against herpes simplex virus. *Antiviral Research*, *83*(3), 282–289.

Haroun-Bouhedja, F., Ellouali, M., Sinquin, C., Boisson-Vidal, C., Colliec, S., Boisson-Vidal, C., et al., (2000). Relationship between sulfate groups and biological activities of fucans. *Thrombosis Research, 100*(5), 453–459.

Hayashi, S., Itoh, A., Isoda, K., Kondoh, M., Kawase, M., & Yagi, K., (2008). Fucoidan partly prevents CCl₄-induced liver fibrosis. *European Journal of Pharmacology, 580*(3), 380–384.

Hemmingson, J. A., Falshaw, R., Furneaux, R. H., & Thompson, K., (2006a). Structure and antiviral activity of the galactofucan sulfates extracted from *Undaria pinnatifida* (Phaeophyta). *J. Appl. Phycol., 18*, 185–193.

Hernández-Corona, D. M., Martínez-Abundis, E., & González-Ortiz, M., (2014). Effect of fucoidan administration on insulin secretion and insulin resistance in overweight or obese adults. *Journal of Medicinal Food, 17*(7), 830–832.

Hidari, K. I. P. J., Takahashi, N., Arihara, M., Nagaoka, M., Morita, K., & Suzuki, T., (2008). Structure and anti-dengue virus activity of sulfated polysaccharide from a marine alga. *Biochemical and Biophysical Research Communications, 376*(1), 91–95.

Holdt, S. L., & Kraan, S., (2011). Bioactive compounds in seaweed: Functional food applications and legislation. *Journal of Applied Phycology, 23*(3), 543–597.

Honya, M., Mori, H., Anzai, M., Araki, Y., & Nisizawa, K., (1999). Monthly changes in the content of fucans, their constituent sugars, and sulfate in cultured *Laminaria japonica. Hydrobiologia, 398/399*, 411–416.

Hsu, H. Y., Lin, T. Y., Hwang, P. A., Tseng, L. M., Chen, R. H., Tsao, S. M., & Hsu, J., (2013). Fucoidan induces changes in the epithelial to mesenchymal transition and decreases metastasis by enhancing ubiquitin-dependent TGFβ receptor degradation in breast cancer. *Carcinogenesis, 34*(4), 874–884.

Hsu, H. Y., Lin, T. Y., Wu, Y. C., Tsao, S. M., Hwang, P. A., Shih, Y. W., et al., (2014). Fucoidan inhibition of lung cancer *in vivo* and *in vitro*: Role of the Smurf2-dependent ubiquitin proteasome pathway in TGFβ receptor degradation. *Oncotarget, 5*(17), 7870–7885.

Iwamoto, K., Hiragun, T., Takahagi, S., Yanase, Y., Morioke, S., Mihara, S., et al., (2011). Fucoidan suppresses IgE production in peripheral blood mononuclear cells from patients with atopic dermatitis. *Archives of Dermatological Research, 303*(6), 425–431.

Jia, Y., Sun, Y., Weng, L., Li, Y., Zhang, Q., Zhou, H., & Yang, B., (2016). Low molecular weight fucoidan protects renal tubular cells from injury induced by albumin overload. *Scientific Reports, 6*, 31759.

Jung, S. H., Ku, M. J., Moon, H. J., Yu, B. C., Jeon, M. J., & Lee, Y. H., (2009). Inhibitory effects of fucoidan on melanin synthesis and tyrosinase activity. *Journal of Life Science, 19*(1), 75–80.

Kang, S. M., Kim, K. N., Lee, S. H., Ahn, G., Cha, S. H., Kim, A. D., et al., (2011). Anti-inflammatory activity of polysaccharide purified from AMG-assistant extract of *Ecklonia cava* in LPS-stimulated RAW 264.7 macrophages. *Carbohydrate Polymers, 85*(1), 80–85.

Kasai, A., Arafuka, S., Koshiba, N., Takahashi, D., & Toshima, K., (2015). Systematic synthesis of low-molecular weight fucoidan derivatives and their effect on cancer cells. *Organic and Biomolecular Chemistry, 13*(42), 10556–10568.

Kim, E. J., Park, S. Y., Lee, J. Y., & Park, J. H. Y., (2010). Fucoidan present in brown algae induces apoptosis of human colon cancer cells. *BMC Gastroenterology, 10*, 96.

Kim, H. H., Shin, C. M., Park, C. H., Kim, K. H., Cho, K. H., Eun, H. C., & Chung, J. H., (2005a). Eicosapentaenoic acid inhibits UV-induced MMP-1 expression in human dermal fibroblasts. *Journal of Lipid Research, 46*(8), 1712–1720.

Kim, K. J., & Lee, B. Y., (2012). Fucoidan from the sporophyll of *Undaria pinnatifida* suppresses adipocyte differentiation by inhibition of inflammation-related cytokines in 3T3-L1 cells. *Nutrition Research, 32*(6), 439–447.

Kim, K. T., Rioux, L. E., & Turgeon, S. L., (2014). Alpha-amylase and alpha-glucosidase inhibition is differentially modulated by fucoidan obtained from *Fucus vesiculosus* and *Ascophyllum nodosum*. *Phytochemistry, 98*, 27–33.

Kim, M. H., & Joo, H. G., (2008). Immunostimulatory effects of fucoidan on bone marrow-derived dendritic cells. *Immunology Letters, 115*(2), 138–143.

Klettner, A., Westhues, D., Lassen, J., Bartsch, S., & Roider, J., (2013). Regulation of constitutive vascular endothelial growth factor secretion in retinal pigment epithelium/choroid organ cultures: P38, nuclear factor κB, and the vascular endothelial growth factor receptor-2/phosphatidylinositol 3 kinase pathway. *Molecular Vision, 19*, 281–291.

Koyanagi, S., Tanigawa, N., Nakagawa, H., Soeda, S., & Shimeno, H., (2003). Oversulfation of fucoidan enhances its anti-angiogenic and antitumor activities. *Biochemical Pharmacology, 65*(2), 173–179.

Kuznetsova, T. A., Besednova, N. N., Mamaev, A. N., Momot, A. P., Shevchenko, N. M., & Zvyagintseva, T. N., (2003). Anticoagulant activity of Fucoidan from brown algae *Fucus evanescens* of the Okhotsk Sea. *Byulleten' Eksperimental'noi Biologii I Meditsiny, 136*(11), 532–534.

Lee, J. B., Hayashi, K., Hashimoto, M., Nakano, T., & Hayashi, T., (2004). Novel antiviral fucoidan from sporophyll of *Undaria pinnatifida* (Mekabu). *Chem. Pharm. Bull. (Tokyo), 52*(9), 1091–1094.

Lee, M. H., & Yang, H. Y., (2001). Contributions in the domain of cancer research: Negative regulators of cyclin-dependent kinases and their roles in cancers. *Cellular and Molecular Life Sciences, 58*(12), 1907–1922.

Lee, S. H., Ko, C. I., Ahn, G., You, S., Kim, J. S., Heu, M. S., et al., (2012). Molecular characteristics and anti-inflammatory activity of the fucoidan extracted from *Ecklonia cava*. *Carbohydrate Polymers, 89*(2), 599–606.

Li, B., Lu, F., Wei, X., & Zhao, R., (2008). Fucoidan: Structure and bioactivity. *Molecules, 13*(8), 1671–1695.

Li, C., Gao, Y., Xing, Y., Zhu, H., Shen, J., & Tian, J., (2011). Fucoidan, a sulfated polysaccharide from brown algae, against myocardial ischemia–reperfusion injury in rats via regulating the inflammation response. *Food and Chemical Toxicology, 49*(9), 2090–2095.

Li, J., Chen, K., Li, S., Liu, T., Wang, F., Xia, Y., et al., (2016). Pretreatment with fucoidan from *Fucus vesiculosus* protected against ConA-induced acute liver injury by inhibiting both intrinsic and extrinsic apoptosis. *PLoS One, 11*(4), e0152570.

Li, X., Zhao, H., Wang, Q., Liang, H., & Jiang, X., (2015). Fucoidan protects ARPE-19 cells from oxidative stress via normalization of reactive oxygen species generation through the Ca^{2+}-dependent ERK signaling pathway. *Molecular Medicine Reports*, https://doi.org/10.3892/mmr.2015.3224.

Lieber, C. S., (1999). Prevention and treatment of liver fibrosis based on pathogenesis. *Alcoholism: Clinical and Experimental Research, 23*(5), 944–949.

Lim, J., Lee, S., Kim, T., Jang, S. A., Kang, S., Koo, H., et al., (2015). Fucoidan from *Fucus vesiculosus* protects against alcohol-induced liver damage by modulating inflammatory mediators in mice and HepG2 cells. *Marine Drugs, 13*(2), 1051–1067.

Liu, F., Luo, G., Xiao, Q., Chen, L., Luo, X., Lv, J., & Chen, L., (2016). Fucoidan inhibits angiogenesis induced by multiple myeloma cells. *Oncology Reports, 36*(4), 1963–1972.

Løvstad, H. S., & Kraan, S., (2011). Bioactive compounds in seaweed: Functional food applications and legislation. *Journal of Applied Phycology*, *23*(3), 543–597.

Makarenkova, I. D., Logunov, D. Y., Tukhvatulin, A. I., Semenova, I. B., Besednova, N. N., & Zvyagintseva, T. N., (2012). Interactions between sulfated polysaccharides from sea brown algae and Toll-like receptors on HEK293 eukaryotic cells *in vitro*. *Bulletin of Experimental Biology and Medicine*, *154*(2), 241–244.

Mandal, P., Mateu, C. G., Chattopadhyay, K., Pujol, C. A., Damonte, E. B., & Ray, B., (2007). Structural features and antiviral activity of sulfated fucans from the brown seaweed *Cystoseira indica*. *Antivir. Chem. Chemother.*, *18*(3), 153–162.

Manzo-Silberman, S., Louedec, L., Meilhac, O., Letourneur, D., Michel, J. B., & Elmadbouh, I., (2011). Therapeutic potential of fucoidan in myocardial ischemia. *Journal of Cardiovascular Pharmacology*, *58*(6), 626–632.

Marinval, N., Saboural, P., Haddad, O., Maire, M., Letourneur, D., Charnaux, N., & Hlawaty, H., (2015). Angiogenesis potentialized by highly sulfated fucoidan: Role of the chemokines and the proteoglycans. *Archives of Cardiovascular Diseases Supplements*, *7*(2).

Matou, S., Helley, D., Chabut, D., Bros, A., & Fischer, A. M., (2002). Effect of fucoidan on fibroblast growth factor-2-induced angiogenesis *in vitro*. *Thrombosis Research*, *106*(4), 213–221.

Menshova, R. V., Shevchenko, N. M., Imbs, T. I., Zvyagintseva, T. N., Malyarenko, O. S., Zaporoshets, T. S., Besednova, N. N, & Ermakova, S. P., (2016) Fucoidans from brown alga *Fucus evanescens:* Structure and biological activity. *Front. Mar. Sci. 3*:129. doi: 10.3389/fmars.2016.00129.

Moon, H. J., Lee, S. R., Shim, S. N., Jeong, S. H., Stonik, V. A., Rasskazov, V. A., et al., (2008). Fucoidan inhibits UVB-induced MMP-1 expression in human skin fibroblasts. *Biological and Pharmaceutical Bulletin*, *31*(2), 284–289.

Moon, H. J., Park, K. S., Ku, M. J., Lee, M. S., Jeong, S. H., Imbs, T. I., et al., (2009). Effect of *Costaria costata* fucoidan on expression of matrix metalloproteinase-1 promoter, mRNA, and protein. *Journal of Natural Products*, *72*(10), 1731–1734.

Moulard, M., Lortat-Jacob, H., Mondor, I., Roca, G., Wyatt, R., Sodroski, J., et al., (2000). Selective interactions of polyanions with basic surfaces on human immunodeficiency virus type 1 gp120. *J. Virol.*, *74*, 1948–1960.

Mourao, P., (2004). Use of sulfated fucans as anticoagulant and antithrombotic agents: Future perspectives. *Current Pharmaceutical Design*, *10*(9), 967–981.

Nader, H. B., Pinhal, M. A. S., Baú, E. C., Castro, R. A. B., Medeiros, G. F., Chavante, S. F., et al., (2001). Development of new heparin-like compounds and other antithrombotic drugs and their interaction with vascular endothelial cells. *Brazilian Journal of Medical and Biological Research*, *34*(6), 699–709.

Nagamine, T., Hayakawa, K., Kusakabe, T., Takada, H., Nakazato, K., Hisanaga, E., & Iha, M., (2009). Inhibitory effect of fucoidan on Huh7 hepatoma cells through downregulation of CXCL12. *Nutrition and Cancer*, *61*(3), 340–347.

Nakashima, H., Kido, Y., Kobayashi, N., & Motoki, Y., (1987). Purification and characterization of an avian myeloblastosis and human immunodeficiency virus reverse transcriptase inhibitor, sulfated polysaccharides extracted from sea algae. *Antimicrob Agents Chemother*, *31*(10), 1524–1528.

Nakazato, K., Takada, H., Iha, M., & Nagamine, T., (2010). Attenuation of N-nitrosodiethylamine-induced liver fibrosis by high-molecular-weight fucoidan derived from *Cladosiphon okamuranus*. *Journal of Gastroenterology and Hepatology*, *25*(10), 1692–1701.

Namkoong, S., Kim, Y. J., Kim, T., & Sohn, E. H., (2012). Immunomodulatory effects of fucoidan on NK cells in ovariectomized rats. *Korean J. Plant Res. 25*(3), 317–322.

Narazaki, M., Segarra, M., & Tosato, G., (2008). Sulfated polysaccharides identified as inducers of neuropilin-1 internalization and functional inhibition of VEGF165 and semaphorin3A. *Blood, 111*(8), 4126–4136.

Nardella, A., Chaubet, F., Boisson-Vidal, C., Blondin, C., Durand, P., & Jozefonvicz, J., (1996). Anticoagulant low molecular weight fucans produced by radical process and ion exchange chromatography of high molecular weight fucans extracted from the brown seaweed *Ascophyllum nodosum. Carbohydrate Research, 289*, 201–208.

Ngo, D. H., & Kim, S. K., (2013). Sulfated polysaccharides as bioactive agents from marine algae. *International Journal of Biological Macromolecules, 62*, 70–75.

Oomizu, S., Yanase, Y., Suzuki, H., Kameyoshi, Y., & Hide, M., (2006). Fucoidan prevents Ce germline transcription and NFjB p52 translocation for IgE production in B cells. *Biochemical and Biophysical Research Communications, 350*(3), 501–507.

Park, H. S., Hwang, H. J., Kim, G. Y., Cha, H. J., Kim, W. J., Kim, N. D., et al., (2013). Induction of apoptosis by fucoidan in human leukemia U937 cells through activation of p38 MAPK and modulation of Bcl-2 family. *Marine Drugs, 11*(7), 2347–2364.

Park, H. Y., Choi, I. W., Kim, G. Y., Kim, B. W., Kim, W. J., & Choi, Y. H., (2015). Fucoidan induces G1 arrest of the cell cycle in EJ human bladder cancer cells through down-regulation of pRB phosphorylation. *Revista Brasileira de Farmacognosia, 25*(3), 246–251.

Park, H. Y., Han, M. H., Park, C., Jin, C. Y., Kim, G. Y., Choi, I. W., et al., (2011). Anti-inflammatory effects of fucoidan through inhibition of NF-κB, MAPK and Akt activation in lipopolysaccharide-induced BV2 microglia cells. *Food and Chemical Toxicology, 49*(8), 1745–1752.

Paternot, S., Bockstaele, L., Bisteau, X., Kooken, H., Coulonval, K., & Roger, P., (2010). Rb inactivation in cell cycle and cancer: The puzzle of highly regulated activating phosphorylation of CDK4 versus constitutively active CDK-activating kinase. *Cell Cycle, 9*(4), 689–699.

Pereira, M. S., Mulloy, B., & Mourão, P. A., (1999). Structure and anticoagulant activity of sulfated fucans. Comparison between the regular, repetitive, and linear fucans from echinoderms with the more heterogeneous and branched polymers from brown algae. *The Journal of Biological Chemistry, 274*(12), 7656–7667.

Ponce, N. M. A., Pujol, C. A., Damonte, E. B., Flores, M. L., & Stortz, C. A., (2003). Fucoidans from the brown seaweed *Adenocystis utricularis:* Extraction methods, antiviral activity, and structural studies. *Carbohydrate Research, 338*(2), 153–165.

Potente, M., Gerhardt, H., Carmeliet, P., Adams, R. H., Alitalo, K., Adams, R. H., et al., (2011). Basic and therapeutic aspects of angiogenesis. *Cell, 146*(6), 873–887.

Qiu, X., Amarasekara, A., & Doctor, V., (2006). Effect of over sulfation on the chemical and biological properties of fucoidan. *Carbohydrate Polymers, 63*(2), 224–228.

Queiroz, K. C. S., Medeiros, V. P., Queiroz, L. S., Abreu, L. R. D., Rocha, H. A. O., Ferreira, C. V., et al., (2008). Inhibition of reverse transcriptase activity of HIV by polysaccharides of brown algae. *Biomedicine and Pharmacotherapy, 62*(5), 303–307.

Rajapakse, N., & Kim, S. K., (2011). Nutritional and digestive health benefits of seaweed. *Adv. Food Nutr. Res., 64*, 17–28.

Riou, D., Colliec-Jouault, S., Pinczon, D. S. D., Bosch, S., Siavoshian, S., Le Bert, V., et al., (1995). Antitumor and antiproliferative effects of a fucan extracted from *Ascophyllum nodosum* against a non-small-cell bronchopulmonary carcinoma line. *Anticancer Research*, *16*(3A), 1213–1218.

Rioux, L. E., Turgeon, S. L., & Beaulieu, M., (2007). Characterization of polysaccharides extracted from brown seaweeds. *Carbohydrate Polymers*, *69*(3), 530–537.

Ritter, L. S., Copeland, J. G., & McDonagh, P. F., (1998). Fucoidin reduces coronary microvascular leukocyte accumulation early in reperfusion. *The Annals of Thoracic Surgery*, *66*(6), 2063–2071.

Rocha, D. S. M. C., Marques, C. T., Guerra, D. C. M., Ferreira, D. S. F. R., Oliveira, R. H. A., & Leite, E. L., (2006). Antioxidant activities of sulfated polysaccharides from brown and red seaweeds. *J. Appl. Phycol.*, *19*(2), 153–160.

Ryu, M. J., & Chung, H. S., (2015). Anti-inflammatory activity of fucoidan with blocking NF-B and STAT1 in human keratinocytes cells. *Natural Product Sciences*, *21*(3), 205–209.

Saito, A., Yoneda, M., Yokohama, S., Okada, M., Haneda, M., & Nakamura, K., (2006). Fucoidan prevents concanavalin A-induced liver injury through induction of endogenous IL-10 in mice. *Hepatol. Res.*, *35*(3), 190–198.

Schneider, T., Ehrig, K., Liewert, I., & Alban, S., (2015). Interference with the CXCL12/CXCR4 axis as potential antitumor strategy: Superiority of a sulfated galactofucan from the brown alga *Saccharina latissima* and Fucoidan over heparins. *Glycobiology*, *25*(8), 812–824.

Senni, K., Gueniche, F., Foucault-Bertaud, A., Igondjo-Tchen, S., Fioretti, F., Colliec-Jouault, S., et al., (2006). Fucoidan a sulfated polysaccharide from brown algae is a potent modulator of connective tissue proteolysis. *Archives of Biochemistry and Biophysics*, *445*(1), 56–64.

Senthilkumar, K., Manivasagan, P., Venkatesan, J., & Kim, S. K., (2013). Brown seaweed fucoidan: Biological activity and apoptosis, growth signaling mechanism in cancer. *International Journal of Biological Macromolecules*, *60*, 366–374.

Serravallo, M., Jagdeo, J., Glick, S. A., Siegel, D. M., & Brody, N. I., (2013). Sirtuins in dermatology: Applications for future research and therapeutics. *Archives of Dermatological Research*, *305*(4), 269–282.

Simpson, E. L., & Hanifin, J. M., (2006). Atopic dermatitis. *Med. Clin. North Am. 90*(1), 149–167.

Smit, A. J., (2004). Medicinal and pharmaceutical uses of seaweed natural products: A review. *Journal of Applied Phycology*, *16*, 245–262.

Streiff, M. B., Bockenstedt, P. L., Cataland, S. R., Chesney, C., Eby, C., Fanikos, J., et al., (2011). Venous thromboembolic disease. *Journal of the National Comprehensive Cancer Network*, *9*(7), 714–777.

Suwan, J., Zhang, Z., Li, B., Vongchan, P., Meepowpan, P., Zhang, F., et al., (2009). Sulfonation of papain-treated chitosan and its mechanism for anticoagulant activity. *Carbohydrate Research*, *344*(10), 1190–1196.

Synytsya, A., Bleha, R., Synytsya, A., Pohl, R., Hayashi, K., Yoshinaga, K., et al., (2014). Mekabu fucoidan: Structural complexity and defensive effects against avian influenza A viruses. *Carbohydrate Polymers*, *111*, 633–644.

Synytsya, A., Kim, W. J., Kim, S. M., Pohl, R., Synytsya, A., Kvasnička, F., et al., (2010). Structure and antitumour activity of fucoidan isolated from sporophyll of Korean brown seaweed *Undaria pinnatifida. Carbohydrate Polymers*, *81*, 41–48.

Teng, H., Yang, Y., Wei, H., Liu, Z., Liu, Z., Ma, Y., et al., (2015). Fucoidan suppresses hypoxia-induced lymphangiogenesis and lymphatic metastasis in mouse hepatocarcinoma. *Marine Drugs*, *13*(6), 3514–3530.

Thomes, P., Rajendran, M., Pasanban, B., & Rengasamy, R., (2010). Cardioprotective activity of *Cladosiphon okamuranus* fucoidan against isoproterenol induced myocardial infarction in rats. *European Journal of Integrative Medicine, 18*, 52–57.

Thuy, T. T. T., Ly, B. M., Van, T. T. T., Van Quang, N., Tu, H. C., Zheng, Y., et al., (2015). Anti-HIV activity of fucoidans from three brown seaweed species. *Carbohydrate Polymers, 115*, 122–128.

Trejo-Avila, L. M., Morales-Martínez, M. E., Ricque-Marie, D., Cruz-Suarez, L. E., Zapata-Benavides, P., Morán-Santibañez, K., & Rodríguez-Padilla, C., (2014). *In vitro* anti-canine distemper virus activity of fucoidan extracted from the brown alga *Cladosiphon okamuranus. Virus Disease, 25*(4), 474–480.

Trinchero, J., Ponce, N. M. A., Córdoba, O. L., Flores, M. L., Pampuro, S., Stortz, C. A., et al., (2009). Antiretroviral activity of fucoidans extracted from the brown seaweed *Adenocystis utricularis. Phytother Res, 23*(5), 707–712.

Ustyuzhanina, N. E., Bilan, M. I., Ushakova, N. A., Usov, A. I., Kiselevskiy, M. V., & Nifantiev, N. E., (2014). Fucoidans: Pro- or antiangiogenic agents? *Glycobiology, 24*(12), 1265–1274.

Wang, J., Liu, H., Li, N., Zhang, Q., & Zhang, H., (2014). The protective effect of fucoidan in rats with streptozotocin-induced diabetic nephropathy. *Marine Drugs, 12*(6), 3292–3306.

Wang, J., Wang, F., Yun, H., Zhang, H., & Zhang, Q., (2012). Effect and mechanism of fucoidan derivatives from *Laminaria japonica* in experimental adenine-induced chronic kidney disease. *J. Ethnopharmacol., 139*(3), 807–813.

Wang, P., Liu, Z., Liu, X., Teng, H., Zhang, C., Hou, L., et al., (2014). Anti-metastasis effect of fucoidan from *Undaria pinnatifida* sporophylls in mouse hepatocarcinoma Hca-F cells. *PLoS One, 9*(8), e106071.

Wang, X., Pei, L., Liu, H., Qv, K., Xian, W., Liu, J., & Zhang, G., (2016). Mice through inhibition of inflammation and oxidative stress. *Int. J. Clin. Exp. Pathol., 9*(7), 6896–6904.

Wang, Y., Nie, M., Lu, Y., Wang, R., Li, J., Yang, B., et al., (2015). Fucoidan exerts protective effects against diabetic nephropathy related to spontaneous diabetes through the NF-κB signaling pathway *in vivo* and *in vitro. International Journal of Molecular Medicine, 35*(4), 1067–1073.

Waris, G., & Ahsan, H., (2006). Reactive oxygen species: Role in the development of cancer and various chronic conditions. *J. Carcinog, 5*, 14.

Warkentin, T. E., Levine, M. N., Hirsh, J., Horsewood, P., Roberts, R. S., Gent, M., & Kelton, J. G., (1995). Heparin-induced thrombocytopenia in patients treated with low-molecular-weight heparin or unfractionated heparin. *New England Journal of Medicine, 332*(20), 1330–1336.

Witvrouw, M., Este, J. A., Mateu, M. Q., Reymen, D., Andrei, G., Snoeck, R., et al., (1994). Activity of a sulfated polysaccharide extracted from the red seaweed *Aghardhiella tenera* against human immunodeficiency virus and other enveloped viruses. *Antiviral Chemistry and Chemotherapy, 5*(5), 297–303.

Wu, L., Sun, J., Su, X., Yu, Q., Yu, Q., & Zhang, P., (2016). A review about the development of fucoidan in antitumor activity: Progress and challenges. *Carbohydrate Polymers, 154*, 96–111.

Wu, M., Xu, L., Zhao, L., Xiao, C., Gao, N., Luo, L., et al., (2015). Structural analysis and anticoagulant activities of the novel sulfated fucan possessing a regular well-defined repeating unit from sea cucumber. *Marine Drugs, 13*(4), 2063–2084.

Xue, M., Ge, Y., Zhang, J., Liu, Y., Wang, Q., Hou, L., & Zheng, Z., (2013). Fucoidan inhibited 4T1 mouse breast cancer cell growth *in vivo* and *in vitro* via downregulation of Wnt/β-Catenin signaling. *Nutrition and Cancer*, *65*(3), 460–468.

Xue, M., Ge, Y., Zhang, J., Wang, Q., Hou, L., Liu, Y., et al., (2012). Anticancer properties and mechanisms of fucoidan on mouse breast cancer *in vitro and in vivo*. *PLoS One*, *7*(8), e43483.

Yang, L., Wang, P., Wang, H., Li, Q., Teng, H., Liu, Z., et al., (2013). Fucoidan derived from *Undaria pinnatifida* induces apoptosis in human hepatocellular carcinoma SMMC-7721 cells via the ROS-mediated mitochondrial pathway. *Marine Drugs*, *11*(6), 1961–1976.

Yang, M., Ma, C., Sun, J., Shao, Q., Gao, W., Zhang, Y., et al., (2008). Fucoidan stimulation induces a functional maturation of human monocyte-derived dendritic cells. *International Immunopharmacology*, *8*, 1754–1760.

Ye, J., Li, Y., Teruya, K., Katakura, Y., Ichikawa, A., Eto, H., et al., (2005). Enzyme-digested fucoidan extracts derived from seaweed Mozuku of *Cladosiphon novae-caledoniae* Kylin inhibit invasion and angiogenesis of tumor cells. *Cytotechnology*, *47*(1–3), 117–126.

Yu, X., Zhang, Q., Cui, W., Zeng, Z., Yang, W., Zhang, C., et al., (2014). Low molecular weight fucoidan alleviates cardiac dysfunction in diabetic Goto-Kakizaki rats by reducing oxidative stress and cardiomyocyte apoptosis. *Journal of Diabetes Research*, 1–13.

Zen, K., Liu, Y., Cairo, D., & Parkos, C. A., (2002). CD11b/CD18-dependent interactions of neutrophils with intestinal epithelium are mediated by fucosylated proteoglycans. *Journal of Immunology 169*(9), 5270–5278.

Zhang, Q., Li, Z., Xu, Z., Niu, X., & Zhang, H., (2003). Effects of fucoidan on chronic renal failure in rats. *Planta Medica*, *69*(6), 537–541.

Zhang, Q., Yu, P., Li, Z., Zhang, H., Xu, Z., & Li, P., (2003). Antioxidant activities of sulfated polysaccharide fractions from *Porphyra haitanesis. Journal of Applied Phycology*, *15*(4), 305–310.

Zhang, W., Du, J. Y., Jiang, Z., Okimura, T., Oda, T., Yu, Q., & Jin, J. O., (2014). Ascophyllan purified from *Ascophyllum nodosum* induces Th1 and Tc1 immune responses by promoting dendritic cell maturation. *Marine Drugs*, *12*(7), 4148–4164.

Zhang, Z., Teruya, K., Eto, H., & Shirahata, S., (2011). Fucoidan extract induces apoptosis in MCF-7 cells via a mechanism involving the ROS-dependent JNK activation and mitochondria-mediated pathways. *PLoS One*, *6*(11), e27441.

Zhang, Z., Teruya, K., Yoshida, T., Eto, H., & Shirahata, S., (2013). Fucoidan extract enhances the anti-cancer activity of chemotherapeutic agents in MDA-MB-231 and MCF-7 breast cancer cells. *Marine Drugs*, *11*(1), 81–98.

Zhu, C., Cao, R., Zhang, S. X., Man, Y. N., Wu, X. Z., Zhu, C., et al., (2013). Fucoidan inhibits the growth of hepatocellular carcinoma independent of angiogenesis. *Evidence-Based Complementary and Alternative Medicine*, https://doi.org/10.1023/A:1025137728525.

CHAPTER 8

CYANOBACTERIA AS A PROMISING SOURCE OF THERAPEUTIC AGENTS AGAINST VARIOUS HUMAN DISEASES

DURDANA YASIN, NAZIA AHMAD, MOSHAHID ALAM RIZVI, and TASNEEM FATMA

Department of Biosciences, Jamia Millia Islamia,
New Delhi – 110025, India

8.1 INTRODUCTION

The discovery of new bioactive compounds has observed a boom in recent years due to the advent of high throughput screening systems, genomics, and bioinformatics techniques, rational drug design, and combinatorial chemistry. Development in the medicinal chemistry laid the foundation stone for the successful hit-to-lead investigation and optimization of new drugs (Vijayakumar and Menakha, 2015). Natural products are one of the major sources of new drugs with unique structures that are used to cure many diseases. Of all the drugs, 60% of anti-cancer drugs and 75% of drugs against infectious diseases are of natural origin (Newman et al., 2003; Lam, 2007). Many products from the microbial sources like bacteria and fungi have been used since long as the pharmaceuticals. Initially, only the terrestrial microbes were being studied but recently scientists have realized that aquatic organisms mainly from oceans particularly the microbes also possess the structurally and functionally diverse compounds that could be utilized as the pharmacological entities. The number of new drugs reaching the market has declined from 53 in 1996 to only 26 in 2005 and only one drug originated from a combinatorial chemistry approach (Vijayakumar and Menakha, 2015). There are many reasons for this, firstly, the investments that pharmaceutical firm has to make in order to develop a new drug is exceptionally high and exerts a financial barrier to the drug research and availability of it to the patients, secondly, the rapid rise in the diseases like cancer, HIV-AIDS (human immunodeficiency virus-acquired immunodeficiency syndrome), various hematological and autoimmune disorders along with multiple drug resistance in pathogens/cancer cells brings new challenges to the researchers. All these concerns compelled the scientists to work for the identification of new biologically active moieties that could be developed into new drugs. They are trying to look forward to the organisms especially microbes like cyanobacteria (CB) that are easily available and require a low cost for drug development. The use of CB as medicine dates back to 1500 BC where it was being used to treat diseases like gout, fistula, and some form of cancer (Liu and Chen, 2003). In the late 1990s, remarkable work in the discovery of drug candidates from CB has been done.

CB possess different types of enzymes that perform special functions like methylations, oxidations, tailoring, and other alterations in the molecules (Jones et al., 2010) which result in the formation of these chemically diverse natural compounds including secondary metabolites. These enzymes may belong to the non-ribosomal peptide synthetase (NRPS) and polyketide synthetase (PKS) genes. The presence of these genes itself unveils the potential for finding the novel natural product from these organisms (Ehernich et al., 2005). The NRPS-PKS system produces many compounds having pharmacological properties like vancomycin (an antibiotic), cyclosporine (an immunosuppressant), and bleomycin (anticancer drug) (Schwarjer et al., 2003). CB produce secondary metabolites that are synthesized as a part of their defense mechanism against predators and various environmental stresses. Secondary metabolite spectrum of CB includes linear peptides (Simmons et al., 2006), cyclic peptides (Sisay et al., 2009), linear lipopeptides (Nogle et al., 2001), depsipeptides (Han et al., 2005a), cyclic depsipeptides (Soria-Mercado et al., 2009), fatty acid amides (Chang et al., 2011), swinholides (Andrianasolo et al., 2005), glicomacrolides (Teruya et al., 2009), or macrolactones (Salvador et al., 2010). Recently, secondary metabolites have been also produced in heterologous hosts like *Escherichia coli* to enhance the production. It has been observed that complex ribosomal metabolites and simple cyanobacterial metabolites, e.g., mycosporine-like amino acids, mycosporine-ornithine, and mycosporine-lysine-the UV (ultraviolet) protectants can be produced using heterologous hosts (Ongley et al., 2013; Katoch et al., 2016) but due to the size and multifunctional character of PKS/NRPS biosynthesis complexes, the heterologous expression of non-ribosomal cyanobacterial peptides is still a challenge.

Nowadays, the scientific community is focusing on harvesting pharma-cologically important compounds/metabolites from the CB to treat various human diseases. It is estimated that among the commercially available marine biomedical compounds, about 24% are of cyanobacterial origin (Tan, 2013). The spectrum of activity of these compounds include antimicrobial, immunosuppressant, anti-HIV, anticoagulant, anti-inflammatory, antiproto-zoal, antimalarial, antiviral, anti-tuberculosis, and antitumor (Mayer et al., 2005; Gademann and Portmann, 2008; Wase and Wright, 2008).

8.2 CYANOBACTERIAL THERAPEUTIC ACTIVITIES

It was reported that of all the new cyanobacterial compounds that were discovered between 2007 and 2008, 21% had antiprotozoal effect, 18.4% were antibacterial, 5.2% had antiviral properties, 15.7% were anti-prolifer-ative, 18.4% were the protease inhibitors, 2.6% inhibited Ca^{2+} channels and the rest were observed to be ineffective (Figure 8.1) (Singh et al., 2011). The biological activities of some CB derived therapeutic compounds are discussed below.

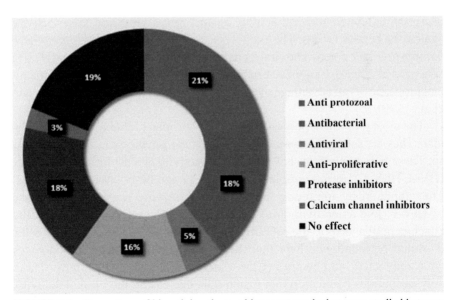

FIGURE 8.1 Percentage of bioactivity observed by compounds that were studied between 2007 and 2008 (adapted from Singh et al., 2011).

8.2.1 ANTIVIRAL

There are large numbers of human disorders including cancer that are caused by viruses. Diseases like Alzheimer's disease (AD), type 1 diabetes, and hepatocellular carcinoma have found to be associated with viral infections (Ball et al., 2013; Morgan et al., 2013). Appearance and re-appearance of viruses due to epidemic outbreaks possess a serious threat to the health of humans, especially when prophylactic vaccines and proper treatment to the viral diseases are unavailable. Despite the critical need for the treatment of viral diseases, there is huge unavailability of effective immunization and only a few antiviral drugs are approved for clinical use. Scientists have identified various antiviral compounds from CB based on their chemical nature, which are categorized in Table 8.1.

Polysaccharides are the polymeric natural resources having many applications, of which exopolysaccharides produced by CB is of greater significance. Most of the cyanobacterial polysaccharides have components like uronic acids, pentoses, a polypeptide moiety, or other monosaccharides like acetyl, pyruvyl, sulfate, and phosphate. These polysaccharides of varying types are being used as the antiviral agent. It was discovered by Mansour et al. (2011) that the polysaccharides from *Gloeocapsa turgidus* and *Synechococcus cedrorum* exhibited antiviral activity against rabies virus and lesser activity against the herpes-1 virus. The exopolysaccharide isolated from *Aphanothece halophytica* has potent antiviral activity against influenza virus A (H1N1) (Zheng et al., 2006). Cyanobacterial polysaccharides have reduced anticoagulant properties as an advantage over other sulfated polysaccharides (Feldmann et al., 1999).

The most prominent polysaccharides are spirulan (Table 8.1) and Ca-spirulan. They are sulfated polysaccharides isolated from the extracts of *Spirulina* sp. These polysaccharides have significant activity against influenza, HIV-1, HIV-2, HSV-1 (Herpes simplex virus type 1) and other enveloped viruses. They function by inhibiting the activity of enzyme reverse transcriptase of HIV-1 (like azidothymidine). The sulfated polysaccharides prevent the attachment of a virus to the cell, thereby preventing its fusion with host cells. It should be mentioned here that the fusion of HIV-infected lymphocytes with uninfected lymphocytes (CD4[+]) increases the viral infectivity. These polysaccharides could stop this fusion thus decreasing the virus infectivity. Aqueous extract of *Arthrospira platensis* also inhibited HIV-1 replication in T-cell *in vitro*, peripheral blood mononuclear cells and Langerhans cells (Ayehunie et al., 1998). Several spirulan like molecules/polysaccharides were isolated from *Arthrospira platensis* with antiviral activity against human cytomegalovirus and HSV-1 (Rechter et al., 2006).

TABLE 8.1   List of Antiviral Compounds from Cyanobacteria

Antiviral Compounds	Chemical Structure	Sources
Polysaccharides		
Spirulan	Sulfated polysaccharide of O-rhamnosyl-acofriose and O-hexuronosylrhamnose	*Spirulina* sp.
Ca-spirulan	Sulfated polysaccharide having rhamnose, 3-O-methylrhamnose (acofriose), 2,3-di-*O*-methylrhamnose, 3-*O*-methylxylose, uronic acid	*Spirulina* sp.
Nostoflan	→4)-D-Glcp-(1→, →6,4)-D-Glcp-(1→, →4)-D-Galp-(1→, →4)-D-Xylp-(1→, D-GlcAp-(1→, D-Manp-(1→ with a ratio of ca. 1:1:1:1:0.8:0.2.	*Nostoc flagilliforme*
Sulfolipids		
Sulfoglycolipid	Sulfoquinovosyl diacylglycerols	*Scytonema* sp.
Sulfolipid-1	Fatty acid (sulpho)	*Lyngbya lagerheimini, Phormidium tenue*
Lectins		
Cyanovirin-N	Carbohydrate binding protein [(NH2) Leu-Gly-Lys-Phe-Scr-Ghs-Thr-Cys-Tyr-Asn-Ser-Ala-Ile-Gln-Gly-Ser-Val-Len-Thr-Ser-The-Cys-Glu-Arg-Thr-Asn-Gly-Gly-Thr-Ser-The-Ser-Ser-Ilg-Asp-Leu-Asn-Ser-Val-Lie-Glu-Asn-Val-Asp-Gly-Ser-Len-Lys-Trp-Gln-Leu-Ata-Gly-Ser-Ser-Glu-Ler-Ala-Ala-Glu-Cys-Lys-Thr-Arg-Ala-Glu-Gln-Phe-Val-Ser-Thr-Lys-Lie-Asn--Leu-Asp-Asp-His-Lie-Ala-Asn-Ile-Asp-Gly-Tia-leu-Lys-Thr-Gla(COOH)]	*Nostoc ellipsosporum*
Scytovirin N	Carbohydrate binding protein [Domain-I (NH2) Ala-Ala-Ala-His-Gly-Ala-Thr-Gly-Gln-Cys-Phe-Gly-Ser-Ser-Ser-Cys-Arg-Asn-Pro-Gly-Gly-Pro-Asn-Lys-Ala-Glu-asp-Trp-Cys-Tyr-Thr-Pro-Gly-Lys-Pro- Domain-II Gly-Pro-Asp-Pro-Lys-Arg-Ser-Thr-Gly-Gln-Cys-Phe-Gly-Ser-Ser-Ser-Cys-Thr-Arg-Ala-Gly-asp-Cys-Gln-Lys-Asn-Asn-Ser-Cys-Arg-Asn-Pro-Gly-Gly-Pro-Asn-Asn-Ala-Glu-Asn-Trp-Cys-Tyr-Thr-Pro-Gly-Ser-Gly(COOH)]	*Scytonema varium*

TABLE 8.1 *(Continued)*

Antiviral Compounds	Chemical Structure	Sources
Microvirin	Mannan-binding lectin [(NH$_2$) Met-Pro-Asn-Phe-Ser-His-Thr-Cys-Ser-Ser-Ile-Asn-Tyr-Asp-Pro-Asp-Ser-Thr-Ile-Leu-Ser-Ala-Glu—Cys-Gln-Ala-Arg-Asp-Gly-Glu-Trp-Leu-Pro-Thr-Glu-Leu-Arg-Leu-Ser-Asp-His-Ile-Gly-Asn-Ile-Asp-Gly-Glu-Leu-Gln-Phe-Gly-Asp-Gln-Asn-Phe-Gln-Glu-Thr-Cys-Gln-Asp-Cys-His-Leu-Glu-Phe-Gly-Asp-Gly-Glu-Gln-Ser-Val-Trp-Leu-Val-Cys-Thr-Cys-Gln-Thr-Met-Asp-Gly-Glu-Trp-Lys-Ser-Thr-Gln-Ile-Leu-Leu-Asp-Ser-Gln-Ile-Asp-Asn-Asn-Asp-Ser-Gln-Leu-Glu-Ile-Gly(COOH)]	*Microcystis aeruginosa*
Others		
Ichthyopeptins A & B	Cyclic depsipeptide [A:cyclo[N(Unk)Gln-Tyr-Nva(Ph(4-OH))-Val-N(Me)Phe-Ile-Thr; B:cyclo[N(Unk)Asn-Leu-Nva(Ph(4-OH))-Ile-N(Me)Phe-Val-Thr	*Microcystis ichthyoblabe*

Nostoflan (Table 8.1) is an acidic polysaccharide isolated from *Nostoc flagelliform* and have virucidal activity against HSV-1 (Kanekiyo et al., 2005), human cytomegalovirus, and influenza A virus, who have carbohydrates as cellular receptors. Sulfoglycolipids (Table 8.1; Figure 8.2a) are structurally similar to acylated diglycolipids, are obtained from *Oscillatoria* and *Phormidium,* also possess antiviral properties. Sulfolipid-1 which was isolated from the CB *Lyngbya, Phormidium, Scytonema, Anabaena, Calothrix,* and *Oscillatoria* have shown inhibition of the HIV-1 virus. The mechanism of action involves the inhibition of HIV-1 reverse transcriptase enzyme's function as DNA polymerase (Gustafson et al., 1989; Reshef et al., 1997; Loya et al., 1998). The sulfoglycolipids exhibited the IC$_{50}$ values in the nano-molar range and found to be more effective than related glycolipids that lack sulfonic acid group (Reshef et al., 1997). The presence of the fatty acid chains in the sulfoglycolipids are mandatory for its antiviral activity.

Lectins are carbohydrate-binding proteins of non-immune origin. The interactions of lectins and carbohydrate target the adherence stage of viruses and can aid in viral elimination. Lectins from CB that possess high specificity for mannose or complex glycans are more potent as anti-viral agents. It is speculated that these cyanobacterial lectins can cure diseases caused by enveloped viruses like HIV, hepatitis, herpes, influenza, and ebolaviruses that are rich in cell surface-attached mannose (Singh et al., 2017). Lectins like cyanovirin-N and scytovirin (Table 8.1) are currently being developed as the virucidal drugs. These proteins manifest their antiviral activity by interfering with the viral fusion process.

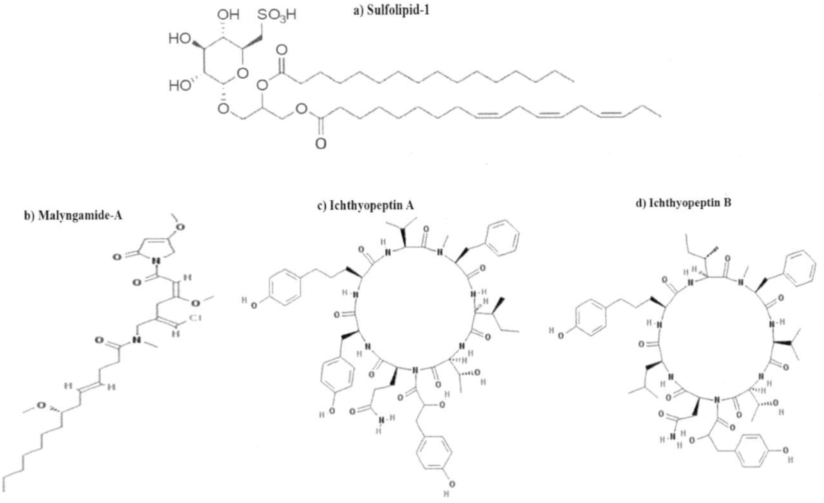

FIGURE 8.2 (a–d) Structure of some antiviral compounds from cyanobacteria.

Cyanovirin-N has been isolated from *Nostoc ellipsosporum*. It is an 11 kDa protein, having 101 amino acids. It has shown promising activity against HIV and Lentiviruses both *in vitro* (Boyd et al., 1997). Mechanism of action involves interference of the binding of gp120 proteins of HIV with CD4$^+$ receptors and the chemokine CCR5 or CXCR4 co-receptors of target cells (Figure 8.3), eventually inhibiting the fusion of HIV virus with the CD4 cell membrane. It is effective at nano-molar concentration and does not have any cytotoxicity even at thousand-fold higher concentration. Cellegy Pharmaceuticals (USA) had developed it as a vaginal gel to inhibit the sexual transmission of HIV. This has reduced the global spread of HIV-1 (Klasse et al., 2008). It is also found to inhibit HSV-6 and measles virus *in vitro* (Dey et al., 2000). It is also effective against influenza A and B, respiratory syncytial virus, enteric viruses, and several corona viruses (O'Keefe et al., 2003; Van der Meer et al., 2007). Bioengineered lactic acid bacteria (LAB) have been made to secrete cyanovirin (LAB-CV-N). These were introduced in a food product, e.g., bioengineered yoghurt. When this yoghurt having LAB-CV-N was given to pigtail macaques, they expressed lower levels of peak viral replication (Li et al., 2011).

Scytovirin has been isolated from the *Scytonema varium*, thus so named (Bokesch et al., 2003). It is a 9.7 kDa protein having 95 amino acids with five disulphide bonds. Scytovirin possess two domains, first domain (48 amino acids) exhibits antiviral property as the whole protein, thus, it can be inferred that it is responsible for the biological activity of scytovirin (Xiong et al., 2006). It has a binding ability to the glycoproteins (gp120, gp160, and gp41) of HIV envelop and thus, it can inactivate the virus at very low concentrations. Another lectin, microvirin, with comparable activity against HIV-1 and having better safety profile, has been isolated from *Microcystis aeruginosa* (Huskens et al., 2010).

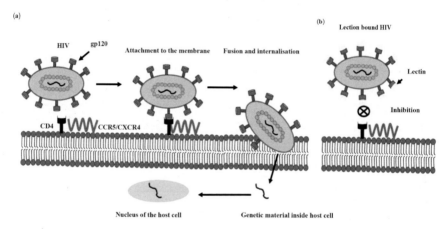

FIGURE 8.3   (a) HIV infection mechanism, (b) Role of cyanobacterial lectins in inhibition of HIV by binding to viral glycoprotein inhibiting T cell and macrophage infection (adapted from Singh et al., 2017).

Other compounds like cyclic depsipeptides, ichthyopeptins A and B (Table 8.1; Figure 8.2c, d) from *Microcystis ichthyoblabe* also have antiviral activity (against influenza A virus) but at higher concentrations (Zainuddin et al., 2007). Various Serinol-derived malyngamides (Figure 8.2b) (e.g., derivative 19), from an ambiguous cyanobacterium, have been reported to have weak anti-HIV activity (Wan and Erickson, 1999). Some β-carbolines bauerines A–C and the indolocarbazoles from *Dichothrix baueriana* (Larsen et al., 1994) and *Nostoc sphaericum* respectively (Knübel et al., 1990) possess activity against HSV type 2.

8.2.2   ANTICANCER

Cancer is the second major killer disease of mankind, first being the cardiovascular diseases (CVDs) (Jemal et al., 2007). Cancer involves the uncontrolled growth of cells that form benign tumors which ultimately lead to the malignant growth that invade the nearby cells. Radiotherapy, surgery, and chemotherapy are the currently available treatments for the disease, among them chemotherapy is a widely used method. Despite the availability of so many anticancer drugs in the market, there is still a serious requirement of new anticancer drugs because tumor cells are very rapidly evolving resistance against conventional drugs (doxorubicin, cisplatin, taxanes, and vinca alkaloids) rendering them ineffective, and causing serious life-threatening side effects. In addition, new types of cancers (e.g., glioblastoma) are increasing at a very fast rate. A search for new anticancer drugs is in its full swing and the natural products including microbes like cyanobacteria are being explored. Few examples of potential anticancer drug candidates from CB are mentioned below.

Cryptophycins (Figure 8.4a) belong to macrolide and depsipeptide family of compounds and show anticancer activities. Twenty-six cryptophycins forms from *Nostoc* sp. GSV 224 were isolated (Golakoti et al., 1994, 1995; Chaganty et al., 2004). Cryptophycin 1 was isolated from *Nostoc* sp. GSV224 and showed anticancer activity against KB human nasopharyngeal cancer cells, LoVo human colorectal cancer cells, against adriamycin-resistant M 17 breast cancer and DMS 273 lung cancer cell lines. It was found more effective than taxol or vinblastine which is the conventional anticancer drugs (Patterson et al., 1991). The mechanism of action was found to be the suppression of microtubule dynamics (Smith et al., 1994). It affects the cytoskeleton, hindering cell cycle and blocking it at G2/M phase. Various analogs of cryptophycin have been naturally isolated or chemically synthesized. One such chemical analog is Cryptophycin-52 (LY 355073) but it produced marginal activity in Phase II clinical trials (Shin et al., 2001) along with some side effects (Edelman et al., 2003; D'Agostino et al., 2006). Other analogs were Cryptophycins 249 and 309, have improved stability and water solubility. They are often known as second-generation clinical candidates (Liang et al., 2005).

Curacin A (Figure 8.4b) is also a potential anticancer compound isolated from an organic extract of a Curacao collection of *Lyngbya majuscula* (hence named so) (Gerwick et al., 1994). It is derived from hybrid PKS-NRPS system of enzymes. It is characterized by the presence of an unusual terminal alkene, thiazoline ring and a cyclopropyl moiety. These structures are essential for the activity of the compound (Gu et al., 2007). It showed activity against breast cancer. Its mechanism of action involves inhibition of cell growth by interacting with cytoskeleton tubulin division (Verdier-Pinard et al., 1998). But this compound has very poor solubility and thus, it did not exhibit any activity *in vivo*. Various semi-synthetic derivatives have been synthesized using combinatorial chemistry so that its solubility could be enhanced. These compounds are being studied in preclinical trials.

Borophycin is a boron-containing compound obtained from *Nostoc spongiaeforme* var. *tenue*. It has effective antiproliferative efficacy for human carcinoma (Davidson, 1995; Banker and Carmeli, 1998; Gupta et al., 2012). Its analog borophycin-8 is isolated from *Nostoc linckia* (Arai et al., 2004). It is composed of two identical parts with their structure similar to boron-containing antibiotics.

Dolastatin-10 was first isolated from the sea hare *Dolabella auricularia*. Later on, it was found out that *D. auricularia* actually obtains it from its diet that included CB *Symploca* sp. (Luesch et al., 2001). Dolastatin-10 (Figure 8.4c) is a pentapeptide made up of amino acid valine and four unique amino acids, dolavaline, dolaproline, dolaisoleucine, and dolaphenine. They show anti-cancerous activity by binding to rhizoxin binding site of tubulin and thus inhibiting microtubule assembly. It has been found to induce apoptosis and Bcl-2 phosphorylation in many malignant cells (Maki et al., 1995). National Cancer Institute of US conducted its phase II clinical trials but it was withdrawn due to its side effects like the development of peripheral neuropathy in many patients and lack of activity in individuals with hormone-refractory metastatic adenocarcinoma (Vaishampayan et al., 2000) and platinum-sensitive adenocarcinoma (Hoffmann et al., 2003). Many of dolastatin analogs are currently in clinical trials. About sixteen dolastatin forms were isolated till date. Symplostatin A (Figure 8.4d) is an analog of dolastatin that is isolated from *Symploca hydnoides* (Harrigan et al., 1998). It also inhibits the microtubule and found to be effective against a drug-resistant mammary tumor and colon tumor. Soblidotin (TZT-1027 or auristatin) (Figure 8.4e) is a synthetic analog of dolastatin 10. It lacks a thiazoline ring on the dolaphenine residue. It has better pharmacological and pharmacokinetic properties. It showed its effects in two human xenograft models, i.e., MX-1 breast carcinoma and LX-1 lung carcinoma in mice (Kobayashi et al., 1997). It also showed efficacy against both p53 normal and mutant cell lines (Natsume et al., 2003). It had successfully cleared phase I and II trials and has reached phase III clinical trials under the supervision of Aska Pharmaceuticals, Tokyo, Japan (Bhatnagar and Kim, 2010). It showed better efficacy than that of the existing anticancer drugs, like paclitaxel and vincristine (Watanabe et al., 2006). Cemadotin (Figure 8.4f) and tasidotin (Figure 8.4g) are the third generation dolastatin 15 analogs. They have cleared phase I trials (Mita et al., 2006) but did not proceed beyond phase II trials either due to toxicity or a lack of efficacy or both (Newman and Cragg, 2017). Only one analog of the dolastatins is developed into a marketed drug, viz. brentuximab vedotin that is used against Hodgkin's lymphoma (Niedermeyer and Brönstrup, 2012).

Apratoxin-A (Figure 8.4h) is a cyclodepsipeptide derivative which was isolated from a *Lyngbya* sp. of Guam (Thornburg et al., 2013; Masuda et al., 2014). It has a mixed peptide-polyketide cyclic structure derived from PKS-NRPS pathway. In 60 tumor cell lines, it was found to induce G1 phase cell arrest and apoptosis. In *in-vivo* studies, only little activity was reported against early-stage adenocarcinoma (Chen et al., 2011; Doi, 2014). Apratoxin-B and -C from *Lyngbya* sp. has shown the effect against KB oral epidermoid cancer and LoVo colon cancer cells (Luesch et al., 2002). Apratoxin-D produced by *L. majascula* is cytotoxic against human lung cancer cells (Gutierrez et al., 2008a). The exact mechanism of action of Apratoxins is not known till date.

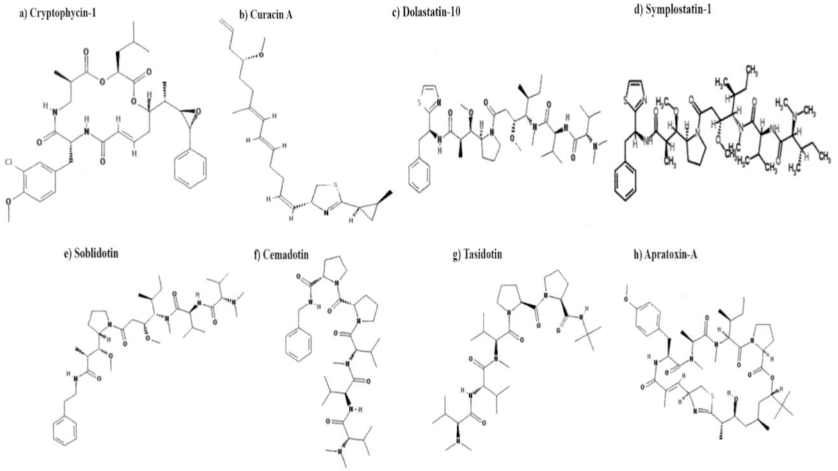

FIGURE 8.4   (a–h) Structure of anticancer compounds from cyanobacteria.

Calothrixins are antiproliferative compounds which possess distinct pentacyclic indolophenanthridine structure having quinoline, quinone, and indole pharmacophore. They are first isolated from *Calothrix* by Rickards et al. (1999). There are two main forms of it, Calothrixins-A, and -B (Figure 8.5a, b). They can kill the human HeLa cancer cells at nano-molar concentrations in a dose-dependent manner. Calothrixin-A can induce apoptosis in human Jurkat cancer cells (Chen et al., 2003) and can arrest cells in S-phase (Khan et al., 2009).

Lyngbyabellins are potential antiproliferative compounds. Lyngbya-bellin-A (Figure 8.5c) which is a peptolide - a depsipeptide, was isolated from the *Lyngbya majuscula* from Guam (Luesch et al., 2000). The cyto-toxicity of this compound depends on the disruption of the microfilament network of cells. Therefore, it could induce G2/M phase arrest in human Burkitt lymphoma cells. Lyngbyabellin-B was also isolated from *Lyngbya majuscula* and has weaker activity than lyngbyabellin-A. The 5 new lyng-byabellins, viz. lyngbyabellins-E-I were isolated by Han et al. (2005b) from *Lyngbya majuscula* from Papua New Guinea. They all had cytotoxicity to NCI-H460 human lung tumor and neuro-2A mouse neuroblastoma cell lines.

Tolyporphin (Figure 8.5d) was extracted from *Tolypothrix nodosa*. It showed the reversal of multiple drug resistance in a human ovarian adeno-carcinoma cell line resistant to vinblastine. It is observed to enhance the cytotoxicity of Adriamycin or vinblastine against SK-VLB cells (Prinsep et al., 1992). It also has a strong photosensitizing activity against tumor cells.

Somocystinamide-A (Figure 8.5e) was also isolated from *L. majuscula*. It is an unusual disulfide dimer formed by PKS-NRPS pathway. It manifests its anticancer activities by inhibiting angiogenesis, cell proliferation and inducing apoptosis. It could affect the proliferation and tubule formation in endothelial cells (Wrasidlo et al., 2008). It possesses cytotoxicity against mouse neuro-2a neuroblastoma cells as well (Nogle and Gerwick, 2002).

Scytonemin (Figure 8.5f) was isolated from *Scytonema* sp. (Proteau et al., 1993) as UV protectant pigment. Later on, it was found that it is able to inhibit the proliferation of three myeloma cells. Scytonemin was the first described small molecule that inhibited the human polo-like kinase (Plk) (Stevenson et al., 2002). Scytonemin hinders the cell growth and it arrests the cell cycle in multiple myeloma cells by decreasing Plk1 activity. Thus, it is a novel agent for the treat-ment of multiple myelomas. Reduced-scytonemin from *Nostoc commune* can suppress the human T-lymphoid Jurkat cell growth (Itoh et al., 2013).

Phycocyanin (C-PC) (Figure 8.5g) is a photosynthetic pigment that belongs to phycobiliproteins and found in the CB. It has various pharma-cological properties that include anti-inflammatory and anticancer activities because of its β-subunit. It was reported that the recombinant β-subunit of C-PC showed anticancer properties on four cell lines. It revealed appre-ciable inhibition of growth and induction of apoptosis. This recombinant protein was found to interact with membrane-associated β-tubulin and glyceraldehyde-3-phosphate dehydrogenase (Wang et al., 2007). Beside all these compounds, there are various other compounds that are being studied for their anticancer properties and possess activity *in vitro*.

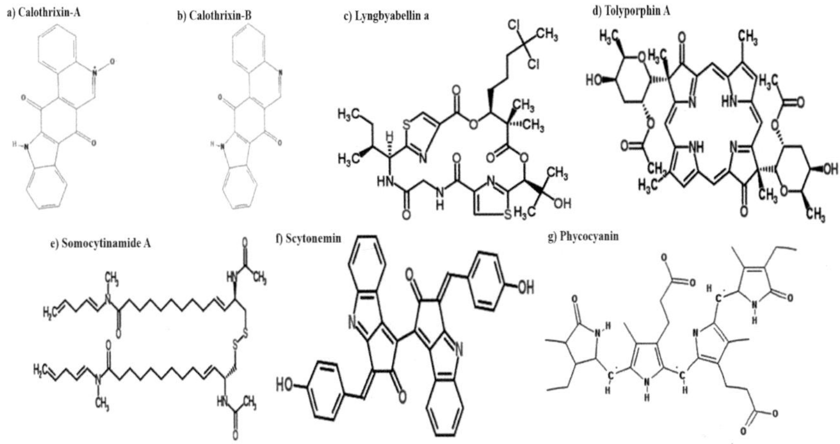

FIGURE 8.5   (a–g) Structure of some more anticancer compounds from cyanobacteria.

8.2.3   ANTIBACTERIAL

Bacteria develop resistance at higher rates than other pathogens because of which many diseases are becoming untreatable. Many species of *Staphylococcus aureus* have become methicillin-resistant, enterococci became vancomycin resistant, and there are AmpC β-lactamase producing *Enterobacteriaceae* became resistant to cephalothin, cefazolin, cefoxitin, most penicillins and β-lactamase inhibitor-β-lactam combinations. All the above-mentioned bacteria can cause nosocomial infections and pose a great therapeutic challenge. Thus, there is an immediate need to search for newer antibiotics.

Anabaena extracts have been reported to have shown antibacterial activity against vancomycin-resistant *S. aureus* (Bhateja et al., 2006). Various cyanobacterial metabolites have been isolated and studied for their antibacterial activities as well, but at times in many cases, antimicrobial activity is found to have associated with general cytotoxicity, thus they are of a little use. Only a few compounds have been characterized till date, some of them are discussed here.

The first antibiotic from CB was Malyngolide (Figure 8.6a), which was isolated from *Lyngbya majuscula*. This compound was effective against gram-positive bacteria like *Mycobacterium smegmatis*, *Streptococcus pyogenes*, *S. aureus,* and *Bacillus subtillis* (Cardellina et al., 1979). Hapalindoles (Figure 8.6b) are indole alkaloids from *Hapalosiphon* sp., *Fischerella*, and *Westiollopsis* (Moore et al., 1984, 1987; Klein et al., 1995). It was found to be very much potent against *S. aureus, B. subtillis, Salmonella gallinarum,* and *E. coli* even in sub-micromolar concentrations. However, these compounds were moderately cytotoxic too (Asthana et al., 2006; Kim et al., 2012). Noscomin (Figure 8.6c) was isolated from *Nostoc commune* and has antibacterial activity against *Bacillus cereus, Staphylococcus epidermidis,* and *E. coli.* This activity was found to be comparable to that of the conventional drugs (Mundt et al., 2003). 6-cyano-5-methoxy-12-methylindolo(2,3-a)carbazole (Figure 8.6d) was also isolated from cyanobacterium *Nostoc sphaericum EX-5-1.* It was found to inhibit *Bacillus anthracis* (Guo et al., 2009).

Carbamidocyclophanes which are paracyclophanes (Figure 8.6e) from *Nostoc* sp. CAVN 10 exhibit intermediate activity against *S. aureus* (Bui et al., 2007). Around nine ambiguines-hapalindole type alkaloids were isolated from *Fischerella* sp., which had antimicrobial activity. Ambiguine-I isonitrile (Figure 8.6f) was found to be more effective than streptomycin (a conventional antibiotic) against *Bacillus subtilis* and *Staphylococcus albus* (Raveh and Carmeli, 2007). Similarly, two norbietane (Figure 8.6g) compounds were extracted from *Micrococcus lacustris* with activity against *E. coli, B. cereus, B. subtilis, Salmonella typhi, S. aureus, S. epidermidis, Vibrio cholarae,* and *Klebsiela pneumonia* (Gutierrez et al., 2008b).

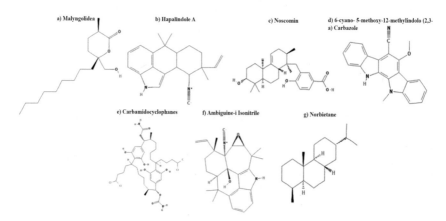

FIGURE 8.6   (a–g) Structure of some antibacterial compounds from cyanobacteria.

8.2.4   ANTIPROTOZOAL

Many protozoans are responsible for a number of serious diseases such as leishmaniasis, amoebiasis, trypanosomiasis, and malaria. The occurrence of these diseases are increasing constantly, moreover, many conventional drugs have some serious side effects. Therefore, new drugs are required for the treatment of these diseases. CB are also investigated for their ability to produce anti-protozoan compounds.

Aerucyclamide B is a cyclic peptide isolated from *Microcyctis aeruginosa.* It is among the most active anti-malarial compound discovered from CB so far (Portmann et al., 2008). Calothrixins A and B (Figure 8.5a, b) also show antimalarial activity against the FAF6 strain of *Plasmodium falciparum* (Rickards et al., 1999). Venturamides A and B (Figure 8.7b, c) are the first cyclic peptides that were extracted from marine species of *Oscillatoria* sp. (Linington et al., 2007) that were active against malarial parasite *Plasmodium falciparum* W2. Symplocamide A from *Symploca* sp. was also effective against *P. falciparum* W2 (malaria), *Trypanosoma cruzi* (Chagas diseases), and *Leishmania donovani* (leishmaniasis) (Linington et al., 2008). Nostocarboline (Figure 8.7d) from *Nostoc* sp. was found to be active against *P. falciparum* and *L. donovani* (Barbaras et al., 2008).

Extracts from *L. majuscula* had activity against even chloroquine-resistant *P. falciparum.* Later on two lipopeptides dragomabin and dragonamide B (Figure 8.7e, f), and metabolites like carmabin A of species of *L. majuscula* Gomont (Panama) were found to have significant antimalarial activity (McPhail et al., 2007). Protease inhibitors and proteases from CB (e.g., from *Symploca* sp.) also possess anti-protozoal as well as antiviral potential (Turk, 2006; Niedermeyer et al., 2012) and can prove to be an effective therapeutic agent in future.

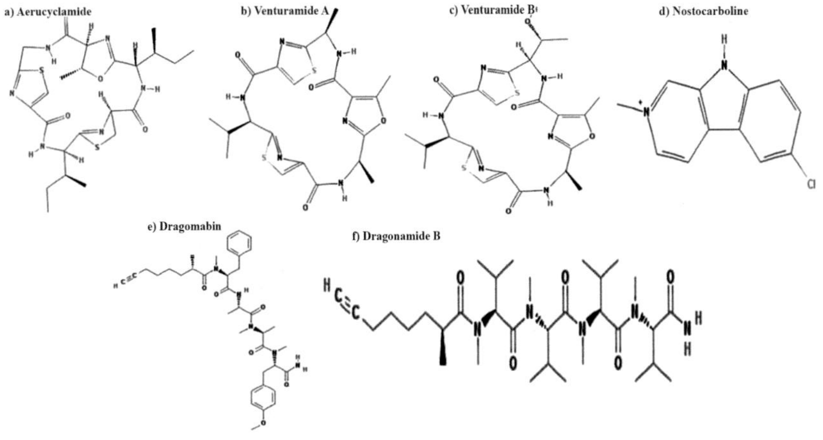

FIGURE 8.7   (a–f) Structure of some antiprotozoal compounds from cyanobacteria.

8.2.5 ANTIFUNGAL

Newer types of human fungal infections are increasing due to the increased rate of diseases like cancer, AIDS, and increased number of immune-compromised patients. Furthermore, many fungal strains have become resistant to the available drugs. Thus, natural sources including CB are being worked on to discover new classes of antifungal compounds.

Ghasemi et al. (2003) from Iran found that species of *Fischerella, Nostoc, Stigonema,* and *Hapalosiphon* had bioactive compounds with potential antifungal activity against *Candida krusei, Candida kefyr* and *Cryptococcus neoformans* and *Aspergillus* sp. They observed that the members of class Stigonemataceae like *Stigonema* sp. and *Fischerella* usually exhibit antifungal activity. Similarly, De Caire et al. (1993) found that members of *Nostocaceae, Microchaetaceae,* and *Scytonemataceae* present in Argentinian paddy fields make compounds which are active against *Candida albicans.* Also, CB like *Phormidium fragile* showed antifungal activities against *C. albicans* and *Trichoderma viride* (Senthil et al., 2013). *Oscillatoria subuliformis* was also found to have activity against *Penicillium, Acremonium,* and *Aspergillus flavus* (Priyadarshini et al., 2013). Some promising antifungal agents of cyanobacterial origin are discussed below.

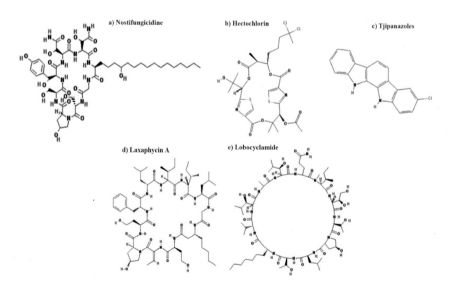

FIGURE 8.8   (a–e) Structure of antifungal compounds from cyanobacteria.

Cryptophycin (Figure 8.4a) also have the fungicidal activity (Burja et al., 2001). Nostifungicidine (Figure 8.8a) is a lipopeptide obtained from *Nostoc commune*. It is also antifungal in its effects (Kajiyama et al., 1998). Hectochlorin (Figure 8.8b), a peptide from *Lyngbya majuscula* has antifungal activity (Koglin and Walsh, 2009). It is currently being studied for its anticancer effects (Ramaswamy et al., 2007). Tjipanazoles (Figure 8.8c) which are N-glycosides of indolo[2,3-a]carbazole obtained from CB *Tolypothrix tjipanasensis* have significant fungicidal activity (Bonjouklian et al., 1991).

Ambiguine isonitriles (Figure 8.6f) are also fungicidal in their effects (Smitka et al., 1992). Laxaphycins (Figure 8.8d) that belong to the family of cyclic undeca- and dodecapeptides have remarkable antifungal activity. They are isolated from *Anabaena laxa*. They have also been isolated from the CB *Hormothamnion enteromorphoides* (Gerwick et al., 1989). A mixture of different laxaphycins showed synergistic effect (Frankmölle et al., 1991).

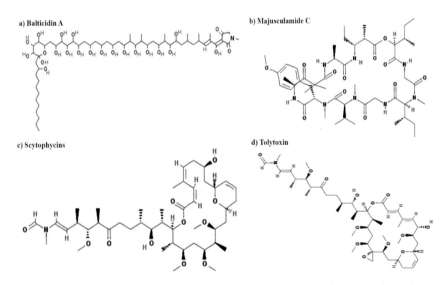

a) Balticidin A

b) Majusculamide C

c) Scytophycins

d) Tolytoxin

FIGURE 8.9   (a–d) Structure of some more antifungal compounds from cyanobacteria.

Lobocyclamides (Figure 8.8e) are the β-amino-containing cyclic lipopeptides isolated from *Lyngbya confervoides* (MacMillan et al., 2002). These lipopeptides had shown intermediate antifungal effects against *C. albicans* and *C. glabrata*. Hassallidins are the non-ribosomal cyclic depsipeptides, first obtained from *Tolypothrix* (Neuhof et al., 2005, 2006b) and found in various filamentous CB (Vestola et al., 2014). They have a sugar and a dihydroxy fatty acid that could also be glycosylated. They are effective against many *Candida* species like *C. albicans* with MICs being in micro-molar ranges (Neuhof et al., 2006a; Vestola et al., 2014). Balticidins A–D is structurally related compounds isolated from *Anabaena cylindrica* and possesses comparable antifungal activity (Bui et al., 2014) (Figure 8.9a). Majusculamide C (Figure 8.9b) from *L. majuscula* was found to be effective against various plant pathogenic fungi like *Rhizoctonia solani*, *Pythium aphanidermatum*, *Aphanomyces euteiches*, and *Phytophthora infestans* at micomolar concentration (Carter et al., 1984; Moore and Mynderse, 1982). But later on, it was found to be toxic even at nano-molar concentrations (Pettit et al., 2008). Scytophycins (Figure 8.9c) and tolytoxin (Figure 8.9d) from *Scytonema* and *Tolypothrix*, respectively, are among the most effective antifungal agents from CB till date (Ishibashi et al., 1986; Carmeli et al., 1990; Patterson and Carmeli, 1992; Patterson et al., 1993; Smith et al., 1993; Patterson and Bolis, 1997).

8.3 CONCLUSION AND FUTURE PERSPECTIVES

We can conclude that CB can serve as the source of many therapeutic compounds that could be effective against various human diseases. It is only in the past three decades that therapeutic potential of CB has been realized like its anti-cancer/anti-tumor, antibacterial, antifungal, antiviral, and antiprotozoal effects. Many of the metabolites of CB have entered clinical trials, but hardly any of these cyanobacterial derived compounds have been approved by the Food and Drug Administration. On the other hand, most of these compounds were cytotoxic beside their respective activities. This has limited the direct use of these agents. But with the help of natural product chemistry, these compounds can serve as the lead for the development of synthetic derivatives that may have lower toxicity and thus, needed to be studied more exhaustively. Moreover, more of the cyanobacterial genera, species, and strains are yet to be discovered and studied for their useful applications in the field of drug discovery.

CB do not require carbon or energy sources in their growth media and can thrive in inorganic salt solutions, which is an economical process. Moreover, the isolation of the secondary metabolites is comparatively easier, as the products do not require to be separated from a complex organic medium. Furthermore, the yield, productivity, and cost of production can be enhanced by heterologous expression of metabolites or by the developing large-scale photo-bioreactors that could work at the lower cost.

Thus, we can say that CB holds a great potential to serve as a source of pharmacologically important metabolites, whose production and activities can be enhanced in future by exploiting various scientific advancements, so that, the human kind may receive the much needed therapeutic drugs to overcome the havoc created by many frightening diseases.

KEYWORDS

- **antibacterial**
- **anticancer**
- **antiprotozoal**
- **cyanobacteria**
- **lectins**
- **NRPS/PKS synthetase**
- **polysaccharides**
- **secondary metabolites**

REFERENCES

Andrianasolo, E. H., Gross, H., Goeger, D., Musafija-Girt, M., McPhail, K., Leal, R. M., et al., (2005). Isolation of swinholide A and related glycosylated derivatives from two field collections of marine cyanobacteria. *Org. Lett., 7*(7), 1375–1378.

Arai, M., Koizumi, Y., Sato, H., Kawabe, T., Suganuma, M., Kobayashi, H., et al., (2004). Boromycin abrogates bleomycin induced G2 checkpoint. *J. Antibiot., 57*(10), 662–668.

Asthana, R. K., Srivastava, A., Singh, A. P., Singh, S. P., Nath, G., Srivastava, R., & Srivastava, B. S., (2006). Identification of an antimicrobial entity from the cyanobacterium *Fischerella* sp. isolated from bark of *Azadirachta indica* (Neem) tree. *J. Appl. Phycol., 18*(1), 33–39.

Ayehunie, S., Belay, A., Baba, T. W., & Ruprecht, R. M., (1998). Inhibition of HIV-1 replication by an aqueous extract of *Spirulina platensis* (*Arthrospira platensis*). *J. Acquir. Immune Defic. Syndr., 18*(1), 7–12.

Ball, M. J., Lukiw, W. J., Kammerman, E. M., & Hill, J. M., (2013). Intracerebral propagation of Alzheimer's disease: Strengthening evidence of a herpes simplex virus etiology. *Alzheimer's Dement., 9*(2), 169–175.

Banker, R., & Carmeli, S., (1998). Tenuecyclamides A-D, cyclic hexapeptides from the cyanobacterium *Nostoc spongiaeforme* var. *tenue*. *J. Nat. Prod., 61*(10), 1248–1251.

Barbaras, D., Kaiser, M., Brunb, R., & Gademann, K., (2008). Potent and selective antiplasmodial activity of the cyanobacterial alkaloid nostocarboline and its dimers. *Bioorg. Med. Chem. Lett., 18*(15), 4413–4415.

Bhateja, P., Mathur, T., Pandya, M., Fatma, T., & Rattan, A., (2006). Activity of blue-green microalgae extracts against *in vitro* generated *Staphylococcus aureus* with reduced susceptibility to vancomycin. *Fitoterpia, 77*(3), 233–235.

Bhatnagar, I., & Kim, S. K., (2010). Immense essence of excellence: Marine microbial bioactive compounds. *Mar. Drugs, 8*(10), 2673–2701.

Bokesch, H. R., O'Keefe, B. R., McKee, T. C., Pannell, L. K., Patterson, G. M., Gardella, R. S., et al., (2003). A potent novel anti-HIV protein from the cultured cyanobacterium *Scytonema varium. Biochem., 42*(9), 2578–2584.

Bonjouklian, R., Smitka, T. A., Doolin, L. E., Molloy, R. M., Debono, M., Shaffer, S. A., et al., (1991). Tjipanazoles, new antifungal agents from the blue-green alga *Tolypothrix tjipanasensis. Tetrahedron, 47*(37), 7739–7750.

Boyd, M. R., Gustafson, K. R., McMahon, J. B., Shoemaker, R. H., O'Keefe, B. R., Mori, T., et al., (1997). Discovery of cyanovirin-N, a novel human immunne deficiency virus inactivating protein that binds viral surface envelope glycorotein gp120: Potential application to microbicide development. *Antimicrob. Agents Chemother., 41*(7), 1521–1530.

Bui, T. H., Wray, V., Nimtz, M., Fossen, T., Preisitsch, M., Schröder, G., et al., (2014). Balticidins A–D, antifungal hassallidin-like lipopeptides from the Baltic Sea cyanobacterium *Anabaena cylindrica* bio33. *J. Nat. Prod., 77*(6), 1287–1296.

Bui, T. N., Jansen, R., Pham, T. L., & Mundt, S., (2007). Carbamidocyclophanes A-E, chlorinated paracyclophanes with cytotoxic and antibiotic activity from the Vietnamese cyanobacterium *Nostoc* sp. *J. Nat. Prod., 70*(4), 499–503.

Burja, A. M., Banaigs, B., Abou-Mansour, E., Grant, B. J., & Wright, P. C., (2001). Marine cyanobacteria - a prolific source of natural products. *Tetrahedron, 57*(46), 9347–9377.

Cardellina, J. H. II., Moore, R. E., Arnold, E. V., & Clardy, J., (1979). Structure and absolute configuration of malyngolide, an antibiotic from the marine blue-green alga *Lyngbya majuscula* Gomont. *J. Org. Chem., 44* (23), 4039–4042.

Carmeli, S., Moore, R. E., & Patterson, G. M. L., (1990). Tolytoxin and new scytophycins from three species of *Scytonema. J. Nat. Prod., 53*(6), 1533–1542.

Carter, D. C., Moore, R. E., Mynderse, J. S., Niemczura, W. P., & Todd, J. S., (1984). Structure of majusculamide C, a cyclic depsipeptide from *Lyngbya majuscula. J. Org. Chem., 49*(2), 236–241.

Chaganty, S., Golakoti, T., Heltzel, C., Moore, R. E., & Yoshida, W. Y., (2004). Isolation and structure determination of cryptophycins 38, 326, and 327 from the terrestrial cyanobacterium *Nostoc* sp. GSV 224. *J. Nat. Prod., 67*(8), 1403–1406.

Chang, T. T., More, S. V., Lu, I. H., Hsu, J. C., Chen, T. J., Jen, Y. C., et al., (2011). Isomalyngamide, A., A-1 and their analogs suppress cancer cell migration *in vitro. Eur. J. Med. Chem., 46*(9), 3810–3819.

Chen, Q. Y., Liu, Y., & Luesch, H., (2011). Systematic chemical mutagenesis identifies a potent novel apratoxin A/E hybrid with improved *in vivo* antitumor activity. *ACS Med. Chem. Lett., 2*(11), 861–865.

Chen, X. X., Smith, G. D., & Waring, P., (2003). Human cancer cell (Jurkat) killing by the cyanobacterial metabolite calothrixin A. *J. Appl. Phycol., 15*(4), 269–277.

D'Agostino, G., Del Campo, J., Mellado, B., Izquierdo, M. A., Minarik, T., Cirri, L., et al., (2006). A multicenter phase II study of the cryptophycin analog LY355703 in patients with platinum-resistant ovarian cancer. *Int. J. Gynecol. Cancer, 16*(1), 71–76.

Davidson, B. S., (1995). New dimensions in natural products research: Cultured marine microorganisms. *Curr. Opin. Biotechnol., 6*(3), 284–291.

De Caire, G. Z., De Cano, M. M. S., De Mule, M. C. Z., & De Halperin, D. R., (1993). Screening of cyanobacterial bioactive compounds against human pathogens. *Phyton., 54*, 59–65.

Dey, B., Lerner, D. L., Lusso, P., Boyd, M. R., Elder, J. H., & Berger, E. A., (2000). Multiple antiviral activities of cyanovirin-N: Blocking HIV type1 gp120 interaction with CD4 co receptor and inhibition of diverse enveloped viruses. *J. Virol., 74*(10), 4562–4569.

Doi, T., (2014). Synthesis of the biologically active natural product cyclodepsipeptides apratoxin A and its analogs. *Chem. Pharm. Bull., 62*(8), 735–743.

Edelman, M. J., Gandara, D. R., Hausner, P., Israel, V., Thornton, D., DeSanto, J., & Doyle, L. A., (2003). Phase 2 study of cryptophycin 52 (LY355703) in patients previously treated with platinum based chemotherapy for advanced non-small cell lung cancer. *Lung Cancer, 39*(2), 197–199.

Ehernich, I. M., Waterbury, J. B., & Webb, E. A., (2005). Distribution and diversity of natural product genes in marine and fresh water cyanobacterial cultures and genomes. *App. Env. Microbiol., 71*(11), 7401–7403.

Feldmann, S. C., Reynaldi, S., Stortz, C. A., Cerezo, A. S., & Damont, E. B., (1999). Antiviral properties of fucoidan fractions from *Leathesia difformis*. *Phytomedicine, 6*(5), 335–340.

Frankmölle, W. P., Larsen, L. K., Caplan, F. R., Patterson, G. M. L., Knübel, G., Levine, I. A., & Moore, R. E., (1991). Antifungal cyclic peptides from the terrestrial bluegreen alga *Anabaena laxa*. I. Isolation and biological properties. *J. Antibiot. (Tokyo), 45*(9), 1451–1457.

Gademann, K., & Portmann, C., (2008). Secondary metabolites from cyanobacteria: Complex structure and powerful bioactivities. *Curr. Org. Chem., 12*(4), 326–341.

Gerwick, W. H., Mrozek, C., Moghaddam, M. F., & Agarwal, S. K., (1989). Novel cytotoxic peptides from the tropical marine cyanobacterium *Hormothamnion enteromorphoides*. 1. Discovery, isolation, and initial chemical and biological characterization of the hormothamnins from wild and cultured material. *Experientia., 45*(2), 115–121.

Gerwick, W. H., Proteau, P. J., Nagle, D. G., Hamel, E., Blokhin, A., & Slate, D. L., (1994). Structure of curacin A, a novel antimitotic, antiproliferative and brine shrimp toxic natural product from the marine cyanobacterium *Lyngbya majuscula*. *J. Org. Chem., 59*(6), 1243–1245.

Ghasemi, Y., Yazdi, M. T., Shokravi, S., Soltani, N., & Zarrini, G., (2003). Antifungal and antibacterial activity of paddy-fields cyanobacteria from the north of Iran. *J. Sci. IRI, 14*(3), 203–209.

Golakoti, T., Ohtani, I., Patterson, G. M. L., Moore, R. E., Corbett, T. H., Valeriote, F. A., & Demchik, L., (1994). Total structures of cryptophycins, potent antitumor depsipeptides from the blue-green-alga *Nostoc* sp. strain GSV-224. *J. Am. Chem. Soc., 116*(11), 4729–4737.

Gu, L., Geders, T. W., Wang, B., Gerwick, W. H., Håkansson, K., Smith, J. L., & Sherman, D. H., (2007). GNAT-like strategy for polyketide chain initiation. *Science, 318*(5852), 970–974.

Guo, S., Tipparaju, S. K., Pegan, S. D., Wan, B., Mo, S., Orjala, J., Mesecar, A. D., et al., (2009). Natural product leads for drug discovery: Isolation, synthesis and biological evaluation of 6-cyano-5-methoxyindolo [2,3-a]carbazole based ligands as antibacterial agents. *Bioorg. Med. Chem., 17*(20), 7126–7130.

Gupta, C., Dhan, P., Amar, P., & Gupta, S., (2012). Why proteins: A novel source of bioceuticals. *Middle East J. Sci. Res., 12*(3), 365–375.

Gustafson, K. R., Cardellina, J. H., Fuller, R. W., Weislow, O. S., Kiser, R. F., Snader, et al., (1989). AIDS-antiviral sulfolipids from cyanobacteria (blue-green algae). *J. Natl. Cancer Inst., 81*(16), 1254–1258.

Gutierrez, M., Suyama, T. L., Engene, N., Wingerd, J. S., Matainaho, T., & Gerwick, W. H., (2008a). Apratoxin, D., a potent cytotoxic cyclodepsipeptide from Papua New Guinea collections of the marine cyanobacteria *Lyngbya majuscula* and *Lyngbya sordida. J. Nat. Prod., 71*(6), 1099–1103.

Gutierrez, R. M. P., Flores, A. M., Solis, R. V., & Jimenez, J. C., (2008b). Two new antibacterial norbietane diterpenoids from cyanobacterium. *Micrococcus lacustris. J. Nat. Med., 62*(3), 328–331.

Han, B. N., McPhail, K. L., Gross, H., Goeger, D. E., Mooberry, S. L., & Gerwick, W. H., (2005b). Isolation and structure of five lyngbyabellin derivatives from a Papua New Guinea collection of the marine cyanobacterium *Lyngbya majuscula. Tetrahedron, 61*(49), 11723–11729.

Han, B., Goeger, D., Maier, C. S., & Gerwick, W. H., (2005a). The wewakpeptins, cyclic depsipeptides from a Papua New Guinea collection of the marine cyanobacterium *Lyngbya semiplena. J. Org. Chem., 70*(8), 3133–3139.

Harrigan, G. G., Luesch, H., Yoshida, W. Y., Moore, R. E., Nagle, D. G., Paul, V. J., et al., (1998). Symplostatin 1: a dolastatin 10 analog from the marine cyanobacterium *Symploca hydnoides. J. Nat. Prod., 61*(9), 1075–1077.

Hoffmann, M., Blessing, J., & Lentz, S., (2003). A phase II trial of dolastatin-10 in recurrent platinum-sensitive ovarian carcinoma: A gynecologic oncology group study. *Gynecol. Oncol., 89*(1), 95–98.

Huskens, D., Férir, G., Vermeire, K., Kehr, J., Balzarini, J., Dittmann, E., & Schols, D., (2010). Microvirin, a novel alpha (1,2)-mannose-specific lectin isolated from *Microcystis aeruginosa*, has anti-HIV-1 activity comparable with that of cyanovirin-N but a much higher safety profile. *J. Biol. Chem., 285*(32), 24845–24854.

Ishibashi, M., Moore, R. E., Patterson, G. M. L., Xu, C., & Clardy, J., (1986). Scytophycins, cytotoxic and antimycotic agents from the cyanophyte *Scytonema pseudohofmanni. J. Org. Chem., 51*(26), 5300–5306.

Itoh, T., Tsuzuki, R., Tanaka, T., Ninomiya, M., Yamaguchi, Y., Takenaka, et al., (2013). Reduced scytonemin isolated from *Nostoc commune* induces autophagic cell death in human T-lymphoid cell line Jurkat cells. *Food Chem. Toxicol., 60*, 76–82.

Jemal, A., Seigal, R., Ward, E., Murray, T., Xu, J., & Thun, M. J., (2007). Cancer statistics. CA *Cancer. J. Clin., 57*(1), 43–66.

Jones, A. C., Monroe, E. A., Eisman, E. B., Gerwick, L., Sherman, D. H., & Gerwick, W. H., (2010). The unique mechanistic transformations involved in the biosynthesis of modular natural products from marine cyanobacteria. *Nat. Prod. Rep., 27*(7), 1048–1065.

Kajiyama, S., Kanzaki, H., Kawazu, K., & Kobayashi, A., (1998). Nostifungicidine, an antifungal lipopeptide from the field-grown terrestrial blue-green alga *Nostoc commune. Tetrahedron Lett., 39*(22), 3737–3740.

Kanekiyo, K., Lee, J. B., Hayashi, K., Takenaka, H., Hayakawa, Y., Endo, S., & Hayashi, T., (2005). Isolation of an antiviral polysacharide, nostoflan, from a terrestrial cyanobacteium, *Nostoc flagilliforme. J. Nat. Prod., 68*(7), 1037–1041.

Katoch, M., Mazmouz, R., Chau, R., Pearson, L. A., Pickford, R., & Neilan, B. A., (2016). Heterologous production of cyanobacterial mycosporine-like amino acids mycosporine-ornithine and mycosporine-lysine in *E. coli. Appl. Environ. Microbiol., 82*(20), 6167–6173.

Khan, Q. A., Lu, J., & Hecht, S. M., (2009). Calothrixins, a new class of human DNA topoisomerase I poisons. *J. Nat. Prod., 72*(3), 438–442.

Kim, H., Lantvit, D., Hwang, C. H., Kroll, D. J., Swanson, S. M., Franzblau, S. G., & Orjala, J., (2012). Indole alkaloids from two cultured cyanobacteria, *Westiellopsis* sp. and *Fischerella muscicola*. *Bioorg. Med. Chem., 20*(17), 5290–5295.

Klasse, P. J., Shattock, R., & Moore, J. P., (2008). Antiretroviral drug-based microbicides to prevent HIV-1 sexual transmission. *Ann. Rev. Med., 59*, 455–471.

Klein, D., Daloze, D., Braekman, J. C., Hoffmann, L., & Demoulin, V., (1995). New hapalindoles from the cyanophyte *Hapalosiphon laingii*. *J. Nat. Prod., 58*(11), 1781–1785.

Knübel, G., Larsen, L. K., Moore, R. E., Levine, I. A., & Patterson, G. M. L., (1990). Cytotoxic, antiviral indolocarbazoles from a blue-green alga belonging to the Nostocaceae. *J. Antibiot., 43*(10), 1236–1239.

Kobayashi, M., Natsume, T., Tamaoki, S., Watanabe, J. I., Asano, H., Mikami, T., et al., (1997). Antitumor activity of TZT-1027, a novel dolastatin 10 derivative. *Jpn. J. Cancer Res., 88*(3), 316–327.

Koglin, A., & Walsh, C. T., (2009). Structural insights into nonribosomal peptide enzymatic assembly lines. *Nat. Prod. Rep., 26*(8), 987–1000.

Lam, K. S., (2007). New aspects of natural products in drug discovery. *Trends Microbiol., 15*(6), 279–289.

Larsen, L. K., Moore, R. E., & Patterson, G. M. L., (1994). Beta-carbolines from the bluegreen alga *Dichothrix baueriana*. *J. Nat. Prod., 57*(3), 419–421.

Li, M., Patton, D. L., Cosgrove-Sweeney, Y., Ratner, D., Rohan, L. C., Cole, A. M., et al., (2011). Incorporation of the HIV-1 microbicide cyanovirin-N in a food product. *J. Acquir. Immune Defic. Syndr., 58*(4), 379–384.

Liang, J., Moore, R. E., Moher, E. D., Munroe, J. E., Al-Awar, R. S., Hay, D. A., et al., (2005). Cryptophycin-309249 and other cryptophycins analogs: Preclinical efficacy studies with mouse and human tumors. *Invest. New Drugs, 23*(3), 213–224.

Linington, R. G., Edwards, D. J., Shuman, C. F., McPhail, K. L., Matainaho, T., & Gerwick, W. H., (2008). Symplocamide, A., a potent cytotoxin and chymotrypsin inhibitor from the marine cyanobacterium *Symploca* sp. *J. Nat. Prod., 71*(1), 22–27.

Linington, R. G., Gonzalez, J., Urena, L. D., Romero, L. I., Ortega-Barria, E., & Gerwick, W. H., (2007). Venturamides A and B: Antimalarial constituents of the Panamanian marine cyanobacterium *Oscillatoria* sp. *J. Nat. Prod., 70*(3), 397–401.

Liu, X. J., & Chen, F., (2003). Cell differentiation and colony alteration of *Nostoc flagelliforme*, an edible terrestrial cyanobacterium in different liquid suspension culture. *Folia Microbiol., 48*, 619–625.

Loya, S., Reshef, V., Mizrachi, E., Silberstein, C., & Rachamim, Y., (1998). The inhibition of the reverse transcriptase of HIV-1 by the natural sulfoglycolipids from cyanobacteria: Contribution of different moieties to their high potency. *J. Nat. Prod., 61*(7), 891–895.

Luesch, H., Moore, R. E., Paul, V. J., Mooberry, S. L., & Corbett, T. H., (2001). Isolation of dolastatin 10 from the marine cyanobacterium *Symploca* sp. VP642 and total stereochemistry and biological evaluation of its analog symplostatin 1. *J. Nat. Prod., 64*(7), 907–910.

Luesch, H., Yoshida, W. Y., Moore, R. E., & Paul, V. J., (2002). New apratoxins of marine cyanobacterial origin from Guam and Palau. *Bioorg. Med. Chem., 10*(6), 1973–1978.

Luesch, H., Yoshida, W. Y., Moore, R. E., Paul, V. J., & Mooberry, S. L., (2000). Isolation, structure determination, and biological activity of lyngbyabellin a from the marine cyanobacterium *Lyngbya majuscula*. *J. Nat. Prod., 63*(5), 611–615.

MacMillan, J., Ernst-Russell, M. A., De Ropp, J. S., & Molinski, T. F., (2002). Lobocyclamides A-C, lipopeptides from a cryptic cyanobacterial mat containing *Lyngbya confervoides*. *J. Org. Chem., 67*(23), 8210–8215.

Maki, A., Diwakaran, H., Redman, B., Al-Asfar, S., Pettit, G. R., Mohammad, R. M., & Al-Katib, A., (1995). The bcl-2 and p53 oncoproteins can be modulated by bryostatin 1 and dolastatins in human diffuse large cell lymphoma. *Anticancer Drugs, 6*(3), 392–397.

Mansour, H. A., Shoman, S. A., & Kdodier, M. H., (2011). Antiviral effect of edaphic cyanophytes on rabies and herpes-1 viruses. *Acta Biol. Hung., 62*(2), 194–203.

Masuda, Y., Suzuki, J., Onda, Y., Fujino, Y., Yoshida, M., & Doi, T., (2014). Total synthesis and conformational analysis of apratoxin C. *J. Org. Chem., 79*(17), 8000–8009.

Mayer, A. M. S., Rodríguez, A. D., Berlinck, R. G. S., & Hamann, M. T., (2005). Marine pharmacology in 2005–6: Marine compounds with anthelmintic, antibacterial, anticoagulant, antifungal, anti-inflammatory, antimalarial, antiprotozoal, antituberculosis, and antiviral activities; affecting the cardiovascular, immune and nervous systems, and other miscellaneous mechanisms of action. *Biochem. Biophys. Acta, 1790*(5), 283–308.

McPhail, K. L., Correa, J., Linington, R. G., Gonzalez, J., Ortega-Barria, E., & Capson, T. L., (2007). Antimalarial linear lipopeptides from a Panamanian strain of the marine cyanobacterium *Lyngbya majuscula*. *J. Nat. Prod., 70*(6), 984–988.

Mita, A. C., Hammond, L. A., Bonate, P. L., Weiss, G., McCreery, H., Syed, S., et al., (2006). Phase I and pharmacokinetic study of tasidotin hydrochloride (ILX651), a third generation dolastatin-15 analogs, administered weekly for 3 weeks every 28 days in patients with advanced solid tumors. *Clin. Cancer Res., 12*(17), 5207–5215.

Moore, R. E., & Mynderse, J. S., (1982). *Majusculamide C*. US Patent 4342751.

Moore, R. E., Cheuk, C., & Patterson, G. M. L., (1984). Hapalindoles: New alkaloids from the blue-green alga *Hapalosiphon fontinalis*. *J. Am. Chem. Soc., 106*(21), 6456–6457.

Moore, R. E., Cheuk, C., Yang, X. Q. G., Patterson, G. M. L., Bonjouklian, R., Smitka, T. A., et al., (1987). Hapalindoles, antibacterial and antimycotic alkaloids from the cyanophyte *Hapalosiphon fontinalis*. *J. Org. Chem., 52*(6), 1036–1043.

Morgan, R. L., Baack, B., Smith, B. D., Yartel, A., Pitasi, M., & Falck-Ytter, Y., (2013). Eradication of hepatitis C virus infection and the development of hepatocellular carcinoma: A meta-analysis of observational studies. *Ann. Intern. Med., 158*, 329–337.

Mundt, S., Kreitlow, S., & Jansen, R., (2003). Fatty acids with antibacterial activity from the cyanobacterium *Oscillatoria redekei* HUB051. *J. Appl. Phycol., 15*(2–3), 263–267.

Natsume, T., Watanabe, J. I., Koh, Y., Fujio, N., Ohe, Y., & Horiuchi, T., (2003). Antitumor activity of TZT-1027 (Soblidotin) against vascular endothelial growth factor secreting human lung cancer *in vivo*. *Cancer Sci., 94*(9), 826–833.

Neuhof, T., Dieckmann, R., Von, D. H., Preussel, K., Seibold, M., & Schmieder, P., (2006a). *Lipopeptides Having Pharmaceutical Activity*. Patent WO 2006/092313 A1.

Neuhof, T., Schmieder, P., Preussel, K., Dieckmann, R., Pham, H., Bartl, F., & Von, D. H., (2005). Hassallidin A, a glycosylated lipopeptide with antifungal activity from the cyanobacterium *Hassallia* sp. *J. Nat. Prod., 68*(5), 695–700.

Neuhof, T., Schmieder, P., Seibold, M., Preussel, K., & Von, D. H., (2006b). Hassallidin B-second antifungal member of the Hassallidin family. *Bioorg. Med. Chem. Lett., 16*(16), 4220–4222.

Newman, D. J., & Cragg, G. M., (2017). Current status of marine-derived compounds as warheads in anti-tumor drug. *Mar. Drugs, 15*(4), 99.

Newman, D. J., Cragg, G. M., & Snader, K. M., (2003). Natural products as sources of new drugs over the period 1981–2002. *J. Nat. Prod., 66*(7), 1022–1037.

Niedermeyer, T., & Brönstrup, M., (2012). Natural-product drug discovery from microalgae. In: Posten, C., & Walter, C., (eds.), *Microalgal Biotechnology: Integration and Economy* (pp. 169–200). De Gruyter, Berlin, Boston.

Nogle, L. M., & Gerwick, W. H., (2002). Somocystinamide, A., a novel cytotoxic disulfide dimer from a Fijian marine cyanobacterial mixed assemblage. *Org. Lett., 4*(7), 1095–1098.

Nogle, L. M., Okino, T., & Gerwick, W. H., (2001). Antillatoxin, B., a neurotoxic lipopeptide from the marine cyanobacterium *Lyngbya majuscula. J. Nat. Prod., 64*(7), 983–9985.

O'Keefe, B. R., Smee, D. F., Turpin, J. A., Saucedo, C. J., Gustafson, K. R., Mori, T., et al., (2003). Potent anti-influenza activity of cyanovirin-N and interactions with viral hemagglutinin. *Antimicrob. Agents Chemother., 47*(8), 2518–2525.

Ongley, S. E., Bian, X., Zhang, Y., Chau, R., Gerwick, W. H., Müller, R., & Neilan, B. E., (2013). High-titer heterologous production in *E. coli* of lyngbyatoxin, a protein kinase C activator from an uncultured marine cyanobacterium. *ACS Chem. Biol., 8*(9), 1888–1893.

Patterson, G. M. L., & Bolis, C. M., (1997). Fungal cell-wall polysaccharides elicit an antifungal secondary metabolite (phytoalexin) in the cyanobacterium *Scytonema ocellatum. J. Phycol., 33*(1), 54–60.

Patterson, G. M. L., & Carmeli, S., (1992). Biological effects of tolytoxin (6-hydroxy-7-O-methyl-scytophycin b), a potent bioactive metabolite from cyanobacteria. *Arch. Microbiol., 157*(5), 406–410.

Patterson, G. M. L., Baldwin, C. L., Bolis, C. M., Caplan, F. R., Karuso, H., & Larsen, L. K., (1991). Antineoplastic activity of cultured blue-green algae (Cyanophyta). *J. Phycol., 27*(4), 530–536.

Patterson, G. M. L., Smith, C. D., Kimura, L. H., Britton, B. A., & Carmeli, S., (1993). Action of tolytoxin on cell morphology, cytoskeletal organization, and actin polymerization. *Cell Motil. Cytoskeleton, 24*(1), 39–48.

Pettit, G. R., Hogan, F., Xu, J. P., Tan, R., Nogawa, T., Cichacz, Z., et al., (2008). Antineoplastic agents. 536. New sources of naturally occurring cancer cell growth inhibitors from marine organisms, terrestrial plants, and microorganisms. *J. Nat. Prod., 71*(3), 438–444.

Portmann, C., Blom, J. F., Kaiser, M., Brun, R., Jüttner, F., & Gademann, K., (2008). Isolation of aerucyclamides C and D and structure revision of microcyclamide 7806A: Heterocyclic ribosomal peptides from *Microcystis aeruginosa* PCC 7806 and their antiparasite evaluation. *J. Nat. Prod., 71*(11), 1891–1896.

Prinsep, M. R., Caplan, F. R., Moore, R. E., Patterson, G. M., & Smith, C. D., (1992). Tolyporphin, a novel multidrug resistance-reversing agent from the blue-green alga *Tolypothrix nodosa. J. Am. Chem. Soc., 114*(1), 385–387.

Priyadharshini, R., Ambikapathy, V., & Pavai, T., (2013). *In vitro* antimicrobial activity of *Oscillatoria angustissima. Int. J. Adv. Res., 1*(4), 60–68.

Proteau, P. J., Gerwick, W. H., Garcia-Pichel, F., & Castenholz, R., (1993). The structure of scytonemin, an ultraviolet sunscreen pigment from the sheaths of cyanobacteria. *Experientia, 49*(9), 825–829.

Ramaswamy, A. V., Sorrels, C. M., & Gerwick, W. H., (2007). Cloning and biochemical characterization of the hectochlorin biosynthetic gene cluster from the marine cyanobacterium *Lyngbya majuscula. J. Nat. Prod., 70*(12), 1977–1986.

Raveh, A., & Carmeli, S., (2007). Antimicrobial ambiguines from the cyanobacterium *Fischerella* sp. collected in Israel. *J. Nat. Prod., 70*(2), 196–201.

Rechter, S., König, T., Auerochs, S., Thulke, S., Walter, H., Dörnenburg, H., et al., (2006). Antiviral activity of *Arthrospira*-derived spirulan-like substances. *Antiviral Res., 72*(3), 197–206.

Reshef, V., Mizrachi, E., Maretzki, T., Silberstein, C., Loya, S., Hizi, A., & Carmeli, S., (1997). New acylated sulfoglycolipids and digalactolipids and related known glycolipids from cyanobacteria with a potential to inhibit the reverse transcriptase of HIV-1. *J. Nat. Prod., 60*(12), 1251–1260.

Rickards, R. W., Rothschild, J. M., Willis, A. C., De Chazal, N. M., Kirk, J., Kirk, K., et al., (1999). Calothrixins A and B., novel pentacyclic metabolites from *Calothrix* cyanobacteria with potent activity against malaria parasites and human cancer cells. *Tetrahedron, 55*(47), 13513–13520.

Salvador, L. A., Paul, V. J., & Luesch, H., (2010). Caylobolide, B., a macrolactone from symplostatin 1-producing marine cyanobacteria *Phormidium* spp. from Florida. *J. Nat. Prod., 73*(9), 1606–1609.

Schwarjer, D., Finking, R., & Marachiel, M. A., (2003). Nonribosomal peptides: From genes to products. *J. Nat. Prod. Rep., 20*(3), 275–287.

Senthil, K. N. S., Sivasubramanian, V., & Mukund, S., (2013). Antimicrobial and antifungal activity of extracts of *Phormidium fragile* Gomont. *J. Algal Biomass Utln., 4*(1), 66–71.

Shin, C., & Teicher, B. A., (2001). Cryptophycins: A novel class of potent antimititotic antitumor depsipeptide. *Curr. Pharm. Des., 13*, 1259–1276.

Simmons, T. L., McPhail, K. L., Ortega-Barria, E., Mooberry, S. L., & Gerwick, W. H., (2006). Belamide, A., a new antimitotic tetrapeptide from a Panamanian marine cyanobacterium. *Tetrahedron Lett., 47*(20), 3387–3390.

Singh, R. K., Tiwari, S. P., Rai, A. K., & Mohapatra, T. M., (2011). Cyanobacteria: An emerging source for drug discovery. *J. Antibiot., 64*(6), 401–412.

Singh, R. S., Walia, A. K., Khattar, J. S., Singh, D. P., & Kennedy, J. F., (2017). Cyanobacterial lectins characteristics and their role as antiviral agents. *Int. J. Biol. Macromol., 102*, 475–496.

Sisay, M. T., Hautmann, S., Mehner, C., Konig, G. M., Bajorath, J., & Gutschow, M., (2009). Inhibition of human leukocyte elastase by brunsvicamides A–C: Cyanobacterial cyclic peptides. *Chem. Med. Chem. 4*(9), 1425–1429.

Smith, C. D., Carmeli, S., Moore, R. E., & Patterson, G. M. L., (1993). Scytophycins, novel microfilament-depolymerizing agents which circumvent P-glycoprotein-mediated multidrug resistance. *Cancer Res., 53*(6), 1343–1347.

Smith, C. D., Zhang, X., Mooberry, S. L., Patterson, G. M. L., & Moore, R. E., (1994). Cryptophycin: A new antimicrotubule agent active against drug-resistant cells. *Cancer Res., 54*(14), 3779–3784.

Smitka, T. A., Bonjouklian, R., Doolin, L., Jones, N. D., Deeter, J. B., Yoshida, W. Y., et al., (1992). Ambiguine isonitriles, fungicidal hapalindole-type alkaloids from three genera of blue-green algae belonging to Stigonemataceae. *J. Org. Chem., 57*(3), 857–861.

Soria-Mercado, I. E., Pereira, A., Cao, Z., Murray, T. F., & Gerwick, W. H., (2009). Alotamide, A., a novel neuropharmacological agent from the marine cyanobacterium *Lyngbya bouillonii*. *Org. Lett., 11*(20), 4704–4707.

Stevenson, C. S., Capper, E. A., Roshak, A. K., Marquez, B., Grace, K., Gerwick, W. H., et al., (2002). Scytonemin - a marine natural product inhibitor of kinases key in hyper proliferative inflammatory diseases. *Inflamm. Res., 51*(2), 112–114.

Tan, L. T., (2013). Marine cyanobacteria: A prolific source of bioactive natural products as drug leads. In: Kim, S. K., (ed.), *Marine Microbiology: Bioactive Compounds and*

Biotechnological Applications (pp. 59–81). Wiley-VCH Verlag GmbH and Co., KGaA, Weinheim.

Teruya, T., Sasaki, H., Kitamura, K., Nakayama, T., & Suenaga, K., (2009). Biselyngbyaside, a macrolide glycoside from the marine cyanobacterium *Lyngbya* sp. *Org. Lett., 11*(11), 2421–2424.

Thornburg, C. C., Cowley, E. S., Sikorska, J., Shaala, L. A., Ishmael, J. E., Youssef, D. T., & McPhail, K. L., (2013). Apratoxin H and apratoxin A sulfoxide from the red sea cyanobacterium *Moorea producens. J. Nat. Prod., 76*(9), 1781–1788.

Turk, B., (2006). Targeting proteases: Successes, failures and future prospects. *Nat. Rev. Drug. Discov., 5*(9), 785–799.

Vaishampayan, U., Glode, M., Du, W., Kraft, A., Hudes, G., Wright, J., & Hussain, M., (2000). Phase II Study of dolastatin 10 in patients with hormone refractory metastatic prostate adenocarcinoma. *Clin. Cancer Res., 6*(11), 4205–4208.

Van, D. M. F. J. U. M., De Haan, C. A. M., Schuurman, N. M. P., Haijema, B. J., Peumans, W. J., Van, D. E. J. M., et al., (2007). Antiviral activity of carbohydrate-binding agents against Nidovirales in cell culture. *Antiviral Res., 76*(1), 21–29.

Verdier-Pinard, P., Lai, J. Y., Yoo, H. D., Yu, J., Marquez, B., Nagle, D. G., et al., (1998). Structure-activity analysis of the interaction of curacin A, the potent colchicine site antimitotic agent, with tubulin and effects of analogs on the growth of MCF-7 breast cancer cells. *Mol. Pharmacol., 53*(1), 62–76.

Vestola, J., Shishido, T. K., Jokela, J., Fewer, D. P., Aitio, O., Permi, P., et al., (2014). Hassallidins, antifungal glycolipopeptides, are widespread among cyanobacteria and are the end-product of a nonribosomal pathway. *Proc. Natl. Acad. Sci. USA, 111*(18), E1909–E1917.

Vijayakumar, S., & Menakha, M., (2015). Pharmaceutical applications of cyanobacteria: A review. *J. Acute Med., 5*(1), 15–23.

Wan, F., & Erickson, K. L., (1999). Serinol-derived malyngamides from an Australian cyanobacterium. *J. Nat. Prod., 62*(12), 1696–1699.

Wang, H., Liu, Y., Gao, X., Carter, C. L., & Liu, Z. R., (2007). The recombinant b subunit of c-phycocyanin inhibits cell proliferation and induces apoptosis. *Cancer Lett., 247*(1), 150–158.

Wase, N. V., & Wright, P. C., (2008). Systems biology of cyanobacterial secondary metabolite production and its role in drug discovery. *Expert Opin. Drug Discov., 3*(8), 903–929.

Watanabe, J., Minami, M., & Kobayashi, M., (2006). Antitumor activity of TZT-1027 (soblidotin). *Anticancer Res., 26*(3A), 1973–1981.

Wrasidlo, W., Mielgo, A., Torres, V. A., Barbero, S., Stoletov, K., Suyama, T. L., et al., (2008). The marine lipopeptide somocystinamide A triggers apoptosis via caspase 8. *Proc. Natl. Acad. Sci. USA, 105*(7), 2313–2318.

Xiong, C., O'Keefe, B. R., Byrd, R. A., & McMohan, J. B., (2006). Potent anti-HIV activity of scytovirin domain 1 peptide. *Peptides, 27*(7), 1668–1675.

Zainuddin, E. N., Mentel, R., Wray, V., Jansen, R., Nimtz, M., Lalk, M., & Mundt, S., (2007). Cyclic depsipeptides, ichthyopeptins A and, B., from *Microcystis ichthyoblabe. J. Nat. Prod., 70*(7), 1084–1088.

Zheng, W., Chen, C., Cheng, Q., Wang, Y., & Chu, C., (2006). Oral administration of exopolysaccharide from *Aphanothece halophytica* (Chroococcales) significantly inhibits influenza virus (H1N1)-induced pneumonia in mice. *Int. Immunopharmacol., 6*(7), 1093–1099.

CHAPTER 9

EVALUATION OF METHODS OF BIOMASS RECOVERY AND LIPID EXTRACTION FOR MICROALGAE

MARIANA LARA MENEGAZZO,[1,2]
JANE MARY LAFAYETTE NEVES GELINSKI,[3] and
GUSTAVO GRACIANO FONSECA[2,3]

[1]Faculty of Engineering, Federal University of Grande Dourados, Dourados, MS, Brazil

[2]Laboratory of Bioengineering, Faculty of Biological and Environmental Sciences, Federal University of Grande Dourados, Dourados, MS, Brazil

[3]Center of Biotechnology, Postgraduate Program in Science and Biotechnology, University of West of Santa Catarina (UNOESC), Videira-SC, Brazil

9.1 INTRODUCTION

Microalgae are single-celled organisms that are found in aquatic systems with a great diversity of forms, traits, and ecological features. They can be economically exploited beneath numerous aspects, including food production, medicinal drugs, and biofuels (Markou and Nerantzis, 2013; Bellou et al., 2014; Klok et al., 2014). Microalgae can develop in autotrophic, heterotrophic, or mixotrophic processes. The latter utilizes both luminosity and organic compounds as energy sources, as well as CO_2 and organic compounds as carbon sources (Liu et al., 2011; Abreu et al., 2012).

Microalgae is as a potential energy source due to the reduced use of water in the cultivation; the high biomass yield per area in relation to agricultural cultures; the opportunity of the usage of non-agricultural areas for cultivation; use of agro-industrial residues as a supply of nutrients (Demirbas et al., 2010; Iyovo et al., 2010; Scott et al., 2010).

The microalgae biomass contains carbohydrates, proteins, lipids, pigments, among others (Behrens and Kyle, 1996; Silva et al., 2014). Each species of microalga synthetizes different degrees of these compounds and has the ability to adjust its metabolism according to the shifts in the chemical composition of the growth medium and other culture conditions (Chisti, 2007).

Due to its capability to synthesize lipids, it has been evaluated as raw material for the generation of products with high added value (Mata et al., 2010; Scott et al., 2010; Silva et al., 2014). The use of microalgae biomass represents as great defiance the choice of an efficient strategy for biomass recuperation and lipid extraction, since the scheduling of these processes can be crucial, requiring the development of an energetically advantageous, ecofriendly, and economically feasible process (Chisti, 2007; Halim et al., 2012a).

The major physical and chemical factors that have an effect on microalgae development are luminosity, temperature, pH, salinity, and availability of nutrients disponible (Piorreck and Pohl, 1984; Richmond, 2004; Makareviciene et al., 2011; Markou and Nerantzis, 2013; Klok et al., 2014). No single growth medium is appropriate for all species of microalgae, since each one has its particular needs. Thus, the biochemical composition of microalgae depends on the employed conditions (Stefanov et al., 1988; Richmond, 2004; Ryckebosch et al., 2014c; Soares et al., 2014; Viegas et al., 2015), including temperature, photoperiod, nutrients, amongst others. Here we reviewed the principal biomass recovery processes and the cellular disruption methods for lipid extraction in order to evaluate their influence on the nature, concentration, and yields of the obtained lipids and fatty acids.

9.2   CULTURE CONDITIONS AND THEIR INFLUENCE ON LIPID CONTENT

The indispensable nutrients for microalgae development are carbon, nitrogen, phosphorus, and microelements (Lourenço, 2006; Abreu et al., 2012; Benemann, 2013), besides temperature, luminosity, agitation/aeration, among others (Raven and Geider, 1988; Richmond, 2004; Breuer et al., 2013b). Carbon is critical for the synthesizing of organic molecules by the cell, including carbohydrates, proteins, lipids, nucleic acids, and vitamins (Lourenço, 2006). Nitrogen also displays an important contribution in the constitution of proteins. It augments the contents of proteins, carotenoids, and chlorophyll, but it increases the lipid content of the microalga if under limitation (Mata et al., 2010; Chen et al., 2011; Abreu et al., 2012; Breuer et al., 2012; Benemann, 2013; Rawat et al., 2013; Viegas et al., 2015).

Temperature straightly affects microalgae metabolism (Breuer et al., 2013b). The response depends on the strain (Raven and Geider, 1988; Breuer et al., 2013b). Increased temperature may augment protein content and diminish carbohydrate and lipid contents. The amounts of saturated fatty acids tend to increase and the unsaturated ones to decrease in the lipid fraction (Mortensen et al., 1988).

Microalgae growth also depends on the intensity of light, wavelength, and photoperiod to which the cells are exposed. The photosynthetic activity increases if irradiation augments to certain values, before the onset of cell growth inhibition. The synthesis of polar lipids can be triggered out by low light intensity. The intensity of light affects the nature of fatty acids: high luminous intensity favors the formation of saturated and mono-unsaturated fatty acids (Vega and Voltolina, 2007).

Aeration and agitation homogenize the growth medium, by diffusing the carbon source, avoiding self-flocculation. The presence of light induces the consumption of the CO_2, increasing the pH. However, the occurrence of CO_2 can diminish the pH and inhibit the growth of some microalgae strains (Richmond, 2004; Vega and Voltolina, 2007).

Some cultivation conditions affect straightly on the lipid composition and fatty acids profile of microalgae, especially in relation to their nitrogen content, temperature, and luminosity (Thompson, 1996; Frumento et al., 2013; Sibi et al., 2015). These stress conditions are utilized to induce the production of lipids or other molecules of commercial interest (Chen et al., 2011; Wang et al., 2014).

Nutrient deficiency may reduce growth rates, provoking a continuous and active synthesis of fatty acids by some strains (Thompson, 1996) that if not utilized in the formation of membrane lipids would result in the accumulation of triglycerides, as energy storage (Guschina and Harwood, 2006). Fatty acids' content and the conversion into triglycerides depend on the microalgae strain, and the growth conditions, including temperature, aeration, luminosity, nutrients, and culture age (Dunstan et al., 1993; Guschina and Harwood, 2006) (Figure 9.1).

9.3 SEPARATION OF MICROALGAE BIOMASS

Separation of microalgae biomass begins with the harvesting from their growth medium (Salim et al., 2011; Granados et al., 2012). The separation of the biomass is difficulted by the microscopic size; the low cell concentration (Grima et al., 2003; Derner et al., 2006); the density similar to water; the low surface charge

(Banerjee et al., 2013); the low ionic strength (Ndikubwimana et al., 2016); and the growth phase (Vandamme et al., 2010). The biomass recovery straightly influences on the cost and quality of the bioproducts (Borges et al., 2016).

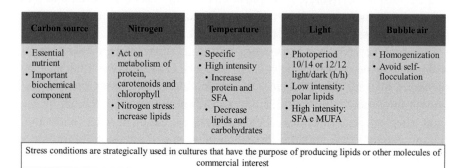

Carbon source	Nitrogen	Temperature	Light	Bubble air
• Essential nutrient • Important biochemical component	• Act on metabolism of protein, carotenoids and chlorophyll • Nitrogen stress: increase lipids	• Specific • High intensity • Increase protein and SFA • Decrease lipids and carbohydrates	• Photoperiod 10/14 or 12/12 light/dark (h/h) • Low intensity: polar lipids • High intensity: SFA e MUFA	• Homogenization • Avoid self-flocculation
Stress conditions are strategically used in cultures that have the purpose of producing lipids or other molecules of commercial interest				

FIGURE 9.1   Microalgae cultivation: elements and influences.

Thickening methods, including gravimetric sedimentation, coagulation/flocculation, flotation or electroflotation, and biomass dewatering methods, including filtration and centrifugation, are utilized to augment biomass concentration and diminish the volume to be processed, respectively (Chen et al., 2011; Christenson and Sims, 2011; Milledge and Heaven, 2013). Drying methods, including solar drying, greenhouse drying, lyophilization, or spray drying, are required to obtain a dried biomass.

It does not exist an ordinary, simple, and low-cost method to be utilized on a large scale (Salim et al., 2011; Liu et al., 2013; Borges et al., 2016; Chatsungnoen and Chisti, 2016a). So, the development of separation processes to intensify recovery efficiencies of microalgal biomass is fundamental to attain the economic viability of the obtained bioproducts (Figure 9.2).

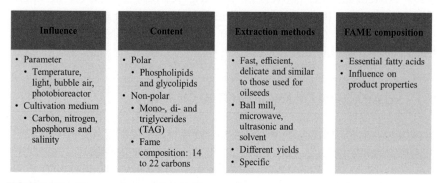

Influence	Content	Extraction methods	FAME composition

• Parameter
  • Temperature, light, bubble air, photobioreactor
• Cultivation medium
  • Carbon, nitrogen, phosphorus and salinity
	• Polar
	  • Phospholipids and glycolipids
	• Non-polar
	  • Mono-, di- and triglycerides (TAG)
	  • Fame composition: 14 to 22 carbons
		• Fast, efficient, delicate and similar to those used for oilseeds
		• Ball mill, microwave, ultrasonic and solvent
		• Different yields
		• Specific
			• Essential fatty acids
			• Influence on product properties

FIGURE 9.2   Process of microalgae lipid extraction.

9.3.1 BIOMASS THICKENING METHODS

Thickening methods augment the biomass concentration and diminish the volume to be processed, saving energy (Barros et al., 2015). Gravimetric sedimentation is a process of high energy efficiency. It is determined by the size and density of the cells allied to the sedimentation rate. Despite slow, this method is the most employed due to the large volumes treated and the low value of the formed biomass (Shelef and Sukenik, 1984; Brennan and Owende, 2010; Pragya et al., 2013). Microalgae debris with low density do not settle properly, therefore they are scarcely removed by sedimentation and can dissipate (Papazi et al., 2010; Uduman et al., 2010; Morioka et al., 2014). However, the method can be enhanced with the use of lamella separators and sedimentation ponds (Barros et al., 2015). Despite trustworthy and less expensive, industries do not widely use sedimentation ponds because it is a leisurely method to concentrate cells (Milledge and Heaven, 2013; Barros et al., 2015).

Self-flocculation is the natural aggregation of the debris before microalgae sedimentation (Şirin et al., 2012; González-Fernández and Ballesteros, 2013). Limitation of carbon or of certain abiotic components may arouse self-flocculation (Pragya et al., 2013; Vandamme et al., 2013). It is not found in all microalgae strains. It is restricted by the sluggish redispersion of the flakes that may arise (Papazi et al., 2010; González-Fernández and Ballesteros, 2013; Morioka et al., 2014). The pH increase may trigger out sedimentation. This pH shift is less subject to interfere in the growth medium compared to the flocculating agents, permitting the medium to be reused (Uduman et al., 2010; Chen et al., 2011). This process is likewise considered self-flocculation.

In general, higher recovery of microalgae is obtained with the action of a pH-increasing agent and the sedimentation time, turning the choice of this method strongly dependent on the specificity of the microalgae, their interaction among the debris, and the pH setting threshold. Literature reports, for example, biomass recovery by self-flocculation with pH modify for *Chlorella vulgaris* (Vandamme et al., 2012) and *Chaetoceros calcitrans*, and sedimentation for *C. calcitrans* (Harith et al., 2009), *C. vulgaris*, and *Neochloris oleoabundans* (Salim et al., 2012). The chemical agent affects the lipid concentration and the nature of fatty acids, with lipid decrease and reduction of polyunsaturated fatty acids (PUFAs) (Borges et al., 2011).

Sedimentation and self-flocculation methods are not adequate to all microalgae species since, for example, cells that form large colonies tend to sediment (Brennan and Owende, 2010; Pragya et al., 2013). Self-flocculation has a higher influence on the recuperation of marine compared to freshwater microalgae (Salim et al., 2012; González-Fernández and Ballesteros, 2013; Barros et al., 2015).

Flotation is frequently described as inverted sedimentation or flocculation in the opposite. It treats vast volumes and demands little space, time, and equipment (Pragya et al., 2013; Barros et al., 2015). Some species are able to float naturally as the lipid content augments. This technique is promising for freshwater microalgae whilst for marine microalgae; flotation can be prejudiced because salinity is a determining for cell adhesion to the bubble (Shelef and Sukenik, 1984; Coward et al., 2014; Barros et al., 2015). This system retains microalgae more successfully than by sedimentation, but large bubbles of dissolved air can disrupt the flakes formed (Shelef and Sukenik, 1984; Christenson and Sims, 2011; Pragya et al., 2013). The dispersed bubbles are produced by an air injection system equipped with a high-speed stirrer, or via continuous air passing through a porous material (Pragya et al., 2013; Yap et al., 2014; Barros et al., 2015; Laamanen et al., 2016). In electrolytic flotation, the bubbles are small and produced by electrolysis. This method requires severe energetic use (Shelef and Sukenik, 1984), therefore is usually not the first choice for the microalgae recovery (Uduman et al., 2010; Barros et al., 2015).

In the flotation, the chemical coagulation can be utilized through surfactants proceeded by flotation of air to augment the yields of the process (Brennan and Owende, 2010; Uduman et al., 2010; Barros et al., 2015). The most often utilized surfactants are aluminum sulfate ($Al_2(SO_4)_3$), iron sulfate ($Fe_2(SO_4)_3$), cetyltrimethylammonium bromide (CTAB), chitosan, and ferric chloride ($FeCl_3$) (Gao et al., 2010; Hanotu et al., 2012; Zhang et al., 2016).

The flotation performance diminishes with high ionic strength. The electrostatic interaction between the collector and the cellular surface displays a crucial role in the separation strategies. Based on bubble generation techniques and process ratios, various units were built to trigger out flotation (Liu et al., 1999; Ndikubwimana et al., 2016). The advantages of flotation include considerable recovery efficiency of microalgae using several chemical conditions and/or agents combined with airflow to guarantee recovery efficiency (Gao et al., 2010; Hanotu et al., 2012; Ndikubwimana et al., 2016; Zhang et al., 2016). The facility of operation and the capability to process large amounts of microalgae cultures at minimum cost turn flotation a potential technology for microalgae recovery (Shelef and Sukenik, 1984; Liu et al., 1999; Hanotu et al., 2012; Ndikubwimana et al., 2016).

Studies and applications of flotation processes in microalgae harvesting are still incipient. Further researches are essential to optimize operational variables so one can conduct process scale-up (Liu et al., 1999; Hanotu et al., 2012; Ndikubwimana et al., 2016). Flocculation is broadly used in varied industrial processes, including beer production, mining, and water and effluent treatments (Chen et al., 2011). It takes place when smaller particles assemble into larger particles through the interaction of coagulant or floccu-lating agents that decant by sedimentation with the time. It can be executed by using traditional methods, including bioflocculation and chemical floc-culation, and by novel technologies, for example, magnetic nanoparticles (Lee et al., 2009; Zhang and Chen, 2015; Chatsungnoen and Chisti, 2016a). Flocculation for microalgae recovery and for water treatment are similar processes (Brennan and Owende, 2010; Banerjee et al., 2013; Borges et al., 2016). The choice of the coagulating/flocculant agent must take into consideration its degree of interference in the processing, the purpose of the biomass obtained, the efficiency at low concentration, and its cost.

In chemical flocculation, the main agents are metal salts, including $Al_2(SO_4)_3$ and $FeCl_3$ and organic polymers including chitosan and cationic starch (Papazi et al., 2010; Granados et al., 2012; Wyatt et al., 2012). Metal salts can be applied for microalga recovery, but their utilization may result in elevated concentrations of metals in the obtained biomass and remain in the cellular residue after extraction of lipids or carotenoids (Rwehumbiza et al., 2012). Nevertheless, contamination of lipid and fatty acid contents is not detected (Salim et al., 2011; Borges et al., 2016). Chitosan and modi-fied starch are organic polymer flocculants reported to have an acceptable recovery of microalgae with a reduced dosage and lower impact, when compared to metal salts (Vandamme et al., 2010; Kim et al., 2013a; Roselet et al., 2016).

Commercial flocculants are an alternative to the recovery of micro-algae biomass. To guarantee the maximum efficiency of the flocculants, it is frequently necessary to adjust the pH by using alkalinizing agents, for example, NaOH, which favors the biomass recovery by flocculation simi-larly to the cationic and anionic flocculants, with increased costs (Coward et al., 2014; Borges et al., 2016). Furthermore, there is also the influence of the species and the growth parameters on the lipid content. The floccula-tion accomplished only with the pH adjustment, i.e., by the alkalinization to pH 10, becomes somewhat more advantageous. The pH is neutralized by washing cells with a NH_4HCO_2 solution after the coagulation, returning to 8. So, the inorganic salts can be removed without altering osmotic pressure, maintaining the flakes, and the lipid content (Borges et al., 2011). Bioflocculation is the assistance of some microorganism, including bacteria or microalgae able to flocculate (Vandamme et al., 2010), without flocculant or coagulant addition. The bioflocculation with bacteria needs the addition of a substrate for bacterial growth (Salim et al., 2011, 2012).

Flocculation is better than the conventional microalgae recovery methods because it permits the treatment of huge volumes of a wide range of microalgae species. Literature reports, for example, flocculation followed by sedimentation of *Chlorella minutissima* (Papazi et al., 2010), *Chlorella sorokiniana* (Xu et al., 2013), *C. vulgaris* (Ras et al., 2011; Rashid et al., 2013), *Conticribra weissflogi* (König et al., 2014), and self-flocculation of *C. vulgaris, Neocloris oleoabundans* (Salim et al., 2012), and *Nannochloropsis oceanica* (Wan et al., 2013). Recovery efficiency can be augmented when flocculation is combined with gravimetric methods (Shelef and Sukenik, 1984; Uduman et al., 2010; Pragya et al., 2013; Barros et al., 2015).

The level of interference in the processing and utilization of the accumulated biomass, the effectiveness at low concentration, and its cost are dependent of the microalga specificity, and are directly evolved in the choice of the flocculating agent (Vandamme et al., 2010). In this sense, the goal is to find a recovery system in consonance with the requirements for the utilization of a high-quality biomass using a few energetic resources.

9.3.2   METHODS OF BIOMASS DEWATERING

The dewatering of densed microalgae biomass can be performed by using mechanical processes, including centrifugation and filtration. After the dewatering, the obtained biomass is generally dried to enhance the efficiency of the downstream processes, including extraction of lipids or carbohydrates (Shelef and Sukenik, 1984; Barros et al., 2015). These processes can be carried out for thickening, but the efficiency is diminished.

Centrifugation is considered a derivation of gravity sedimentation where gravitational acceleration is substituted by centrifugal acceleration, moving the cells through the liquid before settle to the bottom or sides of the vessel (Shelef and Sukenik, 1984; Pragya et al., 2013; Japar et al., 2017). This process efficiently recovers the biomass from most of the microalgae species. However, it requires high energy consumption and costs (Shelef and Sukenik, 1984; Dassey and Theegala, 2013; Pragya et al., 2013; Chatsungnoen and Chisti, 2016b; Japar et al., 2017), which may restrict its use to high added value products, including unsaturated fatty acids, pharmaceuticals, and others, or laboratorial experiments (Barros et al., 2015). Centrifuging is utilized as a dewatering method to the microalgae-thickened biomass (Uduman et al., 2010; Grima et al., 2003; Milledge and Heaven, 2013). Then, the water is separated by draining the supernatant medium.

Centrifugation can be carried out by using two types of equipment: (a) fixed wall systems, for example, hydrocyclone, or (b) rotary wall systems, including centrifugal decanters, tubular centrifuges, and disc centrifuges. The hydrocyclone comprises of a cylinder where the microalgae culture is feed upper tangentially, causing a downward spiral movement that drags the larger and heavier particles to the lower outlet of the equipment. The disc centrifuge comprises of a shallow cylindrical vessel containing a stack of closely spaced rotating metal cones (disks) by where the microalgae culture flows from the center of the disc stack, in such a way that the biomass moves outward, at the bottom, while the aqueous phase moves to the center (Shelef and Sukenik, 1984; Milledge and Heaven, 2013). Tubular centrifuges are used to recover microalgae at small scales. As they do not have a draining system, it is necessary a process interruption for release the liquid. The decanter centrifuges consist of a horizontal conical vessel where the separation occurs by specific weight difference, with the biomass being dislocated to the sides and removed by a helical thread (Shelef and Sukenik, 1984). This configuration of centrifuge consumes more energy than the disk one (Milledge and Heaven, 2013).

Depending on the microalgae strain, and the type and speed of the centrifuge utilized, the biomass recovery may be fast and high (Heasman et al., 2000). Literature reports studies of biomass recovery by centrifugation including, for example, the species *Chlorella* sp. (Ahmad et al., 2014), *Nannochloris* sp. (Dassey and Theegala, 2013), *Nannochloropsis oculata*, *Pavlova lutheri* (Heasman et al., 2000), and *Scenedesmus obliquus* (Anthony et al., 2013). This method is commonly utilized as a second treatment for biomass recovery, mainly due to the energy expenses that it provides, which represents, for example, 50–75 kW for a biomass recovery of 12–25%. In fact, the energy required to operate a centrifuge may be greater than the energy generated to produce microalgae biodiesel (Milledge and Heaven, 2013), thus it this method is indicated only for high added-value products.

Filtration is a separation method that utilizes a permeable or semipermeable medium, whereby the microalgae biomass is concentrated by retaining the solids and liberating the liquid (Show and Lee, 2014; Barros et al., 2015). It is one of the most favorable approaches due to its ability to retain low-density microalgae (Japar et al., 2017). However, cell clogging may occur, leading to an increase in the operational costs and in the process time (Grima et al., 2003; Barros et al., 2015). It is frequently performed after coagulation, flocculation, or flotation to increase biomass recovery yield (Barros et al., 2015).

Filtering methods can be divided into two types: (a) dead-end and (b) tangential flow (or cross filtration). The first one utilizes filter cartridge, filter press, and vacuum filter, while the second one consists of the microfiltration, ultrafiltration, nanofiltration, and reverse osmosis systems.

The filter cartridge system utilizes cartridge filters in a continuous operation, where the cleaning is automatic (Shelef and Sukenik, 1984; Japar et al., 2017). The filter press presents low design and maintenance costs, beyond flexibility in the operation. However, it requires manual and labor dismantling. The vacuum filtration system utilizes vacuum, causing the liquid culture to be sucked and the biomass to be retained in the tissue/membrane (Shelef and Sukenik, 1984; Danquah et al., 2009; Barros et al., 2015).

Membrane filtration is a method for biomass and liquid separation that occurs during the tangential flow of the cultures. It is simple and highly efficient, allowing continuous separation. It does not require the use of chemicals, which brings advantages to this biomass dewatering system (Kim et al., 2013a). The membranes are permeable and selective barriers made from organic and inorganic materials. They are classified according to their porosity and differentiated by the size of the retained compounds. Non-porous membranes are utilized for reverse osmosis and nanofiltration while the porous membranes for microfiltration and ultrafiltration (Shelef and Sukenik, 1984; Ahmad et al., 2014; Barros et al., 2015). The size range of the particles retained are 10–0.01 µm for microfiltration, 0.1–0.01 µm for nanofiltration, and 0.01–0.001 µm for ultrafiltration (Brennan and Owende, 2010; Christenson and Sims, 2011; Japar et al., 2017). However, microfiltration and ultrafiltration are reported to be more efficient (Ahmad et al., 2014).

These methods present diverse differences (Shelef and Sukenik, 1984; Danquah et al., 2009; Barros et al., 2015). Dead-end filtrations are efficient in the recovery of cells with a diameter above 70 µm while the tangential flow filtration is more adequate for the recovery of smaller cells due to reduced fouling problems (Shelef and Sukenik, 1984; Milledge and Heaven, 2013; Barros et al., 2015). Literature reports studies of membrane filtration of *Chlorella* sp. (Ahmad et al., 2014), vacuum filtration of *Coelastrum proboscideum* (Mohn, 1980), microfiltration of *C. vulgaris* and *Phaeodactylum tricornutum* (Bilad et al., 2012), and ultrafiltration of *P. tricornutum*

(Ríos et al., 2012). The costs of filtration are associated to the exchange of membranes, pumping, and energy (Grima et al., 2003; Barros et al., 2015).

9.3.3 SELECTION OF BIOMASS THICKENING AND DRAINAGE METHOD

The microalga specificity straightly affects the biomass recovery process (Vandamme et al., 2010) in terms of energy expenditure and biomass quality. The different methods present advantages and disadvantages regarding thickening and drainage for biomass recovery. Flocculation is simple and fast, and requires low energy, despite that some flocculants could be toxic or expensive (Granados et al., 2012; Borges et al., 2016; Chatsungnoen and Chisti, 2016a).

Auto-flocculation and bioflocculation present low cost and are non-toxic, but are slow and may favor contamination (Uduman et al., 2010; Salim et al., 2012). Sedimentation is simple and present low cost, but it has low efficiency and the biomass may be inadequate for some applications (Shelef and Sukenik, 1984; Japar et al., 2017). Flotation presents low costs and is fast and of easy scales up, but needs chemical flotation agents and display low efficiency for seawater microalgae (Coward et al., 2014; Ndikubwimana et al., 2016). Filtration presents high efficiency and permits the separation of species sensitive to shear, but clogging the pores increases operating costs and demands membrane change (Grima et al., 2003; Barros et al., 2015). Centrifugation is fast and indicated for obtaining high added value products, but is expensive, demands high energy and may shear some cells (Grima et al., 2003; Knuckey et al., 2006; Dassey and Theegala, 2013).

The efficiency of these methods is related to the microalgae species, including strain, size, morphology, and composition of the growth medium. Moreover, important aspects for the selection of a microalgae biomass recovery method are efficiency, viability, cell composition, and sustainability. All available technologies should be evaluated prior to choosing the most appropriate method to be employed.

It is crucial to underline that there is no universal method that can be utilized to recover all strains of microalgae with the same efficacy (Shelef and Sukenik, 1984; Brennan and Owende, 2010; Pragya et al., 2013). In this sense, some industries are working on the development of recovery systems combining mechanical and biological processes (Christenson and Sims, 2011).

9.3.4   EFFECT OF BIOMASS RECOVERY METHODS ON LIPID CONTENT AND FATTY ACIDS PROFILE

The influence of methods for biomass recovery on lipid content is not well understood up to now. Most of the studies from literature usually utilize centrifugation for biomass recovery and evaluate the lipid extraction at small scale, in order to investigate the potential for biofuel production (Gouveia and Oliveira, 2009; Doan et al., 2011; Sydney et al., 2011; Nascimento et al., 2013). The knowledge of the effects of biomass recovery on biomass quality and composition is notably important when biomolecules must meet quality standards for further biomass processing, including, for example, lipids for biodiesel production (Knothe, 2005; Gouveia and Oliveira, 2009).

Common coagulant/flocculating agents, including $FeCl_3$ and $Fe_2(SO_4)_3$, broadly used in systems for the treatment of water and effluents, have an elevated potential for flocculation and biomass recovery (Harith et al., 2009; Gerde et al., 2014). The lipids extracted from the recovered biomass by using these salts undergo slight variations, mostly insignificant, with the recovered lipids highly dependent on the microalga species and on the growth parameters (Chatsungnoen and Chisti, 2016a, b). The recovery by centrifugation usually presents better lipid results if compared to flocculation or filtration, while the changes in fatty acids are insignificant (Borges et al., 2011; Ahmad et al., 2014; Coward et al., 2014; Chatsungnoen and Chisti, 2016b), being this system the most employed for bench studies. The nature and content of fatty acids suffer interference when biomass recovery is carried out with alkalinizing agent, such as the rise of PUFAs and/or the loss of some specific fatty acids (Borges et al., 2011, 2016).

9.4   LIPIDS IN MICROALGAE

The lipids in microalgae can achieve 75% of their biomass (König et al., 2014; Chatsungnoen and Chisti, 2016b) and can be used as food, biofuels, or biomaterials (Thompson, 1996; Chisti, 2007; Lee et al., 2010). These applications are dependent on the cultivation conditions and strain/species chosen, on the physiology of the microalgae, their growth age and the cultivation conditions, including temperature, luminosity, salinity, and nutrients. The biomolecules extraction yield depending on the method used, especially for lipids (Chen et al., 2011; Abreu et al., 2012). Microalgae lipids could be up to 20 times upper when compared to oilseed (Thompson, 1996; Chisti, 2007; Lee et al., 2010).

Microalgae such as *Chlorella* sp., *Nannochloropsis* sp., and *Scenedesmus* sp. are promising candidates for biofuel production because of their total lipid content and rapid growth (Moazami et al., 2012; Nascimento et al., 2013; Oncel, 2013; Milano et al., 2016).

The lipids can be categorized into membrane lipids (polar) and reserve lipids (neutral and nonpolar) and play for many roles (Harwood and Jones, 1989; Guschina and Harwood, 2006; Christie and Han, 2010). Polar or complex lipids, including phospholipids and glycolipids, are majority in most microalgae and in the total lipid; they are associate by hydrogen bonds and electrostatic forces. The nonpolar and neutral lipids are those which do not include charged groups and contain triacylglycerols (TAGs), glycerides, carotenoids, sterols, and a limited range of high molecular weight hydrocarbons; they are associated by van der Waals forces (Molina et al., 1999; Basova, 2005; Petkov and Garcia, 2007; Ryckebosch et al., 2014b). The TAGs are considerate as energy storage products and are favored to produce biodiesel (Breuer et al., 2013b), while phospholipids and glycolipids are lipids present in the cell wall (Halim et al., 2012a; Breuer et al., 2013a).

These interactions (hydrogen bonds, electrostatic forces, and van der Waals forces) must be cracked for their effective extraction. Polar organic solvents destroy the hydrogen bonds between polar lipids while nonpolar organic solvents are mostly used to break up hydrophobic interactions between neutral and nonpolar lipids. Therefore, the selection of solvent is directly connected to the microalgae strain/species and its lipid class. Price, toxicity, volatility, polarity, and selectivity must be considered when selecting the solvent (Li et al., 2014; Ryckebosch et al., 2014a; Chatsungnoen and Chisti, 2016b).

Microalgae lipids are naturally composed of glycerol, sugars or bases esterified to fatty acids, containing between 12 and 24 carbons, whereas medium-chain (C10–C14), long-chain (C16–C18) and very-long-chain fatty acids (C20–C24) (Shahzad et al., 2010). Saturated fatty acids represent the largest fraction of lipids and, in some species, unsaturated fatty acids can be between 20 and 60% of the total lipids (Basova, 2005; Petkov and Garcia, 2007; Breuer et al., 2013a). Unsaturated fatty acids usually have a *cis* configuration, since most *trans* fatty acids are not found in nature, but in fats obtained by artificial processes (Thompson, 1996; Christie and Han, 2010; Bellou et al., 2014).

9.4.1 FATTY ACIDS

The microalgae fatty acids composition is diversified according to species / strain and culture conditions (Figure 9.2). The components of lipid molecules are fatty acids, which may be free or esterified. They are based on a total number of carbon atoms in the chain (usually 12 and 24 carbons) and the number of double bonds in the hydrocarbon chain. (Christie and Han, 2010; Pereira et al., 2012). When the carboxyl terminus of the fatty acid molecule is attached to a glycerol group, then a neutral lipid molecule is formed, for example, the glyceride. When the association of a fatty acid molecule occurs with a phosphate group, then a polar lipid is formed, for example, a phospholipid (Thompson, 1996; Bellou et al., 2014).

These fatty acids differing in the profile and composition of saturated, monounsaturated, and PUFAs (Hu et al., 2008; D'Oca et al., 2011). The quantity of saturated fatty acids in microalgae usually fluctuates between 13% and 58% (Basova, 2005; Guschina and Harwood, 2006). The main fatty acids are C16: 0, C16: 1, C20: 5ω3 and C22: 6ω3 in *Bacillariophyta*; C16: 0, C18: 1, C20: 3 and C20: 4ω3 *Eustigmatophyta*, C16: 0, C18: 1, C18: 2 and C18: 3ω3 in *Chlorophyte*; C16: 0, C20: 1, C18: 3ω3, C18: 4 and C20: 5 *Cryptophyte*, C16: 0, C18: 5ω3 and C22: 6ω3 in *Dinophyte*, and C16: 0, C16: 1, C18: 1, C18: 2 and C18: 3ω3 in *Cyanophyte* (Basova, 2005; Hu et al., 2008; Nascimento et al., 2013; Sahu et al., 2013). At the higher plants, the composition of fatty acids has variation, and the same occurs in microalgae, have the ability to synthesize medium-chain fatty acids, while others produce long-chain fatty acids (Hu et al., 2008; Shahzad et al., 2010). Lipid and fatty acids metabolism have been scarcely studied in microalgae when compared to higher plants. Based on the gene sequence homology and on some similar biochemical characteristics, microalgae, and higher plants are involved in the equivalent lipid metabolism (Hu et al., 2008; Brown and Sharpe, 2016).

The composition of fatty acids is essential for the biodiesel production, as it directly impacts the quality and properties of biodiesel. A high quantity of PUFAs can positively affect viscosity, fog point, cold filter plugging point, but may negatively affect oxidative stability while large amounts of saturated fatty acids concede suitable combustion properties (Knothe, 2005; Ramos et al., 2009; Knothe, 2012).

9.5 LIPID EXTRACTION IN MICROALGAE: PROCESSES

The extraction of lipids and fatty acids in microalgae biomass varies from vegetable oils and foods, due to the existence of rigid cell wall, and multiplicity of lipid classes and fatty acids (Ryckebosch et al., 2012; Breuer et al., 2013a). Specific methods must be used to break the rigid cell wall and release the lipids. The traditional methods of lipid extraction established by Folch (1957) and Bligh and Dyer (1959) use a mixture of chloroform and methanol to release all classes of lipids, but, these methods may not be scaled-up from biomasses of microalgae.

Using different approaches of cell disruption, associated to solvents, the lipid yield tends to be different (D'Oca et al., 2011; Ryckebosch et al., 2012, 2014a; Breuer et al., 2013a; Silva et al., 2014). In an incomplete or selective extraction, the efficiency of recovery the different lipid classes can vary and consequently impacts the types and composition of fatty acids.

The method for lipid extraction should be fast, efficient, and delicate to avoid incomplete extraction, lipid degradation, or oxidation and be economically feasible (Grima et al., 2003; Chisti, 2007). Extraction begins with the microalgae cell wall rupture and then, the lipids can be extracted by different approaches. The cellular disruption process is a precondition for an effectiveness extraction of lipids (Ryckebosch et al., 2014a). There is a large of studies being conducted to improve these processes and maximize the extraction at an even lower cost (Lee et al., 2010; Kim et al., 2013a; Silva et al., 2014). There are a lot of methods for lipid extraction from microalgae, usually separated by several procedures until removal of the residual solvent and debris of biomass (Brennan and Owende, 2010; Niraj et al., 2011; Halim et al., 2012a; Ryckebosch et al., 2014a; Mubarak et al., 2015).

9.5.1 METHODS OF CELL WALL DISRUPTION

The effectiveness of cell disruption methods depends on the microalgae strain/species and on the characteristics of the cell membrane, including composition and morphology. The rupture and extraction costs can be expressively reduced using the suitable method (Kim et al., 2013a; D'Alessandro et al., 2016). These can be categorized in mechanical and non-mechanical. The mechanical methods are pressing, ball mill, high pressure homogenization (HPH), ultrasonic, microwave, while the non-mechanical methods are osmotic shock, chemical breakdown, and enzymes. Mechanical methods can be scaled up industrial, fast, easily controlled, and monitorable, however, their energy consumption is high (Kim et al., 2013a; Mubarak et al., 2015; D'Alessandro et al., 2016). The efficiency of the cell disruption method is measured by the release of biomolecules, ultraviolet (UV) absorbance, turbidity, particle sizing, or cell counting (Dong et al., 2016).

The utilization of ultrasound for cell disruption and lipid release of microalgae has been applied in last years (Adam et al., 2012; Ehimen et al., 2012; Araujo et al., 2013; Balasubramanian et al., 2013; Kim et al., 2013b).

The propagation of the ultrasonic wave's results in cavitation, which breaks the cell structure, permitting the extraction of the lipids (Neto et al., 2013; Mubarak et al., 2015).

Ultrasonic cavitation is more penetrating at low frequency (18–40 kHz) than at high frequency (400–800 kHz) and is affected by cell wall type, viscosity, reaction time, and medium temperature. Besides, the scaling-up ultrasound is tough because the cavitation occurs in regions close the ultrasonic probes (Halim et al., 2012a; Dong et al., 2016).

Ultrasound presents like a pretreatment method and/or associated with solvent for lipid extraction and is suitable for some microalgae species/strain. In a study using *Chlorella* sp., *Nostoc* sp., and *Tolypothrix* sp. where several methods of cell disruption were experienced for lipid extraction, an ultrasound performed the best system (Prabakaran and Ravindran, 2011). For *Chlorella pyrenoidosa* there was no significant difference when using agitation and ultrasound-assisted by the mixture of 2:1 chloroform: methanol (D'Oca et al., 2011). However, in other studies with *Botryococcus* sp., *C. vulgaris,* and *Schizochytrium* sp. S34, the ultrasound showed itself less efficient at breaking down cell walls to allow the release of lipids (Lee et al., 2010; Byreddy et al., 2015).

HPH is known as the French press. This process utilizes hydraulic shear force generated when the biomass at high pressure is sprayed through a thin tube (Kim et al., 2013a; Dong et al., 2016). In this system, heat can be transfer and there is a risk of thermal degradation, however, operational cost is low, average energy consumption when compared to ultrasound, and the possibility of scaling up (Lee et al., 2010). The effectiveness of HPH in microalgal cells differs between species and may decrease depending on the structure of cell walls (Halim et al., 2013; Ursu et al., 2014). Although promising, more evaluation of HPH is essential in an industrial scale for biofuel production.

The biomass cell disruption efficiency of *Chlorococcum* sp. with the methods of HPH, ultrasound, ball mill, assisted with sulfuric acid showed that HPH destroyed 73.8% of the total cells (Halim et al., 2012b). It was reported elsewhere that the recovery of *Scenedesmus acutus* did not reach 80% in fatty acids (Dong et al., 2016). For the extraction of intracellular components of *Nannochloropsis* sp., HPH had the highest efficiency and energy consumption (Grimi et al., 2014).

The cellular breakup of the microalgae cell wall can be performed by pressing or expeller, using mechanical force to rupture the cells and release the lipid content (Mubarak et al., 2015). The mechanical extraction decreases the contamination of the biomass from outside sources and preserves the integrity of the biomolecules (Halim et al., 2012a). Mechanical pressing is a simple method commonly used for the industrial extracting of oilseeds. There are different types of mechanical technologies for extracting microalgae oil; it includes screw press or piston, extruder, and biomass spraying (Rawat et al., 2013).

The extraction of oil from microalgae wet biomass using pressing is not easily reached because part of the biomass can be missed if flowed in the moisture. Pressing can be applied for small and large scales to obtain microalgae oil for the production of biodiesel, however, it is slow and requires a large amount microalgae biomass (Niraj et al., 2011; Rawat et al., 2013; Mubarak et al., 2015). This method is scarcely treated in the literature for the extraction of lipids from microalgae, but it should be considered as a feasible method for the industrial applications as view as oilseed industries already operate with this system.

The ball mill or bead mill consists of a rotating cylinder with metallic or quartz beads, which act as grinding frame. This method causes a direct impact and damage to the cell wall, by collision of the beads. This system is usually in biological samples to extract DNA (Kim et al., 2013a; Mubarak et al., 2015). Damage caused by beads promptly disrupts the wall cell without any preparation to biomass. The ball mill system has been used associated with solvent at research labs and industries (Lee et al., 2010).

The effectiveness and energy consumption of ball mill are affected by the shape of the container, the stirring speed, the size, type, and quantity of spheres. However, this method can cause a thermal degradation of the lipids, being necessary an intensive cooling system (Prabakaran and Ravindran, 2011; Halim et al., 2012b; Kim et al., 2013a; Mubarak et al., 2015).

The fulfillment of the ball mill as a pretreatment method for lipid extraction is a benefit for some species/strain of microalgae. Literature reports an extraction of 28% of lipids from *Botryococcus* sp. when using the ball mill system, which was 20% superior if compared to the ultrasound assisted by solvent. However, the ball mill was less effective than the microwave for lipid extraction of *Scenedesmus* sp. (Lee et al., 2010). In another study with *Tolypothrix* sp. and *Chlorella* sp. there was no difference in lipid extraction when compared to ultrasound (Prabakaran and Ravindran, 2011).

Microwaves are electromagnetic radiations of frequency from 0.3 to 300 GHz (Kim et al., 2013a; Dong et al., 2016). The frequency for laboratorial scale microwaves ovens is approximately 2,450 MHz for cell disruption by induction of heat and interacts with molecules thus releasing lipids.

However, due to the high temperature reached, lipids can be degraded during the process, it is necessary a cooling system or a reduced process time to avoid oxidation and/or degradation (Balasubramanian et al., 2013; Kim et al., 2013a; Pragya et al., 2013; Mubarak et al., 2015).

This method is highly selective and preferred for polar solvents, for example, water, which produces steam and breaks the cell wall, releasing lipids (Dong et al., 2016). The utilization of microwave can be a method with reduced extraction time and less demand for solvents, but it has a high energy consumption considering its scaling-up (Halim et al., 2012a; Pohndorf et al., 2016).

In a study using microwave-assisted by hexane, it was recognized a higher recovery of lipids and fatty acids compared to the solo solvent extraction (Balasubramanian et al., 2013). In another study to release lipids with a mix of microalgae, the microwave system was the fastest and most efficient (33.7%) followed by electroflotation and autoclave methods (Silva et al., 2014). For experiments using different solvent-assisted cell disruption methods, the best results for *Botryococcus* sp., *C. vulgaris*, and *Scenedesmus* sp. were evidenced with the use of microwave (Lee et al., 2010).

Osmotic shock happens when suddenly rise or fall the salt concentration of the medium, and disturbs the balance of the osmotic pressure between the inside and the outside of the cells, causing damage to the cell wall and release the lipids (Kim et al., 2013a; Dong et al., 2016). Hyperosmotic stress causes cell contraction when the salt concentration is higher on the outside. However, when the salt concentration is lower on the outside, it occurs hypo-osmotic stress. In this case, water flows into the cells to balance the osmotic pressure, swelling up to burst, releasing the lipids. The inconvenience is that the process scale-up is unviable due to the large amount of water required for the dilution of the liquid medium (Prabakaran and Ravindran, 2011; Kim et al., 2013a).

Osmotic shock uses low-cost chemical compounds, including sorbitol and NaCl through a simple process, but its performance results in effluents with high salinity. Moreover, this method is specific for microalgae with a cell wall penetrable to this solution (Sharma et al., 2012; Yoo et al., 2012; Byreddy et al., 2015). In order to extract lipids, it is necessary to more steps with solvents (Mandal et al., 2013; Dong et al., 2016).

In a study with wet biomass of *Chlamydomonas reinhardtii* and NaCl-osmotic shock, assisted by solvent, showed yields of 23.81 and 34.50% of lipids during stationary and post-stationary phases respectively (Yoo et al., 2012). Using grind, ultrasound, microwave, and NaCl-osmotic shock, assisted by solvent to extract lipids from *Schizochytrium* sp. S35 and

Thraustochytrium sp. the results showed that osmotic shock showed the highest efficiency compared the other methods like ultrasound and microwave (Byreddy et al., 2015). However, in a study with *Botryococcus* sp., *C. vulgaris*, and *Scenedesmus* sp., the effectiveness of osmotic shock was lower compared to microwaves and ball mill systems (Lee et al., 2010). This means that the use of osmotic shock should be focused to species/strain that tends to disrupt their cell wall from saline solutions.

Microalgae cells can be disrupted by using acids, alkali, or surfactants. These compounds can damage chemical bonds in the cell wall, and induce the release of intracellular biomolecules (Kim et al., 2013a). The energy consumption is lower, since the breakage does not require heat or electricity. It is necessary the use of solvents to release the lipids and transport them to the micelle (Brennan and Owende, 2010; Lee et al., 2010; Ranjan et al., 2010; Pragya et al., 2013). Acids and alkalis may corrode or encrust the surface of the photobioreactors (PBR). The neutralization of acids and alkalis doubles the cost of this system.

The chemical breakdown or solvent-assist method simplifies the lipid extraction of microalgae since it breaks the cell walls and the bonds between the lipids, turning them available to the solvent (Dong et al., 2016). In an extraction using acids and alkalis to break down the wet biomass of *Chlorella* sp. and *Scenedesmus* sp., the cell disruption was carried out with 1 M H_2SO_4 and 5 M NaOH respectively at 90°C for 30 min. The free fatty acids were converted using 0.5 M H_2SO_4 1M solution after chlorophyll dissolution (Sathish and Sims, 2012).

The direct transesterification of microalgae can diminish the cost of biodiesel production and augment the yield of fatty acids because the extraction and transesterification succeed simultaneously in the same step (D'Oca et al., 2011; Liu et al., 2015) with an acid catalysis and lipid conversion in fatty acids. This method is considered as a method of breaking cell by the chemical breakdown.

The breakdown of the microalgae cell wall through enzymes is considered a biological method of disruption cell. Enzymes are favored because of their commercial accessibility and the process is more easily controlled than autolysis and degraded a specific chemical bond. However, it is more expensive than other method; this is a limiting factor for the scale-up (Kim et al., 2013a, 2016). The mixture of enzymes does not always improve results because the inhibition of the reaction can occur if they compete on the same substrates (Zheng et al., 2011). Enzymes should be chosen carefully for an effective cell disruption.

There are two ways to decrease the cost of an enzymatic process: the immobilization of enzymes or the association of this process with other methods (Kim et al., 2013a). In a study with *Chlorella salina*, the yeast *Rhodotorula mucilaginosa* was immobilized in sugar cane bagasse and utilized to breakdown the microalgae cell wall and releases the lipids (Surendhiran et al., 2014). The enzymes lysozyme and cellulase were used to disrupt the biomass of *Scenedesmus* sp. for extraction lipids using solvents, reaching yields of 16.6% and 16.0% respectively. Immobilized enzymes can efficiently degrade the cell walls of *C. pyrenoidosa* and increase the lipid extraction yield by 75% (Kim et al., 2013a).

9.5.2 SELECTION OF METHODS OF CELLULAR WALL DISRUPTION

The pretreatment of biomass exhibit lipid to improve their extraction. However, some of this pretreatment are energy consumption, high cost, and takes time. The high energy consumption is due to the temperature and pressure conditions of the extraction process, the cost of distillation associated with the separation of lipids from solvents, and the cost of drying biomass. These factors are associated to the cell wall of the microalgae, composed of a thick and rigid layer (Halim et al., 2012b; Kim et al., 2013a; Mubarak et al., 2015; D'Alessandro et al., 2016; Dong et al., 2016). The research for extraction lipids should consider a process with less energetic and high efficiency to break the cell wall and release the lipids, utilization of nontoxic solvents to bioproducts and health, processing time, and the scale-up procedures.

The lipid output is affected by the disruption cell method as reported *C. vulgaris* (Lee et al., 2010; Zheng et al., 2011), *Botryococcus* sp., *Scenedesmus* sp. (Lee et al., 2010), *Spirulina* sp. (Pohndorf et al., 2016), *Chlorella* sp., *Nostoc* sp., and *Tolypothrix* sp. (Prabakaran and Ravindran, 2011), for example. There is no single method that can also be profitable for all microalgae species/strain, however, it is demonstrated that the ultrasound, microwave, and osmotic shock methods showed the highest lipid yield efficiencies when assisted by solvents in laboratory scale.

9.6 PROCESS OF DRYING BIOMASS

After the breakdown cell wall, the wet biomass should be dried for the next processing steps. Several studies use biomass drying methods after their recovery (Lee et al., 2010; Balasubramanian et al., 2013; Silva et al., 2014; Chatsungnoen and Chisti, 2016b), including oven drying, spray-drying, freeze-drying (or lyophilization), and solar drying (Grima et al., 2003; Mata et al., 2010; Rawat et al., 2013).

Drying in greenhouses is the most common, simple, and cheaper method used for removing water by heating. It is a slow method that can take from 3 to 24 h at temperatures of 60 or 105°C (Grima et al., 2003; Mata et al., 2010). Solar drying consists of the outside exposing of the biomass to take advantage of the light incidence and heat to evaporate the water. This method is a low cost, but highly dependent on the weather conditions.

Spray drying is a procedure, which consists of spraying the biomass into a chamber subjected to a controlled hot airflow, thereby achieving evaporation of the water, resultant in an ultrafast separation of the biomass with the degradation of the product, resulting in the recovery of the powder product. Freeze drying removes water by sublimation in a process involving biomass freezing before applying vacuum. The temperature is gradually increased, reducing the pressure, allowing the frozen water passes from the solid to the gas, without altering or degrading the properties of the biomass (Richmond, 2004).

The drying temperature influences the lipid composition and the lipid yield of any matrix (Brum et al., 2009; Widjaja et al., 2009; Menegazzo et al., 2016). Drying temperatures below 60°C still maintains a high TAG content in the lipids and only slightly decreases total lipid yield, but with higher temperatures, decreases both the TAG and the lipid yield are affected (Widjaja et al., 2009).

In a study with *Scenedesmus* sp., ultrasound, and microwave cell disruption methods and three different biomass drying methods were tested, without significant differences between the drying methods, but a significant difference between ultrasound and microwave methods (Guldhe et al., 2014). It was reported elsewhere that the drying method associated with cell disruption methods did not have a significant difference in cold lipid extraction while that in the hot method, lipid yields decreased by 60% for *Spirulina* sp. (Pohndorf et al., 2016). In a study comparing freeze-drying, greenhouse, and solar drying, no effect was observed for lipid extraction, although there was a rise in free fatty acids when solar drying was utilized (Balasubramanian et al., 2013).

The methods for production dry biomass in reduced particles are similar to those used in the food or pharmaceutical industry. In laboratory scale the dry biomass can be crushed using mortar and pistil or by grinding with sieves. Spray-drying and freeze-drying generate microparticles of biomass.

9.7   PROCESS OF LIPID EXTRACTION

The lipid yield depends on the method for lipid extraction and the wet or dry biomass. The type of lipids extracted is related with system of disrupt cell wall, the solvent chosen, temperature, and pressure of the procedure. These methods should be fast, scalable, and do not damage biomolecules and bioproducts (Brennan and Owende, 2010; Pragya et al., 2013; Mubarak et al., 2015; Dong et al., 2016). Most of the studies report the use of biomass dried in a greenhouse or lyophilized, however, there are several studies addressing lipid extraction using wet biomass (Lee et al., 2010; Araujo et al., 2013; Nascimento et al., 2013; Silva et al., 2014).

During extraction, the lipids are usually released from the cell by an extraction solvent. The lipids have to be separated from the cellular debris, the solvent and any residual water must be removed, and then utilized according to the chosen lipid class. The debris should be removed by separation techniques, which are commonly filtration and centrifugation (Halim et al., 2012a).

The miscible lipids in the solvent must be separated by distillation, vacuum evaporation, or solid-phase absorption practices and it is often possible to recover the solvent and to reuse them. The remaining biomass may be inadequate for animal feed in cases of contamination by solvents (Rawat et al., 2013).

The common technologies for the extraction of lipids utilize organic solvents and supercritical fluid (Halim et al., 2011, 2012a; Li et al., 2014; Baumgardt et al., 2016). Methods for disrupt cell wall usually are associated with solvent or combination of solvents. One of them are direct trans-esterification in an ultrasound bath, which has emerged as a method of lipid extraction and transesterification for the production of fatty acids in a single process (Menezes et al., 2013).

A supercritical fluid is a clean technology of high selectivity that utilizes substances that have properties of liquids and gases when exposed to high temperatures and pressures (Halim et al., 2011; Pragya et al., 2013; Li et al., 2014; Baumgardt et al., 2016). However, some species/strain of microalgae may not be adequate to use this method. When the temperature and pressure of a fluid achieve its critical values, the fluid enters the supercritical region, allowing them to act as a solvent extraction.

Lipids extraction using supercritical carbon dioxide ($SC\text{-}CO_2$) has the potential to replace some methods of extraction assisted by solvents (Mendes et al., 1995). The process involves a system for compressing and transporting liquid CO_2 to an extraction vessel, and a heating valve to depressurize the input SC-CO_2. Once the oven is heated, the compressed CO_2 enters a supercritical state and extracts the lipid from the microalgae. Then CO_2 evaporates, precipitating the extracted lipid (Halim et al., 2011; Pragya et al., 2013; Taher et al., 2014b; Mubarak et al., 2015). SC-CO_2 has high solvation power and low toxicity. The intermediate diffusion and viscosity properties of the fluid lead to an advantageous mass transfer equilibrium and this process produces solvent-free lipids. However, high infrastructure and operating costs associated with this process are its main disadvantages (Mubarak et al., 2015).

Lipid extraction from microalgae *Chlorococcum* sp. with SC-CO_2 had higher lipid yields and presented a suitable composition of fatty acids for biofuel production, when compared to Soxhlet and hexane extraction methods (Halim et al., 2013). The extraction of lipids from *C. vulgaris* using SC-CO_2 had the lipid extraction yields improved with increasing pressure (Mendes et al., 1995).

The assisted-solvent extraction of microalgae biomass is extensively used to extract metabolites including astaxanthin, β-carotene, and mainly lipids. Organic solvents may extract different lipid classes. Most popular nonpolar solvents are hexane, benzene, diethyl ether, chloroform, and polar solvents are methanol, ethyl acetate, and ethanol. The polarity of the solvents affects the classes of lipids they can release (Grima et al., 1994). With the use of mixtures of nonpolar and polar solvents may increase lipid yields (D'Oca et al., 2011; Silva et al., 2014; Byreddy et al., 2015).

The process for extraction of lipids by non-polar solvent follows the steps: mixture microalgae and solvent; infiltration of the solvent through the already broken cell wall; interaction of the solvent with the neutral/non-polar lipid, breaking out the Van der Walls force; formation of a solvent-lipid complex; the passage of the complex through the cell membrane by the difference of concentration gradient; and formation of a micelle. The extracted lipids remain dissolved in the solvent (Mubarak et al., 2015).

Some lipids are found in the cytoplasm as polar complex lipids. This complex is strongly bound via hydrogen bonds and proteins in the cell wall. Then a polar solvent disrupts the associations, releasing the polar lipids. The utilization of a mixture of solvent would have the advantage of extracting all lipid classes. This mixture is commonly used for quantification of total lipids. The traditional methods of Bligh and Dyer (1959), using a mixture of chloroform, methanol, and water, and Folch et al. (1957) with a combination of chloroform and methanol.

Solvent extraction can be done using Soxhlet equipment (Soxhlet, 1879). This device is based on the solid-liquid extraction, where the solvent extracts, evaporates, and condenses, making it always in interaction with the biomass (Halim et al., 2012a; Mubarak et al., 2015). In some species of microalgae, this method has high efficiency when associate with disruptive methods of cell wall (Cavalcante et al., 2011; Kim et al., 2013b). This method is hard to scale up due to complexity of equipment and moreover, the fatty acids can degraded occur due to high temperature to distillated the solvent (Kim et al., 2013a).

Solvents must be low cost, volatile, low toxicity, pure, water-immiscible, and selective. Organic solvents are commonly used in lipid extraction, but because the antioxidant potential and polarity of the compounds, the yield depends on the type of solvents utilized (Halim et al., 2012a; Ryckebosch et al., 2014a; Hidalgo et al., 2016). The methods of lipid extraction by solvents are usually assisted by ultrasound and microwave. These processes are faster and with higher yields (Halim et al., 2011; Niraj et al., 2011; Ryckebosch et al, 2014c; Mubarak et al., 2015).

In several studies polar, nonpolar, and mixtures of solvents for lipid extraction were used. In studies with *C. pyrenoidosa* (D'Oca et al., 2011), *C. vulgaris* (Araujo et al., 2013), *Botrycoccus braunii* (Lee et al., 2010) and *Schizochytrium* sp. S31 (Byreddy et al., 2015), it was observed that the highest lipid yields were obtained for the methods using mixture of polar and non-polar solvents. *Chlorella* sp. and the *Botrycoccus braunii* presented higher yields with polar solvents. However, for *Schizochytrium* sp. the nonpolar solvents presented better yields than the polar solvent. Different methods of cellular disruption assisted by solvent combination for lipid extraction have influenced yield (Lee et al., 2010; Prabakaran and Ravindran, 2011; Byreddy et al., 2015; Pohndorf et al., 2016). This means that the choice for the most suitable system to extract the lipids is specific for each species/ strain of microalgae, growth conditions, biomolecule purpose, factors that influence the procedure and scale-up process.

To choose the most efficient method to preserve the lipid content and the fatty acid composition suitable for the intended use, it is essential to know the performance of the microalgae biomass against the methods. For scale-up, some parameters must be considered, including energy consumption, operating costs, productivity in lipid extraction, solvent toxicity, and process time, among others (Figure 9.3).

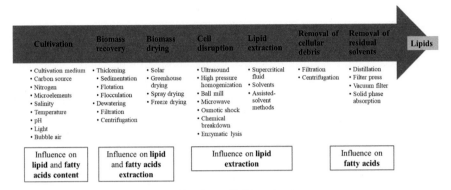

FIGURE 9.3   Microalgae: from cultivation to lipid production.

The effect of the cellular disruption method assisted by single or combination of polar and non-polar solvent is significant and specific on the lipid class and extraction yield, as described for *Tetraselmis* sp. (Li et al., 2014), *Schizochytrium* sp., *Thraustochytrium* sp. (Byreddy et al., 2015), *Nannochloropsis* sp. (Balasubramanian et al., 2013), *C. pyrenoidosa* (D'Oca et al., 2011), and *C. vulgaris* (Araujo et al., 2013). The combination of polar and nonpolar solvents (chloroform and methanol, hexane, and ethanol, dichloromethane, and methanol) generally results in greater results when compared to the extraction sole solvent. The chosen method for disrupt cell walls affect significantly the extraction lipids, even with the use of the combination of solvents.

In order to measure the total lipids in microalgae biomass, the conventional methods as Bligh and Dyer (1959) and Folch (1957) show the total lipid content because these methods are combined with polar and nonpolar solvents. Although the disrupt cell wall method should be considered. To scale up the process of lipid extraction it is necessary to evaluate the best system of cellular disruption and solvents, and the technological process complete. Regarding the profile and composition of fatty acids, slight variations occur (D'Oca et al., 2011; Halim et al., 2011; Byreddy et al., 2015; Hidalgo et al., 2016).

9.8   CONCLUSION

The lipid yield is affected by the choice of the strain/species, growth conditions, as light intensity, salinity, nitrogen, and carbon sources, pH, air, and methods for biomass harvesting and lipid extraction. The most common method for harvesting is centrifugation. This method has minimal influence on lipids extraction, but it is high energy consumption and hard to scale up. Flocculation has a low influence on the lipid content, energy consumption, and easy scale-up. The effectiveness of the lipid extraction method is affected by several factors; including a method for disrupt cell wall, type of lipids (polar, neutral, or non-polar), solvents, operating systems, and energy costs. Solvent-assisted ultrasound and microwave are evidenced as efficient for cell disruption and lipid extraction but are highly energy consuming and hard to scale-up. More research is necessary to develop processes that allow obtaining microalgae lipids at competitive costs, through the selection of the best strain/species and cultivation conditions that enable maximum lipid productivity with a fatty acids profile indicated for the intended application.

KEYWORDS

- **bioflocculation**
- **biomass recovery**
- **electroflotation**
- **fatty acids**
- **high pressure homogenization**
- **lipid extraction**
- **lyophilization**
- **filtration**
- **proteins**
- **reverse osmosis systems**
- **self-flocculation**
- **solar drying**

REFERENCES

Abreu, A. P., Fernandes, B., Vicente, A. A., Teixeira, J., & Dragone, G., (2012). Mixotrophic cultivation of *Chlorella vulgaris* using industrial dairy waste as organic carbon source. *Bioresour. Technol., 118*, 61–66.

Adam, F., Abert-Vian, M., Peltier, G., & Chemat, F., (2012). "Solvent-free" ultrasound-assisted extraction of lipids from fresh microalgae cells: A green, clean, and scalable process. *Bioresour. Technol., 114*, 457–465.

Ahmad, A. L., Yasin, N. H. M., Derek, C. J. C., & Lim, J. K., (2014). Comparison of harvesting methods for microalgae *Chlorella* sp. and its potential use as a biodiesel feedstock. *Environ. Technol., 35*, 2244–2253.

Ahmad, A. L., Yasin, N. H. M., Derek, C. J. C., & Lim, J. K., (2011). Microalgae as a sustainable energy source for biodiesel production: A review. *Renew. Sustain. Energy Rev., 15*, 584–593.

Anthony, R. J., Ellis, J. T., Sathish, A., Rahman, A., Miller, C. D., & Sims, R. C., (2013). Effect of coagulant/flocculants on bioproducts from microalgae. *Bioresour. Technol., 149*, 65–70.

Araujo, G. S., Matos, L. J. B. L., Fernandes, J. O., Cartaxo, S. J. M., Gonçalves, L. R. B., Fernandes, F. A. N., & Farias, W. R. L., (2013). Extraction of lipids from microalgae by ultrasound application: Prospection of the optimal extraction method. *Ultrason. Sonochem., 20*, 95–98.

Balasubramanian, R. K., Yen, D. T. T., & Obbard, J. P., (2013). Factors affecting cellular lipid extraction from marine microalgae. *Chem. Eng. J., 215, 216*, 929–936.

Banerjee, C., Ghosh, S., Sen, G., Mishra, S., Shukla, P., & Bandopadhyay, R., (2013). Study of algal biomass harvesting using cationic guar gum from the natural plant source as flocculant. *Carbohydr. Polym., 92*, 675–681.

Barros, A. I., Gonçalves, A. L., Simões, M., & Pires, J. C. M., (2015). Harvesting techniques applied to microalgae: A review. *Renew. Sustain. Energy Rev., 41*, 1489–1500.

Basova, M. M., (2005). Fatty acid composition of lipids in microalgae. *Int. J. Algae., 7*, 33–57.

Baumgardt, F. J. L., Zandoná, F. A., Brandalize, M. V., Da Costa, D. C., Antoniosi, F. N. R., Abreu, P. C. O. V., et al., (2016). Lipid content and fatty acid profile of *Nannochloropsis oculata* before and after extraction with conventional solvents and/or compressed fluids. *J. Supercrit. Fluids, 108*, 89–95.

Behrens, P. W., & Kyle, D. J., (1996). Microalgae as a source of fatty acids. *J. Food Lipids, 3*, 259–272.

Bellou, S., Baeshen, M. N., Elazzazy, A. M., Aggeli, D., Sayegh, F., & Aggelis, G., (2014). Microalgal lipids biochemistry and biotechnological perspectives. *Biotechnol. Adv., 32*, 1476–1493.

Benemann, J., (2013). Microalgae for biofuels and animal feeds. *Energies, 6*, 5869–5886.

Bilad, M. R., Vandamme, D., Foubert, I., Muylaert, K., & Vankelecom, I. F. J., (2012). Harvesting microalgal biomass using submerged microfiltration membranes. *Bioresour. Technol., 111*, 343–352.

Bligh, E. G., & Dyer, W. J., (1959). A rapid method of total lipid extraction and purification. *Natl. Res. Counc. Canada, 37*, 911–917.

Borges, L., Caldas, S., D'Oca, M. G. M., & Abreu, P. C., (2016). Effect of harvesting processes on the lipid yield and fatty acid profile of the marine microalga *Nannochloropsis oculata*. *Aquac. Reports, 4*, 164–168.

Borges, L., Morón-Villarreyes, J. A., D'Oca, M. G. M., & Abreu, P. C., (2011). Effects of flocculants on lipid extraction and fatty acid composition of the microalgae *Nannochloropsis oculata* and *Thalassiosira weissflogii*. *Biomass Bioenergy, 35*, 4449–4454.

Brennan, L., & Owende, P., (2010). Biofuels from microalgae: A review of technologies for production, processing, and extractions of biofuels and co-products. *Renew. Sustain. Energy Rev., 14*, 557–577.

Breuer, G., Evers, W. A. C., De Vree, J. H., Kleinegris, D. M. M., Martens, D. E., Wijffels, R. H., & Lamers, P. P., (2013a). Analysis of fatty acid content and composition in microalgae. *J. Vis. Exp.,* 80. doi: 10.3791/50628.

Breuer, G., Lamers, P. P., Martens, D. E., Draaisma, R. B., & Wijffels, R. H., (2013b). Effect of light intensity, pH, and temperature on triacylglycerol (TAG) accumulation induced by nitrogen starvation in *Scenedesmus obliquus. Bioresour. Technol., 143*, 1–9.

Breuer, G., Lamers, P. P., Martens, D. E., Draaisma, R. B., & Wijffels, R. H., (2012). The impact of nitrogen starvation on the dynamics of triacylglycerol accumulation in nine microalgae strains. *Bioresour. Technol., 124*, 217–226.

Brown, A. J., & Sharpe, L. J., (2016). *Biochemistry of Lipids, Lipoproteins and Membranes* (4th edn., p. 612). Elsevier, Nova Scotia.

Brum, A. A. S., Arruda, L. F., Regitano-D'Arce, M. A. B., (2009). Extraction methods and quality of the lipid fraction of vegetable and animal samples. *Quim. Nova, 32*, 849–854.

Byreddy, A. R., Gupta, A., Barrow, C. J., & Puri, M., (2015). Comparison of cell disruption methods for improving lipid extraction from thraustochytrid strains. *Mar. Drugs, 13*, 5111–5127.

Cavalcante, A. K., de Sousa, L. B., & Hamawaki, O. T., (2011). Determination and evaluation of oil content in soybean seeds by nuclear magnetic resonance methods and soxhlet. *Biosci. J., 27*, 8–15.

Cecchi, H. M., (2003). *Fundamentos Teóricos e Práticos Em Análise De Alimentos* (p. 208). Unicamp. Campinas.

Chatsungnoen, T., & Chisti, Y., (2016a). Harvesting microalgae by flocculation–sedimentation. *Algal Res., 13*, 271–283.

Chatsungnoen, T., & Chisti, Y., (2016b). Oil production by six microalgae: Impact of flocculants and drying on oil recovery from the biomass. *J. Appl. Phycol., 28*, 2697–2705.

Chen, C. Y., Yeh, K. L., Aisyah, R., Lee, D. J., & Chang, J. S., (2011). Cultivation, photobioreactor design, and harvesting of microalgae for biodiesel production: A critical review. *Bioresour. Technol., 102*, 71–81.

Chisti, Y., (2007). Biodiesel from microalgae. *Biotechnol. Adv., 25*, 294–306.

Christenson, L., & Sims, R., (2011). Production and harvesting of microalgae for wastewater treatment, biofuels, and bioproducts. *Biotechnol. Adv., 29*, 686–702.

Christie, W. W., & Han, X., (2010). *Lipid Analysis: Isolation, Separation, Identification and Lipidomic Analysis* (p. 446). Oily Press, Bridgwater.

Coward, T., Lee, J. G. M., & Caldwell, G. S., (2014). Harvesting microalgae by CTAB-aided, foam flotation increases lipid recovery and improves fatty acid methyl ester characteristics. *Biomass Bioenergy, 67*, 354–362.

D'Alessandro, E. B., & Antoniosi-Filho, N. R., (2016). Concepts and studies on lipid and pigments of microalgae: A review. *Renew. Sustain. Energy Rev., 58*, 832–841.

D'oca, M. G. M., Viêgas, C. V., Lemões, J. S., Miyasaki, E. K., Morón-Villarreyes, J. A., Primel, E. G., & Abreu, P. C., (2011). Production of FAMEs from several microalgal lipidic extracts and direct transesterification of the *Chlorella pyrenoidosa. Biomass Bioenergy, 35*, 1533–1538.

Danquah, M. K., Ang, L., Uduman, N., Moheimani, N., & Forde, G. M., (2009). Dewatering of microalgal culture for biodiesel production: Exploring polymer flocculation and tangential flow filtration. *J. Chem. Technol. Biotechnol., 84*, 1078–1083.

Dassey, A. J., & Theegala, C. S., (2013). Harvesting economics and strategies using centrifugation for cost effective separation of microalgae cells for biodiesel applications. *Bioresour. Technol., 128*, 241–245.

Demirbas, A., (2010). Use of algae as biofuel sources. *Energy Convers. Manag., 51*, 2738–2749.

Derner, R. B., Ohse, S., Villela, M., Carvalho, S. M., & Fett, R., (2006). Microalgas, produtos e aplicações. *Ciência Rural, 36*, 1959–1967.

Doan, T. T. Y., Sivaloganathan, B., & Obbard, J. P., (2011). Screening of marine microalgae for biodiesel feedstock. *Biomass Bioenergy, 35*, 2534–2544.

Dong, T., Knoshaug, E. P., Pienkos, P. T., & Laurens, L. M. L., (2016). Lipid recovery from wet oleaginous microbial biomass for biofuel production: A critical review. *Appl. Energy, 177*, 879–895.

Dunstan, G. A., Volkman, J. K., Barrett, S. M., Leroi, J. M., & Jeffrey, S. W., (1993). Essential polyunsaturated fatty acids from 14 species of diatom (*Bacillario phyceae*). *Phytochemistry, 35*, 155–161.

Ehimen, E. A., Sun, Z., & Carrington, G. C., (2012). Use of ultrasound and co-solvents to improve the *in-situ* transesterification of microalgae biomass. *Procedia Environ. Sci., 15*, 47–55.

Folch, J., Lees, M., & Stanley, G. H. S., (1957). A simple method for the isolation and purification of total lipids from animal tissues. *J. Biol. Chem., 226*, 497–509.

Frumento, D., Casazza, A. A., Al Arni, S., & Converti, A., (2013). Cultivation of *Chlorella vulgaris* in tubular photobioreactors: A lipid source for biodiesel production. *Biochem. Eng. J., 81*, 120–125.

Gao, S., Yang, J., Tian, J., Ma, F., Tu, G., & Du, M., (2010). Electro-coagulation-flotation process for algae removal. *J. Hazard. Mater., 177*, 336–343.

Gerde, J. A., Yao, L., Lio, J., Wen, Z., & Wang, T., (2014). Microalgae flocculation: Impact of flocculant type, algae species and cell concentration. *Algal. Res., 3*, 30–35.

González-Fernández, C., & Ballesteros, M., (2013). Microalgae autoflocculation: An alternative to high-energy consuming harvesting methods. *J. Appl. Phycol., 25*, 991–999.

Gouveia, L., & Oliveira, A. C., (2009). Microalgae as a raw material for biofuels production. *J. Ind. Microbiol. Biotechnol., 36*, 269–274.

Granados, M. R., Acién, F. G., Gómez, C., Fernández-Sevilla, J. M., & Grima, E. M., (2012). Evaluation of flocculants for the recovery of freshwater microalgae. *Bioresour. Technol., 118*, 102–110.

Grima, E. M., Belarbi, E. H., Fernández, F. G. A., Medina, A. R., & Chisti, Y., (2003). Recovery of microalgal biomass and metabolites: Process options and economics. *Biotechnol. Adv., 20*, 491–515.

Grima, E. M., Medina, A. R., Giménez, A. G., Sánchez, P. J. A., Camacho, F. G., García, S. J. L., (1994). Comparison between extraction of lipids and fatty acids from microalgal biomass. *J. Am. Oil Chem. Soc., 71*, 955–959.

Grimi, N., Dubois, A., Marchal, L., Jubeau, S., Lebovka, N. I., & Vorobiev, E., (2014). Selective extraction from microalgae *Nannochloropsis* sp. using different methods of cell disruption. *Bioresour. Technol., 153*, 254–259.

Guldhe, A., Singh, B., Rawat, I., Ramluckan, K., & Bux, F., (2014). Efficacy of drying and cell disruption techniques on lipid recovery from microalgae for biodiesel production. *Fuel, 128*, 46–52.

Guschina, I. A., & Harwood, J. L., (2006). Lipids and lipid metabolism in eukaryotic algae. *Prog. Lipid. Res., 45*, 160–186.

Halim, R., Danquah, M. K., & Webley, P. A., (2012a). Extraction of oil from microalgae for biodiesel production: A review. *Biotechnol. Adv., 30*, 709–732.

Halim, R., Gladman, B., Danquah, M. K., & Webley, P. A., (2011). Oil extraction from microalgae for biodiesel production. *Bioresour. Technol., 102*, 178–185.

Halim, R., Harun, R., Danquah, M. K., & Webley, P. A., (2012b). Microalgal cell disruption for biofuel development. *Appl. Energy, 91*, 116–121.

Halim, R., Rupasinghe, T. W. T., Tull, D. L., & Webley, P. A., (2013). Mechanical cell disruption for lipid extraction from microalgal biomass. *Bioresour. Technol., 140*, 53–63.

Hanotu, J., Bandulasena, H. C. H., & Zimmerman, W. B., (2012). Microflotation performance for algal separation. *Biotechnol. Bioeng., 109*, 1663–1673.

Harith, Z. T., Ariff, A. B., Yusoff, F. M., Mohamed, M. S., Shariff, M., & Din, M., (2009). Effect of different flocculants on the flocculation performance of microalgae, *Chaetoceros calcitrans* cells. *Afr. J. Biotechnol., 8*, 5971–5978.

Harwood, J. L., & Jones, A. L., (1989). Lipid metabolism in algae. *Adv. Bot. Res., 16*, 1–53.

Heasman, M., Diemar, J., O'Connor, W., Sushames, T., & Foulkes, L., (2000). Development of extended shelf-life microalgae concentrate diets harvested by centrifugation for bivalve mollusks: A summary. *Aquac. Res., 31*, 637–659.

Hidalgo, P., Ciudad, G., & Navia, R., (2016). Evaluation of different solvent mixtures in esterifiable lipids extraction from microalgae *Botryococcus braunii* for biodiesel production. *Bioresour. Technol., 201*, 360–364.

Hu, Q., Sommerfeld, M., Jarvis, E., Ghirardi, M., Posewitz, M., Seibert, M., & Darzins, A., (2008). Microalgal triacylglycerols as feedstock's for biofuel production: Perspectives and advances. *Plant J., 54*, 621–639.

Iyovo, G. D., Du, G., & Chen, J., (2010). Sustainable bioenergy bioprocessing: Biomethane production, digestate as biofertilizer and as supplemental feed in algae cultivation to promote algae biofuel commercialization. *J. Microb. Biochem. Technol., 2*, 100–106.

Japar, A. S., Takriff, M. S., & Yasin, N. H. M., (2017). Harvesting microalgal biomass and lipid extraction for potential biofuel production: A review. *J. Environ. Chem. Eng., 5*, 555–563.

Kim, D. Y., Vijayan, D., Praveenkumar, R., Han, J. I., Lee, K., Park, J. Y., Chang, W. S., Lee, J. S., & Oh, Y. K., (2016). Cell-wall disruption and lipid/astaxanthin extraction from microalgae: *Chlorella* and *Haematococcus*. *Bioresour. Technol., 199*, 300–310.

Kim, J., Yoo, G., Lee, H., Lim, J., Kim, K., Kim, C. W., Park, M. S., & Yang, J. W., (2013a). Methods of downstream processing for the production of biodiesel from microalgae. *Biotechnol. Adv., 31*, 862–876.

Kim, Y. H., Park, S., Kim, M. H., Choi, Y. K., Yang, Y. H., Kim, H. J., Kim, H., Kim, H. S., Song, K. G., & Lee, S. H., (2013b). Ultrasound-assisted extraction of lipids from *Chlorella vulgaris* using [Bmim][MeSO$_4$]. *Biomass Bioenergy, 56*, 99–103.

Klok, A. J., Lamers, P. P., Martens, D. E., Draaisma, R. B., & Wijffels, R. H., (2014). Edible oils from microalgae: Insights in TAG accumulation. *Trends Biotechnol., 32*, 521–528.

Knothe, G., (2005). Dependence of biodiesel fuel properties on the structure of fatty acid alkyl esters. *Fuel Process Technol., 86*, 1059–1070.

Knothe, G., (2012). Fuel properties of highly polyunsaturated fatty acid methyl esters. prediction of fuel properties of algal biodiesel. *Energy Fuels, 26*, 5265–5273.

Knuckey, R. M., Brown, M. R., Robert, R., & Frampton, D. M. F., (2006). Production of microalgal concentrates by flocculation and their assessment as aquaculture feeds. *Aquac. Eng., 35*, 300–313.

König, R. B., Sales, R., Roselet, F., & Abreu, P. C., (2014). Harvesting of the marine microalga *Conticribra weissflogii* (Bacillariophyceae) by cationic polymeric flocculants. *Biomass Bioenergy, 68*, 1–6.

Laamanen, C. A., Ross, G. M., & Scott, J. A., (2016). Flotation harvesting of microalgae. *Renew. Sustain. Energy Rev., 58*, 75–86.

Lee, A. K., Lewis, D. M., & Ashman, P. J., (2009). Microbial flocculation, a potentially low-cost harvesting technique for marine microalgae for the production of biodiesel. *J. Appl. Phycol., 21*, 559–567.

Lee, J. Y., Yoo, C., Jun, S. Y., Ahn, C. Y., & Oh, H. M., (2010). Comparison of several methods for effective lipid extraction from microalgae. *Bioresour. Technol., 101*, 575–577.

Li, Y., Ghasemi, N. F., Garg, S., Adarme-Vega, T. C., Thurecht, K. J., Ghafor, W. A., Tannock, S., & Schenk, P. M., (2014). A comparative study: The impact of different lipid extraction methods on current microalgal lipid research. *Microb. Cell Fact*, p. 13. doi: https://doi.org/10.1186/1475-2859-13-14.

Liu, J. C., Chen, Y. M., & Ju, Y. H., (1999). Separation of algal cells from water by column flotation. *Sep. Sci. Technol., 34*, 2259–2272.

Liu, J., Huang, J., Sun, Z., Zhong, Y., Jiang, Y., & Chen, F., (2011). Differential lipid and fatty acid profiles of photoautotrophic and heterotrophic *Chlorella zofingiensis*: Assessment of algal oils for biodiesel production. *Bioresour. Technol., 102*, 106–110.

Liu, J., Liu, Y., Wang, H., & Xue, S., (2015). Direct transesterification of fresh microalgal cells. *Bioresour. Technol., 176*, 284–287.

Liu, J., Zhu, Y., Tao, Y., Zhang, Y., Li, A., Li, T., Sang, M., & Zhang, C., (2013). Freshwater microalgae harvested via flocculation induced by pH decrease. *Biotechnol. Biofuels, 6*. doi: https://doi.org/10.1186/1754-6834-6-98.

Lourenço, S. O., (2006). *Cultivo De Microalgas Marinhas-Princípios E Aplicações* (p. 606). RIMA. São Carlos.

Makareviciene, V., Andrulevičiūtė, V., Skorupskaitė, V., & Kasperovičienė, J., (2011). Cultivation of microalgae *Chlorella* sp. and *Scenedesmus* sp. as a potential biofuel feedstock. *Environ. Res. Eng. Manag., 57*, 21–27.

Mandal, S., Patnaik, R., Singh, A. K., & Mallick, N., (2013). Comparative assessment of various lipid extraction protocols and optimization of transesterification process for microalgal biodiesel production. *Environ. Technol., 34*, 2009–2018.

Markou, G., & Nerantzis, E., (2013). Microalgae for high-value compounds and biofuels production: A review with focus on cultivation under stress conditions. *Biotechnol. Adv., 31*, 1532–1542.

Mata, T. M., Martins, A. A., & Caetano, N. S., (2010). Microalgae for biodiesel production and other applications: A review. *Renew. Sustain. Energy Rev., 14*, 217–232.

Mendes, R. L., Coelho, J. P., Fernandes, H. L., Marrucho, I. J., Cabral, J. M. S., Novais, J. M., & Palavra, A. F., (1995). Applications of supercritical CO_2 extraction to microalgae and plants. *J. Chem. Technol. Biotechnol., 62*, 53–59.

Menegazzo, M. L., Petenucci, M. E., & Fonseca, G. G., (2016). Quality assessment of Nile tilapia and hybrid sorubim oils during low temperature storage. *Food Biosci., 16*, 1–4.

Menezes, R. S., Leles, M. I. G., Soares, A. T., Franco, P. I. B. M., Antoniosi, F. N. R., Sant, A. C. L., & Vieira, A. A. H., (2013). Evaluation of the potentiality of freshwater microalgae as a source of raw material for biodiesel production. *Quim. Nova, 36*, 10–15.

Milano, J., Ong, H. C., Masjuki, H. H., Chong, W. T., Lam, M. K., Loh, P. K., & Vellayan, V., (2016). Microalgae biofuels as an alternative to fossil fuel for power generation. *Renew. Sustain. Energy Rev., 58*, 180–197.

Milledge, J. J., & Heaven, S., (2013). A review of the harvesting of micro-algae for biofuel production. *Rev. Environ. Sci. Bio. Technology, 12*, 165–178.

Moazami, N., Ashori, A., Ranjbar, R., Tangestani, M., Eghtesadi, R., & Nejad, A. S., (2012). Large-scale biodiesel production using microalgae biomass of *Nannochloropsis*. *Biomass Bioenergy, 39*, 449–453.

Mohn, F. H., (1980). Experiences and strategies in the recovery of biomass from mass cultures of microalgae. In: Shelef, G., & Soeder, C. J., (eds.), *Algae Biomass* (pp. 547–571). Amsterdam.

Molina, G. E., Medina, R. A., & Gimenez, G. A., (1999). Recovery of algal PUFAs. In: Cohen, Z., (ed.), *Chemicals from Microalgae* (p. 154). London: Taylor & Francis.

Morioka, L. R. I., Matos, Â. P., Olivo, G., & Sant, A. E. S., (2014). Evaluation of flocculation and lipid extraction of *Chlorella* sp. cultivated in concentrated desalination. *Quim. Nova, 37*, 44–49.

Mortensen, S. H., Borsheim, K. Y., Rainuzzo, J. R., & Knutsen, G., (1988). Fatty acid and elemental composition of the marine diatom *Chaetoceros gracilis* Schutt. Effects of silicate deprivation, temperature and light intensity. *J. Exp. Mar. Bio. Ecol., 122*, 173–185.

Mubarak, M., Shaija, A., & Suchithra, T. V., (2015). A review on the extraction of lipid from microalgae for biodiesel production. *Algal Res., 7*, 117–123.

Nascimento, I. A., Marques, S. S. I., Cabanelas, I. T. D., Pereira, S. A., Druzian, J. I., De Souza, C. O., Vich, D. V., et al., (2013). Screening microalgae strains for biodiesel production: Lipid productivity and estimation of fuel quality based on fatty acids profiles as selective criteria. *Bioenergy Res. 6*, 1–13.

Ndikubwimana, T., Chang, J., Xiao, Z., Shao, W., Zeng, X., Ng, I. S., & Lu, Y., (2016). Flotation: A promising microalgae harvesting and dewatering technology for biofuels production. *Biotechnol. J., 11*, 315–326.

Neto, A. M. P., De Souza, R. A. S., Leon-Nino, A. D., Da Costa, J. D. A., Tiburcio, R. S., Nunes, T. A., et al., (2013) Improvement in microalgae lipid extraction using a sonication-assisted method. *Renew. Energy, 55*, 525–531.

Niraj, S. J. R., Tapare, S., Renge, V. C., Khedka, S. V., Chavan, Y. P., & Bhagat, S. L., (2011). Extraction of oil from algae by solvent extraction and oil expeller method. *Int. J. Chem. Sci., 9*, 1746–1750.

Noraini, M. Y., Ong, H. C., Badrul, M. J., & Chong, W. T., (2014). A review on potential enzymatic reaction for biofuel production from algae. *Renew. Sustain. Energy Rev., 39*, 24–34.

Oncel, S. S., (2013). Microalgae for a macroenergy world. *Renew. Sustain. Energy, Rev., 26*, 241–264.

Papazi, A., Makridis, P., & Divanach, P., (2010). Harvesting *Chlorella minutissima* using cell coagulants. *J. Appl. Phycol., 22*, 349–355.

Park, J. Y., Park, M. S., Lee, Y. C., & Yang, J. W., (2015). Advances in direct transesterification of algal oils from wet biomass. *Bioresour. Technol., 184*, 267–275.

Peña, E. H., Medina, A. R., Callejón, M. J. J., Sánchez, M. D. M., Cerdán, L. E., Moreno, P. A. G., & Grima, E. M., (2015). Extraction of free fatty acids from wet *Nannochloropsis gaditana* biomass for biodiesel production. *Renew. Energy, 75*, 366–373.

Pereira, C. M. P., Hobuss, C. B., Maciel, J. V., Ferreira, L. R., Del, P. F. B., Mesko, M. F., Jacob-Lopes, E., & Colepicolo, N. P., (2012). Biodiesel derived from microalgae: Advances and perspectives. *Quim. Nova, 35*, 2013–2018.

Petkov, G., & Garcia, G., (2007). Which are fatty acids of the green alga *Chlorella*? *Biochem. Syst. Ecol., 35*, 281–285.

Piorreck, M., & Pohl, P., (1984). Formation of biomass, total protein, chlorophylls, lipids, and fatty acids in green and blue-green algae during one growth phase. *Phytochemistry, 23*, 217–223.

Pohndorf, R. S., Camara, Á. S., Larrosa, A. P. Q., Pinheiro, C. P., Strieder, M. M., & Pinto, L. A. A., (2016). Production of lipids from microalgae *Spirulina* sp.: Influence of drying, cell disruption and extraction methods. *Biomass Bioenergy, 93*, 25–32.

Prabakaran, P., & Ravindran, A. D., (2011). A comparative study on effective cell disruption methods for lipid extraction from microalgae. *Lett. Appl. Microbiol., 53*, 150–154.

Pragya, N., Pandey, K. K., & Sahoo, P. K., (2013). A review on harvesting, oil extraction and biofuels production technologies from microalgae. *Renew. Sustain. Energy Rev., 24*, 159–171.

Ramos, M. J., Fernández, C. M., Casas, A., Rodríguez, L., & Pérez, A., (2009). Influence of fatty acid composition of raw materials on biodiesel properties. *Bioresour. Technol., 100*, 261–268.

Ranjan, A., Patil, C., & Moholkar, V. S., (2010). Mechanistic assessment of microalgal lipid extraction. *Ind. Eng. Chem. Res., 49*, 2979–2985.

Ras, M., Lardon, L., Bruno, S., Bernet, N., & Steyer, J. P., (2011). Experimental study on a coupled process of production and anaerobic digestion of *Chlorella vulgaris*. *Bioresour. Technol., 102*, 200–206.

Rashid, N., Rehman, M. S. U., & Han, J. I., (2013). Use of chitosan acid solutions to improve separation efficiency for harvesting of the microalga *Chlorella vulgaris*. *Chem. Eng. J., 226*, 238–242.

Raven, J. A., & Geider, R. J., (1988). Temperature and algal growth. *New Phytol., 110*, 441–461.

Rawat, I., Ranjith, K. R., Mutanda, T., & Bux, F., (2013). Biodiesel from microalgae: A critical evaluation from laboratory to large scale production. *Appl. Energy, 103*, 444–467.

Richmond, A., (2004). *Handbook of Microalgal Culture Biotechnology and Applied Phycology* (p. 566). Wiley-Blackwell, New Jersey.

Ríos, S. D., Salvadó, J., Farriol, X., & Torras, C., (2012). Antifouling microfiltration strategies to harvest microalgae for biofuel. *Bioresour. Technol., 119*, 406–418.

Roselet, F., Burkert, J., & Abreu, P. C., (2016). Flocculation of *Nannochloropsis oculata* using a tannin-based polymer: Bench scale optimization and pilot scale reproducibility. *Biomass Bioenergy, 87*, 55–60.

Rwehumbiza, V. M., Harrison, R., & Thomsen, L., (2012). Alum-induced flocculation of preconcentrated *Nannochloropsis salina*: Residual aluminum in the biomass, FAMEs and its effects on microalgae growth upon media recycling. *Chem. Eng. J., 200–202*, 168–175.

Ryckebosch, E., Bermúdez, S. P. C., Termote-Verhalle, R., Bruneel, C., Muylaert, K., Parra-Saldivar, R., & Foubert, I., (2014a). Influence of extraction solvent system on the extractability of lipid components from the biomass of *Nannochloropsis gaditana*. *J. Appl. Phycol., 26*, 1501–1510.

Ryckebosch, E., Bruneel, C., Termote-Verhalle, R., Goiris, K., Muylaert, K., & Foubert, I., (2014b). Nutritional evaluation of microalgae oils rich in omega-3 long chain polyunsaturated fatty acids as an alternative for fish oil. *Food Chem., 160*, 393–400.

Ryckebosch, E., Bruneel, C., Termote-Verhalle, R., Muylaert, K., & Foubert, I., (2014c). Influence of extraction solvent system on extractability of lipid components from different microalgae species. *Algal. Res., 3*, 36–43.

Ryckebosch, E., Muylaert, K., & Foubert, I., (2012). Optimization of an analytical procedure for extraction of lipids from microalgae. *J. Am. Oil. Chem. Soc., 89*, 189–198.

Sahu, A., Pancha, I., Jain, D., Paliwal, C., Ghosh, T., Patidar, S., Bhattacharya, S., & Mishra, S., (2013). Fatty acids as biomarkers of microalgae. *Phytochemistry, 89*, 53–58.

Salim, S., Bosma, R., Vermuë, M. H., & Wijffels, R. H., (2011). Harvesting of microalgae by bio-flocculation. *J. Appl. Phycol., 23*, 849–855.

Salim, S., Vermuë, M. H., & Wijffels, R. H., (2012). Ratio between auto flocculating and target microalgae affects the energy-efficient harvesting by bio-flocculation. *Bioresour. Technol., 118*, 49–55.

Sathish, A., & Sims, R. C., (2012). Biodiesel from mixed culture algae via a wet lipid extraction procedure. *Bioresour. Technol., 118*, 643–647.

Scott, S. A., Davey, M. P., Dennis, J. S., Horst, I., Howe, C. J., Lea-Smith, D. J., & Smith, A. G., (2010). Biodiesel from algae: Challenges and prospects. *Curr. Opin. Biotechnol., 21*, 277–286.

Shahzad, I., Hussain, K., Nawaz, K., Nisar, M. F., & Bhatti, K. H., (2010). Algae as an alternative and renewable resource for biofuel production. *BIOL: E-J. Life Sci., 1*, 16–23.

Sharma, K. K., Schuhmann, H., & Schenk, P. M., (2012). High lipid induction in microalgae for biodiesel production. *Energies, 5*, 1532–1553.

Shelef, G., Sukenik, A., & Green, M., (1984). *Microalgae Harvesting and Processing: A Literature Review* (p. 70). Report, Solar Energy Research Institute, Golden Colorado, SERI/STR-231-2396.

Show, K. Y., & Lee, D. J., (2014). Algal biomass harvesting. In: Pandey, A., Lee, D. J., Chisti, Y., & Soccol, C., (eds.), *Biofuels from Algae* (pp. 85–110). Elsevier. Nova Scotia.

Sibi, G., Shetty, V., & Mokashi, K., (2015). Enhanced lipid productivity approaches in microalgae as an alternate for fossil fuels: A review. *J. Energy Inst., 89*, 330–334.

Silva, A. P. F. S., Costa, M. C., Lopes, A. C., Abdala, N. E. F., Leitão, R. C., Mota, C. R., & Dos, S. A. B., (2014). Comparison of pretreatment methods for total lipids extraction from mixed microalgae. *Renew. Energy, 63*, 762–766.

Silva, G. S., Corazza, M. L., Ramos, L. P., & ZandonáFilho, A., (2014). Oil extraction of microalgae for biodiesel production. *Espaço. Energ., 21*, 12–19.

Şirin, S., Trobajo, R., Ibanez, C., & Salvadó, J., (2012). Harvesting the microalgae *Phaeodactylum tricornutum* with polyaluminum chloride, aluminum sulfate, chitosan and alkalinity-induced flocculation. *J. Appl. Phycol., 24*, 1067–1080.

Soares, A. T., Costa, D. C., Silva, B. F., Lopes, R. G., Derner, R. B., & AntoniosiFilho, N. R., (2014). Comparative analysis of the fatty acid composition of microalgae obtained by different oil extraction methods and direct biomass transesterification. *Bio. Energy Res., 7*, 1035–1044.

Soxhlet, F., (1879). Soxhlet, übergewichts analytische bestimmung des milchfettes. *Polytech. J., 232*, 461–465.

Stefanov, K., Konaklieva, M., Brechany, E. Y., & Christie, W. W., (1988). Fatty acid composition of some algae from the black sea. *Phytochemistry, 27*, 3495–3497.

Surendhiran, D., Vijay, M., & Sirajunnisa, A. R., (2014). Biodiesel production from marine microalga *Chlorella salina* using whole cell yeast immobilized on sugarcane bagasse. *J. Environ. Chem. Eng., 2*, 1294–1300.

Sydney, E. B., Da Silva, T. E., Tokarski, A., Novak, A. C., DE Carvalho, J. C., Woiciecohwski, A. L., Larroche, C., & Soccol, C. R., (2011). Screening of microalgae with potential for biodiesel production and nutrient removal from treated domestic sewage. *Appl. Energy, 88,* 3291–4329.

Taher, H., Al-Zuhair, S., Al-Marzouqi, A. H., Haik, Y., & Farid, M., (2014a). Effective extraction of microalgae lipids from wet biomass for biodiesel production. *Biomass Bioenergy, 66,* 159–167.

Taher, H., Al-Zuhair, S., Al-Marzouqi, A. H., Haik, Y., & Farid, M., (2014b). Mass transfer modeling of *Scenedesmus* sp. lipids extracted by supercritical CO_2. *Biomass Bioenergy, 70,* 530–541.

Thompson, G. A., (1996). Lipids and membrane function in green algae. *Biochim. Biophys. Acta-Lipids Metab., 1302,* 17–45.

Uduman, N., Qi, Y., Danquah, M. K., Forde, G. M., & Hoadley, A., (2010). Dewatering of microalgal cultures: A major bottleneck to algae-based fuels. *J. Renew. Sustain. Energy, 2,* 012701. doi: https://doi.org/10.1063/1.3294480.

Ursu, A. V., Marcati, A., Sayd, T., Sante-Lhoutellier, V., Djelveh, G., & Michaud, P., (2014). Extraction, fractionation, and functional properties of proteins from the microalgae *Chlorella vulgaris. Bioresour. Technol., 157,* 134–139.

Vandamme, D., Foubert, I., & Muylaert, K., (2013). Flocculation as a low-cost method for harvesting microalgae for bulk biomass production. *Trends Biotechnol., 31,* 233–239.

Vandamme, D., Foubert, I., Fraeye, I., Meesschaert, B., & Muylaert, K., (2012). Flocculation of *Chlorella vulgaris* induced by high pH: Role of magnesium and calcium and practical implications. *Bioresour. Technol., 105,* 114–119.

Vandamme, D., Foubert, I., Meesschaert, B., & Muylaert, K., (2010). Flocculation of microalgae using cationic starch. *J. Appl. Phycol., 22,* 525–530.

Vega, B. O. A., & Voltolina, D., (2007). *Metodos y Herramientas Analiticas En La Evaluacion De La Biomassa Microalgal* (p.97). Centro de Investigaciones Biológicas del Noroeste, Mexico.

Viegas, C. V., Hachemi, I., Mäki-Arvela, P., Smeds, A., Aho, A., Freitas, S. P., Gorgônio, C. M. S., et al., (2015). Algal products beyond lipids: Comprehensive characterization of different products in direct saponification of green alga *Chlorella* sp. *Algal. Res., 11,* 156–164.

Wan, C., Zhao, X. Q., Guo, S. L., Asraful-Alam, M., & Bai, F. W., (2013). Bioflocculant production from *Solibacillus silvestris* W01 and its application in cost-effective harvest of marine microalga *Nannochloropsis oceanica* by flocculation. *Bioresour. Technol., 135,* 207–212.

Wang, S. K., Hu, Y. R., Wang, F., Stiles, A. R., & Liu, C. Z., (2014). Scale-up cultivation of *Chlorella ellipsoidea* from indoor to outdoor in bubble column bioreactors. *Bioresour. Technol., 156,* 117–122.

Widjaja, A., Chien, C. C. C., & Ju, Y. H. H., (2009). Study of increasing lipid production from fresh water microalgae *Chlorella vulgaris. J. Taiwan Inst. Chem. Eng., 40,* 13–20.

Wyatt, N. B., Gloe, L. M., Brady, P. V., Hewson, J. C., Grillet, A. M., Hankins, M. G., & Pohl, P. I., (2012). Critical conditions for ferric chloride-induced flocculation of freshwater algae. *Biotechnol. Bioeng., 109,* 493–501.

Xu, Y., Purton, S., & Baganz, F., (2013). Chitosan flocculation to aid the harvesting of the microalga *Chlorella sorokiniana. Bioresour. Technol., 129,* 296–301.

Yap, R. K. L., Whittaker, M., Diao, M., Stuetz, R. M., Jefferson, B., Bulmus, V., Peirson, W. L., et al., (2014). Hydrophobically-associating cationic polymers as micro-bubble surface modifiers in dissolved air flotation for cyanobacteria cell separation. *Water Res., 61,* 253–262.

Yoo, G., Park, W. K., Kim, C. W., Choi, Y. E., & Yang, J. W., (2012). Direct lipid extraction from wet *Chlamydomonas reinhardtii* biomass using osmotic shock. *Bioresour. Technol., 123*, 717–722.

Zhang, B., & Chen, S., (2015). Effect of different organic matters on flocculation of *Chlorella sorokiniana* and optimization of flocculation conditions in swine manure wastewater. *Bioresour. Technol., 192*, 774–780.

Zhang, J., & Hu, B., (2012). A novel method to harvest microalgae via co-culture of filamentous fungi to form cell pellets. *Bioresour. Technol., 114*, 529–535.

Zhang, X., Wang, L., Sommerfeld, M., & Hu, Q., (2016). Harvesting microalgal biomass using magnesium coagulation-dissolved air flotation. *Biomass Bioenergy, 93*, 43–49.

Zheng, H., Yin, J., Gao, Z., Huang, H., Ji, X., & Dou, C., (2011). Disruption of *Chlorella vulgaris* cells for the release of biodiesel-producing lipids: A comparison of grinding, ultrasonication, bead milling, enzymatic lysis, and microwaves. *Appl. Biochem. Biotechnol., 164*, 1215–1224.

CHAPTER 10

DNA REARRANGEMENTS IN CYANOBACTERIAL NITROGEN FIXING GENES DURING HETEROCYST DEVELOPMENT

NEHA SAMI and TASNEEM FATMA

Department of Biosciences, Jamia Millia Islamia, New Delhi, India

10.1 INTRODUCTION

The most abundant gas in the Earth's atmosphere is nitrogen (78%, 3.9×10^{15} tonnes) but paradoxically, it remains unexploited until it gets fixed into usable forms like ammonia, nitrites, and nitrates. The molecular nitrogen gets reduced either by natural or industrial processes. Natural processes include both non-biological and biological methods. The biological process includes soil bacteria and blue-green algae, that fixes about 122×10^6 tonnes of nitrogen per year and is the major contributor to the global nitrogen fixation, the second being the industrial method that contributes approximately 50×10^6 nitrogen while other natural non-biological processes, like combustion and lightning, fixes only 30×10^6 tonnes of nitrogen per year globally. The industrial process employs Haber's process that derives energy from the fossil fuels and results in the emission of carbon dioxide that adversely affects the environment and human health in the long run. The assimilation of atmospheric nitrogen into nitrite, nitrate, and finally ammonia by free-living, symbiotic, and associative bacterial groups holds an immense significance in sustainable agriculture and environmental management (Staley and Reysenbach, 2002; Raymond et al., 2004; Ruffing, 2011). Nitrogen-fixers or diazotrophs have been used in agriculture as biofertilizer since antiquity. Among them, cyanobacteria (CB), the diverse group of gram-negative prokaryotes has evolved to carry out photosynthesis and nitrogen fixation simultaneously and contributes maximally to the global nitrogen cycle (LeBauer and Treseder, 2008). Free-living CB such as *Nostoc* and those existing in symbiotic associations with higher plants like *Anabaena* with the water fern *Azolla* have been known for their nitrogen-fixing ability. There are many genes that regulate nitrogen fixation like nitrogen fixation genes (*nifH, nifD,* and *nifK*), ferredoxin gene (*fdxN*), and hydrogen uptake gene (*hupL*). However, these genes in the case of CB are interrupted by direct repeats that get excised during heterocysts development and nitrogen fixation. The code for different components of the unique enzyme, nitrogenase that fixes nitrogen (Thiel and Pratte, 2001).

Nitrogenase (EC 1.18.6.1) is a highly conserved metalloenzyme among all nitrogen fixers and constitutes around 10% of the entire proteins present in the cyanobacterial cell. It is made up of two components, i.e., the dinitrogenase (Component I), an iron-molybdenum protein (Fe-Mo protein), and dinitrogenase reductase (Component II), an iron protein (Fe protein) (Figure 10.1). Based on the metal present, there are three forms of nitrogenase that include vanadium nitrogenase having iron and vanadium (Fe/V), e.g., *Azotobacter*, molybdenum nitrogenase having iron and molybdenum (Fe/Mo), e.g., *Anabaena variabilis* and nitrogenase having only iron is found in *Rhodobacter capsulatus*. Among the three forms, Molybdenum nitrogenase (Fe/Mo type) is the most common which has been comprehensively studied and therefore the most characterized (Mazur and Chui, 1982; Kentemich et al., 1991).

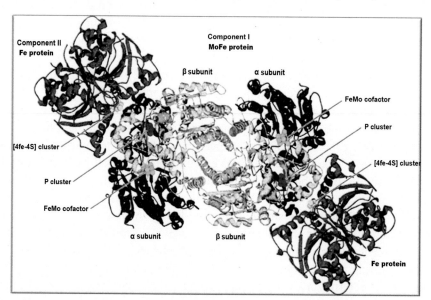

FIGURE 10.1 Structure of nitrogenase. It is made up of two components, i.e., the dinitrogenase (Component I), an iron-molybdenum protein (Fe-Mo protein), and dinitrogenase reductase (Component II), an iron protein (Fe protein).

Nitrogenase reduces molecular nitrogen through many electron transfer reactions into ammonia, with the release of hydrogen as given in the following equation. As seen below in the equation, nitrogen fixation is an energy-consuming process (Saikia and Jain, 2007).

$$N_2 + 8H^+ + 8e^-v + 16ATP \rightarrow 2NH_3 + H_2 + 16ADP + P_i$$

The three basic *nif* genes of the *nifHDK* operon, i.e., *nifH, nifD,* and *nifK* genes play an important role in formation of hetero-tetrameric complex of dinitrogenase (Component I) or molybdoferredoxin, which reduces the atmopsheric nitrogen. Genes *nifD* and *nifK* code for the α-subunit and β-subunit (α₂β₂) of the dinitrogenase (Bohme and Haselkorn, 1998). The first operon *nifBSU-fdxN,* present before *nifHDK* operon consists of *nifB, nifS, nifU,* and *fdxN* genes which play an important role in synthesis of Fe-Mo cofactors (FeMo-Co) that forms the Magnetic center, i.e., M-center (Fe$_7$MoS$_9$C) of dinitrogenase (Mazur and Chui, 1982). In case of *Nostoc commune, glbN* gene is present between *nifS* and *nifU* that codes for a membrane-associated protein, the prokaryotic Myoglobin called Cyano-globin to scavenge oxygen. *nifN* and *nifE* genes forms a heterotetramer like *nifK* and *nifD* respectively and function as scaffolds for assembly of FeMo-Co. *nifU* and *nifS* are required for the mobilization of sulphur and iron and further in the fabrication of [4Fe-4S] clusters. These clusters are then transferred to NifB, encoded by *nifB* gene, and finally transformed into NifB-cofactor, which is a precursor for FeMo-Co biosynthesis. The other component dinitrogen reductase (Component II), also known as azofer-redoxin, is a homodimer of α-subunits coded by *nifH* gene (Figure 10.1). Dinitrogenase reductase is known to transfer electrons from electron donors like flavodoxin or ferredoxin to dinitrogenase. The basic function of dinitro-genase reductase is to gain an electron from flavodoxin and ferredoxin and transfer it to dinitrogenase. Furthermore, it also helps in synthesis and proper insertion of FeMo cofactor into dinitrogenase.

CB has evolved strategies to protect this enzyme and subsist with the oxygen produced within the cell during photosynthesis that supplies energy for every cellular processes including fixation of nitrogen. They harmonize nitrogen fixation with photosynthesis through temporal and spatial separa-tion. In case of non-heterocystous, CB like *Lyngbya* and *Symploca* and the unicellular strains *Cyanothece species* (ATCC 51142), *Synechococcus elongatus* and *Gloeothece* fixation of nitrogen occurs during the night and carbon-dioxide gets fixed through photosynthesis during the day (Gallon and Hamadi, 1984; Colon-Lopez et al., 1997; Steunou et al., 2008; Toepel et al., 2008; Stal et al., 2010; Egli, 2017). This temporal separation is regulated by the cyanobacterial biological clock, the KaiABC clock, or post-translational oscillator (PTO) that evolved 3.5 billion years ago. It is composed of three proteins called KaiA, KaiB, and KaiC (Nakajima et al., 2005; Tomita et al., 2005). The concentration of KaiA, B, C as well as KaiAC, KaiBC, and KaiABC complexes concentration generates a circadian oscillation in CB and maintains the circadian rhythm (Mori et al., 2007). The protein KaiB binds to KaiC in night (darkness) to repudiate the effect of KaiA. By morning (light) the complex dissociate from each other and thus no nitrogen fixation takes place. Other proteins that help in maintaining the circadian rhythm are circadian input kinase (CikA) and its cognate-response regulator (CikR) and *Synechococcus* adaptive sensor (SasA) (Cohen and Golden, 2015).

Filamentous heterocystous CB like *Anabaena* species PCC 7120 show spatial separation between carbon and nitrogen fixation to protect the nitrogenase. As soon as the cell senses nitrogen scarcity, the genes responsible for scavenging combined nitrogen gets activated. It has been recorded by light microscopy that after 6–12 hours of nitrogen deprivation, 5–10% of the vegetative cells first differentiate into pro-heterocyst, which then matures to form heterocysts within 20 hours at 30°C. Heterocysts are morphologically and metabolically very different from vegetative cells and provide a micro-oxic environment that protects the oxygen labile nitrogenase enzyme. They are larger and more rounded than vegetative cells. The cell possesses thicker envelopes composed of superimposed layers of polysaccharides and glycolipids. The polysaccharide layer protects the delicate glycolipid layer that decreases the diffusion of oxygen (Hoiczyk and Hansel, 2000). During heterocyst differentiation, photosystem II (PSII) gets knocked down which results in an augmented rate of respiration (Wolk et al., 1994).

Development of heterocysts in CB under nitrogen limiting conditions has been divided into three stages that include Early Proheterocysts (early stage), Proheterocysts (middle stage), and Mature heterocysts (late stage) (Figure 10.2). The early stage of heterocyst differentiation comprises the initiation of a complex gene regulatory cascade that is triggered by a metabolic signal, i.e., nitrogen deprivation and simultaneous accumulation of 2-oxoglutarate (2-OG), due to lack of the key enzyme 2-OG dehydrogenase and an incomplete Kreb's cycle (Figure 10.2) (Lee et al., 1999; Laurent et al., 2005; Sandh et al., 2009; Kumar et al., 2010). NtcA, a catabolite activator protein (CAP) family transcriptional regulator functions as the global regulator of nitrogen metabolism which senses this increased level of 2-OG and further activates the expression of the master regulator of heterocyst differentiation HetR via NrrA (Herrero et al., 2001; Ehira and Ohmori., 2006; Valladares et al., 2008). The induction of *hetR* produces a transient increase of NtcA levels in a HetR-dependent manner, and this mutual up-regulation under the nitrogen deficiency status is essential for the subsequent induction of other heterocyst development genes (Muro-Pastor et al., 2002; Olmedo-Verd et al., 2005). "Early proheterocysts" loses phycocyanin (PC) and oxygen evolving capacity (Apte and Nareshkumar, 1996; Nicolaisen et al., 2009). It has also been found that the concentration of glutamate and glutamine decreases, as the nitrogen-containing compounds are first reduced to ammonium and then assimilated into 2-OG by glutamine synthetase (GS) and glutamine synthase (GOGAT) respectively in the GS-GOGAT cycle (Moreno-Vivián and Flores, 2005). The keto group of 2-OG serves as the carbon skeleton for the assimilation of ammonia which consequently activates the *nif* genes that help in nitrogen fixation (Chen et al., 2006). Following the nitrogen deprivation, nitrogen reserves degradation occurs due to inducible specific proteases in the committed. The gene encoding nitrate reductase (*narB*), nitrite reductase (*nirA*) and a nitrate/nitrite uptake transporter get activated as soon as ammonium concentration decreases, i.e., within 30 minutes (Flores et al., 2005).

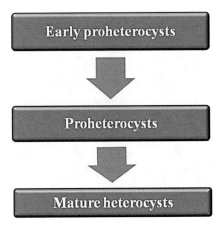

FIGURE 10.2 Developmental stages of heterocyst. It has been divided into three stages: early, middle, and late stages. Early stage of differentiation results in formation of early proheterocysts from vegetative cells. These cells undergo morphological and structural changes to form proheterocysts. After DNA rearrangements in *nif* genes, nitrogen gets fixed and also results in maturation of heterocysts.

The middle stage of heterocyst differentiation is marked by structural and physiological changes that are regulated by many genes (Table 10.1) (Kumar et al., 2010). Morphogenesis of the heterocyst envelope results in the deposition of the outer polysaccharide and inner glycolipid layer that restricts the entry of oxygen by 12 hours after the nitrogen step-down. The outer polysaccharide layer further divides into a homogeneous inner layer and an external fibrous layer that protects the delicate glycolipid layer and inhibits the oxygen diffusion (Fay, 1992; Nicolaisen et al., 2009). In addition to this, pattern formation (patterning) also takes place that is regulated by *pat* genes (Table 10.2). It gets influenced by the physiology of individual cells, cell cycle, nutrients, or signals from vegetative cells and the products of nitrogen fixation (Aldea et al., 2008). Differentiation

TABLE 10.1 Genes Regulating Structural and Physiological Changes During Middle Stage of Differentiation in Heterocysts

Genes	Protein Formed	Function
hanA	HanA	Initiates heterocyst differentiation by regulating transcription of *hetR*
hetR	HetR	Regulation of heterocyst differentiation and patterning
hetF	HetF	Positively affects localized expression of *hetR* and influences the development of heterocysts
hetP	HetP	Produces multiple heterocysts phenotype
hetC	HetC	Regulate the expression of later-acting heterocyst specific genes
hetL	HetL	Regulates the development of heterocysts
all2874	All2874	Normal development of heterocysts under high light intensity conditions
asr1734	Asr1734	Over expression inhibits the development of heterocysts

TABLE 10.2 Genes Regulating Patterning; These Genes Maintain Regular Pattern of Differentiation with Approximately One Cell in Ten Becoming Heterocyst

Genes	Protein Formed	Function
patA	PatA	Attenuate the negative effects of the PatS and HetN, thereby facilitating pattern formation in heterocyst
patB	PatB	It acts as a sensor of redox potential, the iron concentration within the cell and regulates the transcription of heterocyst-specific genes
patS	PatS	Inhibit the DNA-binding activity of HetR and inhibits formation of heterocyst

is reversible till this time. The late stage that occurs between 18 and 24 hours after nitrogen deprivation culminates with fully developed mature heterocysts and expression of the nitrogen-fixation (*nif*) genes (Kumar et al., 2010) (Figure 10.3).

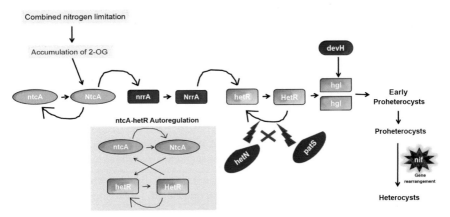

FIGURE 10.3 Signaling network regulating heterocyst development. Deprivation of nitrogen results in accumulation of 2-oxoglutarate that triggers a complex gene regulatory cascade. *ntcA* gets activated that further activates *hetR* via *nrrA*. Induction of *hetR* produces a transient increase of NtcA levels in a HetR dependent manner. Other genes regulating development of heterocysts like *devH*, *hgl* gets activated that causes morphological changes and result in formation of early proheterocysts and proheterocysts. After DNA rearrangements, *nif* genes get activated that results in nitrogen fixation.

10.2   DNA REARRANGEMENTS

The final stage of differentiation is accompanied by DNA rearrangements that take place within the *nifD, fdxN,* and *hupL* genes and involve the deletion of *nifD, fdxN,* and *hupL* elements respectively (Figure 10.4 and Table 10.3). DNA rearrangements through site-specific recombination results in the formation of mature heterocyst and activation of genes involved in nitrogen fixation. In many prokaryotic organisms, it has been seen that site-specific recombination accompanies cellular differentiation (Kunkel et al., 1990; Haselkorn, 1992). In heterocystous CB like *Anabena* sp. strain PCC 7120, DNA rearrangements are developmentally regulated and are coupled with differentiation of heterocysts in response to nitrogen-deficient conditions. The *nif* operon in vegetative cells is interrupted by large discontinuities and their excision is essential prior to the expression of the operon (Böhme, 1998; Adams and Duggan., 1999). These

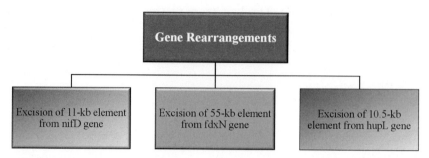

FIGURE 10.4 DNA rearrangements in heterocystous cyanobacteria. Three genetic rearrangements take place in late stages of heterocysts development that result in activation of *nif* genes. Active *nif* genes code for nitrogenase enzyme that fixes atmospheric nitrogen.

TABLE 10.3 Developmentally Regulated DNA Rearrangements During the Late Stage of Heterocyst Development that Activates *nif*-Genes and Forms the Nitrogenase Enzyme

Specifications	First DNA Rearrangement	Second DNA Rearrangement	Third DNA Rearrangement
DNA element rearranged	nif D element	fdxN element	hup L element
Size	11 kb	55 kb	10.5 kb
Interrupted element	*nifD*	*fdxN*	*hupL*
Direct repeats	11bp	5 bp	16 bp
Recombination site	GGATTACTCCG	TATTC	GGTATATTGACGACAC
Recombinase	Excisase A	Excisase B	Excisase C
Coded by gene	*xis A*	*xis F*	*xis C*
Length of ORF	1062 bp	1545 bp	9435 bp
Fate of rearrangement	Development of HC	Development of HC	Development of HC
Result	Functional *nif K,D,H* operon and nitrogen fixation	Functional *nifB-fdxN-nif S-nifU* operon and nitrogen fixation	Functional *hupL* operon and hydrogen fixation
Protein formed	α and β-subunit of nitrogenase reductase	Heterocyst-specific ferredoxin	Membrane-bound (NiFe), uptake hydrogenase

elements look like introns but no RNA transcripts have been detected from them. The evolutionary origin of these genetic elements is not clear but it has been proposed that these elements represent ancient viruses that have been maintained as they have come under the control of the host and can be excised when required (Haselkorn, 1992). All the elements get precisely excised independently from its respective gene during the late stage of heterocyst differentiation simultaneously, when the *nif*-genes also begin to be transcribed. The excision occurs due to site-specific recombination between the direct repeats that flank each of the elements (Figure 10.5). The length and sequence of the nucleotide in these direct repeats is different for each of the element, and each of them encodes its own site-specific recombinase that is responsible for its excision (Golden et al., 1985; Brusca et al., 1990). The three functional *nif*-operons codes for the different components of nitrogenase enzyme. The *nifHDK* operon encodes the molybdenum-containing nitrogenase α-subunit of the nitrogenase enzyme complex. Another operon is *nifB-fdxN-nifS-nifU*, located upstream of the previous operon and codes for the bacterial type ferredoxin. The last operon is the *hupSL* operon that codes for a heterocyst-specific uptake hydrogenase that recovers the reductant from hydrogen produced by nitrogenase.

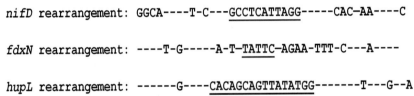

```
nifD rearrangement: GGCA----T-C---GCCTCATTAGG-----CAC-AA----C

fdxN rearrangement: ----T-G-----A-T-TATTC-AGAA-TTT-C---A----

hupL rearrangement: ------G----CACAGCAGTTATATGG-------T---G--A
```

FIGURE 10.5   Recombination sites (shown in underline). Recombinase acts at these sites present in the nifD, fdxN, and hupL genes. Excision takes place at the recombination sites and the direct repeats get removed in form of circles. The gene then get contiguous and the respective operon become functional and codes for different subunits of nitrogenase (nifD and fdxN) and hydrogenase (in case of hupL DNA rearrangement).

10.3   FIRST DNA REARRANGEMENT

The *nifHDK* operon in *Anabena* 7120 is interrupted by 11bp direct repeats present between *nifK* and *nifD* that needs to be excised. The site-specific recombinase involved in the excision of 11kb *nifD* element between two 11bp direct repeats present at the terminals of the element is called Excisase A (XisA) encoded by *xisA* (Lammers et al., 1986) (Figure 10.6). Excisase A is a member of the tyrosine family of site-specific recombinases but does not contain the conserved active site tetrad. It has a tyrosine in place of histidine (His) R-Y-R-Y in its active site (Nunes-Duby et al., 1998). The specific nucleotides present within and surrounding the direct repeats help in excision (Voziyanov et al., 1999). It has also been found that the distance between the direct repeats affect recombination and gets inhibited when more than one nucleotides in the direct repeats is mutated. It has been reported that null mutant of *xisA* forms heterocyst but could not excise the *nifD* element or grow on medium that lacks nitrogen (Golden and Wiest, 1988). However, in case of its over expression, the excision of *nifD* occurs even in vegetative cells (Brusca et al., 1990).

The excision of *nifD* element is a four-step process. In the first step, the direct repeats come closer to each other, followed by exchange of DNA strands formation of Holiday junction, branch migration and finally, excision occurs. It occurs between the nucleotides D14 and D21 on the proximal side of *nifD* and K14 and K21 on the proximal side of *nifK*.

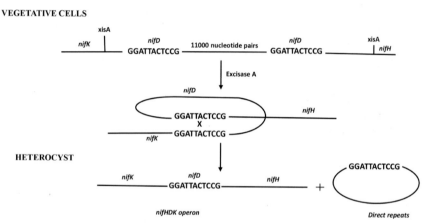

FIGURE 10.6 *nifD* rearrangement. *xisA* encodes XisA that excises the interrupting gene of 11kb in between the two nifD genes. The genes get excised and removed in form of direct repeats. Recombination results are active nifHDK operon that encodes the polypeptide components of nitrogen reductase.

10.4   SECOND DNA REARRANGEMENT

The rearrangement of *fdxN* element within the *nifB-fdxN-nifS-nifU* operon is the second DNA rearrangement. It involves the excision of 55kb *fdxN* element within the *fdxN* gene. The site-specific recombinase is known as Excisase F (XisF) encoded by *xisF* gene present on the element (Carrasco et al., 1994; Golden and Wiest, 1988; Mulligan and Haselkorn., 1989) (Figure 10.7). Active *fdxN* encodes for a heterocyst specific ferredoxin (Bohme, 1988). The recombinase belongs to the resolvase class of the serine family of site-specific recombinases. The knock out mutant of *xisF* gene forms heterocyst but does not excise the *fdxN* element or grow on medium without nitrogen as it is not able to form functional nitrogenase due to polar effects on *nifS* and *nifU* expression (Stark et al., 1992). The N-terminal of XisF exhibit end-to-end similarity to ORF469 and *spoIVCA* which are required for excision of the actinophage R4 (infects *Streptomyces parvulus* 2297) and 48kb skin element (present in *Bacillus subtilis*) respectively. So, it has been concluded that *fdxN* element is a remnant of an ancestral phage or even a lysogen of an active phage (Matsuura et al., 1995). Complementation analysis of KSR9, the null mutant of *xisF* suggests that along with XisF, other elements are also required for excision. It has been observed that no excision occurs in vegetative cells even in presence of XisF. Thus, there are other heterocyst specific factors that play an important role in this excision (Carrasco et al., 1994). Moreover, the deletion of 3.2 kb region containing two overlapping genes *xisH* and *xisI* separated by 13 bases, present 368 bases downstream of *xisF* inhibits the excision of *fdxN* element. But, their over expression (*xisHI*) results in the excision even in vegetative cells. Thus, *xisHI* effect the cell-type specificity of excision of *fdxN* element. *xisH* and *xisI* encodes polypeptide of 138 and 112 amino acids respectively. The conserved Shine-Dalgarno sequence (SD) is present within the *xisH* gene upstream of the *xisI* start codon. XisH (MW=15578 Da) and XisI (MW= 13135) has an isoelectric point (pI) of 4.6 and 6.27 respectively. Thus, *fdxN* element rearrangement requires all the three genes *xisF, xisH,* and *xisI* (Ramaswamy et al., 1997).

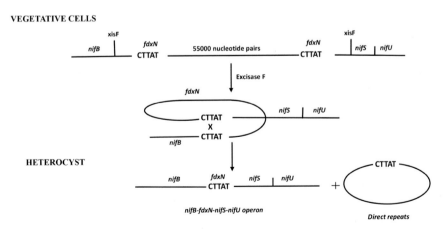

FIGURE 10.7 *fdxN* rearrangement. *xisF* encodes XisF that excises the interrupting gene of 55kb in between the two *fdxN* genes. The genes get excised and removed in form of direct repeats. Recombination results in active *nifB-fdxN-nifS-nifU* operon that encodes a heterocyst specific ferredoxin.

10.5   THIRD REARRANGEMENT

Recently, the third programmed DNA rearrangement has been reported in *Anabena* 7120 through pulsed-field gel electrophoresis of vegetative cell and heterocyst DNA. It involves the excision of a DNA element from within a hydrogenase gene (*hupL* gene) by site-specific recombination. *hupL* element of 10.5kb is excised by Excisase C (XisC) encoded by *xisC* (Matveyev et al., 1994; Carrasco et al., 1995). After excision, the stretch of 16 bp direct repeats gets removed as a circular molecule (Figure 10.8). Consequently, the *hupSL* operon becomes contiguous and the *hupL* gene encodes the large subunit of membrane-bound (Ni-Fe) hydrogenases which utilizes molecular hydrogen, i.e., fixes hydrogen, a byproduct of nitrogen fixation, for the energy-conserving reduction of electron acceptors. This improves the efficiency of nitrogen fixation in heterocysts (Przybyla et al., 1992; Wu and Mandrand, 1993). The HupL poly-peptide shows homology to many hydrogen uptake hydrogenases (Ni-Fe). It is 55% similar and 31% identical to the membrane-bound HupL of *Rhodobacter capsulatus* and 53% similar and 31% identical to the HydB of *Desulfovibrio fructosovorans* (Leclerc et al., 1988). HupL contains the nickel-binding motifs as found at the amino and carboxy termini of large subunits of (Ni-Fe) hydrog-enases but contains serine in place of proline. The 3' end of this open reading frame is similar to *hupS* gene that codes for the small subunit of the hydrogenase. It is present upstream of *hupL*. The *hupL* gene is regulated at the transcriptional level as seen in case of *nifH* gene (Carrasco et al., 1995). XisC belongs to the tyrosine family of site-specific recombinase (Carrasco et al., 1994). Any muta-tion in expression of genes does not block fixation of nitrogen in heterocyst or growth on nitrogen-free medium but certainly decreases its efficiency.

10.6   FACTORS REGULATING DEVELOPMENT OF HETEROCYSTS

10.6.1   CALCIUM CONCENTRATION

After deprivation of nitrogen, the intracellular levels of calcium ions increases as detected by using aequorin, the calcium-binding luminescent photoprotein (Torrecilla et al., 2004). In addition to this, obelin, the calcium ions reporter also confirmed the higher concentration of calcium ions in heterocysts as compared with vegetative cells. The increase in the concentration of calcium ions has been attributed to the reduced calcium sequestering protein, i.e., calcium-binding protein (CcbP) (Zhao et al., 2005). Its inactivation results in a multiple-contiguous-heterocyst formation (Mch) phenotype and its over expression inhibit the development of heterocysts. Further, it has also been hypothesized that there exists a regulatory pathway involving NtcA, HetR, and CcbP proteins that controls the concentration of intracellular free calcium ions. HetR specifically degrades CcbP in a calcium-dependent manner. This down-regulation is also assisted by the binding of NtcA to its promoter which is dependent on 2-OG. The increased concentration of calcium induces HetR's calcium-dependent serine protease and other proteolytic activities (Shi et al., 2006).

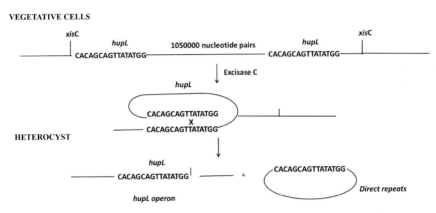

FIGURE 10.8   *hupL* rearrangement. *xisC* encodes XisC that excises the interrupting gene of 10.5kb in between the two *hupL* genes. The genes get excised and removed in form of direct repeats. Recombination results in active *hupL* operon that encodes the large subunit of membrane-bound (Ni-Fe) hydrogenases which utilize molecular hydrogen, i.e., fixes hydrogen, a byproduct of nitrogen fixation, for the energy-conserving reduction of electron acceptors.

10.6.2   *IRON CONCENTRATION*

Many iron-containing substances are needed for important biological pathways like catabolic and metabolic pathways, assimilation of nitrogen, and transport of electrons and production of chlorophyll a (Morel and Price, 2003). Perhaps, the most "iron-expensive" process is nitrogen fixation (Kupper et al., 2008). Under diazotrophic conditions, the demand for iron increases but decreases in the presence of nitrogen due to down-regulation of nitrogenase activity. Formation and development of heterocyst is regulated by iron. The master regulator of iron homeostasis is FurA as found in *Anabena* sp. PCC 7120 that controls the expression of the global regulator of nitrogen metabolism NtcA. *furA* expression gets activated by NtcA and induced in proheterocysts during the first 15 hours after nitrogen deprivation and gets stably expressed even in mature heterocysts. However, over expression of *furA* inhibits the differentiation of heterocysts at an early stage (Narayan et al., 2011). This results in the formation of a large number of filaments that exhibit semi-regularly spaced early proheterocysts suggesting that morphogenesis of heterocysts has been arrested. Consequently, this also increases the expression of *ntcA* that generally occurs after the limitation of nitrogen. Thus, FurA acts as a key connector between nitrogen metabolism and iron homeostasis (González et al., 2013).

10.7   CONCLUSION AND FUTURE PERSPECTIVES

Nitrogen fixation is the most important task attributed to CB. It is catalyzed by an oxygen-sensitive enzyme, nitrogenase. Therefore, temporal, and spatial mechanisms have been evolved in CB to protect it. DNA rearrangements in *nif* genes in heterocystous CB involve excision of interrupting elements including *nifD* element (11kb), fdxN element (55 kb), and hupL element. After excision, the operon becomes contiguous and codes for *nif* genes which then encodes nitrogenase. It occurs only after the morphological and biochemical changes have taken place in the vegetative cells.

Nitrogen is an important nutrient that affects the yield of crops. Biologically reactive nitrogen is hence regularly supplied to crops in the form of artificial nitrogen fertilizer. In the developed world, the extensive use of these fertilizers in agriculture has added a significant financial and environmental burden. On the contrary, in the developing world, the lack of fertilizer has resulted in decreased crop yields that have led to malnutrition and even deaths due to hunger. These problems could be circumvented if prokaryotic organisms and higher plant symbionts are engineered to fix more atmospheric nitrogen. This will minimize a load of fertilizers and their harmful effects on the environment. The dream for developing genetically modified crops, fruits, and vegetables with active "*nif* genes" is yet to be achieved for sustainable agricultural practices needed for organic farming.

KEYWORDS

- *Anabaena* sp. PCC 7120
- cyanobacteria
- direct element

- **DNA rearrangement**
- **excisase**
- *fdxN*
- **heterocysts**
- *hupL*
- **hydrogenase**
- **interrupting genes**
- *nifD*
- **nitrogen fixation**
- **nitrogenase**
- **operon**
- **serine family**
- **site-specific recombinase**
- **site-specific recombination**
- **tyrosine family**

REFERENCES

Adams, D. G., & Duggan, P. S., (1999). Heterocyst and akinete differentiation in cyanobacteria. *New Phytol.*, *144*(1), 3–33.

Aldea, M., Kumar, K., & Golden, J., (2008). Heterocyst development and pattern formation, In: Winans, S., & Bassler, B., (eds.), *Chemical Communication among Bacteria* (pp. 75–90). ASM Press., Washington, DC.

Apte, S. K., & Nareshkumar, G., (1996). A model for cell type-specific differential gene expression during heterocyst development and the constitution of aerobic nitrogen fixation ability in *Anabaena* sp. strain PCC 7120. *J. Biosci.*, *21*(3), 397–411.

Blair, D. F., (1995). How bacteria sense and swim. *Annu. Rev. Microbiol.*, *49*(1), 489–522.

Bohme, H., & Haselkorn, R., (1988). Molecular cloning and nucleotide sequence analysis of the gene coding for heterocyst ferredoxin from the cyanobacterium *Anabaena* sp. strain PCC 7120. *Mol. Gen. Genet.*, *214*(2), 278–285.

Böhme, H., (1998). Regulation of nitrogen fixation in heterocyst-forming cyanobacteria. *Trends Plant Sci.*, *3*(9), 346–351.

Brusca, J. S., Chastain, C. J., & Golden, J. W., (1990). Expression of the *Anabaena* sp. strain PCC 7120 xisA gene from a heterologous promoter results in excision of the nifD element. *J. Bacteriol.*, *172*(7), 3925–3931.

Budrene, E. O., & Berg, H. C., (1995). Dynamics of formation of symmetrical patterns by chemotactic bacteria. *Nature*, *376*(6535), 49–53.

Burris, R. H., (1991). Nitrogenases. *J. Biol. Chem.*, *266*(15), 9339–9342.

Carrasco, C. D., Buettner, J. A., & Golden, J. W., (1995). Programmed DNA rearrangement of a cyanobacterial hupL gene in heterocysts. *Proc. Natl. Acad. Sci. USA*, *92*(3), 791–795.

Carrasco, C. D., Ramaswamy, K. S., Ramasubramanian, T. S., & Golden, J. W., (1994). *Anabaena* xisF gene encodes a developmentally regulated site-specific recombinase. *Genes Dev.*, *8*(1), 74–83.

Chen, H., Laurent, S., Bedu, S., Ziarelli, F., Chen, H. L., Cheng, Y., et al., (2006). Studying the signaling role of 2-oxoglutaric acid using analogs that mimic the ketone and ketal forms of 2-oxoglutaric acid. *Chem. Biol.*, *13*(8), 849–856.

Cohen, S. E., & Golden, S. S., (2015). Circadian rhythms in cyanobacteria. *Microbiol Mol Biol Rev., 79*(4), 373–385.

Colon-Lopez, M. S., Sherman, D. M., & Sherman, L. A., (1997). Transcriptional and translational regulation of nitrogenase in light-dark- and continuous-light-grown cultures of the unicellular cyanobacterium *Cyanothece* sp. strain ATCC 51142. *J. Bacteriol.*, *179*(13), 4319–4327.

Egli, M., (2017). Architecture and mechanism of the central gear in an ancient molecular timer. *J. R. Soc. Interface*, *14*(128). doi: 10.1098/rsif.2016.1065.

Ehira, S., & Ohmori, M., (2006). NrrA directly regulates expression of hetR during heterocyst differentiation in the cyanobacterium *Anabaena* sp. strain PCC 7120. *J. Bacteriol.*, *188*(24), 8520–8525.

Fay, P., (1992). Oxygen relations of nitrogen fixation in cyanobacteria. *Microbiol. Rev.*, *56*(2), 340–373.

Flores, E., Frias, J. E., Rubio, L. M., & Herrero, A., (2005). Photosynthetic nitrate assimilation in cyanobacteria. *Photosynth. Res., 83*(2), 117–133.

Gallon, J. R., & Hamadi, A. F., (1984). Studies on the effects of oxygen on acetylene reduction (nitrogen fixation) in *Gloeothece* sp. ATCC 27152. *Microbiology, 130*(3), 495–503.

Golden, J. W., & Wiest, D. R., (1988). Genome rearrangement and nitrogen fixation in *Anabaena* blocked by inactivation of xisA gene. *Science*, *242*(4884), 1421–1423.

Golden, J. W., Robinson, S. J., & Haselkorn, R., (1985). Rearrangement of nitrogen fixation genes during heterocyst differentiation in the cyanobacterium *Anabaena*. *Nature*, *314*(6010), 419.

González, A., Valladares, A., Peleato, M. L., & Fillat, M. F., (2013). FurA influences heterocyst differentiation in *Anabaena* sp. PCC 7120. *FEBS Letters*, *587*(16), 2682–2690.

Haraldsen, J. D., & Sonenshein, A. L., (2003). Efficient sporulation in *Clostridium difficile* requires disruption of the sigmaK gene. *Mol. Microbiol.*, *48*(3), 811–821.

Haselkorn, R., (1992). Developmentally regulated gene rearrangements in prokaryotes. *Annu. Rev. Genet.*, *26*(1), 113–130.

Henson, B. J., Pennington, L. E., Watson, L. E., & Barnum, S. R., (2008). Excision of the nifD element in the heterocystous cyanobacteria. *Arch. Microbiol.*, *189*(4), 357–366.

Herrero, A., Muro-Pastor, A. M., & Flores, E., (2001). Nitrogen control in cyanobacteria. *J. Bacteriol.*, *183*(2), 411–425.

Hoiczyk, E., & Hansel, A., (2000). Cyanobacterial cell walls: News from an unusual prokaryotic envelope. *J. Bacteriol.*, *182*(5), 1191–1199.

Kentemich, T., Haverkamp, G., & Bothe, H., (1991). The expression of a third nitrogenase in the cyanobacterium *Anabaena variabilis*. *Z. Natur. Forsch. C., 46*(3–4), 217–222.

Kitts, P. A., & Nash, H. A., (1988). Bacteriophage lambda site-specific recombination proceeds with a defined order of strand exchanges. *J. Mol. Biol.*, *204*(1), 95–107.

Kumar, K., Mella-Herrera, R. A., & Golden, J. W., (2010). Cyanobacterial heterocysts. *Cold Spring Harb. Perspect. Biol.*, *2*(4), a000315. doi: 10.1101/cshperspect.a000315.

Kunkel, B., Losick, R., & Stragier, P., (1990). The *Bacillus subtilis* gene for the development transcription factor sigma K is generated by excision of a dispensable DNA element containing a sporulation recombinase gene. *Genes Dev.*, *4*(4), 525–535.

Kupper, H., Setlik, I., Seibert, S., Prasil, O., Setlikova, E., Strittmatter, M., et al., (2008). Iron limitation in the marine cyanobacterium *Trichodesmium* reveals new insights into regulation of photosynthesis and nitrogen fixation. *New Phytol.*, *179*(3), 784–798.

Lammers, P. J., Golden, J. W., & Haselkorn, R., (1986). Identification and sequence of a gene required for a developmentally regulated DNA excision in *Anabaena*. *Cell*, *44*(6), 905–911.

Laurent, S., Chen, H., Bedu, S., Ziarelli, F., Peng, L., & Zhang, C. C., (2005). Nonmetabolizable analog of 2-oxoglutarate elicits heterocyst differentiation under repressive conditions in *Anabaena* sp. PCC 7120. *Proc. Natl. Acad. Sci. USA, 102*(28), 9907–9912.

LeBauer, D. S., & Treseder, K. K., (2008). Nitrogen limitation of net primary productivity in terrestrial ecosystems is globally distributed. *Ecology*, *89*(2), 371–379.

Leclerc, M., Colbeau, A., Cauvin, B., & Vignais, P. M., (1988). Cloning and sequencing of the genes encoding the large and the small subunits of the H_2 uptake hydrogenase (hup) of *Rhodobacter capsulatus*. *Mol. Gen. Genet.*, *214*(1), 97–107.

Lee, H. M., Vazquez-Bermudez, M. F., & De Marsac, N. T., (1999). The global nitrogen regulator NtcA regulates transcription of the signal transducer PII (GlnB) and influences its phosphorylation level in response to nitrogen and carbon supplies in the Cyanobacterium *Synechococcus* sp. strain PCC 7942. *J. Bacteriol.*, *181*(9), 2697–2702.

Matsuura, M., Noguchi, T., Aida, T., Asayama, M., Takahashi, H., & Shirai, M., (1995). A gene essential for the site-specific excision of actinophage R4 prophage genome from the chromosome of a lysogen. *J. Gen. Appl. Microbiol.*, *41*(1), 53–61.

Matveyev, A. V., Rutgers, E., Söderbäck, E., & Bergman, B., (1994). A novel genome rearrangement involved in heterocyst differentiation of the cyanobacterium *Anabaena* sp. PCC 7120. *FEMS Microbiol. Lett.*, *116*(2), 201–207.

Mazur, B. J., & Chui, C. F., (1982). Sequence of the gene coding for the β-subunit of dinitrogenase from the blue-green alga *Anabaena*. *Proc. Natl. Acad. Sci. USA*, *79*(22), 6782–6786.

Morel, F. M., & Price, N. M., (2003). The biogeochemical cycles of trace metals in the oceans. *Science*, *300*(5621), 944–947.

Moreno-Vivián, C., & Flores, E., (2007). Nitrate assimilation in bacteria. In: *Biology of the Nitrogen Cycle* (pp. 263–282). Elsevier Science, USA.

Mori, T., Williams, D. R., Byrne, M. O., Qin, X., Egli, M., McHaourab, H. S., et al., (2007). Elucidating the ticking of an *in vitro* circadian clockwork. *PLoS Biol.*, *5*(4), e93.

Mulligan, M. E., & Haselkorn, R., (1989). Nitrogen fixation (nif) genes of the cyanobacterium *Anabaena* species strain PCC 7120. The nifB-fdxN-nifS-nifU operon. *J. Biol. Chem.*, *264*(32), 19200–19207.

Muro-Pastor, A. M., Valladares, A., Flores, E., & Herrero, A., (2002). Mutual dependence of the expression of the cell differentiation regulatory protein HetR and the global nitrogen regulator NtcA during heterocyst development. *Mol. Microbiol.*, *44*(5), 1377–1385.

Nakajima, M., Imai, K., Ito, H., Nishiwaki, T., Murayama, Y., Iwasaki, H., et al., (2005). Reconstitution of circadian oscillation of cyanobacterial KaiC phosphorylation *in vitro*. *Science*, *308*(5720), 414–415.

Narayan, O. P., Kumari, N., & Rai, L. C., (2011). Iron starvation-induced proteomic changes in *Anabaena* (*Nostoc*) sp. PCC 7120: Exploring survival strategy. *J. Microbiol. Biotechnol., 21*(2), 136–146.

Nicolaisen, K., Hahn, A., & Schleiff, E., (2009). The cell wall in heterocyst formation by *Anabaena* sp. PCC 7120. *J. Basic. Microbiol., 49*(1), 5–24.

Nunes-Duby, S. E., Kwon, H. J., Tirumalai, R. S., Ellenberger, T., & Landy, A., (1998). Similarities and differences among 105 members of the Int family of site-specific recombinases. *Nucleic Acids Res., 26*(2), 391–406.

Olmedo-Verd, E., Flores, E., Herrero, A., & Muro-Pastor, A. M., (2005). HetR-dependent and independent expression of heterocyst-related genes in an *Anabaena* strain overproducing the NtcA transcription factor. *J. Bacteriol., 187*(6), 1985–1991.

Przybyla, A. E., Robbins, J., Menon, N., & Peck, Jr. H. D., (1992). Structure-function relationships among the nickel-containing hydrogenases. *FEMS Microbiol. Lett., 88*(2), 109–136.

Ramaswamy, K. S., Carrasco, C. D., Fatma, T., & Golden, J. W., (1997). Cell-type specificity of the *Anabaena* fdxN-element rearrangement requires xisH and xisI. *Mol. Microbiol., 23*(6), 1241–1249.

Raymond, J., Siefert, J. L., Staples, C. R., & Blankenship, R. E., (2004). The natural history of nitrogen fixation. *Mol. Biol. Evol., 21*(3), 541–554.

Ruffing, A. M., (2011). Engineered cyanobacteria: Teaching old bug new tricks. *Bioeng. Bugs, 2*(3), 136–149.

Saikia, S., & Jain, V., (2007). Biological nitrogen fixation with non-legumes: An achievable target or a dogma? *Curr. Sci., 92*(3), 317–322.

Sandh, G., El-Shehawy, R., Diez, B., & Bergman, B., (2009). Temporal separation of cell division and diazotrophy in the marine diazotrophic cyanobacterium *Trichodesmium erythraeum* IMS101. *FEMS Microbiol. Lett., 295*(2), 281–288.

Sato, T., Samori, Y., & Kobayashi, Y., (1990). The cisA cistron of *Bacillus subtilis* sporulation gene spoIVC encodes a protein homologous to a site-specific recombinase. *J. Bacteriol., 172*(2), 1092–1098.

Shi, Y., Zhao, W., Zhang, W., Ye, Z., & Zhao, J., (2006). Regulation of intracellular free calcium concentration during heterocyst differentiation by HetR and NtcA in *Anabaena* sp. PCC 7120. *Proc. Natl. Acad. Sci. USA, 103*(30), 11334–11339.

Stal, L. J., Severin, I., & Bolhuis, H., (2010). The ecology of nitrogen fixation in cyanobacterial mats. *Adv. Exp. Med. Biol., 675*, 31–45.

Staley, J. T., & Reysenbach, A. L., (2002). Biodiversity of microbial life. *Curr. Opin. Biotechnol., 8*(3), 340–345.

Stark, W. M., Boocock, M. R., & Sherratt, D. J., (1992). Catalysis by site-specific recombinases. *Trends Genet, 8*(12), 432–439.

Steunou, A. S., Jensen, S. I., Brecht, E., Becraft, E. D., Bateson, M. M., Kilian, O., et al., (2008). Regulation of *nif* gene expression and the energetics of N_2 fixation over the diel cycle in a hot spring microbial mat. *ISME J., 2*(4), 364–378.

Thiel, T., & Pratte, B., (2001). Effect on heterocyst differentiation of nitrogen fixation in vegetative cells of the cyanobacterium *Anabaena variabilis* ATCC 29413. *J. Bacteriol., 183*(1), 280–286.

Toepel, J., Welsh, E., Summerfield, T. C., Pakrasi, H. B., & Sherman, L. A., (2008). Differential transcriptional analysis of the cyanobacterium *Cyanothece* sp. strain ATCC 51142 during light-dark and continuous-light growth. *J. Bacteriol., 190*(11), 3904–3913.

Tomita, J., Nakajima, M., Kondo, T., & Iwasaki, H., (2005). No transcription-translation feedback in circadian rhythm of KaiC phosphorylation. *Science, 307*(5707), 251–254.

Torrecilla, I., Leganés, F., Bonilla, I., & Fernández-Piñas, F., (2004). A calcium signal is involved in heterocyst differentiation in the cyanobacterium *Anabaena* sp. PCC7120. *Microbiology, 150*(11), 3731–3739.

Valladares, A., Flores, E., & Herrero, A., (2008). Transcription activation by NtcA and 2-oxoglutarate of three genes involved in heterocyst differentiation in the cyanobacterium *Anabaena* sp. strain PCC 7120. *J. Bacteriol., 190*(18), 6126–6133.

Voziyanov, Y., Pathania, S., & Jayaram, M., (1999). A general model for site-specific recombination by the integrase family recombinases. *Nucleic Acids Res., 27*(4), 930–941.

Wolk, C. P., Ernst, A., & Elhai, J., (1994). Heterocyst metabolism and development. In: *The Molecular Biology of Cyanobacteria* (pp. 769–823). Springer, Dordrecht.

Wu, L. F., & Mandrand, M. A., (1993). Microbial hydrogenases: Primary structure, classification, signatures, and phylogeny. *FEMS Microbiol. Rev., 10*(3–4), 243–269.

Zhao, Y., Shi, Y., Zhao, W., Huang, X., Wang, D., Brown, N., et al., (2005). CcbP, a calcium-binding protein from *Anabaena* sp. PCC 7120, provides evidence that calcium ions regulate heterocyst differentiation. *Proc. Natl. Acad. Sci. USA, 102*(16), 5744–5748.

CHAPTER 11

DETERMINATION OF KINETIC PARAMETERS FOR BIOTECHNOLOGICAL APPLICATIONS WITH MICROALGAE

NATHANYA NAYLA DE OLIVEIRA,[1] MARIANA LARA MENEGAZZO,[1,2] and GUSTAVO GRACIANO FONSECA[1,3]

[1]Laboratory of Bioengineering, Faculty of Biological and Environmental Sciences, Federal University of Grande Dourados, Dourados, MS, CEP – 79804-970, Brazil

[2]Faculty of Engineering, Federal University of Grande Dourados, Dourados, MS, CEP – 79804-970, Brazil

[3]Center of Biotechnology, Postgraduate Program in Science and Biotechnology, the University of West of Santa Catarina (UNOESC), Videira-SC, CEP – 89600-000, Brazil

11.1 INTRODUCTION

In the present day, our society has demonstrated a relevant concern about non-renewable raw materials. Thus, it was started a search for alternative sources of renewable raw materials that meet the requirements of a sustainable bio-economy. In this scenario, microalgae may present themselves as possible candidates to solve much of this problem. From its growth and development, it achieves the synthesis and bioaccumulation of interesting biotechnological products in the industrial context.

These bioproducts can be polysaccharides, proteins, lipids, enzymes, among other metabolites derived from microalga metabolism. Consequently, there are several industrial applications for each of these compounds, e.g., for biofuel, food, animal feed, cosmetic, and pharmaceutical applications (Hariskos and Posten, 2014).

Several parameters can influence growth, development, and synthesis of bioproducts by microalgae, such as temperature, carbon source (mixotrophic, autotrophic or heterotrophic), presence, or absence of some macro- and micronutrients, luminosity, pH, agitation/aeration, time of cultivation, type of bioreactor, and mode of cultivation among others. Therefore, these variables have been studied over the years in order to optimize the culture conditions for the various species of microalgae.

In this sense, it is important to evaluate the kinetic parameters of cultivations. They are obtained from simple mathematical models, providing responses on microbial behavior in different growing conditions, which results in relevant information about the physiology and metabolism of the same (Hiss, 2001). One of the main advantages of evaluating these parameters is that it does not require a complex analytical infrastructure.

11.2 MICROALGAE

The term microalga is designated to name a diverse and heterogeneous group of microorganisms including prokaryotes (cyanobacteria (CB)) and eukaryotes (most microalgae) (Castro-Puyana et al., 2013) that mostly have the ability to use carbon dioxide combined with sunlight, macro, and micronutrients present in aquatic environments to perform their photosynthetic process (Tramontin et al., 2018). One of the main characteristics of microalgae is the fact that they perform photosynthesis, i.e., they are primary producers that present a remarkable adaptability to adverse situations.

Microalgae can be found in aquatic (freshwater, saline, or brackish) environments, among others. In addition, they can be used to perform bioremediation of environments contaminated with heavy metals, inorganic, and organic matter, among others (Torres et al., 2017; De Mendonça et al., 2018).

Studies with microalgae have been intensified with years due to the fact that they represent a heterogeneous group of microorganisms that present interesting characteristics of biosynthesis and bioaccumulation of products, which are excellent sources of raw material for the production of several bioproducts (Zhan et al., 2017).

It is important to highlight the many applications in which they fit (Table 11.1), e.g., the production of pigments, such as beta-carotene (Srinivasan et al., 2015), which may present as a supplement in human nutrition. Another example is the fatty acids, such as polyunsaturated fatty acids (PUFAs), which have antioxidant activity and can be destined to the pharmaceutical industry, as well as many others that can be destined to biodiesel production

(Miazek et al., 2017). There are also mycosporine-producing microalgae, which can be targeted to sunscreens (Sathasivam et al., 2017). Other applications include, e.g., sterols, proteins, bioethanol, and bioplastics (Table 11.1).

TABLE 11.1   Microalgae Bioproducts and Applications by Species

Bioproduct	Application	Microalgae	References
Pigments	Nutritional and medical	*Neochloris oleoabundans*	Castro-Puyana et al., 2013
		Dunaliella salina, Arthrospira maxima, Synechococcus sp.	Mäki-Arvela et al., 2014
Fatty acids	Pharmaceutical and cosmetics	*Chlorella sorokiniana*	Kumar et al., 2014
		Scenedesmus obliquus	Salama et al., 2014
		Chlorella sp.	Soares et al., 2012; Kumar et al., 2014
		Schizochytrium limacinum	Nelson and Viamajala, 2016
Micosporines	Sunscreen	*Chlorella sorokiniana, Scenedesmus* spp.	Sathasivam et al., 2017
Sterols	Animal feed	*Navicula incerta, Asterionella glacialis, Haslea ostrearia*	Volkman, 2003
Proteins	Human and animal nutrition	*Tetraselmis* sp.	Schwenzfeier et al., 2011
		Chlorella vulgaris	Coustets et al., 2015
		Spirulina platensis, Nannochloropsis sp.	Wang et al., 2017
Lipids in general	Biodiesel	*Scenedesmus obliquus*	Salama et al., 2014
		Chlorella sp., *Schizochytrium limacinum*	Nelson and Viamajala, 2016
Carbohydrates	Bioethanol	*Nannochloropsis* sp.	Wang et al., 2017
		Chlorella sorokiniana	Freitas et al., 2017

11.3   KINETIC PARAMETERS

Kinetic parameters are obtained from mathematical models in order to understand the physiological and biochemical aspects of the microorganisms during cultivation. In the same way, they can help in the optimization of the culture conditions through the results obtained in a productive system and also into provide conclusions about the cost-benefit ratio (Tramontin et al., 2018).

One of the models most accepted to explain the growth of microorganisms, which is also used for microalgae, was adapted from the Monod equation and is based on the nutrient consumption (Hiss, 2001; Table 11.1).

$$\mu = \mu_{max} \frac{S}{ks + S} \tag{11.1}$$

where: μ = specific growth rate; μ_{max} = maximum specific growth rate; K_s = substrate affinity constant; S = substrate concentration.

For that, it is identified as the linear region on a ln (X) vs. time plot for batch cultivation data, which represents the exponential growth phase (EGP). The maximum specific growth rate (μ_{max}) is determined as the slope of this linear region (Nascimento et al., 2016). As growth is determined by a rate, the biomass concentration (X) may be indicated by the dried cell mass concentration (DW), optical density (OD) or cell counting (N), observed during the experiment (Dewan et al., 2012; Demirel et al., 2016; Fazeli Danesh et al., 2018). To calculate μ_{max} all these parameters must be obtained during EGP, where the specific growth rate is constant and maximum (Hiss, 2001; Eqs. 11.2 and 11.3).

$$\mu = \frac{1}{X}\left(\frac{dX}{dt}\right) \tag{11.2}$$

$$\ln \frac{X_f}{X_i} = \mu_{max}\left(t_f - t_i\right) \tag{11.3}$$

where, d = derivative; ln = Napierian logarithm; X_f = final biomass concentration; X_i = initial biomass concentration; t_f = final culture time; and t_i = initial culture time.

From μ_{max} and Eq. (11.3), one calculates the doubling time (DT) also denominated as generation time (GT) (Hiss, 2001) (Eq. 11.4), considering $X_f = 2 X_i$.

$$DT = \frac{\ln 2}{\mu_{max}} \tag{11.4}$$

Analogously, the specific substrate consumption rate (μ_s) can be calculated, according to Eq. (11.5), which combined to Eq. (11.2) gives Eq. (11.6).

$$\mu_S = \frac{1}{X}\left(-\frac{dS}{dt}\right) \qquad (11.5)$$

$$\mu_S = \frac{\mu_{max}}{Y_{X/S}} \qquad (11.6)$$

where, μ_S = specific rate of substrate consumption; d = derivative; X = biomass concentration; S = substrate concentration; t = culture time; and $Y_{X/S}$ = biomass yield.

The yields are calculated based on the consumption of a substrate for the formation of biomass or a specific product. From Eq. (11.5), it observes that the biomass yield ($Y_{X/S}$) was obtained from the quotient of the variation of biomass (X) and substrate (S) (Eq. 11.7). Similarly, it obtains the product yield ($Y_{P/S}$) from the quotient of the variation of product (P) and substrate (S) (Eq. 11.8). They are also named conversion factor of substrate into cell while ($Y_{X/S}$) and a conversion factor of substrate into product $Y_{P/S}$. It is important to highlight that measured absorbance values or number of cells must to be converted into mass values using a linear relationship of OD units (or cells) per gram dry cell mass to calculate biomass yield.

$$Y_{X/S} = \frac{X_f - X_i}{S_i - S_f} \qquad (11.7)$$

$$Y_{P/S} = \frac{P_f - P_i}{S_i - S_f} \qquad (11.8)$$

where, X_f = final biomass concentration; X_i = initial biomass concentration; P_f = final product concentration; P_i = initial product concentration; S_f = final substrate concentration; S_i = initial substrate concentration.

Other parameters that are essential to provide a visualization of the effectiveness and efficiency of a production process of microorganisms are the biomass productivity (P_x; Eq. 11.9) and product productivity (P_p; Eq. 11.10) (Ribeiro and Horii, 1999). Generally, the final biomass and product productivities are utilized, but it is worth mentioning that such productivities can be calculated during any period of the bioprocess. In most of the cases, final or maximum productivities are achieved. The biomass productivity of the microalgae is mostly determined using the dry biomass, so the biomass is centrifuged and subjected to a drying methodology and the productivity is determined after constant weigh, in terms of concentration per time.

$$P_x = \frac{X_2 - X_i}{t} \qquad (11.9)$$

$$P_p = \frac{P_2 - P_i}{t} \qquad (11.10)$$

All kinetic parameters are extremely important and can be used to evaluate growth, substrate consumption, and product formation in different bioprocesses with microalgae, as well as other microorganisms. However, the complete evaluation of these parameters is not applied in most of the studies using microalgae. One of the reasons could be the fact that the authors seek only the parameters that best fit the purpose of their work. Thus, Table 11.2 presents studies with different microalgae, their targets, and the kinetic parameters evaluated.

TABLE 11.2 Kinetic Parameters Determined for Different Species of Microalgae

Microalgae	μ_{max}	D_T	$Y_{x/s}$	$Y_{P/S}$	P_X	P_P	Aim of the study	References
Auxenochlorella prototheoides	+	−	+	−	+	−	Characterization of microalgal growth under heterotrophic conditions using organic acids as carbon source.	Turon et al., 2015
Botryococcus braunii	+	−	−	−	+	+	Investigate the effect of the variation of light intensity and increase of carbon dioxide on microalgae growth.	Huang et al., 2017
Botryococcus braunii	+	−	−	−	+	−	Evaluate the influence of photoperiod on the growth and productivity of microalgae.	Krzemińska et al., 2014
Chlorella minutissima	−	−	−	−	+	+	Use of low cost crude glycerol in heterotrophic cultivation.	Katiyar et al., 2017

TABLE 11.2 *(Continued)*

Microalgae	μ_{max}	D_T	$Y_{x/s}$	$Y_{P/S}$	P_X	P_P	Aim of the study	References
Chlorella sorokiniana	–	–	+	–	+	+	Use of beechwood (*Fagus sylvatica*) for microalgae growth and the production of fatty acids and pigments.	Miazek et al., 2017
Chlorella sorokiniana	+	–	+	–	+	–	Characterization of microalgal growth under heterotrophic conditions using organic acids as carbon source.	Turon et al., 2015
Chlorella vulgaris	+	–	–	–	+		Assess the ability to remove organic and inorganic pollutants from agricultural sources.	Baglieri et al., 2016
Chlorella vulgaris	–		–	–	+	–	Investigate decomposition of algal biomass when exposed to high temperatures.	Ali et al., 2015
Chlorella vulgaris	+	–	–	–	–	–	Verify how growth influences microalgal cell size and investigate possible effect on intracellular content.	Dewan et al., 2012
Chlorella vulgaris	–	–	–	–	+	+	Use of effluent for microalgae cultivation and lipid production.	Ebrahimian et al., 2014
Dunaliella salina	–	–	–	–	+	–	Maximize biomass production.	Khadim et al., 2018
Monoraphidium griffithii	+	–	–	–	+	–	Evaluate the feasibility of using various carbon sources to stimulate biomass growth and productivity.	Yee, 2015

TABLE 11.2   *(Continued)*

Microalgae	μ_{max}	D_T	$Y_{x/s}$	$Y_{P/S}$	P_X	P_P	Aim of the study	References
Nannochloropsis oculta	–	–	–	–	+	–	Investigate decomposition of algal biomass when exposed to high temperatures.	Ali et al., 2015
Neochloris conjuncta	+	–	–	–	+	–	Evaluate the influence of photoperiod on the growth and productivity of microalgae.	Krzemińska et al., 2014
Neochloris terrestris	+	–	–	–	+	–	Evaluate the influence of photoperiod on the growth and productivity of microalgae.	Krzemińska et al., 2014
Neochloris texensis	+	–	–	–	+	–	Evaluate the influence of photoperiod on the growth and productivity of microalgae.	Krzemińska et al., 2014
Scenedesmus abundans	+	–	–	–	+	+	Verify the effect of red and white lights with respect to cell growth, carbon assimilation, cellular respiration, production of pigment and lipids, and the fatty acid profile.	Gupta and Pawar, 2018
Scenedesmus obliquus	+	–	–	–	+	–	Evaluate the influence of photoperiod on the growth and productivity of microalgae.	Krzemińska et al., 2014
Scenedesmus quadricauda	+	–	–	–	+	–	Assess the ability to remove organic and inorganic pollutants from agricultural sources.	Baglieri et al., 2016

TABLE 11.2 *(Continued)*

Microalgae	μ_{max}	D_T	$Y_{x/s}$	$Y_{P/S}$	P_X	P_P	Aim of the study	References
Spirulina sp.	+	+	–	–	+	–	Stimulate the increase of carbohydrates in the microalgae cells.	Braga et al., 2018
Mix of microalgae	+	–	–	–	–	+	Optimization of the culture medium to increase the accumulation of lipids.	Fazeli Danesh et al., 2018

μ_{max} = maximum specific growth rate; D_T: doubling time; $Y_{x/s}$ = biomass yield; $Y_{P/S}$ = product yield; P_X: biomass productivity; P_P: product productivity: +: presence; –: absence.

11.4 EFFECT OF THE CULTIVATION CONDITIONS ON THE PHYSIOLOGY OF MICROALGAE

The main cultivation conditions that affect the microalgae physiology are their environmental and chemical requirements, specifically the sources and concentrations of carbon (Zhan et al., 2017), other nutrients, temperature, photoperiod, pH, among other variables (Kim et al., 2012).

Microalgae present specific needs for better performance in the growth, development, and synthesis of essential compounds. However, to achieve these needs, studies must be carried out in order to optimize the variables to obtain the best possible cultivation conditions. More than this, it is crucial to determine the kinetic parameters during these evaluations at different cultivation conditions to better understand the physiology and metabolism of each strain.

11.4.1 CARBON SOURCE

A requirement that needs to be evaluated before starting a microalgal culture is the carbon source that will be used because it can influence the growth, synthesis, and accumulation of biomolecules. For that, there are different types of carbon metabolism: mixotrophic, autotrophic, and heterotrophic (Abreu et al., 2012; Salati et al., 2017), depending on the nature of the carbon source, which can be inorganic or organic.

The autotrophic cultures are those where the carbon source used is the inorganic atmospheric carbon dioxide, so the microalga harbors the same and uses it for the photosynthetic process, whereby it produces the energy it needs to survive, grow, and reproduce. This system is commonly seen in natural and artificial environments (Zhan et al., 2017).

The advantage of this type of cultivation is the high availability of atmospheric carbon dioxide (Gupta and Pawar, 2018). However, the disadvantages are related to the dependence of the climatic conditions, difficulty to maintain the antiseptic environment, besides the difficulty of implementation in open environments and due bioreactors, often requiring aerators to promote a greater contact of the cell with the atmospheric carbon dioxide, which can trigger a high cost (Zhan et al., 2017).

The heterotrophic cultivations are characterized by the absence of light and supplementation of the culture medium with organic sources of carbon, e.g., glucose, glycerol, acetate, domestic effluents, among others. These cultures have the advantage of not dependence on luminosity, greater control of the production process, greater cell growth, besides the possibility of obtaining a higher content of lipids in the cells, which can be applied, e.g., for biodiesel production (Devi et al., 2012; Zhang et al., 2013; Mohan et al., 2015; Katiyar et al., 2017). The main disadvantage is contamination mainly when cultivating is carried out with glucose (Zhan et al., 2017).

The mixotrophic cultivations are characterized by using an organic carbon source and an inorganic carbon source (Salati et al., 2017), i.e., it presents the advantages of the autotrophic and heterotrophic system in the same culture. In this way, the microalga can first use the organic carbon and later make use of the carbon dioxide.

This system allows the microalga to synthesize the products that would be found separately in each culture system, in addition to a lower energy expenditure that is found in the autotrophic system (Abreu et al., 2012). It has as main advantages the obtaining of a high and fast cellular growth, which results in higher biomass productivities, and lipid accumulation, besides the decrease of carbon dioxide production, the result of the heterotrophic metabolism (Zhan et al., 2017).

11.4.2   OTHER NUTRIENTS

Microalgae nutrition is characterized mainly by its diversity, because there are some microalgae with an increased nutritional demand compared to others. Beyond the carbon source, some macros and micronutrients are part of the needs of microalgae, such as nitrogen and phosphorus. These nutrients directly affect cell metabolism.

Nitrogen is essential because it is necessary in the production of proteins as well as in the synthesis of nucleic acids, being an indispensable nutrient for cell division. As the concentration of nitrogen is increased in the cultivations, it becomes possible to obtain a high protein yield, but when this nutrient is limited, some microalgae go through a stress and, thus, avoid the protein synthesis, thus accumulating lipids, which in some contexts may be interesting (Ji et al., 2014).

Phosphate is important in cell proliferation, since it is also necessary in the synthesis of nucleic acids. They are important to maintain cell integrity, as the phospholipids are present in the cell membrane. This nutrient is associated to the higher growths achieved in shorter times (Park et al., 2014).

There are several culture media for microalgal production, but it is always necessary to pay attention to the nutritional needs of each species (and strain) and, of course, the concentration of each of the reactants. For example, *Dunaliella salina* is a marine microalga; consequently, there is a requirement for the presence of salts in high concentration for its development (Johnson et al., 1968; Morowvat and Ghasemi, 2016). In addition, some media have in their composition organic carbon sources that may stimulate increased cell productivity, e.g., glucose, when compared to some autotrophic cultures, but again it depends on the species/strain chosen.

11.4.3 pH, LUMINOSITY AND PHOTOPERIOD

pH is an important variable to be analyzed before and during a microalgal cultivation, as it can affect the availability of the chemical elements in the medium. Each species of microalga has an ideal pH for its cultivation. If the pH suffers variations, this could result in some effect in the distribution of the carbon dioxide species, being able to modify it. In addition, it affects the availability of macro and micronutrients, and thus directly affects the culture medium and the physiological aspects of the microalga (De Melo et al., 2018).

Microalgae are highly sensitive to environmental and chemical changes, so other factors that can also influence both the growth and the production of biomass and its intracellular contents are the light intensity and the photoperiod cycle (light and dark) chosen. The last one is usually utilized as 12:12 or 24:0 h of light and dark, respectively, depending of the objective.

These parameters end up exerting a greater influence on autotrophic cultures, since most microalgae require light to perform their photosynthetic process (Wahidin et al., 2013). However, it is necessary that the luminous intensity and the photoperiod chosen be sufficient so that the microalga does not suffer from the excess of light or even the lack of luminosity, which characterizes the photo-limitation, i.e., when the luminous intensity or exposure is increased in below to what microalgae supports (Chen et al., 2011).

These factors are considered as primordial in microalga cultivation, since one must be considered the species of microalga, because they present specific needs, the volume of the culture, the cellular density that can be developed throughout the cultivation in order to improve the efficiency of biomass growth and the biomass productivity (Wahidin et al., 2013). The excess of brightness, e.g., can stimulate that the excess of carbohydrate/ sugars be converted into lipids, reducing cell division (Mitra et al., 2012).

11.5   KINETIC PARAMETERS APPLIED TO MICROALGAE

The determining of kinetic parameters of microalgae cultivations is important for the understanding of the cell physiology and the optimization of the production systems. The parameters most evaluated in the studies with microalgae are the maximum specific rate of growth (μ_{max}) and the biomass productivity (Px) (Table 11.2). These two parameters together with the maximum biomass concentration (X_{max}) where compiled from the literature for microalgae from the genus *Chlorella* (Table 11.3), *Spirulina* (Table 11.4), *Scenedesmus* (Table 11.5), beyond other species (Table 11.6) in order to have a comparison in terms of carbon source and concentration, media, temperature, pH, photoperiod, and luminous intensity, once that chemical and physical aspects can influence the microalgal culture (Tables 11.3–11.6).

For *Chlorella* species, it observes results from mixotrophic and heterotrophic cultures, with different carbon sources, from glucose to industrial byproducts and residues, in addition to the autotrophic cultures with carbon dioxide (Table 11.3). For example, it can be observed that the study carried out by Mondal et al. (2016) the carbon source that most demonstrated viability for cultivation of the *Chlorella* sp. was cheese whey, as it resulted in the highest cell concentration of 1.62 g L^{-1} while the autotrophic culture presented a cell concentration of only 0.69 g L^{-1} (Table 11.3).

TABLE 11.3 Kinetic Parameters Obtained from *Chlorella* Cultivations

Microalgae	S (g L^{-1})	Carbon Source	μ_{max} (d^{-1})	PP (h:h)	X_{max} (gL^{-1})	Px (gL^{-1}d^{-1})	T (°C)	pH	Medium	Light Intensity (µmol m^{-2} s^{-1})	References
Chlorella sp.	–	–	0.158 ± 0.016	8:16	1.58	0.089 ± 0.003	22 ± 2	7.6 ± 0.2	Mod. BG11	~250	Andruleviciute et al., 2014
Chlorella sp.	2	Glycerol	0.342 ± 0.029	8:16	1.65	0.227 ± 0.027	22 ± 2	7.6 ± 0.2	BG11	~250	Andruleviciute et al., 2014
Chlorella sp.	5	Glycerol	0.188 ± 0.016	8:16	1.92	0.148 ± 0.009	22 ± 2	7.6 ± 0.2	BG11	~250	Andruleviciute et al., 2014
Chlorella sp.	10	Glycerol	0.122 ± 0.003	8:16	1.70	0.138 ± 0.003	22 ± 2	7.6 ± 0.2	BG11	~250	Andruleviciute et al., 2014
Chlorella sp. BTA 9031	*a	Carbon dioxide	0.44	12:12	0.95	0.03	25	6.9	BG11	70	Mondal et al., 2016
Chlorella sp. BTA 9031	5	Sodium acetate	1.21	12:12	1.45	0.06	25	6.9	BG11	70	Mondal et al., 2016
Chlorella sp. BTA 9031	5	Sodium bicarbonate	0.27	12:12	0.71	0.01	25	6.9	BG11	70	Mondal et al., 2016
Chlorella sp. BTA 9031	5	Fructose	1.2	12:12	1.24	0.05	25	6.9	BG11	70	Mondal et al., 2016
Chlorella sp. BTA 9031	5	Cheese whey permeate	1.8	12:12	1.62	0.07	25	6.9	BG11	70	Mondal et al., 2016
Chlorella sp. BTA 9031	5	Molasses	1.47	12:12	1.55	0.07	25	6.9	BG11	70	Mondal et al., 2016

TABLE 11.3 (Continued)

Microalgae	S (g L^{-1})	Carbon Source	μ_{max} (d^{-1})	PP (h:h)	X$_{max}$ (gL^{-1})	Px (gL^{-1}d^{-1})	T (°C)	pH	Medium	Light Intensity (μmol m^{-2} s^{-1})	References
Chlorella emersonii	*b	Carbon dioxide	0.86	–	0.39	0.028	25	–	Watanabe	25	Illman et al., 2000
Chlorella emersonii	*b	Carbon dioxide	0.46	–	1.11	0.079	25	–	Watanabe	25	Illman et al., 2000
Chlorella emersonii	*c	Carbon dioxide	0.10 ± 0.06	–	2.06	–	25	8	Jüttner	200	Borkenstein et al., 2011
Chlorella emersonii	*c	Carbon dioxide from flue gas	0.13 ± 0.08	–	2.00	–	25	8	Jüttner	200	Borkenstein et al., 2011
Chlorella emersonii	–	Livestock wastewater	0.61				35	–	Mod. BG11	160	Kim and Kim, 2017
Chlorella emersonii	–	Livestock wastewater	0.53			–	35	–	Mod. BG11	160	Kim and Kim, 2017
Chlorella emersonii	–	Livestock wastewater	0.51			–	35	–	Mod. BG11	160	Kim and Kim, 2017
Chlorella minutissima	*b	Carbon dioxide	0.43	–	0.46	0.032	25	–	Guillards	25	Illman et al., 2000
Chlorella minutissima	*b	Carbon dioxide	0.43	–	0.22	0.016	25	–	Guillards LN	25	Illman et al., 2000
Chlorella protothe–coides	10	Glycerol	0.91 ± 0.1	24:00	2.67 ± 0.3		24	–	BG11+pep.	100	Sforza et al., 2011
Chlorella protothe–coides	10	Glycerol	0.87 ± 0.01	–	1.10 ± 0.2		24	–	BG11+pep.	Without	Sforza et al., 2011

Microalgae	S (g L⁻¹)	Carbon Source	μ_{max} (d⁻¹)	PP (h:h)	X_{max} (gL⁻¹)	Px (gL⁻¹d⁻¹)	T (°C)	pH	Medium	Light Intensity (μmol m⁻² s⁻¹)	References
Chlorella protothe–coides	*b	Carbon dioxide	0.33	–	0.03	0.002	25	–	Watanabe	25	Illman et al., 2000
Chlorella protothe–coides	*b	Carbon dioxide	0.27	–	0.33	0.023	25	–	Watanabe LN	25	Illman et al., 2000
Chlorella sorokiniana	*b	Carbon dioxide	0.58	–	0.04	0.002	25	–	Watanabe	25	Illman et al., 2000
Chlorella sorokiniana	*b	Carbon dioxide	0.19	–	0.07	0.004	25	–	Watanabe LN	25	Illman et al., 2000
Chlorella sorokiniana	*e	Carbon dioxide	0.05	–	–	6.00	32	7	TBP	180 ± 20	Ojo et al., 2014
Chlorella sorokiniana	*e	Carbon dioxide	0.05	–	–	6.07	32	7	TBP	180 ± 20	Ojo et al., 2014
Chlorella sorokiniana	*e	Carbon dioxide	0.09	–	–	6.05	32	7	TBP	180 ± 20	Ojo et al., 2014
Chlorella sorokiniana	*e	Carbon dioxide	0.11	–	–	6.61	32	7	TBP	180 ± 20	Ojo et al., 2014
Chlorella sorokiniana	*e	Carbon dioxide	0.12	–	–	6.23	32	7	TBP	180 ± 20	Ojo et al., 2014
Chlorella sorokiniana	–	Livestock wastewater	0.34				35	–	Mod. BG11-nitrogen	160	Kim and Kim, 2017
Chlorella sorokiniana	–	Livestock wastewater	0.45				35	–	Mod. BG11-nitrogen	160	Kim and Kim, 2017

TABLE 11.3 (Continued)

Microalgae	S (g L⁻¹)	Carbon Source	μ_{max} (d⁻¹)	PP (h:h)	X_{max} (gL⁻¹)	Px (gL⁻¹d⁻¹)	T (°C)	pH	Medium	Light Intensity (μmol m⁻² s⁻¹)	References
Chlorella sorokiniana	–	Livestock wastewater	0.33				35	–	Mod. BG11-nitrogen	160	Kim and Kim, 2017
Chlorella variabilis	–	–	0.04 ± 0.02	–	3.76 ± 0.02	0.63 ± 0.02	–	10 ± 0.1	Zarrouk *f	1.88–555	De Bhowmick et al., 2014
Chlorella variabilis	–	–	0.09 ± 0.02	–	4.90 ± 0.08	1.07 ± 0.08	–	10 ± 0.1	Zarrouk *f	1.88–555	De Bhowmick et al., 2014
Chlorella variabilis	–	–	0.38 ± 0.04	–	8.11 ± 0.04	1.76 ± 0.04	–	10 ± 0.1	Zarrouk *f	1.88–555	De Bhowmick et al., 2014
Chlorella variabilis	–	–	0.43 ± 0.02	–	6.65 ± 0.04	1.70 ± 0.04	–	10 ± 0.1	Zarrouk *f	1.88–555	De Bhowmick et al., 2014
Chlorella variabilis	–	–	0.31 ± 0.04	–	6.00 ± 0.06	1.32 ± 0.06	–	10 ± 0.1	Zarrouk *f	1.88–555	De Bhowmick et al., 2014
Chlorella variabilis	–	–	0.21 ± 0.02	–	7.20 ± 0.01	1.55 ± 0.01	–	10 ± 0.1	Zarrouk *f	1.88–555	De Bhowmick et al., 2014
Chlorella variabilis	–	–	0.16 ± 0.03	–	5.06 ± 0.06	1.14 ± 0.06	–	10 ± 0.1	Zarrouk *f	1.88–555	De Bhowmick et al., 2014
Chlorella vulgaris	2.2	Cheese whey	–	12:12	2.59 ± 0.04	0.52 ± 0.02	25	8.4	BG11	370	Salati et al., 2017
Chlorella vulgaris	2.2	Digestate ultrafiltrate plus glycerol	–	12:12	1.67 ± 0.03	0.33 ± 0.04	25	8.4	BG11	370	Salati et al., 2017
Chlorella vulgaris	2.2	White wine lees	–	12:12	1.75 ± 0.05	0.35 ± 0.02	25	8.4	BG11	370	Salati et al., 2017

Microalgae	S (g L⁻¹)	Carbon Source	μ_{max} (d⁻¹)	PP (h:h)	X_{max} (gL⁻¹)	Px (gL⁻¹d⁻¹)	T (°C)	pH	Medium	Light Intensity (µmol m⁻² s⁻¹)	References
Chlorella vulgaris	–	Carbon dioxide	–	12:12	1.21 ± 0.02	0.24 ± 0.03	25	8.4	BG11	370	Salati et al., 2017
Chlorella vulgaris	–	Non-hydrolyzed cheese whey powder solution	0.12 ± 0.00	24:0	1.98 ± 0.43	0.32 ± 0.13	30	–	–	70	Abreu et al., 2012
Chlorella vulgaris	–	Mixture of glucose and galactose	0.47 ± 0.05	24:0	2.24 ± 0.34	0.46 ± 0.09	30	–	–	70	Abreu et al., 2012
Chlorella vulgaris	–	Hydrolyzed cheese whey powder solution	0.43 ± 0.00	24:0	3.58 ± 0.12	0.75 ± 0.01	30	–	–	70	Abreu et al., 2012
Chlorella vulgaris	*h	Carbon dioxide	0.13 ± 0.01	24:0	1.22 ± 0.12	0.10 ± 0.01	30	–	–	70	Abreu et al., 2012
Chlorella vulgaris	–	Orange peel extract	–	24:0	2.20	0.183	26	4.4–8.5	Orange peel extract	160–180	Park et al., 2014
Chlorella vulgaris	5	Glucose	–	24:0	0.65	0.052	26	4.4–8.5	Mod. BG11	160–180	Park et al., 2014
Chlorella vulgaris	–	Carbon dioxide	–	24:0	0.44	0.036	26	4.4–8.5	BG11	160–180	Park et al., 2014
Chlorella vulgaris	10	Lactose	0.29 ± 0.02	–	1.60 ± 0.01	0.12 ± 0.01	27 ± 1	6.8	BBM with cheese whey	72 ± 4	De Melo et al., 2018
Chlorella vulgaris	0.1 *g	Corn steep liquor	0.38 ± 0.01	–	2.10 ± 0.01	0.20 ± 0.02	27 ± 1	6.8	BBM with corn steep liquor	72 ± 4	De Melo et al., 2018

TABLE 11.3 (Continued)

Microalgae	S (g L⁻¹)	Carbon Source	μ_{max} (d⁻¹)	PP (h:h)	X_{max} (gL⁻¹)	P_x (gL⁻¹d⁻¹)	T (°C)	pH	Medium	Light Intensity (μmol m⁻² s⁻¹)	References
Chlorella vulgaris	0.2 *g	Vinasse	0.34 ± 0.0	–	0.77 ± 0.03	0.09 ± 0.02	27 ± 1	6.8	BBM with vinasse	72 ± 4	De Melo et al., 2018
Chlorella vulgaris	–	–	0.29 ± 0.0	–	0.91 ± 0.02	0.08 ± 0.02	27 ± 1	6.8	BBM	72 ± 4	De Melo et al., 2018
Chlorella vulgaris	1/25 *d	Monosodium glutamate	–	24:0	0.80 ± 0.05	0.049	25 ± 3	7.0	MSGW	30	Ji et al., 2014
Chlorella vulgaris	1/50 *d	Monosodium glutamate	–	24:0	0.97 ± 0.01	0.058	25 ± 3	7.0	MSGW	30	Ji et al., 2014
Chlorella vulgaris	1/100 *d	Monosodium glutamate	–	24:0	1.02 ± 0.05	0.061	25 ± 3	7.0	MSGW	30	Ji et al., 2014
Chlorella vulgaris	1/200 *d	Monosodium glutamate	–	24:0	0.78 ± 0.00	0.047	25 ± 3	7.0	MSGW	30	Ji et al., 2014
Chlorella vulgaris	1/400 *d	Monosodium glutamate	–	24:0	0.56 ± 0.02	0.034	25 ± 3	7.0	MSGW	30	Ji et al., 2014
Chlorella vulgaris	1/600 *d	Monosodium glutamate	–	24:0	0.47 ± 0.02	0.030	25 ± 3	7.0	MSGW	30	Ji et al., 2014
Chlorella vulgaris	1/800 *d	Monosodium glutamate	–	24:0	0.43 ± 0.01	0.028	25 ± 3	7.0	MSGW	30	Ji et al., 2014
Chlorella vulgaris	–	Carbon dioxide	–	24:0	0.35 ± 0.02	0.022	25 ± 3	7.0	BG11	30	Ji et al., 2014
Chlorella vulgaris	10	Glucose	–	24	7.1	–	28	6.8	MBM	11.87	Mitra et al., 2012
Chlorella vulgaris	–	Soy whey	–	24	5	–	28	6.8	–	11.87	Mitra et al., 2012

Microalgae	S (g L⁻¹)	Carbon Source	μ_max (d⁻¹)	PP (h:h)	X_max (gL⁻¹)	Px (gL⁻¹d⁻¹)	T (°C)	pH	Medium	Light Intensity (µmol m⁻² s⁻¹)	References
Chlorella vulgaris	–	Thin stillage	–	24	8.3	–	28	6.8	–	11.87	Mitra et al., 2012
Chlorella vulgaris	10	Glucose	–	24	5.6	–	28	6.8	MBM	Without	Mitra et al., 2012
Chlorella vulgaris	–	Soy whey	–	24	2.5	–	28	6.8	–	Without	Mitra et al., 2012
Chlorella vulgaris	–	Thin stillage	–	24	5.8	–	28	6.8	–	Without	Mitra et al., 2012
Chlorella vulgaris	–	Glucose *i	–	24	8.0 ± 0.2	2.0 ± 0.05	28	6.8	MBM	11.87	Mitra et al., 2012
Chlorella vulgaris	–	Soy whey *i	–	24	6.3 ± 0.1	1.6 ± 0.03	28	6.8	–	11.87	Mitra et al., 2012
Chlorella vulgaris	–	Thin stillage *i	–	24	9.8 ± 0.3	2.5 ± 0.08	28	6.8	–	11.87	Mitra et al., 2012
Chlorella vulgaris	*b	Carbon dioxide	0.99	–	0.41	0.041	25	–	Watanabe	25	Illman et al., 2000
Chlorella vulgaris	*b	Carbon dioxide	0.77	–	0.52	0.037	25	–	Watanabe LN	25	Illman et al., 2000

S: Concentration of the carbon source; μ_max: maximum specific growth rate; PP: photoperiod; X_max: maximum cellular concentration; Px: cellular productivity; LN: low nitrogen; TBP: Tris-base phosphate medium; Mod.: modified; *a: 3% v/v of CO_2 with a flow of 0.25 L min⁻¹; *b: 5% v/v of CO_2 with a flow of 1.0 L min⁻¹; *c with flue gas the final concentration of CO_2 was 67.8 ± 3.5; and with the pure CO_2 was 71.2 ± 3.5 Kg; *d: dilution described at Ji et al. (2014), without specified; *e: 2% v/v of CO_2 with a flow of 0.2 L min⁻¹; *f: related to the seasons of the year, respectively: Winter, Spring, Summer, Rainy, Autumn, and Winter; *g: concentration in % v/v (mL/L); *h: 2% v/v of CO_2; *i: refers to the use of bioreactor, the other treatments in this study occurred in Erlenmeyer flasks; "–": not mentioned.

TABLE 11.4 Kinetic Parameters Obtained from *Spirulina* Cultivations

Microalgae	S (g L^{-1})	Carbon Source	μ_{max} (d^{-1})	PP (h:h)	X$_{max}$ (g L^{-1})	Px (g L^{-1}d^{-1})	T (°C)	pH	Medium	Light Intensity (μmol m^{-2} s^{-1})	References
Spirulina maxima	–	–	0.11 ± 0.01	16:08	4.0 ± 0.1	0.20 ± 0.01	32	9.3	Schlösser	81	Barrocal et al., 2010
Spirulina maxima	1	Beet vinasse	0.13 ± 0.05	16:08	4.8 ± 0.2	0.24 ± 0.01	32	9.3	Schlösser+vinasse	81	Barrocal et al., 2010
Spirulina maxima	2	Beet vinasse	0.11 ± 0.06	16:08	4.3 ± 0.1	0.16 ± 0.01	32	9.3	Schlösser+vinasse	81	Barrocal et al., 2010
Spirulina maxima	5	Beet vinasse	0.17 ± 0.01	16:08	4.8 ± 0.2	0.15 ± 0.01	32	9.3	Schlösser+vinasse	81	Barrocal et al., 2010
Spirulina maxima	7	Beet vinasse	–	16:08	–	–	32	9.3	Schlösser+vinasse	81	Barrocal et al., 2010
Spirulina maxima	2	Beet vinasse	0.08 ± 0.05	16:08	3.5 ± 0.3	0.16 ± 0.02	32	9.3	Schlösser+alkalized diluted vinasse	81	Barrocal et al., 2010
Spirulina maxima	5	Beet vinasse	0.08 ± 0.05	16:08	3.5 ± 0.3	0.15 ± 0.02	32	9.3	Schlösser+alkalized diluted vinasse	81	Barrocal et al., 2010
Spirulina maxima	–	–	0.97	12:12	–	–	30 ± 3	9.5	AO	70	Dos Santos et al., 2016
Spirulina maxima	0.1 *a	Sugarcane vinasse	0.28	12:12	0.676	–	30 ± 3	9.5	AO	70	Dos Santos et al., 2016
Spirulina maxima	1.0 *a	Sugarcane vinasse	0.14	12:12	0.716	–	30 ± 3	9.5	AO	70	Dos Santos et al., 2016
Spirulina maxima	0.1 *a	Sugarcane vinasse	0.54	–	–	–	30 ± 3	9.5	AO	Without	Dos Santos et al., 2016

Microalgae	S (g L⁻¹)	Carbon Source	μ_{max} (d⁻¹)	PP (h:h)	X_{max} (g L⁻¹)	Px (g L⁻¹ d⁻¹)	T (°C)	pH	Medium	Light Intensity (μmol m⁻² s⁻¹)	References
Spirulina maxima	1.0 *a	Sugarcane vinasse	1.02	–	–	–	30 ± 3	9.5	AO	Without	Dos Santos et al., 2016
Spirulina platensis	0.5	Glucose	0.1272	12:12	4.02 ± 0.07	–	30	–	Zarrouk	24.3	Muliterno et al., 2005
Spirulina platensis	0.5	Glucose	0.1104	12:12	5.35 ± 0.05	–	30	–	Zarrouk	40.5	Muliterno et al., 2005
Spirulina platensis	0.5	Glucose	0.3312	12:12	1.28 ± 0.02	–	30	–	Zarrouk	24.3	Muliterno et al., 2005
Spirulina platensis	0.5	Glucose	0.1512	12:12	5.38 ± 0.09	–	30	–	Zarrouk	40.5	Muliterno et al., 2005
Spirulina platensis	1.0	Glucose	0.0336	12:12	3.18 ± 0.03	–	30	–	Zarrouk	24.3	Muliterno et al., 2005
Spirulina platensis	1.0	Glucose	0.1152	12:12	3.09 ± 0.02	–	30	–	Zarrouk	40.5	Muliterno et al., 2005
Spirulina platensis	1.0	Glucose	0.0984	12:12	2.51 ± 0.04	–	30	–	Zarrouk	24.3	Muliterno et al., 2005
Spirulina platensis	1.0	Glucose	0.2328	12:12	1.50 ± 0.04	–	30	–	Zarrouk	40.5	Muliterno et al., 2005
Spirulina platensis	–	–	0.17	–	–	0.29	30 ± 1	9.5	Schlösser	55	Converti et al., 2006
Spirulina platensis	–	–	1.7	–	–	0.42	30 ± 1	9.5	Schlösser	80	Converti et al., 2006
Spirulina platensis	–	–	0.19	–	–	0.62	30 ± 1	9.5	Schlösser	120	Converti et al., 2006

TABLE 11.4 *(Continued)*

Microalgae	S (g L⁻¹)	Carbon Source	μ_{max} (d⁻¹)	PP (h:h)	X_{max} (g L⁻¹)	Px (g L⁻¹d⁻¹)	T (°C)	pH	Medium	Light Intensity ($\mu mol\ m^{-2}\ s^{-1}$)	References
Spirulina platensis	–	–	0.13	–	–	0.18	30 ± 1	9.5	Schlösser	550	Converti et al., 2006
Spirulina platensis	–	–	0.12 *b	–	–	0.05	30 ± 1	9.5	Schlösser	55	Converti et al., 2006
Spirulina platensis	–	–	0.098 *b	–	–	0.054	30 ± 1	9.5	Schlösser	55	Converti et al., 2006
Spirulina platensis	–	–	0.081 *b	–	–	0.06	30 ± 1	9.5	Schlösser	55	Converti et al., 2006
Spirulina platensis Paracas	16.8	Sodium bicarbonate	0.031	12:12	–	0.019	30	–	Zarrouk	33.75	Reihner and Costa, 2006
Spirulina platensis LEB-52	16.8	Sodium bicarbonate	0.034	12:12	–	0.022	30	–	Zarrouk	33.75	Reihner and Costa, 2006
Spirulina platensis	16.8	Sodium bicarbonate	0.20	12:12	–	–	20	9	Zarrouk	25.65	Costa et al., 2000
Spirulina platensis	16.8	Sodium bicarbonate	0.24	12:12	–	–	20	9	Zarrouk+0.003 g/L sodium nitrate	25.65	Costa et al., 2000
Spirulina platensis	16.8	Sodium bicarbonate	0.28	12:12	–	–	20	9	Zarrouk+0.015 g/L sodium nitrate	25.65	Costa et al., 2000
Spirulina platensis	16.8	Sodium bicarbonate	0.26	12:12	–	–	20	9	Zarrouk+0.030 g/L sodium nitrate	25.65	Costa et al., 2000
Spirulina platensis	16.8	Sodium bicarbonate	0.23	12:12	–	–	20	9	Zarrouk+0.060 g/L sodium nitrate	25.65	Costa et al., 2000

Microalgae	S (g L⁻¹)	Carbon Source	μ_{max} (d⁻¹)	PP (h:h)	X_{max} (g L⁻¹)	Px (g L⁻¹d⁻¹)	T (°C)	pH	Medium	Light Intensity (μmol m⁻² s⁻¹)	References
Spirulina platensis	2.8	Sodium bicarbonate	0.25	12:12	0.75	0.145	30	8.2	Mod. Zarrouk	43.2	Andrade et al., 2008
Spirulina platensis	5	Sodium bicarbonate	0.12	12:12	0.64	0.079	30	8.2	Mod. Zarrouk	43.2	Andrade et al., 2008
Spirulina platensis	10	Sodium bicarbonate	0.16	12:12	0.55	0.088	30	8.2	Mod. Zarrouk	43.2	Andrade et al., 2008
Spirulina platensis	20	Sodium bicarbonate	0.10	12:12	0.5	0.069	30	8.2	Mod. Zarrouk	43.2	Andrade et al., 2008
Spirulina platensis	50	Sodium bicarbonate	0.09	12:12	0.21	0.011	30	8.2	Mod. Zarrouk	43.2	Andrade et al., 2008
Spirulina platensis	–	–	0.71	–	2.105	0.114	30 ± 1	8.7	Paoletti+20% potassium nitrate+15 mM ammonium chloride	156 ± 6	Rodrigues et al., 2011
Spirulina platensis	–	–	0.75	–	2.613	0.135	30 ± 1	8.7	Schlösser+20% potassium nitrate+7.5 mM ammonium chloride	156 ± 6	Rodrigues et al., 2011
Spirulina platensis	–	–	0.30	–	2.252	0.097	30 ± 1	8.7	Schlösser+20% potassium nitrate+7.5 mM ammonium chloride	156 ± 6	Rodrigues et al., 2011

TABLE 11.4 *(Continued)*

Microalgae	S (g L⁻¹)	Carbon Source	μ$_{max}$ (d⁻¹)	PP (h:h)	X$_{max}$ (g L⁻¹)	Px (g L⁻¹d⁻¹)	T (°C)	pH	Medium	Light Intensity (μmol m⁻² s⁻¹)	References
Spirulina platensis	–	–	0.60	–	2.338	0.108	30 ± 1	8.7	Schlösser+20% potassium nitrate+7.5 mM ammonium chloride+1:33 sodium carbonate/ sodium hydrogen carbonate	156 ± 6	Rodrigues et al., 2011

S: Concentration of the carbon source; μ$_{max}$: maximum specific growth rate; PP: photoperiod; X$_{max}$: maximum cellular concentration; Px: cellular productivity; *a: % of v/v (mL/L); *b: Converti et al. (2006) used different initial cell concentrations, respectively: 0.15, 0.30 and 0.50 gL⁻¹; Mod.: modified; "–": not mentioned.

TABLE 11.5 Kinetic Parameters Obtained from *Scenedesmus* Cultivations

Microalgae	S (g L⁻¹)	Carbon Source	μ_{max} (d⁻¹)	PP (h:h)	X_{max} (g L⁻¹)	Px (g L⁻¹d⁻¹)	T (°C)	pH	Medium	Light Intensity (μmol m⁻² s⁻¹)	References
Scenedesmus sp.	10	Glucose	0.23	12:12	0.757	0.079	25 ± 2	–	BG11	150	Pancha et al., 2015
Scenedesmus sp.	–	–	0.124 ± 0.027	8:16	1.67	0.094 ± 0.006	22 ± 2	7.6 ± 0.2	Mod. BG11	~250	Andruleviciute et al., 2014
Scenedesmus sp.	2	Glycerol	0.154 ± 0.023	8:16	1.87	0.141 ± 0.029	22 ± 2	7.6 ± 0.2	BG11	~250	Andruleviciute et al., 2014
Scenedesmus sp.	5	Glycerol	0.220 ± 0.003	8:16	2.15	0.229 ± 0.047	22 ± 2	7.6 ± 0.2	BG11	~250	Andruleviciute et al., 2014
Scenedesmus sp.	10	Glycerol	0.127 ± 0.011	8:16	–	0.092 ± 0.007	22 ± 2	7.6 ± 0.2	BG11	~250	Andruleviciute et al., 2014
Scenedesmus obliquus	5	Food wastewater +10% CO_2	–	–	0.42	–	27	–	–	120	Ji et al., 2015
Scenedesmus obliquus	10	Food wastewater +14% CO_2	–	–	0.44	–	27	–	–	120	Ji et al., 2015
Scenedesmus obliquus	4	Xylose	0.21	24:0	2.2	–	–	–	BG11	160	Yang et al., 2014

TABLE 11.5 *(Continued)*

Microalgae	S (g L⁻¹)	Carbon Source	μ$_{max}$ (d⁻¹)	PP (h:h)	X$_{max}$ (g L⁻¹)	Px (g L⁻¹d⁻¹)	T (°C)	pH	Medium	Light Intensity (μmol m⁻² s⁻¹)	References
Scenedesmus obliquus	4	Glucose	0.30	24:0	–	–	–	–	BG11	160	Yang et al., 2014
Scenedesmus quadricauda	2	Xylose	0.32	24:0	–	0.2113	25 ± 1	7–7.5	BG11	60	Song and Pei, 2018
Scenedesmus quadricauda	4	Xylose	0.35	24:0	–	0.3614	25 ± 1	7–7.5	BG11	60	Song and Pei, 2018
Scenedesmus quadricauda	6	Xylose	–	24:0	–	0.3108	25 ± 1	7–7.5	BG11	60	Song and Pei, 2018
Scenedesmus quadricauda	8	Xylose	–	24:0	–	0.271	25 ± 1	7–7.5	BG11	60	Song and Pei, 2018
Scenedesmus quadricauda	10	Xylose	0.35	24:0	–	0.2581	25 ± 1	7–7.5	BG11	60	Song and Pei, 2018
Scenedesmus quadricauda	0	Xylose	0.1	24:0	–	0.0186	25 ± 1	7–7.5	BG11	60	Song and Pei, 2018

S: Concentration of the carbon source; μ$_{max}$: maximum specific growth rate; PP: photoperiod; X$_{max}$: maximum cellular concentration; Px: cellular productivity; Mod.: modified; "–": not mentioned.

TABLE 11.6 Kinetic Parameters Obtained from the Cultivations of Several Other Strains of Microalgae

Microalgae	S (g L⁻¹)	Carbon Source	μ_{max} (d⁻¹)	PP (h:h)	X_{max} (g L⁻¹)	Px (g L⁻¹d⁻¹)	T (°C)	pH	Medium	Light Intensity (µmol m⁻² s⁻¹)	References
Dunaliella salina	–	–	0.24	–	0.571	0.18	25	–	Johnson	60	Morowvat and Ghasemi, 2016
Dunaliella salina	15.0	Glucose	0.37	–	0.997	0.46	25	–	Optimized Johnson (with glucose)	60	Morowvat and Ghasemi, 2016
Dunaliella salina	5	Sodium bicarbonate	0.70	–	–	–	25	–	Mod. Johnson	340	Kim et al., 2017
Dunaliella salina	5	Sodium bicarbonate	1.05	12:12	–	–	25	–	Mod. Johnson	340	Kim et al., 2017
Haematococcus pluvialis	2.4	Sodium acetate	–	12:12	0.834 ± 0.031	–	24	8	MCM	~57	Sun et al., 2015
Haematococcus pluvialis	2.4	Sodium acetate	–	12:12	0.832 ± 0.027	–	20	8	MCM	~57	Sun et al., 2015
Haematococcus pluvialis	2.4	Sodium acetate	–	12:12	0.863 ± 0.030	–	18	8	MCM	~57	Sun et al., 2015
Haematococcus pluvialis	2.4	Sodium acetate	–	12:12	0.894 ± 0.032	–	16	8	MCM	~57	Sun et al., 2015

TABLE 11.6 (Continued)

Microalgae	S (g L⁻¹)	Carbon Source	μ_{max} (d⁻¹)	PP (h:h)	X_{max} (g L⁻¹)	Px (g L⁻¹d⁻¹)	T (°C)	pH	Medium	Light Intensity (μmol m⁻² s⁻¹)	References
Haematococcus pluvialis	*a	Carbon dioxide	0.236	12:12	2.028 ± 0.09	–	23 ± 2	7	–	202.5–405	Haque et al., 2017
Haematococcus pluvialis	*a	Carbon dioxide	0.280	12:12	2.836 ± 0.03	–	23 ± 2	7	–	202.5–405	Haque et al., 2017
Haematococcus pluvialis	*a	Carbon dioxide	0.317	12:12	4.37 ± 0.07	–	23 ± 2	7	–	202.5–405	Haque et al., 2017
Haematococcus pluvialis	*a	Carbon dioxide	0.013	12:12	0.259 ± 0.02	–	23 ± 2	7	–	202.5–405	Haque et al., 2017
Nannochloropsis salina	–	Atmospheric carbon dioxide	–	12:12	0.61 ± 0.03	–	25 ± 2	7.5	f/2 Guillard solution *c	150	Fakhry and El Maghraby, 2015
Nannochloropsis salina	–	Atmospheric carbon dioxide	–	12:12	0.57 ± 0.06	–	25 ± 2	7.5	f/2 Guillard solution with Low N *b	150	Fakhry and El Maghraby, 2015
Nannochloropsis salina	–	Atmospheric carbon dioxide	–	12:12	0.53 ± 0.05	–	25 ± 2	7.5	f/2 Guillard solution with Low N *b	150	Fakhry and El Maghraby, 2015
Nannochloropsis salina	–	Atmospheric carbon dioxide	–	12:12	0.48 ± 0.02	–	25 ± 2	7.5	f/2 Guillard solution with Low N *b	150	Fakhry and El Maghraby, 2015

Microalgae	S (g L^{-1})	Carbon Source	μ_{max} (d^{-1})	PP (h:h)	X$_{max}$ (g L^{-1})	Px (g L^{-1} d^{-1})	T (°C)	pH	Medium	Light Intensity (μmol m^{-2} s^{-1})	References
Nannochloropsis salina	–	Atmospheric carbon dioxide	–	12:12	0.42 ± 0.05	–	15	7.5	f/2 Guillard solution	150	Fakhry and El Maghraby, 2015
Nannochloropsis salina	–	Atmospheric carbon dioxide	–	12:12	0.45 ± 0.04	–	20	7.5	f/2 Guillard solution	150	Fakhry and El Maghraby, 2015
Nannochloropsis salina	–	Atmospheric carbon dioxide	–	12:12	0.5 ± 0.04	–	25	7.5	f/2 Guillard solution	150	Fakhry and El Maghraby, 2015
Nannochloropsis salina	–	Atmospheric carbon dioxide	–	12:12	0.53 ± 0.02	–	30	7.5	f/2 Guillard solution	150	Fakhry and El Maghraby, 2015
Nannochloropsis salina	–	Atmospheric carbon dioxide	–	12:12	0.51 ± 0.02	–	35	7.5	f f/2 Guillard solution	150	Fakhry and El Maghraby, 2015
Nannochloropsis salina	10	Glycerol	0.36 ± 0.2	24:0	0.43 ± 0.2	–	24	–	f/2 Guillard solution modified	100 ± 10	Sforza et al., 2011
Nannochloropsis salina	10	Glycerol	0.01	–	0.05	–	24	–	Without	100 ± 10	Sforza et al., 2011

S: Concentration of the carbon source; μ_{max}: maximum specific growth rate; PP: photoperiod; X$_{max}$: maximum cellular concentration; Px: cellular productivity; *a: % of CO_2; air, 2.5, 5 and 10 v/v (mL/L); *b: Low N: Low nitrogen, respectively 25, 50 and 75% of the original levels used in the control medium; *c: the control medium.

Another example is the study carried out by Salati et al. (2017), which evaluated different industrial byproducts as a carbon source for the microalga *Chlorella vulgaris*. They reported that the higher cell concentration (2.59 g L^{-1}) and cell productivity (0.52 g L^{-1} d^{-1}) were obtained using cheese whey. There was an increase of 2-fold in relation to the autotrophic cultivation that achieved the cell concentration of 1.21 g L^{-1} and cell productivity of 0.24 g L^{-1} d^{-1} (Table 11.3). According to Abreu et al. (2012), when cheese whey is used as a carbon source, it is possible to increase the cell concentration from 1.6 to 2.9 times compared to autotrophic cultures.

In a study conducted by Barrocal et al. (2010) with *Spirulina maxima*, beet vinasse at various concentrations was used as a carbon source. For mixotrophic cultivation, the highest cell concentration obtained was 4.8 g L^{-1} and the highest cell productivity was 0.24 g L^{-1} d^{-1} while that for the autotrophic cultivation the cell concentration obtained was 4.0 g L^{-1} and the cell productivity was 0.20 g L^{-1} d^{-1} (Table 11.4). From that, it concludes that there are not such differences in growth considering these two substrates.

Another source of carbon widely used in mixotrophic cultivations is glycerol, an industrial residue from biodiesel production. A study performed by Andruleviciute et al. (2014) showed that the highest cell concentration of 1.92 g L^{-1} was obtained for *Chlorella* sp. using 5 g of glycerol per liter of medium (Table 11.3). This concentration was the same that showed the highest cell concentration of 2.15 g L^{-1} and cell productivity of 0.229 g L^{-1} d^{-1} for *Scenedesmus* sp. (Table 11.5). However, considering cell productivity, better results were obtained with glycerol at 2 g L^{-1} for *Chlorella* sp. (0.227 g L^{-1} d^{-1}; Table 11.3). Regarding autotrophic cultivations, cell concentrations of 1.58 and 1.67 g L^{-1} and cell productivities of 0.089 and 0.094 g L^{-1} d^{-1} were obtained for *Chlorella* sp. (Table 11.3) and *Scenedesmus* sp. (Table 11.5), respectively. So, it may be concluded that *Chlorella* sp. has its metabolism more adapted for both mixotrophic cultivations with glycerol and autotrophic cultivations with dioxide carbon when compared to *Scenedesmus* sp.

Considering other microalgae species, Morowvat and Ghasemi (2016) while evaluating the traditional Johnson culture medium for *Dunaliella salina*, showed that glucose at the concentration of 15 g L^{-1} was the most appropriated for increased cell productivity, which was 0.46 g L^{-1} d^{-1} for the glucose-optimized medium, whereas for the traditional medium, without glucose, was 0.18 g L^{-1} d^{-1} (Table 11.6).

In relation to the different carbon sources, it is possible to observe that mixotrophic cultures in general presented higher cell concentrations and cell productivities than the autotrophic cultures for the *Chlorella, Spirulina,*

Scenedesmus, and other microalgae (Tables 11.3–11.6). It is justified by the fact that when an organic carbon source is added and the luminosity maintained, the microalgae grow and develop is induced by making use of both organic and inorganic carbon sources, thus presenting the advantages of autotrophic and heterotrophic metabolisms in the same cultivation (Zhan et al., 2017).

11.6 CONCLUSION

The kinetic parameters are very useful for the evaluation of a productive process in terms of the influence of the variables in a complex system. They are easy to determine by using simple and basic mathematical tools. The kinetic parameters provide quick information about the behavior of microorganisms, in order to more accurately identify the most suitable growing conditions for each potential biotechnological application, taking into consideration each strain. For that, it is important to critically evaluate each one of these parameters. From the economic point of view, the maximal productivities of biomass and product are the most important parameters for defining processes, however maximal specific growth rates, and biomass and product yields are crucial for the comprehension of the physiology and metabolism of the different microalgae strains.

ACKNOWLEDGMENTS

Author Gustavo Graciano Fonseca gratefully acknowledges the Brazilian research funding agency CAPES for the financial support.

KEYWORDS

- **algal biomass**
- **biomass production**
- **bioproducts**
- **carbon source**
- **cultivation condition**
- **growth conditions**

- **mathematical models**
- **microalgae strain**
- **mixotrophic cultivation**
- **specific growth**
- **substrate concentration**

REFERENCES

Abreu, A. P., Fernandes, B., Vicente, A. A., Teixeira, J., & Dragone, G., (2012). Mixotrophic cultivation of *Chlorella vulgaris* using industrial dairy waste as organic carbon source. *Bioresour. Technol., 118*, 61–66.

Ali, S. A. M., Razzak, S. A., & Hossain, M. M., (2015). Apparent kinetics of high temperature oxidative decomposition of microalgal biomass. *Bioresour. Technol., 175*, 569–577.

Andrade, M. R., Camerini, F. V., & Costa, J. A. V., (2008). Chemical carbon losses and growth kinetics in *Spirulina* cultures. *Quím. Nova, 31*, 2031–2034.

Andruleviciute, V., Makareviciene, V., Skorupskaite, V., & Gumbyte, M., (2014). Biomass and oil content of *Chlorella* sp., *Haematococcus* sp., *Nannochloris* sp. and *Scenedesmus* sp. under mixotrophic growth conditions in the presence of technical glycerol. *J. Appl. Phycol., 26*, 83–90.

Baglieri, A., Sidella, S., Barone, V., Fragalà, F., Silkina, A., Nègre, M., & Gennari, M., (2016). Cultivating *Chlorella vulgaris* and *Scenedesmus quadricauda* microalgae to degrade inorganic compounds and pesticides in water. *Environ. Sci. Poll. Res., 23*, 18165–18174.

Barrocal, V. M., García-Cubero, M. T., Gonzáles-Benito, G., & Coca, M., (2010). Production of biomass by *Spirulina maxima* using sugar beet vinasse in growth media. *New Biotechnol., 27*, 851–856.

Borkenstein, C. G., Knoblechner, J., Frühwirth, H., & Schagerl, M., (2011). Cultivation of *Chlorella emersonii* with flue gas derived from a cement plant. *J. Appl. Phycol., 23*, 131–135.

Braga, V. S., Mastrantonio, D. J. S., Costa, J. A. V., & Morais, M. G., (2018). Cultivation strategy to stimulate high carbohydrate content in *Spirulina* biomass. *Bioresour. Technol., 269*, 221–226.

Castro-Puyana, M., Herrero, M., Urreta, I., Mendiola, J. A., Cifuentes, A., Ibáñez, E., & Suárez-Alvarez, S., (2013). Optimization of clean extraction methods to isolate carotenoids from the microalga *Neochloris oleoabundans* and subsequent chemical characterization using liquid chromatography tandem mass spectrometry. *Anal. Bioanal. Chem., 405*, 4607–4616.

Chen, X., Goh, Q. Y., Tan, W., Hossain, I., Chen, W. N., & Lau, R., (2011). Lumostatic strategy for microalgae cultivation utilizing image analysis and chlorophyll a content as design parameters. *Bioresour. Technol., 102*, 6005–6012.

Converti, A., Lodi, A., Del Borghi, A., & Solisio, C., (2006). Cultivation of *Spirulina platensis* in a combined airlift-tubular reactor system. *Biochem. Eng. J., 32*, 13–18.

Costa, J. A. V., Linde, G. A., Atala, D. I. P., Mibiell, G. M., & Krüger, R. T., (2000). Modeling of growth conditions for cyanobacterium *Spirulina platensis* in microcosms. *World J. Microbiol. Biotechnol., 16*, 15–18.

Coustets, M., Joubert-Durigneux, V., Hérault, J., Schoefs, B., Blanckaert, V., Garnier, J. P., & Teissié, J., (2015). Optimization of protein electroextraction from microalgae by a flow process. *Bioelect. Chem., 103*, 74–81.

De Bhowmick, G., Subramanian, G., Mishra, S., & Sen, R., (2014). Raceway pond cultivation of a marine microalga of Indian origin for biomass and lipid production: A case study. *Algal. Res., 6*, 201–209.

De Melo, R. G., Andrade, A. F., Bezerra, R. P., Correia, D. S., Souza, V. C., Brasileiro-Vida, A. C., Marques, D. A. V., & Porto, A. L. F., (2018). *Chlorella vulgaris* mixotrophic growth enhanced biomass productivity and reduced toxicity from agro-industrial by-products. *Chemosphere, 204*, 344–350.

De Mendonça, H. V., Ometto, J. P. H. B., Otenio, M. H., Marques, I. P. R., & Reis, A. J. D., (2018). Microalgae-mediated bioremediation and valorization of cattle wastewater previously digested in a hybrid anaerobic reactor using a photobioreactor: Comparison between batch and continuous operation. *Sci. Total Environ., 633*, 1–11.

Demirel, Z., Demirkaya, C., Imamoglu, E., & Conk, D. M., (2016). Diatom cultivation and lipid productivity for non-cryopreserved and cryopreserved cells. *Agron. Res., 14*, 1266–1273.

Devi, M. P., Subhash, G. V., & Mohan, S., (2012). Heterotrophic cultivation of mixed microalgae for lipid accumulation and wastewater treatment during sequential growth and starvation phases: Effect of nutrient supplementation. *Renew. Energy, 43*, 276–283.

Dewan, A., Kim, J., McLean, R. H., Vanapalli, S. A., & Karim, M. N., (2012). Growth kinetics of microalgae in microfluidic static droplet arrays. *Biotechnol. Bioeng., 109*, 2987–2996.

Dos Santos, R. R., Araújo, O. D. Q. F., De Medeiros, J. L., & Chaloub, R. M., (2016). Cultivation of *Spirulina maxima* in medium supplemented with sugarcane vinasse. *Bioresour. Technol., 204*, 38–48.

Ebrahimian, A., Kariminia, H. R., & Vosoughi, M., (2014). Lipid production in mixotrophic cultivation of *Chlorella vulgaris* in a mixture of primary and secondary municipal wastewater. *Renew. Energy, 71*, 502–508.

Fakhry, E. M., & El Maghraby, D. M., (2015). Lipid accumulation in response to nitrogen limitation and variation of temperature in *Nannochloropsis salina. Bot. Stud., 56*, 6.

Fazeli, D. A., Mooij, P., Ebrahimi, S., Kleerebezem, R., & Van, L. M., (2018). Effective role of medium supplementation in microalgal lipid accumulation. *Biotechnol. Bioeng., 115*, 1152–1160.

Freitas, B. C. B., Cassuriaga, A. P. A., Morais, M. G., & Costa, J. A. V., (2017). Pentoses and light intensity increase the growth and carbohydrate production and alter the protein profile of *Chlorella minutissima. Bioresour. Technol., 238*, 248–253.

Gupta, S., & Pawar, S. B., (2018). Mixotrophic cultivation of microalgae to enhance the quality of lipid for biodiesel application: Effects of scale of cultivation and light spectrum on reduction of α-linolenic acid. *Bioproc. Biosys. Eng., 41*, 531–542.

Haque, F., Dutta, A., Thimmanagari, M., & Chiang, Y. W., (2017). Integrated *Haematococcus pluvialis* biomass production and nutrient removal using bioethanol plant waste effluent. *Proc. Safety Environ. Protec., 111*, 128–137.

Hariskos, I., & Posten, C., (2014). Biorefinery of microalgae-opportunities and constraints for different production scenarios. *Biotechnol. J., 9*, 739–752.

Hiss, H., (2001). Cinética de processos fermentativos. *Biotecnologia. Industrial, 2*, 93–122.

Huang, Y. T., Lai, C. W., Wu, B. W., Lin, K. S., Wu, J. C., Hossain, M. S. A., & Wu, K. C. W., (2017). Advances in bioconversion of microalgae with high biomass and lipid productivity. *J. Taiwan. Inst. Chem. Eng., 79*, 37–42.

Illman, A. M., Scragg, A. H., & Shales, S. W., (2000). Increase in *Chlorella* strains calorific values when grown in low nitrogen medium. *Enz. Microb. Technol., 27*, 631–635.

Ji, M. K., Yun, H. S., Park, Y. T., Kabra, A. N., Oh, I. H., & Choi, J., (2015). Mixotrophic cultivation of a microalga *Scenedesmus obliquus* in municipal wastewater supplemented with food wastewater and flue gas CO_2 for biomass production. *J. Environ. Manag., 159*, 115–120.

Ji, Y., Hu, W., Li, X., Ma, G., Song, M., & Pei, H., (2014). Mixotrophic growth and biochemical analysis of *Chlorella vulgaris* cultivated with diluted monosodium glutamate wastewater. *Bioresour. Technol., 152*, 471–476.

Johnson, M. K., Johnson, E. J., MacElroy, R. D., Speer, H. L., & Bruff, B. S., (1968). Effects of salts on the halophilic alga *Dunaliella viridis. J. Bacteriol., 95*, 1461–1468.

Katiyar, R., Gurjar, B. R., Bharti, R. K., Kumar, A., Biswas, S., & Pruthi, V., (2017). Heterotrophic cultivation of microalgae in photobioreactor using low cost crude glycerol for enhanced biodiesel production. *Renew. Energy, 113*, 1359–1365.

Khadim, S. R., Singh, P., Singh, A. K., Tiwari, A., Mohanta, A., & Asthana, R. K., (2018). Mass cultivation of *Dunaliella salina* in a flat plate photobioreactor and its effective harvesting. *Bioresour. Technol., 270*, 20–29.

Kim, G. Y., Heo, J., Kim, H. S., & Han, J. I., (2017). Bicarbonate-based cultivation of *Dunaliella salina* for enhancing carbon utilization efficiency. *Bioresour. Technol., 237*, 72–77.

Kim, J. Y., & Kim, H. W., (2017). Photoautotrophic microalgae screening for tertiary treatment of livestock wastewater and bioresource recovery. *Water, 9*, 192.

Kim, W., Park, J. M., Gim, H. G., Jeong, S. H., Kang, C. M., Kim, D. J., & Kim, S. W., (2012). Optimization of culture conditions and comparison of biomass productivity of three green algae. *Bioproc. Biosys. Eng., 35*, 19–27.

Krzemińska, I., Pawlik-Skowrońska, B., Trzcińska, M., & Tys, J., (2014). Influence of photoperiods on the growth rate and biomass productivity of green microalgae. *Bioproc. Biosys. Eng., 37*, 735–741.

Kumar, V., Muthuraj, M., Palabhanvi, B., Ghoshal, A. K., & Das, D., (2014). Evaluation and optimization of two stage sequential *in situ* transesterification process for fatty acid methyl ester quantification from microalgae. *Renew. Energy, 68*, 560–569.

Mäki-Arvela, P., Hachemi, I., & Murzin, D. Y., (2014). Comparative study of the extraction methods for recovery of carotenoids from algae: Extraction kinetics and effect of different extraction parameters. *J. Chem. Technol. Biotechnol., 89*, 1607–1626.

Miazek, K., Remacle, C., Richel, A., & Goffin, D., (2017). Beech wood *Fagus sylvatica* dilute-acid hydrolysate as a feedstock to support *Chlorella sorokiniana* biomass, fatty acid and pigment production. *Bioresour. Technol., 230*, 122–131.

Mitra, D., Van, L. J. H., & Lamsal, B., (2012). Heterotrophic/mixotrophic cultivation of oleaginous *Chlorella vulgaris* on industrial co-products. *Algal Res., 1*, 40–48.

Mohan, S. V., Rohit, M. V., Chiranjeevi, P., Chandra, R., & Navaneeth, B., (2015). Heterotrophic microalgae cultivation to synergize biodiesel production with waste remediation: Progress and perspectives. *Bioresour Technol., 184*, 169–178.

Mondal, M., Ghosh, A., Sharma, A. S., Tiwari, O. N., Gayen, K., Mandal, M. K., & Halder, G. N., (2016). Mixotrophic cultivation of *Chlorella* sp. BTA 9031 and *Chlamydomonas* sp. BTA 9032 isolated from coal field using various carbon sources for biodiesel production. *Energy Conver. Manag., 124*, 297–304.

Morowvat, M. H., & Ghasemi, Y., (2016). Culture medium optimization for enhanced β-carotene and biomass production by *Dunaliella salina* in mixotrophic culture. *Biocatal Agric. Biotech., 7*, 217–223.

Muliterno, A., Mosele, P. C., Costa, J. A. V., Hemkemeier, M., Bertolin, T. E., & Colla, L. M., (2005). Mixotrophic growth of *Spirulina platensis* in fed-batch mode. *Ciên. Agrotecnol., 29*, 1132–1138.

Nascimento, V. M., Silva, L. F., Gomez, J. G. C., & Fonseca, G. G., (2016). Growth of *Burkholderia sacchari* LFM 101 cultivated in glucose, sucrose, and glycerol. *Sci. Agric., 73*, 429–433.

Nelson, D. R., & Viamajala, S., (2016). One-pot synthesis and recovery of fatty acid methyl esters (FAMEs) from microalgae biomass. *Catal. Today, 269*, 29–39.

Ojo, E. O., Auta, H., Baganz, F., & Lye, G. J., (2014). Engineering characterization of a shaken, single-use photobioreactor for early stage microalgae cultivation using *Chlorella sorokiniana. Bioresour. Technol., 173*, 367–375.

Pancha, I., Chokshi, K., & Mishra, S., (2015). Enhanced biofuel production potential with nutritional stress amelioration through optimization of carbon source and light intensity in *Scenedesmus* sp. CCNM 1077. *Bioresour. Technol., 179*, 565–572.

Park, W. K., Moon, M., Kwak, M. S., Jeon, S., Cho, G. G., Yang, J. W., & Lee, B., (2014). Use of orange peel extract for mixotrophic cultivation of *Chlorella vulgaris*: Increased production of biomass and FAMEs. *Bioresour Technol., 171*, 343–349.

Reihner, C. O., & Costa, J. A. V., (2006). Repeated batch cultivation of the microalga *Spirulina platensis. World J. Microbiol. Biotechnol., 22*, 937–943.

Ribeiro, C. A. F., & Horii, J., (1999). Potentialities of yeast strains of *Saccharomyces cerevisiae* for sugar cane juice fermentation. *Sci. Agric., 56*, 255–263.

Rodrigues, M. S., Ferreira, L. S., Converti, A., Sato, S., & Carvalho, J. C. M., (2011). Influence of ammonium sulphate feeding time on fed-batch *Arthrospira (Spirulina) platensis* cultivation and biomass composition with and without pH control. *Bioresour. Technol., 102*, 6587–6592.

Salama, E. S., Kabra, A. N., Ji, M. K., Kim, J. R., Min, B., & Jeon, B. H., (2014). Enhancement of microalgae growth and fatty acid content under the influence of phytohormones. *Bioresour. Technol., 172*, 97–103.

Salati, S., D'Imporzano, G., Menin, B., Veronesi, D., Scaglia, B., Abbruscato, P., Mariani, P., & Adani, F., (2017). Mixotrophic cultivation of *Chlorella* for local protein production using agro-food by-products. *Bioresour. Technol., 230*, 82–89.

Sathasivam, R., Radhakrishnan, R., Hashem, A., & Abdallah, E. F., (2017). Microalgae metabolites: A rich source for food and medicine. *Saudi J. Biol. Sci.* https://doi.org/10.1016/j.sjbs.2017.11.003 (accessed on 26 May 2020).

Schwenzfeier, A., Wierenga, P. A., & Gruppen, H., (2011). Isolation and characterization of soluble protein from the green microalgae *Tetraselmis* sp. *Bioresour. Technol., 102*, 9121–9127.

Sforza, E., Cipriani, R., Morosinotto, T., Bertucco, A., & Giacometti, G. M., (2012). Excess CO_2 supply inhibits mixotrophic growth of *Chlorella protothecoides* and *Nannochloropsis salina. Bioresour. Technol., 104*, 523–529.

Soares, B. M., Vieira, A. A., Lemões, J. S., Santos, C. M., Mesko, M. F., Primel, E. G., et al., (2012). Investigation of major and trace element distribution in the extraction–transesterification process of fatty acid methyl esters from microalgae *Chlorella* sp. *Bioresour. Technol., 110*, 730–734.

Song, M., & Pei, H., (2018). The growth and lipid accumulation of *Scenedesmus quadricauda* during batch mixotrophic/heterotrophic cultivation using xylose as a carbon source. *Bioresour. Technol., 263*, 525–531.

Srinivasan, R., Kumar, V. A., Kumar, D., Ramesh, N., Babu, S., & Gothandam, K. M., (2015). Effect of dissolved inorganic carbon on β-carotene and fatty acid production in *Dunaliella* sp. *Appl. Biochem. Biotechnol., 175*, 2895–2906.

Sun, H., Kong, Q., Geng, Z., Duan, L., Yang, M., & Guan, B., (2015). Enhancement of cell biomass and cell activity of astaxanthin-rich *Haematococcus pluvialis*. *Bioresour. Technol., 186*, 67–73.

Torres, E. M., Hess, D., McNeil, B. T., Guy, T., & Quinn, J. C., (2017). Impact of inorganic contaminants on microalgae productivity and bioremediation potential. *Ecotoxicol. Environ. Safety, 139*, 367–376.

Tramontin, D. P., Gressler, P. D., Rörig, L. R., Derner, R. B., Pereira-Filho, J., Radetski, C. M., & Quadri, M. B., (2018). Growth modeling of the green microalga *Scenedesmus obliquus* in a hybrid photobioreactor as a practical tool to understand both physical and biochemical phenomena in play during algae cultivation. *Biotechnol. Bioeng., 115*, 965–977.

Turon, V., Baroukh, C., Trably, E., Latrille, E., Fouilland, E., & Steyer, J. P., (2015). Use of fermentative metabolites for heterotrophic microalgae growth: Yields and kinetics. *Bioresour. Technol., 175*, 342–349.

Volkman, J., (2003). Sterols in microorganisms. *Appl. Microbiol. Biotechnol., 60*, 495–506.

Wahidin, S., Idris, A., & Shaleh, S. R. M., (2013). The influence of light intensity and photoperiod on the growth and lipid content of microalgae *Nannochloropsis* sp. *Bioresour. Technol., 129*, 7–11.

Wang, X., Sheng, L., & Yang, X., (2017). Pyrolysis characteristics and pathways of protein, lipid, and carbohydrate isolated from microalgae *Nannochloropsis* sp. *Bioresour. Technol., 229*, 119–125.

Yang, S., Liu, G., Meng, Y., Wang, P., Zhou, S., & Shang, H., (2014). Utilization of xylose as a carbon source for mixotrophic growth of *Scenedesmus obliquus*. *Bioresour. Technol., 172*, 180–185.

Yee, W., (2015). Feasibility of various carbon sources and plant materials in enhancing the growth and biomass productivity of the freshwater microalgae *Monoraphidium griffithii* NS16. *Bioresour. Technol., 196*, 1–8.

Zhan, J., Rong, J., & Wang, Q., (2017). Mixotrophic cultivation, a preferable microalgae cultivation mode for biomass/bioenergy production, and bioremediation, advances and prospect. *Int. J. Hydrog. Energy, 42*, 8505–8517.

Zhang, T. Y., Wu, Y. H., Zhu, S. F., Li, F. M., & Hu, H. Y., (2013). Isolation and heterotrophic cultivation of mixotrophic microalgae strains for domestic wastewater treatment and lipid production under dark condition. *Bioresour. Technol., 149*, 586–589.

CHAPTER 12

BIOTECHNOLOGY OF MICROALGAE: A GREEN APPROACH TOWARDS EXPLOITATION OF OMEGA-3 FATTY ACIDS

RAMU MANJULA[1] and MANJUNATH CHAVADI[2]

[1]National Institute of Mental Health and Neurosciences, Department of Biophysics, Bangalore – 560029, Karnataka, India

[2]Hassan Institute of Medical Science, Sri Chamarjendra Hospital Campus, Department of Microbiology, Hassan, Karnataka, India

12.1 INTRODUCTION

The only type of fat that cannot be cut back on in our diet is omega-3 fatty acids. Omega-3 fatty acids (also called ω-3 fatty acids) are polyunsaturated fatty acids (PUFAs), which are required for the normal functioning of the body cells. They are the EFA (essential fatty acids) with a double bond at the third carbon position from the end of the carbon tail (Figure 12.1). Foods high in omega-3 fatty acids include salmon, albacore, trout, herring, halibut, sardines, walnut, flaxseed oil, and canola oil. Omega-3 fatty acids are also present in shrimp, clams, light chunk tuna, catfish, cod, and spinach.

There are four types of ω-3 PUFAs important for human physiology are α-linolenic acid (ALA; 18:3: ω3), stearidonic acid (SDA, 18:4: ω3), eicosapentaenoic acid (EPA; 20:5: ω3), and docosahexaenoic acid (DHA; 22:6: ω3). ALA and SDA are the short-chain-PUFAs (SC-PUFAs) which are consisting of 18 or less than 18 carbons in a chain. EPA and DHA are the long-chain-PUFAs (LC-PUFAs) play essential roles in human nutrition, mainly during neonatal development and in adult cardiovascular health. The ALA is mainly plant-derived whereas, LC-PUFAs are marine-derived, as they are abundant in certain fishes.

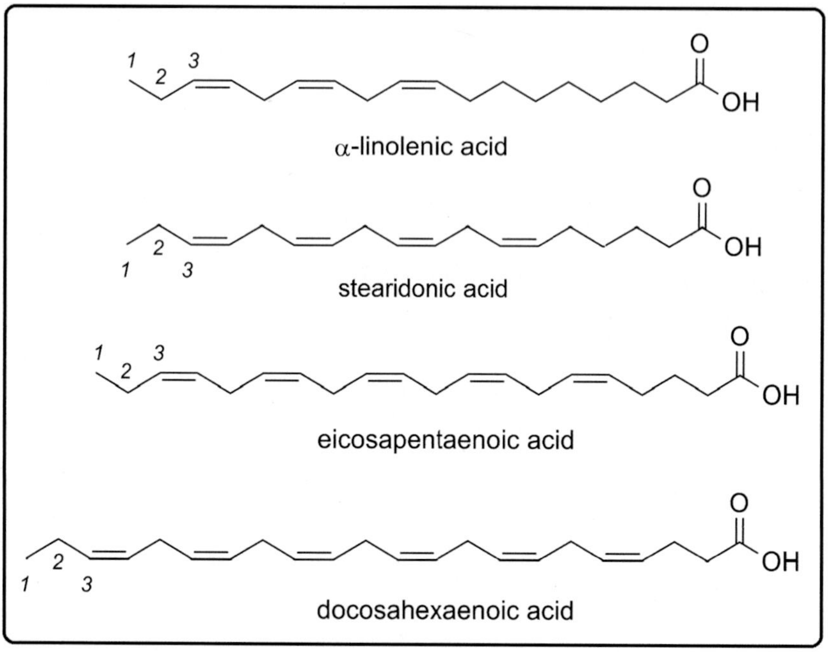

FIGURE 12.1 The structure of omega-3 fatty acids.

12.2 WHAT ARE THE BENEFITS OF OMEGA-3 PUFA?

The risk of malnutrition increases with age, where 90 years were identified with >40% increased malnutrition risk than those of <70 years (Elia et al., 2008). A better quality of life can be achieved by the optimal nutrition that will maintain mental and physical functions intact (Kozlowska et al., 2008). Specific nutrients, particularly LC-PUFAs, might have the potential of preventing and reducing co-morbidities in older adults (Swanson et al., 2012). Omega-3 PUFAs are able to regulate inflammation, hyperlipidemia, platelet aggregation, and hypertension (Ubeda et al., 2012). LC-PUFAs are also the precursors of several metabolites that are potent lipid mediators, which are beneficial in the prevention or treatment of several diseases (Serhan et al., 2008). Different mechanisms contribute to these effects, including conditioning cell membrane function and composition, eicosanoid production, and gene expression. The beneficial functions of omega-3 PUFA in various organs are schematically represented in Figure 12.2.

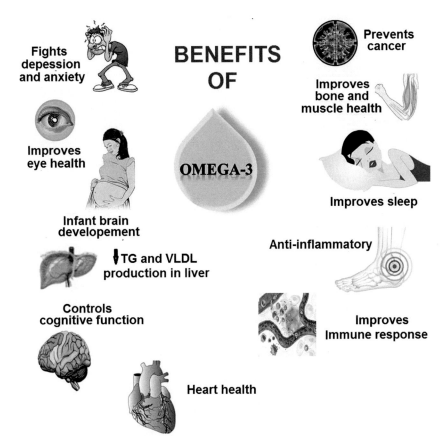

FIGURE 12.2   Organs and functions modulated by omega-3 PUFAs.

12.2.1   OMEGA-3 PUFAs ON BRAIN FUNCTION

Omega-3 brain benefits are derived mostly from neuroprotective DHA, which is found in membrane phospholipids and, to a lesser degree, EPA. EPA is the single vital nutrient that controls communication between nerve cells and the brain. From a recent study on the US population proved that the higher dietary intake of omega-3 LC-PUFAs was associated with lower prevalence of a headache which is supporting the hypothesis that omega-3 PUFAs may prevent or reduce a headache (Sanders et al., 2018). In the peripheral blood of dementia patients, the level of omega-3 PUFAs are significantly decreased (Lin et al., 2012). According to the study by Gu et al. (2010), higher intake of foods with high omega-3 PUFAs and lower intake of foods with low omega-3

PUFAs would reduce the risk of Alzheimer's disease (AD) (Gu et al., 2010). In an image analysis of brain sections of an aged Alzheimer mouse model, a diet enriched with the DHA reduced the amyloid burden (Lim et al., 2005). From another clinical study, it was found that the increased plasma EPA level was associated with a decreased incidence of dementia in a cohort of 1214 older non-demented persons (Samieri et al., 2008).

Both EPA and DHA have their effect on reducing depression in adults. According to a six months randomized controlled trial on 50 people aged >65 years with mild cognitive impairment (MCI), the supplementation of LC-PUFAs has dramatically reduced the depression (Sinn et al., 2012). The administration of phosphatidylserine-containing DHA (PS-DHA) to the elderly with memory complaints has significantly improved the cognitive functions in a double-blind placebo-controlled trial (Vakhapova et al., 2010). Preterm infants are vulnerable to the adverse effect of DHA deficiency on both visual and neuronal development as they cannot synthesize enough LC-PUFAs during last trimester pregnancy.

The limited evidence suggested that supplementation of EPA may be a useful adjunct to antipsychotic therapy in schizophrenic patients. A pilot study of 45 schizophrenic patients showed that the intake of 2 g/day of EPA along with the standard antipsychotic therapy was superior to the intake of a 2 g/day to DHA or a placebo in reducing the residual symptoms. According to another meta-analysis of randomized controlled trials on symptomatic outcome revealed no beneficial effect of EPA augmentation in established schizophrenia (Fusar-Poli and Berger, 2012). However, the published results are conflicting and larger long-term studies addressing clinically relevant outcomes are needed. In summary, the majority of the available evidence supports a positive correlation between omega-3 status/supplementation with brain function.

12.2.2   OMEGA-3 PUFAs ON CARDIOVASCULAR SYSTEM

There is an inverse relationship between fish consumption and sudden cardiac death. In a prospective study, omega-3 fatty acid intakes equivalent to two fatty fish meals per week were associated with a 50% decrease in the risk of primary cardiac arrest. A recent result of randomized controlled trials in individuals with documented coronary heart disease (CHD) suggests a beneficial effect of omega-3 fatty acids.

A 2018 meta-analysis found no correlation between the daily supplementation of omega-3 fatty acid in individuals and prevention of CHD, nonfatal myocardial infarction (MI) or any other vascular event (Aung et al., 2018). However, omega-3 fatty acid supplementation greater than one gram daily for a year might be protective in people who have a history of cardiovascular disease (CVD) against cardiac death, sudden death, and MI (Casula et al., 2013). Fish-oil supplementation has induced the anti-inflammatory gene expression profiles in human blood mononuclear cells. In this study of human blood samples, EPA+DHA supplementation changed the expression profile of 1040 genes and resulted in a decreased expression of genes associated with inflammatory and atherogenesis-related pathways, such as nuclear transcription factor κB signaling, scavenger receptor activity, adipogenesis, eicosanoid synthesis, and hypoxia signaling (Bouwens et al., 2009). A study followed 1822 men for 30 years and claimed that mortality from CHD was 38% lower in men who consumed an average 35 g (1.2 ounces) of fish daily than in men who did not eat fish, while mortality from MI was 67% lower (Daviglus et al., 1997). From a recent meta-analysis of 18 random controlled trials (involving about 93000 individuals) and 16 prospective cohort studies (involving about 732000 individuals), it was found that EPA and DHA supplementation significantly reduced CHD risk in individuals with elevated triglycerides or LDL cholesterol levels (Alexander et al., 2017). The intake of 250 mg/day dietary omega-3 LC-PUFAs has also shown to reduce sudden cardiac death (Musa-Veloso et al., 2011).

The FAs composition analysis of plasma phospholipids showed that DHA and EPA are essential FAs for distinguishing non-cerebral atherosclerotic stenosis (NCAS) and intracranial atherosclerotic stenosis (ICAS) in strokes (Kim et al., 2012). The study also concluded that DHA intake could reduce the risk of ICAS.

12.2.3 OMEGA-3 PUFAs ON CANCER

According to the research higher intake of omega-3 PUFAs would reduce the risk of cancer (Weylandt et al., 2015). The review concludes that the animal and *in vitro* data showed health benefits of omega-3 PUFAs. However, the outcomes of human studies have been so far quite controversial. Few of the studies have shown the association between intake of EPA and DHA with the risk of breast (Gago-Dominguez et al., 2003) and colorectal cancers. According to a study on 35016 female participants aged 50–76 years, the use of fish oil has reduced the risk of breast cancer to 32% (Brasky et al., 2010). However, not much of clinical trials are performed to conclude the association of PUFAs on breast cancer. In the case of colorectal cancer limited studies suggests that higher intake of omega-3 PUFAs is associated with reduced risk of cancer (Gerber, 2012). In summary, the data from observational studies show no consistent relationship between omega-3s and overall cancer risk.

12.2.4   OMEGA-3 PUFAs ON OTHER CONDITIONS

12.2.4.1   RHEUMATOID ARTHRITIS (RA)

Rheumatoid arthritis (RA) is the common systemic inflammatory joint disease. A high-dose fish oil supplement in addition to the triple therapy for RA (methotrexate, sulfasalazine, and hydroxychloroquine), achieved better outcomes in patients. Other studies suggest that omega-3 PUFAs will help RA patients lower their dose of nonsteroidal anti-inflammatory drugs (NSAIDs) (Galarraga et al., 2008). According to the information from NIH, administering fish oil reduces swollen and tender joints in people with RA.

12.2.4.2   DIABETES

The glucose tolerance and symptoms of diabetic neuropathy and nephropathy may improve with the supplementation of fish oil. A number of randomized controlled trials have shown the decreased serum triglyceride levels in diabetic individuals after the fish oil supplementation (Harris, 1997). From one of the studies with more than 3000 older US adults, it was found that over a 10-year period those with the highest blood levels of the omega-3 PUFAs were about one-third less likely to develop diabetes than those adults with the lowest levels.

12.2.4.3   FETAL DEVELOPMENT

Along with protein and caloric supplements, PUFAs are also crucial in maternal nutrition. Supplementation of omega-3 PUFAs during pregnancy was associated with several benefits such as decreased incidence of asthma (Olsen et al., 2008), enhanced infant problem-solving skills (Judge et al., 2007), improved cognitive and language development in offspring during early childhood (Makrides et al., 2010) and decreased risk of food allergy (Furuhjelm et al., 2009).

12.2.4.4   BONE HEALTH

In a study with 45–90 aged subjects, it was found that a higher ratio of omega-6 to omega-3 associated with lower hip bone mineral density. Hence, bone health is significantly correlated with omega-3 status.

12.2.4.5   IMMUNE FUNCTION

The essential FAs regulate T-cell proliferation and inflammatory responses (Han et al., 2012). In the case of hospitalized patients, the supplementation of omega-3 PUFAs through fish oil has increased the immune function. However, due to inconsistent results, there are no clear outcomes from the study. In some cases, DHA intake has inhibited the activity of immune cells. Overall researches on the benefits of omega-3 PUFAs are highly inconsistent in many cases which will suggest the need for further studies.

12.3   SOURCES OF OMEGA-3 PUFA

The human body cannot synthesize the ALA, and other omega-3 fatty acids such as SDA, EPA, and DHA can be synthesized from ALA. The adequate intake of total omega-3s for newborn babies from 0–12 months should be 0.5 g (Trumbo et al., 2002) and for the kids from 1–8 years the ALA intake should be 0.7–0.9 g. Male require higher daily intake compared to females after the age of 9 years. An adult male and female require daily 1.6 and 1.1 g of omega-3 respectively. Plant oils such as flaxseed (linseed), soybean, and canola oils and chia seeds and walnuts contain ALA (Table 12.1). The main dietary source of LC-PUFAs is cold-water fatty fishes such as salmon, herring, mackerel, anchovies, menhaden, tuna, and sardines. However, there is a threat of contaminants like heavy metals, fat-soluble polychlorinated biphenyls, and dioxins. Beef from grass-fed cows found to contain a high amount of omega-3 FAs. The main sources of omega-3 PUFAs used in India include Flax seeds, Sardines, Chia seeds, Walnuts, Eggs, Salmon, Brussel sprouts, Hemp seeds, Soybeans, and fish oil (Figure 12.3).

TABLE 12.1   ALA Sources

Food	Serving	Alpha-Linolenic Acid (g)
Flaxseed oil	1 tablespoon	8.5
Walnuts, English	1 ounce	2.6
Flaxseeds	1 tablespoon	2.2
Walnut Oil	1 tablespoon	1.4
Canola Oil	1 tablespoon	1.2
Mustard Oil	1 tablespoon	0.8
Soybean Oil	1 tablespoon	0.9
Walnuts, Black	1 ounce	0.6
Olive Oil	1 tablespoon	0.1
Broccoli, raw	1cup, chopped	0.1

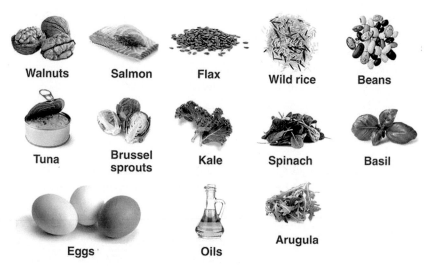

FIGURE 12.3   The primary sources of omega-3 PUFA rich food.

12.3.1   NON-FISH OIL BASED OMEGA-3 POLYUNSATURATED FATTY ACIDS (PUFAs)

For those who do not consume fish, ALA is the source for LC-PUFAs. Some of the ALA sources are mentioned in Table 12.1. Plant foods and vegetable oils lack EPA and DHA; however, they are with high ALA contents. The desaturation and elongation reactions in the body convert the dietary ALA to EPA and DHA. Flax seeds are the powerhouse of nutrients with a higher content of ALA than soybean and canola oil. Cauliflower and chia seeds are also rich in omega-3 FAs along with vitamins and minerals. Brussel sprouts which are used for skin nourishment are having a high content of ALA. The hemp seed is rich in essential FAs such as SDA and Gamma-linolenic acid (GLA).

12.3.2   FISH OIL OMEGA-3 POLYUNSATURATED FATTY ACIDS (PUFAs)

Fish oil omega-3 PUFAs consist mainly of the essential FAs such as EPA and DHA. Some types of omega-3 PUFAs are found in foods such as fatty fish and shellfish. Fish oils from salmon, herring, mackerel, and anchovies will have seven times a high profile of omega-3 than omega-6 FAs. Marine and freshwater fish oils differ in their LC-PUFAs.

12.3.3   THE NEED FOR ALTERNATIVE LC-PUFA SOURCES

Many health organizations recommend daily intake for PUFA in addition to essential nutrients. The World Health Organization (WHO) recommends the intake of 2 portions of oily fish per week to obtain sufficient dietary omega-3 PUFAs (200–500 mg) (Margetts, 2003). Many studies have also recommended a daily dietary intake of 1 g DHA and EPA in seafood which would require the intake of a considerable amount of fish. Although, fish-oil is the primary traditional source for the essential PUFAs there are few limitations associated with the application of fish oil as a food additive. Mass-scale fisheries are not sustainable if we continue with the growing demand for fish-oils. There is the possibility of accumulation of toxins, in particular mercury, lead, nickel, arsenic, and cadmium, in the omega-3 PUFAs derived from fish sources and reliance on these fish-oils is also complicated due to an unpleasant smell, taste, and poor oxidative stability (Spolaore et al., 2006). The process of fish-oil production is also tedious and complicated (Lenihan-Geels et al., 2013).

Thus, there is an urgent need for the alternative sources of LC-PUFA in order to meet the demands and health-related requirements. In 1990s fungi *Mortierella,* yeasts, *and* microalgae were explored as sustainable alternatives to fish oil. The different sources with their yield of omega-3 are shown in Figure 12.4. However, the vegetable oils also served as the better alternative to fish oil in PUFA production. By using the genetically modified *Camelina sativa* plant the high amount of DHA and EPA was synthesized (Napier et al., 2015). Though the GM plants are a better choice as fish oil alternatives, they synthesis less DHA compared to microbes. Microalgae which are present as spacious and relatively unexploited resource of fatty acids thus are used as an alternative source to declining fish stocks (Ratledge, 2004). It is well known that the PUFA yields from microorganism are in the order, microalgae > fungi > bacteria.

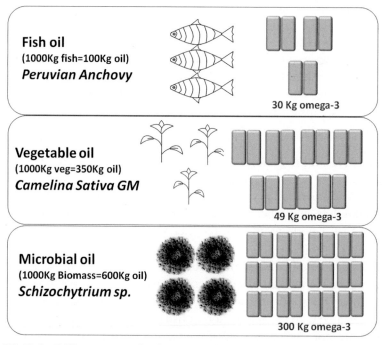

FIGURE 12.4   Different sources for the commercially available omega-3 oil.

12.4   THE BIOTECHNOLOGY OF MICROALGAE

To date, the green algae such as *Chlorella* and cyanobacteria (CB) such as *Spirulina* are commercially used to produce nutraceuticals, dietary supplements, pigments, and functional ingredients. They are fast growing and also rich in amino acids. However, they are low in essential omega-3 FAs. Since only plants can synthesize PUFAs consumers can get them in their diet. The essential dietary nutrients EPA and DHA can also be obtained from the aquatic organisms.

Fishes obtain LC-PUFAs from these organisms such as microalgae which could be used for commercial PUFA production. Many algae are rich in phytosterols which helps in reducing serum cholesterol levels and also has a beneficial effect in AD and other cancers. In practice, while screening for the strains following properties has to be considered: (1) high growth rates, (2) a high proportion of DHA/EPA as a percentage of total lipids in elevated temperature fermentation (30 °C), and (3) growth unaffected by low salinity conditions.

The process of culturing the algae or algae-like microorganism for the production of oil rich in omega-3 fatty acids was called "Omega-3 Biotechnology" (Gupta et al., 2012). The biotechnological production of lipids started in the nineteenth century. In 1992, Martek Corporation began to produce DHA from *Crypthecodinium cohnii*. In 2002, DHASCO™ (for an infant) came into the market with GRAS (generally recognized as safe) from FDA in the USA. The most promising producers of LC-PUFAs currently are *Nannochloropsis* (Eustigmatophyte) (Adarme-Vega et al., 2012). Since Cryptophytes are rich in phytosterol, PUFA, and amino acid concentration, they serve as a potential source for nutraceutical purposes. Cryptophytes are below 500 μm^3 in size and do not possess heavy cell wall hence easier to break during the commercial processes compared to diatoms. The marine members of the Thraustochytriaceae (genera *Schizochytrium* and *Ulkenia*) and Crypthecodiniaceae (genus *Crypthecodinium*) families are found to be the excellent source for the omega-3 fatty acid DHA. Because of the heterotrophic culturing of these families, the production of DHA has gone to the industrial scale (Ren et al., 2010). In 2004, the microbial oil produced by *Schizochytrium limacinum* was granted a GRAS by the FDA. The marine diatom *Phaeodactylum tricornutum* is an excellent choice for the commercial production of EPA. In industries, the microalgae will be grown in controlled conditions to get the biomass with constant biochemical composition. The major companies working in this sector are listed in Table 12.2. In 2009, DSM's Life's DHA™-S algal oil is also approved for the dietary intake. In July 2012, Life's Omega™, a DHA and EPA-rich oil from DSM was approved by European Novel Food Regulation.

The LC-PUFAs can be produced both by aerobic and anaerobic pathways in micro-organisms. Fermentation process development is a necessary step for both strain construction and commercial manufacturing of omega-3 and omega-6 fatty acids. Micro-algal fermentations do not compete for land space and some can be used to fix carbon dioxide (Winwood, 2013). The cultivation process is divided into two stages: (a) biomass growth and (b) lipid accumulation. The first stage facilitates cell reproduction and the second stage leads

TABLE 12.2   The Private Companies Producing the PUFA Across the World

Countries	Number of Companies
USA	76
Norway	15
China	14
Canada	8
Australia	9
Peru	8
Chile	7
Spain	5
Netherlands	4
UK	5
Germany	5
Japan	7
Switzerland	3
Israel	4
Turkey	3
France	3

to the production of fatty acids in the absence of nitrogen. The fermentation may be in batch, fed-batch or continuous processes. Fermentation is followed by the recovery of biomass by centrifugation, filtration, or other means of solid-liquid separation. This biomass is dried and oil is extracted. In an industrial scale extraction mixture of solvents such as acetone, hexane, dichloromethane, trichloroethylene, methyl ethyl ketone, and carbon dioxide are used. By desolventizing process, all these used solvents are removed from the oil. Before the refinement process, the oxidation of PUFA is protected by freezing in an inert atmosphere. The refining steps include degumming, neutralization with NaOH, bleaching, and deodorizing. The final oil is subjected to high pressure to remove odor, color, and undesirable tastes in order to purify it for use in food supplements (Kim et al., 2013). In microbial biomass production by using a better substrate and microorganism, enhanced productivity can be obtained. Some of the sources used for DHA and EPA production along with the carbon source used are listed in Table 12.3. During the process optimum temperature, pH, and salinity have to be maintained.

TABLE 12.3   Culture Mode, the Carbon Source for the Highest Reported EPA and DHA Productivities of Microalgae Cultivated Under Heterotrophic Conditions

Species	Product	Culture Mode	Carbon Source
Aurantiochytrium limacinum MH0186	DHA	Batch	Glucose
Aurantiochytrium limacinum	DHA	Fed-batch	Glucose
Aurantiochytrium limacinum	DHA	Batch	Glycerol
Cryptecodinium cohnii	DHA	Batch	Ethanol
Cryptecodinium cohnii	DHA	Fed-batch	Acetic acid
Cryptecodinium cohnii	DHA	Fed-batch	Glucose
Cryptecodinium cohnii	DHA	Fed-batch	Carob syrup
Cryptecodinium cohnii	DHA	Fed-batch	Glucose
Phaeodactylum tricornutum (transgenic)	DHA	Fed-batch	Glucose
Phaeodactylum tricornutum (transgenic)	EPA	Fed-batch	Glucose
Nitzschia laevis	EPA	Fed-batch	Glucose
Nitzschia laevis	EPA	Perfusion	Glucose
Schizochytrium sp. LU310	DHA	Fed-batch	Glucose
Schizochytrium sp. SR21	DHA	Batch	Glycerol

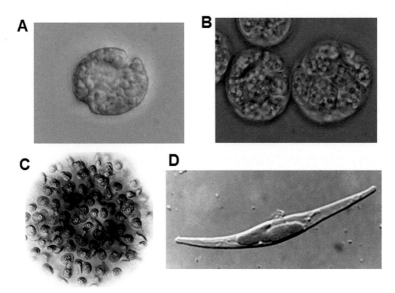

FIGURE 12.5 Photomicrograph of the micro-algae *Crypthecodinium cohnii* (A), *Schizochytrium* sp. (B), *Phaeodactylum tricornutum* (C), and *Nannochloropsis* (D).

Some promising oleaginous microorganisms such as *Yarrowia lipolytica, Mortierella alpina, Schizochytrium limacinum* (high lipid accumulation), *Crypthecodinium cohnii* (Figure 12.5), and *Ulkenia* sp. already have GRAS status. To address the challenges in large-scale land-based EPA production from organisms, DuPont (USA) has performed research on metabolically engineered *Yarrowia lipolytica*. The two-stage continuous fermentation was used as an active process to produce the omega-3 and other fermentation products where non-growth or partially growth associated kinetics characterize the process (Xie et al., 2017).

12.5 CONCLUSION

Omega-3 LC-PUFAs show significant benefits to human health in maintaining the metabolic functions, immune systems, and cardiovascular systems which have been proven by many types of research. The primary dietary source of DHA and EPA, i.e., fish, has reached global production and sustainability is now under the question. The production of these bioactive lipids is a bigger challenge for many food industries across the world. The only better option for replacing the fish oil is the algal oils with excellent sustainability credentials can be produced in an industrial scale. However, the production of algal oil is also expensive, but refinements in the process and economies of scale have attenuated the final cost. To reduce the cost of the substrate during fermentation, it is highly advisable to use agricultural wastes to replace the cultivating media. Scientists have used corn starch for the fermentation to grow *Schizochytrium* sp. to produce LC-PUFAs which was further commercialized by Evonik and DSM. This technology has secured the supply of essential fatty acid without relying on the marine ecosystem. Along with the usage of wastes as new substrates, the research on molecular biology, metabolic engineering, and microbial genetics would help us to produce new lipid producing strains for food and therapeutics.

KEYWORDS

- docosahexaenoic acid
- eicosapentaenoic acid
- fish oil

- **long chain-PUFAs**
- **malnutrition**
- **microalgae**
- **omega-3**
- **sustainability**
- **α-linolenic acid**

REFERENCES

Adarme-Vega, T. C., Lim, D. K., Timmins, M., Vernen, F., Li, Y., & Schenk, P. M., (2012). Microalgal biofactories: A promising approach towards sustainable omega-3 fatty acid production. *Microb. Cell Fact., 11*, 96. doi: 10.1186/1475-2859-11-96.

Alexander, D. D., Miller, P. E., Van, E. M. E., Kuratko, C. N., & Bylsma, L. C., (2017). A meta-analysis of randomized controlled trials and prospective cohort studies of eicosapentaenoic and docosahexaenoic long-chain omega-3 fatty acids and coronary heart disease risk. *Mayo Clin. Proc., 92*, 15–29.

Aung, T., Halsey, J., Kromhout, D., Gerstein, H. C., Marchioli, R., Tavazzi, L., et al., (2018). Associations of omega-3 fatty acid supplement use with cardiovascular disease risks: Meta-analysis of 10 trials involving 77917 individuals. *JAMA Cardiology, 3*, 225–234.

Bouwens, M., Van, D. R. O., Dellschaft, N., Bromhaar, M. G., De Groot, L. C., Geleijnse, J. M., et al., (2009). Fish-oil supplementation induces antiinflammatory gene expression profiles in human blood mononuclear cells. *Am. J. Clin. Nutr., 90*, 415–424.

Brasky, T. M., Lampe, J. W., Potter, J. D., Patterson, R. E., & White, E., (2010). Speciality supplements and breast cancer risk in the vitamins and lifestyle (VITAL) Cohort. *Cancer Epidemiol. Biomarkers Prev., 19*, 1696–1708.

Casula, M., Soranna, D., Catapano, A. L., & Corrao, G., (2013). Long-term effect of high dose omega-3 fatty acid supplementation for secondary prevention of cardiovascular outcomes: A meta-analysis of randomized, placebo controlled trials [corrected]. *Atheroscler. Suppl., 14*, 243–251.

Daviglus, M. L., Stamler, J., Orencia, A. J., Dyer, A. R., Liu, K., Greenland, P., et al., (1997). Fish consumption and the 30-year risk of fatal myocardial infarction. *N. Engl. J. Med., 336*, 1046–1053.

Elia, M., Jones, B., & Russell, C., (2008). Malnutrition in various care settings in the UK: The 2007 nutrition screening week survey. *Clin. Med. (Lond.), 8*, 364–365.

Finco, A. M. O., Mamani, L. D. G., Carvalho, J. C., De Melo Pereira, G. V., Thomaz-Soccol, V., & Soccol, C. R., (2017). Technological trends and market perspectives for production of microbial oils rich in omega-3. *Crit. Rev. Biotechnol., 37*, 656–671.

Furuhjelm, C., Warstedt, K., Larsson, J., Fredriksson, M., Bottcher, M. F., Falth-Magnusson, K., & Duchen, K., (2009). Fish oil supplementation in pregnancy and lactation may decrease the risk of infant allergy. *Acta Paediatr., 98*, 1461–1467.

Fusar-Poli, P., & Berger, G., (2012). Eicosapentaenoic acid interventions in schizophrenia: Meta-analysis of randomized, placebo-controlled studies. *J. Clin. Psychopharmacol., 32*, 179–185.

Gago-Dominguez, M., Yuan, J. M., Sun, C. L., Lee, H. P., & Yu, M. C., (2003). Opposing effects of dietary n-3 and n-6 fatty acids on mammary carcinogenesis: The Singapore Chinese health study. *Br. J. Cancer., 89*, 1686–1692.

Galarraga, B., Ho, M., Youssef, H. M., Hill, A., McMahon, H., Hall, C., et al., (2008). Cod liver oil (n-3 fatty acids) as a non-steroidal anti-inflammatory drug sparing agent in rheumatoid arthritis. *Rheumatology (Oxford), 47*, 665–669.

Gerber, M., (2012). Omega-3 fatty acids and cancers: A systematic update review of epidemiological studies. *Br. J. Nutr., 107*(S2), S228–239.

Gu, Y., Nieves, J. W., Stern, Y., Luchsinger, J. A., & Scarmeas, N., (2010). Food combination and Alzheimer disease risk: A protective diet. *Arch. Neurol., 67*, 699–706.

Gupta, A., Barrow, C. J., & Puri, M., (2012). Omega-3 biotechnology: Thraustochytrids as a novel source of omega-3 oils. *Biotechnol. Adv., 30*, 1733–1745.

Han, S. N., Lichtenstein, A. H., Ausman, L. M., & Meydani, S. N., (2012). Novel soybean oils differing in fatty acid composition alter immune functions of moderately hypercholesterolemic older adults. *J. Nutr., 142*, 2182–2187.

Harris, W. S., (1997). n-3 fatty acids and serum lipoproteins: Human studies. *Am. J. Clin. Nutr., 65*, 1645S–1654S.

Judge, M. P., Harel, O., & Lammi-Keefe, C. J., (2007). Maternal consumption of a docosahexaenoic acid-containing functional food during pregnancy: Benefit for infant performance on problem-solving but not on recognition memory tasks at 9 months of age. *Am. J. Clin. Nutr., 85*, 1572–1577.

Kim, J., Yoo, G., Lee, H., Lim, J., Kim, K., Kim, C. W., et al., (2013). Methods of downstream processing for the production of biodiesel from microalgae. *Biotechnol. Adv., 31*, 862–876.

Kim, Y. J., Kim, O. Y., Cho, Y., Chung, J. H., Jung, Y. S., Hwang, G. S., & Shin, M. J., (2012). Plasma phospholipid fatty acid composition in ischemic stroke: Importance of docosahexaenoic acid in the risk for intracranial atherosclerotic stenosis. *Atherosclerosis, 225*, 418–424.

Kozlowska, K., Szczecinska, A., Roszkowski, W., Brzozowska, A., Alfonso, C., Fjellstrom, C., et al., (2008). Patterns of healthy lifestyle and positive health attitudes in older Europeans. *J. Nutr. Health Aging, 12*, 728–733.

Lenihan-Geels, G., Bishop, K. S., & Ferguson, L. R., (2013). Alternative sources of omega-3 fats: Can we find a sustainable substitute for fish? *Nutrients, 5*, 1301–1315.

Lim, G. P., Calon, F., Morihara, T., Yang, F., Teter, B., Ubeda, O., et al., (2005). A diet enriched with the omega-3 fatty acid docosahexaenoic acid reduces amyloid burden in an aged Alzheimer mouse model. *J. Neurosci., 25*, 3032–3040.

Lin, P. Y., Chiu, C. C., Huang, S. Y., & Su, K. P., (2012). A meta-analytic review of polyunsaturated fatty acid compositions in dementia. *J. Clin. Psychiatry, 73*, 1245–1254.

Makrides, M., Gibson, R. A., McPhee, A. J., Yelland, L., Quinlivan, J., & Ryan, P., (2010). Effect of DHA supplementation during pregnancy on maternal depression and neurodevelopment of young children: A randomized controlled trial. *JAMA, 304*, 1675–1683.

Margetts, B., (2003). Feedback on WHO/FAO global report on diet, nutrition and prevention of chronic diseases (NCD). *Public Health Nutr., 6*(5), 423–424.

Morales-Sánchez, D., Martinez-Rodriguez, O. A., & Martinez, A., (2017). Heterotrophic cultivation of microalgae: Production of metabolites of commercial interest. *Journal of Chemical Technology and Biotechnology, 92*, 925–936.

Musa-Veloso, K., Binns, M. A., Kocenas, A., Chung, C., Rice, H., Oppedal-Olsen, H., et al., (2011). Impact of low v. moderate intakes of long-chain n-3 fatty acids on risk of coronary heart disease. *Br. J. Nutr., 106*, 1129–1141.

Napier, J. A., Usher, S., Haslam, R. P., Ruiz-Lopez, N., & Sayanova, O., (2015). Transgenic plants as a sustainable, terrestrial source of fish oils. *Eur. J. Lipid Sci. Technol., 117*, 1317–1324.

Olsen, S. F., Osterdal, M. L., Salvig, J. D., Mortensen, L. M., Rytter, D., Secher, N. J., & Henriksen, T. B., (2008). Fish oil intake compared with olive oil intake in late pregnancy and asthma in the offspring: 16 y of registry-based follow-up from a randomized controlled trial. *Am. J. Clin. Nutr., 88*, 167–175.

Ratledge, C., (2004). Fatty acid biosynthesis in microorganisms being used for single cell oil production. *Biochimie., 86*, 807–815.

Ren, L. J., Ji, X. J., Huang, H., Qu, L., Feng, Y., Tong, Q. Q., & Ouyang, P. K., (2010). Development of a stepwise aeration control strategy for efficient docosahexaenoic acid production by *Schizochytrium* sp. *Appl. Microbiol. Biotechnol., 87*, 1649–1656.

Samieri, C., Feart, C., Letenneur, L., Dartigues, J. F., Peres, K., Auriacombe, S., et al., (2008). Low plasma eicosapentaenoic acid and depressive symptomatology are independent predictors of dementia risk. *Am. J. Clin. Nutr., 88*, 714–721.

Sanders, A. E., Shaikh, S. R., & Slade, G. D., (2018). Long-chain omega-3 fatty acids and headache in the U.S. population. *Prostaglandins, Leukotrienes and Essential Fatty Acids, 135*, 47–53.

Serhan, C. N., Chiang, N., & Van, D. T. E., (2008). Resolving inflammation: Dual anti-inflammatory and pro-resolution lipid mediators. *Nat. Rev. Immunol., 8*, 349–361.

Sinn, N., Milte, C. M., Street, S. J., Buckley, J. D., Coates, A. M., Petkov, J., & Howe, P. R., (2012). Effects of n-3 fatty acids, EPA v. DHA, on depressive symptoms, quality of life, memory, and executive function in older adults with mild cognitive impairment: A 6-month randomized controlled trial. *Br. J. Nutr., 107*, 1682–1693.

Spolaore, P., Joannis-Cassan, C., Duran, E., & Isambert, A., (2006). Commercial applications of microalgae. *J. Biosci. Bioeng., 101*, 87–96.

Swanson, D., Block, R., & Mousa, S. A., (2012). Omega-3 fatty acids EPA and DHA: Health benefits throughout life. *Adv. Nutr., 3*, 1–7.

Trumbo, P., Schlicker, S., Yates, A. A., & Poos, M., (2002). Dietary reference intakes for energy, carbohydrate, fiber, fat, fatty acids, cholesterol, protein and amino acids. *J. Am. Diet. Assoc., 102*, 1621–1630.

Ubeda, N., Achon, M., & Varela-Moreiras, G., (2012). Omega 3 fatty acids in the elderly. *Br. J. Nutr., 107*(S2), S137–151.

Vakhapova, V., Cohen, T., Richter, Y., Herzog, Y., & Korczyn, A. D., (2010). Phosphatidylserine containing omega-3 fatty acids may improve memory abilities in non-demented elderly with memory complaints: A double-blind placebo-controlled trial. *Dement Geriatr. Cogn. Disord., 29*, 467–474.

Weylandt, K. H., Serini, S., Chen, Y. Q., Su, H. M., Lim, K., Cittadini, A., & Calviello, G., (2015). Omega-3 polyunsaturated fatty acids: The way forward in times of mixed evidence. *Biomed Res. Int.*, 143109. doi:10.1155/2015/143109.

Winwood, R. J., (2013). Recent developments in the commercial production of DHA and EPA rich oils from micro-algae. *OCL, 20*, D604. doi: https://doi.org/10.1051/ocl/2013030.

Xie, D., Miller, E., Sharpe, P., Jackson, E., & Zhu, Q., (2017). Omega-3 production by fermentation of *Yarrowia lipolytica*: From fed-batch to continuous. *Biotechnol. Bioeng., 114*, 798–812.

CHAPTER 13

GENETIC ENGINEERING AS A TOOL FOR IMPROVED APPLICATIONS OF ALGAE

ANNAMALAI JAYSHREE

Center for Environmental Studies, Anna University, Guindy Campus, Chennai – 600025, Tamil Nadu, India

13.1 INTRODUCTION

Algae are diverse groups of photosynthetic organisms ranging from unicellular microalgae to multicellular macroalgae thriving in both marine and freshwater environment. These simple structured chlorophyll-containing organisms photosynthetically convert sunlight, water, and CO_2 to a wide range of metabolites and chemicals in algal biomass (Demirbas, 2010; Bharathiraja et al., 2015; Sambusiti et al., 2015; Raheem et al., 2015). There are about 55,000 species and more than 100,000 strains of marine, brackish, freshwater, and terrestrial algae; however, only a dozen of algae species are commercially cultivated worldwide (Raslavicius et al., 2014; Ullah et al., 2015). Algae are generally categorized based on their variations in photosynthetic pigmentation as: green, blue-green, red, brown, and golden algae (Mohan et al., 2006; Demirbas, 2010; Chen et al., 2015). Pigmentation, growth rate, size, weight, and chemical composition of algae are significantly affected by their habitual environmental conditions such as light, temperature, pH, salinity, nutrient, pollution, and even water motion, particularly depending on their taxonomical classes and species (Jung et al., 2013; Sambusiti et al., 2015).

Microalgae are one of the most primitive forms of plants ranging <400 µm in length and <30 µm in diameter (Huber et al., 2006; Ziolkowska and Simon, 2014). Biologists have categorized microalgae into a variety of classes, mainly distinguished by their pigmentation, life cycle, and basic cellular structure (Demirbas, 2010). The most important classes or categories of microalgae in terms of their abundance are: (1) diatoms (Bacillariophyceae); (2) green (Chlorophyceae); (3) blue and blue-green cyanobacteria (CB) (Cyanophyceae); (4) golden (Chrysophyceae); and (5) red (Rhodophyceae) algae (Demirbas, 2010; Demirbas et al., 2011; Bharathiraja et al., 2015; Sambusiti et al., 2015; Raheem et al., 2015; Gust et al., 2008). Diatoms are the dominant forms of life among phytoplanktons and represent the largest group of biomass producers on Earth (Demirbas, 2010; Demirbas, 2011). CB are referred to as blue-green microalgae even though they combine characteristics of bacteria and algae (Gust et al., 2008; Ziolkowska and Simon, 2014). Microalgae are grouped additionally into photosynthetic, non-photosynthetic, autotrophic, heterotrophic, mixotrophic, prokaryotic, and eukaryotic (Brennan and Owende, 2010; Bharathiraja et al., 2015; Sambusiti et al., 2015). Macroalgae are comparatively large, multicellular, and photoautotrophic organisms and belong to lower group of plants consisting of a leaf-like thallus instead of root, stem, and leaves that grow up to 60 m in length (Jung et al., 2013; Bharathiraja et al., 2015; Sambusiti et al., 2015). Macroalgae are classified mainly into three major groups according to the thallus color derived from photosynthetic pigmentation variations, namely green (Chlorophyta), red (Rhodophyta) and brown (Phaeophyta) (Demirbas, 2010; Jung et al., 2013; Chen et al., 2015; Raheem et al., 2015).

Algae are easily cultured, more readily and rapidly bioengineered than other biomass, and no herbicide or pesticide is recommended in algal cultivation (Mohan et al., 2006; Noraini et al., 2014). The total annual world production of algal biomass is about 9200 t dry basis for microalgae and about 12 million t dry basis (around 16 million t wet basis) for macroalgae, which are harvested from aquaculture farms and wild habitats (Jung et al., 2013; Chen et al., 2015; Raheem et al., 2015). The amount of the mass-cultivated macroalgae has continuously increased over the last 10 years at an average of 10%. About 98% of commercial algal biomass production is currently with open ponds due to its most economical and preferable way of cultivation. Besides the natural environment, microalgae are cultivated in freshwater, seawater, and wastewater within open ponds and closed photo-bioreactors (Jung et al., 2013). Heterotrophic microalgae are more flexible than autotrophic microalgal cultivation as they tend to grow under light free condition and are capable of accumulating higher lipid in the cells (Bharathiraja et al., 2015). Algae are reported to have high productivities when cultivated on nutrient-rich (with N and P) wastewaters such as municipal effluents, domestic sewage, dairy manure effluents and source-separated urine (Mulbry et al., 2008; Abbasi and Abbasi, 2010; Singh et al., 2011).

13.2   ALGAE AS NATURE'S BOON

Algae are one of the biggest oxygen producers, consume most of the CO_2 from the atmosphere, and have a wide range of applications in industries all over the world. They have adapted throughout evolutionary history to hostile environmental conditions and competition for light, nutrients, and space which has resulted in a complex and wide range of secondary metabolites such as carotenoids, terpenoids, vitamins, saturated/unsaturated fatty acids, amino acids, and polysaccharides. Based on these complex compounds, algae have the following applications.

13.2.1   ALGAE AS FOOD

Seaweeds are valuable low-calorie food sources; rich in vitamins, minerals, proteins, polysaccharides, steroids, and dietary fibers. Phycocolloides such as agar-agar, carrageenan, and alginic acids are the main constituents of brown and red algal cell walls, widely used in several food industries, especially in Asian countries such as China, Japan, and Korea (De Almeida et al., 2011). Agar-agar is a vegetarian substitute for gelatin, frequently used as a soup thickener in ice cream and other desserts. Carrageenan also has similar applications; used as gelling, thickening, and stabilizing agents in dairy and meat products. *Porphyra*, commonly known as nori, is the most widely consumed seaweed used to prepare sushi.

Apart from seaweeds, microalgae are also rich sources of carbohydrates, protein, enzymes, and fiber. Besides, many vitamins and minerals like vitamins A, C, B_1, B_2, B_6, niacin, iodine, potassium, iron, magnesium, and calcium are abundantly found in microalgae. Green micro-algae were been used as nutritional supplements or food sources in Asiatic countries for hundreds of years; now in the growing years consumed throughout the world for their nutritional value. Microalgal species such as green algae: *Chlorella vulgaris*, *Haematococcus pluvialis*, *Dunaliella salina,* and CB: *Spirulina maxima* and *Spirulina platensis* are widely commercialized nutritional supplements for both human and animal feed additives (Rangel-Yagui et al., 2004; Colla et al., 2007; Madhyastha and Vatsala, 2007; Ogbonda et al., 2007; Sajilata et al., 2008).

13.2.2   ALGAE AS COSMETICS

Chlorella is the most used genus for the production of cosmetics; other important genera include *Scenedesmus, Anabaena,* and *Spirulina.* These algae are
used in several forms such as algal oils, algal powders, algal flour, and algal
flakes; algal products include: algal soaps, algal clays masks, beauty serums
and oils, scrubs, and shampoos. Microalgal extracts are used mainly in face
and skincare products such as anti-aging cream, refreshing, or regenerate care
products, emollient, and as an anti-irritant in peelers (Stolz and Obermayer,
2005). Microalgae are also used in sun protection and hair care products. Algal
species widely used in cosmetics are *Chondrus crispus*, *Mastocarpus stellatus*,
Ascophyllum nodosum, *Alaria esculenta*, *Spirulina platensis*, *Nannochloropsis
oculata*, *Chlorella vulgaris,* and *Dunaliella salina.*

13.2.3   ALGAE AS FOOD COLORANTS

Beta carotene, a fat-soluble photosynthetic pigment has anticarcinogenic
properties and also effective in controlling cholesterol and reducing risks of
heart disease. Though algal-derived food colorings are not photostable and
tend to bleach with cooking, the potential market for micro-algae-derived food
coloring is vast. Beta-carotene is used as an orange dye and as a vitamin C
supplement where, *Dunaliella salina* serves as a major source (Priyadarshini
and Rath, 2012).

13.2.4   ALGAE AS HIGH-VALUE BIOMOLECULES

Marine microalgae are an important renewable source of bioactive lipids with
a high proportion of polyunsaturated fatty acids (PUFAs), especially *n*-3 PUFA
such as α-linolenic acid (ALA), eicosapentaenoic acid (EPA), docosapentae-
noic acid, and docosahexaenoic acid (DHA). In turn, these are effective in
preventing or treating diseases such as cardiovascular disorders, cancer, type
2 diabetes, inflammatory bowel disorders, asthma, arthritis, kidney, and skin
disorders, depression, and schizophrenia (Priyadarshini and Rath, 2012).

13.2.5   ALGAE AS BIOFUEL

Both macro and microalgae potentially serve as a source for biofuel, an
alternative to fossil fuels. Algae deliver the highest lipid amount among all

biofuels feedstocks available in the market nowadays and their triglyceride production rates are normally 45–220 times higher than terrestrial biomass (Huber et al., 2006; Ziolkowska and Simon, 2014). Oil yields per hectare (on an annual basis) from certain algae strains could be much higher than from corn (300 times), soybean (100–130 times), jatropha (30 times), rapeseed (10–20 times), oil palm (10 times), other traditional oil plants (10–100 times) and terrestrial crops (7–31 times) (Demirbas, 2010, 2011; Ziolkowska and Simon, 2014; Ullah et al., 2015). As a result, about 47,000–308,000 L of oil per hectare could be produced annually using algae (Demirbas, 2010). While a terrestrial crop cycle may take from three months to three years for production, algae can start producing oil within 3–5 days and thereafter oil can be harvested on daily basis (Demirbas, 2011).

13.2.6 ALGAE AS BIOFERTILIZER

Most of the microalgal species are employed in agriculture as biofertilizers and soil conditioners; among which cyanobacterial species are efficiently capable of fixing atmospheric nitrogen as biofertilizers (Song et al., 2005). Blue-green algae of genera *Nostoc*, *Anabaena*, *Tolypothrix*, and *Aulosira* are reported to efficiently fix atmospheric nitrogen and generally used as inoculants for paddy crop grown both under upland and low land conditions. *Anabaena* in association with water fern *Azolla* contributes nitrogen up to 60 kg/ha/season and also enriches soils with organic matter (Priyadarshani and Rath, 2012).

13.2.7 ALGAE AS PHARMACEUTICAL PRODUCTS

Algae are rich sources of novel and biologically active primary and secondary metabolites. These metabolites may be potential bioactive compounds of interest in the pharmaceutical industry (Rania and Hala, 2008). Cell-free extracts of *Chlorella vulgaris* and *Chlamydomonas pyrenoidosa* have been proved to have antibacterial and antioxidant activity (Annamalai et al., 2012; Annamalai and Nallamuthu, 2014). The antibacterial, antifungal, and anti-viral activity of various macroalgae have also been demonstrated (Salvador et al., 2007; Guedes et al., 2012). Fucoxanthin, a xanthophyll found as an accessory pigment in the chloroplasts of brown algae and its metabolite fucoxanthinol are potentially active against Adult T-cell leukemia (Ishikawa et al., 2008). Polysaccharide fractions of the red algae *Gracilaria verrucosa* exhibit antioxidant activity as well as anti-inflammatory activity by stimulating phagocytosis in rats (De Almeida et al., 2011).

13.2.8   ALGAE AS AQUACULTURE FEED

Microalgal feeds are currently used in aquaculture for culturing larvae, juvenile shell, finfish, as well as for raising the zooplankton required for feeding of juvenile animals (Chen, 2003). The most frequently used species in aquaculture as feed are *Chlorella, Tetraselmis, Isochrysis, Pavlova, Phaeodactylum, Chaetoceros, Nannochloropsis, Skeletonema,* and *Thalassiosira. Spirulina* and *Chlorella* are also used in this domain for many types of animals: cats, dogs, aquarium fish, ornamental birds, horses, poultry, cows, and breeding bulls (Spolaore et al., 2006). Microalgae for larval feeds include *Chaetoceros, Thalassiosira, Tetraselmis, Isochrysis,* and *Nannochloropsis* that are fed either directly or indirectly (Da Silva and Barbosa, 2008).

13.2.9   ALGAE AS A CARBON SEQUESTER

Microalgae are capable of mitigating CO_2 10 to 50 times higher than that of terrestrial plants (Raheem et al., 2015). Producing 1 t of algal biomass fixes 1.6–2.0 t of CO_2; approximately half of the dry weight of algal biomass (Abbasi and Abbasi, 2010; Demirbas, 2010; Ziolkowska and Simon, 2014; Bharathiraja et al., 2015).

13.2.10   ALGAE AS A BIOREMEDIATOR

Algae are increasingly used to treat both municipal and industrial wastewater. They play a vital role in aerobic treatment of wastewater, mostly in the removal of nutrients (nitrogen and phosphorous), toxic compounds (heavy metals), and pathogens. The algal-based treatment has some advantages over conventional treatments such as a reduction in the use of chemicals and lower energy costs (Bertagnolli et al., 2014).

13.3   ALGAL GENOMICS

Algae are ubiquitous, highly diverse group of photosynthetic organisms having significant role in maintaining terrestrial and atmospheric conditions. These organisms occur from microscopic to macroscopic range, i.e., survive as tiny picoplankton in open oceans and also as turf meadows and forests in coastal waters (Biegala et al., 2003). The diversity among the algae is enormous, not only based on size and shape of the organisms, but also based on the production of various chemical compounds through novel biosynthetic pathways. Algal pigments that comprise the light-harvesting antennae are visually striking and biochemically diverse. In the green algae, the light-harvesting antennae contain chlorophylls a and b, with a significant level of carotenoids, while pigments of red algae and CB are predominantly the phycobiliproteins. In contrast, diatoms, and dinofagellates use oxygenated carotenoids as their major light-harvesting pigments. The composition of polysaccharides and cell walls also shows enormous diversity among the algae; ranging from microfibrillar walls of cellulose, polysaccharides to proteinaceous and silicacious walls. Thus, a vast diversity among algal genomics exists, where genomes are fundamental element for genetic manipulation. Genomic study not only provide location and distribution of metabolic pathways and enzymes but also aid in the identification of elements that can improve genetic engineering, including *cis*-acting elements, transacting factors and other regulatory elements (Grossman, 2005).

13.3.1 CYANOBACTERIAL GENOMICS

Emerging field of marine algal genomics first began with publications of three genomes of the smallest known oxygen evolving autotroph *Prochlorococcus* (Dufresne et al., 2003; Rocap et al., 2003). To date, over 20 cyanobacterial genomes have been released. CB possess several traits in their genome that are different from other algae such as their outer-membrane light harvesting antenna, two-component signal transduction system and their autotrophic metabolism. Many CB have water-soluble, light-harvesting protein-pigment complex phycobilisomes that reach a width of 40 nm attached to the cytoplasmic surface of thylakoid membrane (Yi et al., 2005). In *Synechococcus* sp., the genes for the metabolism of phycobiliproteins are distributed as several operons or gene clusters. However, in *Prochlorococcus*, most of the genes for phycobiliproteins disappear and only a small set of genes for phycoerythrin type III and their reductases are conserved in the genome, suggesting the genes for phycobilierythrin are being lost through selection in the evolutionary process (Ting et al., 2001).

13.3.2   GUILLARDIA THETA: NUCLEOMORPH GENOMICS

Cryptomonads are among chlorophyll-c containing chromophytic algae and these are the only organisms to retain an enslaved red algal nucleus resulting from secondary endosymbiosis (Cavalier-Smith, 2000; Maier et al., 2000). This reduced nucleus or nucleomorph has an envelope membrane with nuclear pores; however, the genetic content of the nucleomorph is highly reduced relative to a red algal genome. DNA of the nucleomorph from Cryptomonad, *G. theta* contains 3 mini-chromosomes that together constitute 551 kb. This genome is predicted to have 464 genes encoding polypeptides, nearly one-half of which encode proteins of unknown function. The genes are highly compacted in the genome such that only 17 protein-coding genes contain introns and could be removed by spliceosomes. Most of the introns are near the 5' ends of the transcripts, and 11 of these 17 intron-containing genes encode ribosomal proteins (Grossman, 2005). Understanding the steps involved in the biosynthesis and expression of genes encoded on the nuclear, plastid, and nucleomorph genomes would further elucidate the roles of various cellular processes such as communication and protein metabolite exchange.

13.3.3   CYANIDIOSCHYZAN MEROLAE GENOMICS

Cyanidiales is a group of unicellular, asexual red algae that grow at high temperatures and under acidic conditions. This group includes the genera *Cyanidium, Cyanidioschyzon, and Galdieria* (Ciniglia et al., 2004). The first algal nuclear genome to be sequenced was that of a member of *C. merolae* which is the smallest genome is among the smallest genome that occurs in photosynthetic eukaryotes. *C. merolae* is an organism that grows in the hotspring (45°C) at a pH of 1.5 and is considered one of the most primitive algal species (Matsuzaki et al., 2004; Nozaki et al., 2004). Its subcellular structure is relatively simple with a single golgi apparatus, ER, and a relatively small number of internal membrane structures. Plastid genome of this alga is about 150 kb and contains 243 genes (Ohta et al., 2003). There is an overlap between the protein coding sequences for many of these genes (40%), resulting in a highly compacted plastid genome. Analysis of *C. merolae* genomic sequence implicates with respect to the endosymbiont origins of the plastid. Enzymes of the Calvin cycle are revealed to be originated from a combination of genes derived from a cyanobacterial endosymbiont and its eukaryotic host, *Arabidopsis thaliana* suggesting stability of ancestral origin even after separation of lineages.

13.3.4 *THALASSIOSIRA PSEUDONANA GENOMICS*

Diatoms are diverse group of organisms present in marine, freshwater, and terrestrial environments. These organisms have different gross morphologies (pennate, centric, coccoid, and triangular) with precisely patterned and beautifully ornamented silicified cell walls or frustules. Diatoms are usually employed to sophisticated molecular techniques and biolistic procedures (Dunahey et al., 1995; Apt et al., 1996; Zaslavskaia et al., 2000). Research work involves insertion of reporter genes into diatoms to study gene expression and these reporters include: *Escherichia coli uidA* gene encoding β-glucuronidase, Tn9-derived *cat* gene encoding chloramphenicol acetyltransferase, firefly *luc* gene encoding luciferase, a variant of the green fluorescent protein gene (*egfp*), and *aequorin* gene from jellyfish *Aequorea victoria* (Falciatore et al., 1999). One of the major areas of interest in diatom genomics over the last decade concerns cell wall or frustule formation. Frustules are silicified cell walls of the diatoms in which the deposition of the silica creates a precise, nano-scale pattern; these structures have the potential for exploitation as substrates for nanotechnology development. Other areas of interest include: the way diatoms position themselves in the water column, the function and evolution of light-harvesting components, the mechanisms associated with non-photochemical quenching of excess absorbed light energy, carbon metabolism and role of the C_4 pathway in CO_2 fixation, biosynthesis of long-chain PUFAs, role of Ca_{21} in signaling cellular processes, identification, and functional analyses of photoreceptors, development of different cell morphotypes, and the control of morphogenesis (Buchel, 2003; Lebeau and Robert, 2003; Wen and Chen, 2003; Oeltjen et al., 2004; Reinfelder et al., 2004; Tonon et al., 2004).

13.3.5 *CHLAMYDOMONAS REINHARDTII GENOMICS*

Genetic, molecular, physiological, and genomic features of *C. reinhardtii*, a unicellular green alga ideally elucidates the critical of biological processes of both plants and animals relating to photosynthesis, biogenesis, and functioning of the flagella. These haploid algae are capable of growing heterotrophically in dark using acetate as the sole carbon source and maintain normal chloroplast structure which could resume photosynthetic CO_2 fixation upon illumination (Harris, 2001). Indeed, using the genetic manipulations first elegantly demonstrated by Sager (1960), Levine, and his colleagues began to delineate the pathway of photosynthetic electron transport and the regulation of the photosynthetic activity (Gorman and Levine, 1966; Bennoun and Levine, 1967; Givan and Levine, 1967; Lavorel and Levine, 1968; Levine and Goodenough, 1970; Moll and Levine, 1970; Sato et al., 1971). Studies on *C. reinhardtii* mutants in the identification of motility and the biochemical characterization of flagella has made this organism as an ideal for dissecting flagellar function. Polypeptides associated with flagellar assembly are found to be similar to proteins altered in diseased mammalian cells; in turn elucidating the biology of both photosynthetic and non-photosynthetic eukaryotes (Li et al., 2004; Snell et al., 2004).

The genomic sequence of *C. reinhardtii* is about 110 Mb, nearly 95 Mb of the sequence completed; yet sequence information remain dispersed over approximately 3,000 individual scaffolds. Many genes encode proteins that are similar to those of animal cells. A comparison of the *C. reinhardtii* gene models with that of human genome generated 4,348 matches. Pool of these genes encodes number of known flagellar and basal body polypeptides and contains genes associated with human diseases resulting from impairment of cilia or basal body function. Six genes ((Bardet-Beidl syndrome) BBS1, 2, 4, 5, 7, and 8) are found to be associated with Bardet-Biedl syndrome, a human disease characterized by retinal dystrophy, obesity, polydactyly, renal, and genital malformation, and learning disabilities (Li et al., 2004). Hence, genomic studies in *C. reinhardtii* would significantly improve understanding on mechanistic aspects of human diseases associated with defects in centriole and cilia function and assembly.

13.4 METHODS OF GENETIC TRANSFORMATION IN ALGAE

More than thirty different strains of algae have been transformed successfully to date; resulting in stable expression of transgenes, either from the nucleus or plastids.

13.4.1 TRANS-CONJUGATION

Trans-conjugation is the transfer of DNA between a cyanobacterial cell and a bacterial cell (usually *Escherichia coli*) by direct cell-to-cell contact or by a bridge-like connection between two cells. Application of this method is common among freshwater cyanobacterial genetic manipulation, while the versatility of gene transfer by transconjugation in marine CB were also demonstrated in strains of *Synechococcus* sp., *Synechocystis* sp. and *Pseudanabaena* sp. (Sode et al., 1992). Conjugation was implemented using a mobilizable transposon and a broad-host range vector pKT230. This research confirmed the wide applicability of conjugation in CB. A plasmid DNA containing green fluorescent protein (GFP) was transferred into a *Prochlorococcus* strain by interspecific conjugation with *E. coli* and the expression of this protein was detected by Western blotting and cellular fluorescence. This is the first report of GFP expression in oceanic CB (Tolonen et al., 2006).

13.4.2 NATURAL TRANSFORMATION AND INDUCED TRANSFORMATION

Natural transformation and induced transformation may allow cyanobacterial cell to absorb extrinsic DNA directly in the form of natural competence cells or artificially induced to competence cells. By treating with ethidium bromide, a cured strain *Synechococcus* sp. NKBG042902-YG 1116 was successfully transformed under dark incubation conditions for 16 h (Matsunaga and Takeyama, 1995). The mechanism of competence in CB is similar to that of bacteria. However, the transformation efficiencies of the marine *Synechococcus* strains are ten times lower than those of the freshwater *Synechococcus* strains. This might be due polysaccharides surrounding the marine *Synechococcus* cell which would inhibit DNA uptake. Thus, marine cyanobacterial algal genetic transformation is replaced by electroporation.

13.4.3 ELECTROPORATION

Gene transfer by electroporation has been applied in various cells and bacteria for over 30 years because of the simplicity of the procedure and a high efficiency with a small amount of DNA (Neumann et al., 1982). Electroporation potentially transfer extrinsic genes independent of the cell's ability and universally to different genera. The electric field strength required for marine cyanobateria is lower than that for freshwater strains. The decrease in efficiency due to the electric field strength could be compensated for by enhancing the $CaCl_2$ pretreatment of marine strains. Efficient electroporation-mediated transformation has been achieved in both wild-type and cell wall-deficient eukaryotic *Chlamydomonas reinhardtti* strains (Brown et al., 1991). The efficiency of electroporation is two times higher than that obtained with the glass beads method to introduce exogenous DNA to algal cells (Shimogawara et al., 1998). To date, the electroporation transformation is established in marine genera from prokaryotic cells to eukaryotic red algae, green algae, and diatoms. 'Star' marine alga, *Nannochloropsis* sp. has been successfully genetically transformed by electroporation technique and potentially employed in biodiesel production (Kilian et al., 2011). However, the application of the electroporation technology in brown algae remains constraint due to difficulties in viable protoplast preparation and regeneration technologies.

13.4.4   BIOLISTIC TRANSFORMATION

Direct gene transfer by the biolistic/micro-particle bombardment method has been proven to be the most efficient method and is highly reproducible in introducing exogenous DNA into algal cells. This method has been successfully employed for the transformation of many microalgal nuclear and chloroplast expression systems, and it is not surprising that biolistic transformation remains the most useful tool for transgenic studies of macroalgae regardless of their cell walls and life cycle. *Synechococcus* sp. NKBG15041c was transformed successfully using particle bombardment with bacterial magnetic particles purified from one magnetic bacterium known as *Magnetospirillum* sp. AMB-1. This particle is covered with a thin phospholipid layer that could bind larger quantities of DNA rather than gold or tungsten particles (Matsunaga et al., 1991). To date, the available tools for genetically engineering diatoms remain sparse. Biolistic transformation is the only efficient tool to genetically manipulate diatoms. Transformation methods with particle bombardment have been well established for several species of diatoms such as *Thalassiosira pseudonana, Thalassiosira weissflogii, Chaetoceros* sp., *C. cryptica, Navicula saprophila, P. tricornutum,* and *Cylindrotheca fusiformis* (Dunahay et al., 1995; Apt et al., 1996; Falciatore et al., 1999; Zaslavskaia et al., 2000; Poulsen and Kroger, 2005; Poulsen et al., 2006; Miyagawa et al., 2009; Miyagawa-Yamaguchi et al., 2011). A transient transformation system has been successfully established in the genera *Porphyra*; the cultivation of which currently provides a yearly turnover of approximately 1×10^9 US Dollars and forms a mature algal industry for food, cosmetics, and other high value products (Kuang et al., 1998; Pulz and Gross, 2004; Zhang et al., 2010).

13.4.5   GLASS BEADS

Agitation with glass beads has been used to efficiently introduce foreign DNA into microalgae and was first reported in the freshwater alga *C. reinhardtii*

(Kindle, 1990). The advantages of glass beads are their simplicity and independence from expensive and specialized equipments. Genetic transformation using glass beads in *D. salina* that lacks rigid cell wall, by glass beads was successfully established due to more efficient, easily controlled and less physically destructive approach to the cells than electroporation and particle bombardment (Feng et al., 2009). Red seaweed, *Porphyra haitanensis* was transformed using glass beads by agitating the freshly released conchospores that either had thin cell walls or none at all. The maximum number of transformants was more than six out of the 1 million agitated conchospores (Wang et al., 2010). The main drawback of this method is its inability to transfer DNA into thick-walled cells (Coll, 2006). To overcome such constraint, thick-walled cells are pre-treated with cell wall desoluting enzyme yielding protoplasts. These are then mixed membrane fusion agent, polyethylene glycol and agitated in presence of glass beads. However, in some seaweed, cell viability decreases on removal of cell walls and differentiation of callus tends to complicate the isolation of transformants (Reddy et al., 2008; Baweja et al., 2009).

13.4.6   SILICON CARBON WHISKERS METHOD

In contrast to agitation of the cells with glass beads, agitating algal cells with silicon carbon (SiC) whiskers for up to 10 min results in a minor loss in cell viability. The SiC whisker method produce transformants at an efficiency of up to 10^{-5} per cell for walled cells and up to 10^{-4} per cell for cell wall deficient mutant strains (Dunahay, 1995). Ability to overcome cell wall obstruction to exogenous DNA and inexpensive cost of this method has made the stable genetic transformation of marine dinoflagellates; the efficiency range was approximately 5–24 per 10^7 cells (Lohuis and Miller, 1998). Cell viability following SiC whisker agitation is greatly improved; however, unreliable source of SiC whisker and the inhalation hazard restricts it wide usage compared to the glass bead agitation method (Potvin and Zhang, 2010).

13.4.7   MICROINJECTION

As a direct physical method that is able to penetrate intact cell walls, the microinjection method does not necessarily require a protoplast regeneration system. Additionally, microinjection allows the introduction of DNA under microscopical control into specific targets (Schnorf et al., 1991). Theoretically, the receptor cells of microinjection could be defined as either compartments of a single cell or as defined cells within a multicellular structure, from plants, animals, or microbes. High yield and stable nuclear transformation was achieved in a unicellular green alga *Acetabularia mediterranea* by microinjecting SV40 DNA and pSV2neo into the isolated nuclei of algal cells and then implanting the injected nuclei into nucleate cell fragments of the same species (Neuhaus et al., 1986). This established and high yield method was successfully used to tackle the problem of the nuclear transport of algal proteins (Pfeiffer et al., 2009). Regardless of its complicated and delicate procedure, microinjection is considered to be highly efficient and low cost transformation method for algae.

13.4.8   ARTIFICIAL TRANSPOSON METHOD

Transposons, mobile DNA elements originally discovered in maize are strong genetic tools for both prokaryotic and eukaryotic cells (Haapa et al., 1999). Artificial transposons that are extracted and transformed from the essential elements of natural transposons have been broadly developed for *in vitro* mutagenesis and genetic transformation. It is worth emphasizing that this stable genetic transformation method has been used successfully in various algal genera such as *Spirullina platensis* (Kawata et al., 2004). However, this technique has drawbacks as in most cases, the genes are integrated randomly into the genome leading to rearrangement and truncation of DNA sequences, causing unintentional changes or silencing the expression of the foreign gene.

13.4.9   RECOMBINANT EUKARYOTIC ALGAL VIRUSES AS TRANSFORMATION VECTORS

Large dsDNA viruses infecting eukaryotic algae show promising application as genetic vectors in algal biotechnology. The two main groups of eukaryotic algal viruses are *Chlorella* system, which displays high levels of infectivity and complete pathogenesis (complete lysis of the unicellular host), and Brown algal virus system by which only specialized reproductive cells of a multi-cellular free-living organism are infected and lysed. More extensive and comprehensive fundamental studies on eukaryotic algal viruses including genome and mechanism of infection are necessary for effective implementation of this method in future (Van Etten and Meints, 1999; Leon and Fernandez, 2007).

13.4.10   AGROBACTERIUM TUMEFACIENS-MEDIATED GENETIC TRANSFORMATION

This method genetically transforms plants by transferring and integrating a portion of resident Ti-plasmid with a large segment of DNA up to 150 kb into a plant nuclear genome with the assistance of several virulence proteins (Tzfira and Citovsky, 2006). The first report of a stable genetic transformation by *A. tumefaciens* in algae was conducted in the marine red seaweed *Porphyra yezoensis* (Cheney et al., 2001). The transformation frequency of gene transfer to the nuclear genome of the freshwater green alga *C. reinhardtii* by *A. tumefaciens* was 50-fold higher than that of the glass bead transformation (Kumar et al., 2004). The transformation systems in the microalgae *Nannochloropsis* sp. and *Dunaliella bardawil* were also recently established with this method (Anila et al., 2011; Cha et al., 2011). In the *Dunaliella* study, the transformation frequency obtained (41.0 ± 4 CFU 10^{-6} cells) was not higher than those reported for glass beads transformation and electroporation, but the transformants obtained were found to be stable for 18 months (Anila et al., 2011). However, this method is technically challenging because of the large size and low copy number of Ti plasmids leading to difficulties in plasmid isolation and manipulation (Meyers et al., 2010).

13.5   VECTOR CONSTRUCTION AND GENE SELECTION STRATEGIES

13.5.1   VECTOR CONSTRUCTION: PROMOTER SELECTION AND CODON USAGE

Vector element construction is one of the crucial parts in determining the stability and frequency of exogenous DNA expression in algae. Among CB, shuttle vectors constructed from a cyanobacterial chromosome segment and cyanophages are the main possible sources of the vector backbone. In eukaryotic algae, vectors are typically constructed based on their own chromosome segment. Additionally, several vectors from *E. coli* and a few constructed vectors from higher plants are used in algal transformations. Promoter availability and selection is a critical factor in genetic transformation (Liu et al., 2003; Qin et al., 2004; Wang et al., 2010; Anila et al., 2011). Apart from the universal promoters from viruses, the endogenous promoters from specific algae are considered to be the most efficient while constructing a vector. The diatom fucoxanthin-chlorophyll a/c binding protein gene (fcp) promoter is effective in marine diatoms and other marine algae (Apt et al., 1996; Zaslavskaia et al., 2000; Qin et al., 2004; Miyagawa-Yamaguchi et al., 2011). The duplicated carbonic anhydrase 1 (DCY1) promoter is used for stable nuclear transformation in *D. salina* (Lu et al., 2011). An endogenous PyAct1 (5′ upstream region of the actin1 gene from *P. yezoensis*) promoter was also found to be effective in the transient gene expression of 12 red seaweed species (Takahashi et al., 2010; Hirata et al., 2011). In *Nannochloropsis* sp. transformation, the endogenous promoters were developed from two unlinked violaxanthin/chlorophyll α-binding protein (VCP) genes, VCP1 and VCP2. The VCP1 promoter was a unidirectional promoter, while the VCP2 promoter was bidirectional (Kilian et al., 2011).

13.5.2   REPORTER AND MARKER GENES

Reporter genes demarcate transformation, expression, stability, and easy determination of transformed cells; thus, proteins expressed by reporter genes are considered to be sensitive and intuitionistic. The widely used reporter genes in algal transformation are *GUS* and *lacZ*. The *GUS* gene encodes β-glucuronidase and is found to be an effective reporter for transient and stable expression in algae such as *D. salina*, *Amphidinium* sp., *Symbiodinium microadriaticum*, *T. weissflogii*, *Ectocarpus* sp., *Porphyra yezoesis*, *Ulva lactuca*, *L. japonica*, and *Undaria pinnatifida* (Qin et al., 1994; Huang et al., 1996; Lohuis and Miller, 1998; Kuang et al., 1998; Falciatore et al., 1999; Liu et al., 2003; Tan et al., 2005; Hirata et al., 2011). The *Luc* gene encoding luciferase is another reporter gene applied in a freshwater microalgae and marine diatom *P. tricornutum* (Falciatore et al., 1999). The GFP of the jellyfish *Aequorea victoria* is considered to be the universal reporter of gene expression and in subcellular localization analyses of various algae (Poulsen et al., 2006; Hirakawa et al., 2008; Takahashi et al., 2010; Wang et al., 2010; Miyagawa-Yamaguchi et al., 2011; Watanabe et al., 2011).

13.5.3   GENE COPY NUMBER AND HOMOLOGY-DEPENDENT GENE SILENCING

Expression levels of transgenic genes in algae are inconsistent and difficult to predict. The significant reasons for unpredictable variation arise from inconsistencies in the number of integrated transgene copies and the subsequent homology-dependent gene silencing. Silencing increases with a growing number of integrated gene copies that occurs at the transcriptional or post-transcriptional phase as a defense mechanism of plants against viruses (Depicker and VanMontagu, 1997; Baulcombe, 2004; Marenkova and Deineko, 2010; Potvin and Zhang, 2010). Electroporation commonly results in highly variable integrated transgene copy numbers and low copy transformants. *Agrobacterium*-mediated transformation typically leads to low copy numbers and higher single-copy transformants while, the direct DNA-transfer methods such as glass beads and biolistic bombardment leads to a large number of integrated gene copies in the receptor algal genome, in turn increasing silencing effect (Yao et al., 2006; Lowe et al., 2009).

13.5.4 RNA INTERFERENCE (RNAi) TECHNOLOGY

Gene silencing can occur either through repression of transcription, termed transcriptional gene silencing, or through mRNA degradation, termed post-transcriptional gene silencing (Angaji et al., 2010). These silencing effects have been proved to be an invaluable tool for analyzing the biological function of the target gene and sequence-specific knockdowns (Cerutti et al., 2011). A wide range of core RNAi machinery components, which promotes the transient gene silencing and stable gene repression have been found in marine algal species such as red alga *P. yezoensis* (Liang et al., 2010), green alga *D. salina* (Jia et al., 2009), diatom *P. tricornutum* (De Riso et al., 2009), *T. pseudonana* (Armbrust et al., 2004), and brown alga *Ectocarpus siliculosus* (Cock et al., 2010).

13.6 GENETIC TRANSFORMATION IN ALGAE: A MODEL

DNA transformation of the chloroplast was first reported in 1988 using the single-celled green alga, *Chlamydomonas reinhardtii*; since, that time the tools and techniques for chloroplast genetic engineering of *C. reinhardtii* advanced significantly (Boynton et al., 1988; Purton, 2007; Doron et al., 2016). Recently, chloroplast transformation are also reported among other algal species such as green algae *Haematococcus pluvialis* and *Dunaliella tertiolecta*, the red alga *Cyanidioschyzon merolae* and diatom *Phaeodactylum tricornutum* (Gutierrez et al., 2012; Georgianna et al., 2013; Xie et al., 2014; Zienkiewicz et al., 2017). Most of the microalgal species are generally recognized as safe (GRAS) status and are considered to be free of harmful viral or endotoxin contaminants, thereby simplifying the procedures for product purification. Alternatively, whole algae may be used for oral delivery (to animals, if not to humans) of vaccines, enzymes, or hormones (Yan et al., 2016). Typically, transgenic DNA is introduced into the chloroplast by the bombardment of an algal lawn with DNA-coated gold microparticles. Alternative DNA delivery strategies include electroporation or agitation of a DNA/cell suspension in the presence of glass beads. DNA integration into the plastome occurs almost exclusively *via* homologous recombination (HR) between matching sequences on the incoming DNA and plastome sequence (Boynton et al., 1988). Consequently, transgenes are precisely targeted to any locus by flanking the DNA with chloroplast sequences upstream and downstream of the target locus (Figure 13.1).

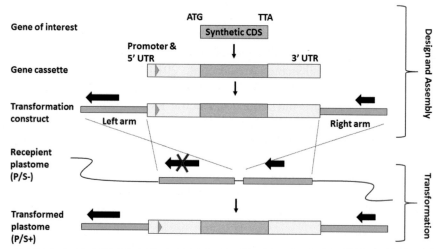

FIGURE 13.1 Marker-free strategy for introducing transgenes into the *C. reinhardtii* chloroplast. The gene of interest (GOI) is codon-optimized to match chloroplast genes and assembled into a transformation construct using a 'one-step' method such as Golden Gate or Gibson assembly. The left and right arms are chloroplast DNA parts that ensure insertion into a specific intergenic region *via* homologous recombination between the arms and the recipient plastome. One of the arms carries a wild-type copy of a gene that is essential for photosynthesis, and selection is based on the repair of a mutated form of this gene (indicated with an 'X') in the photosynthesis-deficient (P/S) recipient strain. The resulting transformant is therefore restored to phototrophy (P/S+), with only the GOI being introduced into the plastome (Redrawn with permission from Duo and Purton, 2018).

Several selection strategies have been developed based on the use of bacterial antibiotic-resistance genes such as aadA and aphA6; however, a superior selection strategy involves the rescue of chloroplast mutant to phototrophy

(Boynton et al., 1988). This results in marker-free transformants in which only the foreign DNA in the plastome is the gene of interest (Bertalan et al., 2015; Wannathong et al., 2016). Expression of gene is achieved by fusing the coding sequence to promoters and untranslated regions (UTRs) from highly expressed endogenous genes, such as the photosynthesis genes psaA and psbA. The efficiency of translation could be significantly improved by using a synthetic coding sequence optimized to match the AT-rich codon bias seen in chloroplast genes (Purton et al., 2013). Biocontainment could also be built into the transgene by replacing several tryptophan codons (UGG) with the UGA stop codon and using an orthogonal tryptophan tRNA to recognize these internal stop codons in the chloroplast (Young and Purton, 2016). Although almost all of the transgenes that have been inserted into the *C. reinhardtii* chloroplast to date were been constituently expressed, regulation of transgene expression was achieved using a vitamin-based system; i.e., by suppressing the nuclear gene encoding a factor essential for the translation of the chloroplast psbD gene. Any transgene fused to the psbD 5' UTR is therefore only translated in the absence of vitamins (Ramundo and Rochaix, 2015).

13.7   SURVIVAL AND PERSISTENCE OF GENETICALLY ENGINEERED (GE) ALGAE

The ability of genetically engineered (GE) algae (or their DNA) to survive outside of open ponds or enclosed bioreactors is crucial baseline information for risk assessment. Proponents of GE algae often suggest that lab-created strains should not be able to survive in the wild, especially if they are domesticated and bred to produce large volumes of industrial by-products (Gressel, 2008). This scenario would obviate many environmental concerns, but ecological research is needed to determine how well weak GE microalgae would survive under a wide range of natural biotic and abiotic conditions. Certain engineered traits, such as tolerance of harsh conditions or enhanced growth, may facilitate the survival and growth of GE algae in unmanaged ecosystems. Unfortunately, much more is known about the baseline performance of microalgae under controlled laboratory conditions than in complex, uncontrolled natural habitats. To fill this gap, new GE lab strains that are intended for large-scale cultivation should be examined in comparison with their natural, non-GE counterparts in contained mesocosm experiments and modeling studies. A recent simulation analysis concluded that depending on the specific nature of genetic modifications, escaping populations of GE algae may quickly be outcompeted by natural forms (Flynn et al., 2010). However, it should also be noted that introduced changes in algal metabolic processes may adversely affect the food value of such organisms for zooplankton, which often provide top-down regulation of algal populations. In some cases, unintended effects of genetic engineering might inadvertently produce new strains of harmful algae that could become persistent (Flynn et al., 2010).

13.8 IMPROVED APPLICATIONS OF GENETICALLY ENGINEERED (GE) ALGAE

13.8.1 GENETICALLY ENGINEERED (GE) ALGAE AS BIOSORBENTS SWEEPING-OFF HEAVY METAL FROM INDUSTRIAL EFFLUENTS

Seaweed, green macroalgae and their alginate derivatives exhibit high affinity for many metal ions (Mani and Kumar, 2014). In the emerging trend, biological materials used in the passive removal of toxic heavy metals are known as biosorbents. Brown macroalgal strains are termed to be potential green biosorbent due to synthesis of extrapolyaccharide, alginate (Davis et al., 2003). Alginate plays a critical role in metal biosorption by participating in ion-exchange and complexation causing bond between heavy metals and this polymer. The adsorption capacity of the brown algae is directly related to the alginate content, availability, and its specific macromolecular conformation. Alginate comprises a significant component up to 40–45% of the dry weight of *Sargassum* biomass (Fourest and Volesky, 1995). The affinity of alginates for divalent cations such as Pb^{2+}, Cu^{2+}, Cd^{2+} and Zn^{2+} donate 227, 51, 79, and 78 mg g^{-1} metal uptake (Davis et al., 2003). Sargassum packed columns was investigated to be used in flow-through column systems. Implementation of such packed bed columns inactively adsorb and detoxify heavy metals bearing industrial wastewater (Bertagnolli et al., 2014).

Exploiting these biological mechanisms at a molecular level to produce engineered organisms with higher biosorption capacity and selectivity for specific metal ions could be used to develop new biosorbents. Many genes are involved in metal-uptake, detoxification, or tolerance (Rosen, 2002). Cysteine-rich peptides such as glutathione, lipopolysaccharides (LPSs), phytochelatins, and metallothioneins bind metal ions such as Cd, Cu, Hg, and many more heavy metals, and enhance metal ion bioaccumulation (Ghosh et al., 1999). To date, little attention has been paid to investigate the recombinant microalgal strains for metal ion biosorption, and it remains highly prospective for engineered algae in achieving higher sorption capacities and specificity for targeted metal ions (Zeraatkar et al., 2016).

13.8.2 GENETICALLY ENGINEERED (GE) ALGAE AS BIOREACTORS TO PRODUCE PHARMACEUTICALS AND THERAPEUTICAL PROTEINS

As a platform for the production of subunit vaccines, transgenic microalgae have attracted considerable attention (Specht and Mayfield, 2014). Microalgae plays versatile role as expression system for various industrial, therapeutic, and diagnostic recombinant proteins; in terms of high growth rate and eukaryotic expression system. In the genetic engineering of algae, potential vaccine candidates have been produced against viruses and bacteria as well as malaria and other communicable diseases.

13.8.2.1 SUBUNIT VACCINES

1. The E2 protein, an antigen for vaccines against classical swine fever virus (CSFV) was produced in *Chlamydomonas* chloroplast. The E2 protein could accumulate 1.5–2% of total soluble protein (TSP), and crude extracts of E2-transformed algae were administered via subcutaneous injection and oral immunization in mice; the former resulted in a significant level of serum antibody against CSFV, whereas the later caused no immune response to E2 (He et al., 2007).
2. Foot-and-mouth disease virus (FMDV) is an important virus that affects livestock. To produce a vaccine against FMDV in *C. reinhardtii*, the VP1 protein was fused with a potent mucosal adjuvant, cholera toxin B subunit (CTB). VP1-CTB accumulation to 3% of TSP and exhibited binding affinity towards GM1-ganglioside, separately acting as an antigen for FMDV VP1 and CTB proteins (Sun et al., 2003).
3. The gene that encodes the hepatitis B virus surface antigen (HBsAg) was successfully transformed into the green alga *D. salina* by electroporation, and stable nuclear transformants were obtained. Western blot analysis demonstrated HBsAg production ranging between 1.6–3.1 ng mg^{-1} of total protein in different transformants with different insertion sites (Geng et al., 2003).

4. White spot syndrome virus (WSSV) cause severe damage to shrimp farms worldwide. The main viral envelops protein of WSSV, viral protein 28 (VP28), is a target for subunit vaccines (Van Hulten et al., 2001). The gene encoding the VP28 protein was successfully transformed into the nuclear genome of *D. salina* and into the chloroplast of *C. reinhardtii*, with an output of 21% of total protein in the former case, and approximately 3 ng mg^{-1} of total protein in the latter. Under the challenge of WSSV, the survival rate of Ds-VP28-vaccinated crayfish was 41%, much higher than the control group, which had a mortality rate of 100%. This indicated that oral administration of VP28-transformed algae protected crayfish against viral pathogens (Surzycki et al., 2009; Feng et al., 2014; Dyo and Purton, 2018).

5. Glutamic acid decarboxylase-65 (GAD65) is a major autoantigen for human type 1 diabetes which is an autoimmune disease resulting from the destruction of insulin-producing beta cells in the pancreas. By immunizing non-obese diabetic (NOD) mice, studies on type 1 diabetes found that GAD65 can prevent or delay diabetes onset (Tisch et al., 1993). The expression level of human GAD65 in the chloroplasts of *C. reinhardtii* reached 0.25% to 0.3% of TSP, and the algal-derived hGAD65 reacted with diabetic sera and has the ability to induce the proliferation of spleen lymphocytes from NOD mice (Wang et al., 2008).

6. Many other vaccines have been expressed in microalgae, such as in *C. reinhardtii* malaria vaccines were produced by fusing algae's granule-bound starch synthase (GBSS) with apical major antigen (AMA1) and major surface protein (MSP1), two clinically relevant malaria antigens, separately. Experimental results showed that immune sera from GBSS-MSP1 immunized mice resulted in strong protection against malarial infection (Dauvillee et al., 2010; Jones et al., 2013).

13.8.2.2 ANTIBODIES, IMMUNOTOXINS, ANTIMICROBIAL AGENTS, AND HORMONES

1. Monoclonal antibody (mAb) is an effective therapy for the treatment of various human diseases. The first mammalian protein produced in the chloroplast of microalgae was a mAb protecting human against glycoprotein D of herpes simplex virus (HSV). This large

single-chain antibody, a fusion protein of the entire IgA heavy chain and the light chain variable region *via* a flexible linker expressed in *C. reinhardtii* chloroplast (Mayfield et al., 2003).

2. Monoclonal human IgG antibody against the Hepatitis B surface protein and the respective antigen was expressed in the diatom *Phaeodactylum tricornutum*. Antibodies were fully-assembled, functional, and were able to accumulate TSP up to 8.7%, which complied with 21 mg antibody per gram algal dry weight (Hempel et al., 2011).

3. Immunotoxins, consisting of a toxin molecule and an antibody domain could bind to the target cells and inhibit its proliferation, in turn acting as anti-cancer therapeutic agent. Both monovalent and divalent immunotoxins could be produced and assembled in the chloroplasts of microalgae. Furthermore, both immunotoxin molecules could significantly inhibit tumor growth in animal experiments, improving mouse survival rate in the assay of tumor-challenge. Similar results were observed when gelonin, a different eukaryotic toxin from *Gelonium multiflorum* was fused to monovalent and divalent anti-CD22 single-chain antibodies (Tran et al., 2013 a, b).

4. Defensins, a kind of small cationic peptides serves to be a promising alternative to antibiotics. Mature rabbit neutrophil peptide 1, an α-defensin with a broad anti-microbial activity that could defend against gram-negative or positive bacteria, certain viruses, and pathogenic fungi were been successfully transformed into *C. ellipsoidea* (Bai et al., 2013).

5. Flounder growth hormone was be expressed in *C. ellipsoidea* and accumulated up to approximately 400 µg L^{-1}. Flounder fry fed transgenic *Chlorella* increased by 25% in length and width within 30 days (Kim et al., 2002).

13.8.3 GENETICALLY ENGINEERED (GE) ALGAE AS ENHANCED BIOFUEL PRODUCER

Economically viable production of biofuel production using microalgae as a source requires higher lipid biosynthesis and accumulation ability. In order to enhance neutral lipid biosynthesis in microalgae, *Chlamydomonas reinhardtii* key enzyme: diacylglycerol acyltransferase (BnDGAT2) from *Brassica napus* was GE. The transformed colonies harboring aph7 gene were screened on hygromycin-supplemented medium and the transformation frequency was ~120 ± 10 colonies/1×10^6 cells. Major class of lipid includes: neutral lipids that constitute about 80% of the total lipids and are the most significant requirement for biodiesel production; among transformed algal cells neutral lipid were remarkably higher than that of wild control cells. Experimental results also revealed 7% decrease in saturated fatty acids and increase in unsaturated fatty acids in transformed algal cells. PUFAs, especially ALA, an essential omega-3 fatty acid were enhanced up to 12% in the transformed line. Expression of BnDGAT2 significantly improved the fatty acids profile in GE algal cells revealing their valuable strategy in biofuel production (Ahmad et al., 2015).

13.8.4 GENETICALLY ENGINEERED (GE) ALGAE AS PESTICIDE RESISTANT NITROGEN FIXATIVE AGENT

GE cyanobacterial species may act more efficiently as the self-renewable constitutive nitrogen (N) supply in wet agriculture, producing ammonia at the sole expense of the plentiful solar energy, atmospheric CO_2 and N_2 and water; irrespective of the pre- or post-treatment of the fields with pesticides and synthetic N fertilizers. These may contribute in considerably reducing the use of synthetic N fertilizers. Further, these may help in making up the significant loss of the available fixed N from the soil, which often happens as a result of the action of many of the naturally occurring denitrifying bacteria (Broadbent and Clark, 1965; Kuhlbusch et al., 1991). Nitrogen-fixing cyanobacterium *Nostoc muscorum* may be used as a gene bank or source material for the agronomically important markers in genetic engineering studies. Thus, pesticide-resistant CB are not only important for thriving in chemicalized agricultural fields and performing constitutive nitrogen fixation (with the help of additional markers producing derepressed nitrogenase), but are also an essential unique source for transforming pesticide-sensitive plants to develop higher tolerance to the synthetic agro-chemicals (Vaishampayan et al., 1998).

13.8.5 GENETICALLY ENGINEERED (GE) ALGAE AS A SAVIOR OF CORAL REEFS

Coral reefs are the most diverse marine habitat per unit area and provide world economies with nearly US$30 billion in net benefits from goods and services annually (Knowlton et al., 2010). Climate change impact models predict that most reefs will be severely damaged or lost in this century unless immediate protection efforts are made prompting calls for the development of novel mitigation and restoration approaches (Hoegh-Guldberg et al., 2007; Rinkevich, 2014; van Oppen et al., 2015, 2017; Mora et al., 2016; Piaggio et al., 2016; Hughes et al., 2017). Environmental bioengineering is an alternative strategy to safeguard against climate change (Piaggio et al., 2016). Microalgae, such as *Symbiodinium* are clear and promising candidates for genetic engineering with the aim of regaining and preserving ecosystem in terms of climate homeostasis (Berkelmans and van Oppen, 2006; Kirk and Weis, 2016; Murray et al., 2016). GE *Symbiodinium* variants having increased stress tolerance and naturally harbored by at-risk corals holds potential to reduce bleaching susceptibility without negatively impacting the fitness of the coral host; in turn *Symbiodinium*-coral partnerships would be preserved. *Fe-sod, Mn-sod Prxd*, and *Hsp70* genes from *Symbiodinium* are standout candidates whose engineered up-regulation may enhance thermal and bleaching tolerance by reducing heat-induced oxidative damage (Levin et al., 2016; Gierz et al., 2017; Goyen et al., 2017). However, a thorough evaluation of artificial up-regulation would be mandatory in contributing long term fitness and *Symbiodinium*-coral symbiosis (Levin et al., 2017).

13.8.6 GENETICALLY ENGINEERED (GE) ALGAE AS AN EFFICIENT DRUG DELIVERING AGENT

The ability to selectively inhibit cancerous cell proliferation while leaving healthy cells unaffected is a key goal in anticancer therapeutics. Though production of nanoporous silica-based materials requires costly and toxic chemicals, use of these materials as drug-delivery vehicles has been recently proved to be successful drug-delivering technology. However, diatom *Thalassiosira pseudonana* was GE to display an IgG-binding domain of protein G on the biosilica surface, enabling attachment of cell-targeting antibodies. Studies report that, neuroblastoma, and B-lymphoma cells are selectively targeted and there proliferation is inhibited by biosilica displaying specific antibodies sorbed onto drug-loaded nanoparticles. Nanoporous biosilica also exhibits regression of tumor growth in a subcutaneous mouse xenograft model of neuroblastoma. These data indicate that GE biosilica frustules may be used as versatile 'backpacks' for the targeted delivery of poorly water-soluble anticancer drugs to tumor sites (Delalalt et al., 2015).

13.9   FUTURE PROSPECTS AND CONCLUSION

A challenging task for algal biologists is to move from morphological, chemical, and geophysical descriptors of algal communities to more molecular descriptors that include both gene content and expression levels. Indeed, our understanding of biological, biophysical, and geochemical processes acquired using a spectrum of biotechnological methods developed over the last 20 years needs to be expanded. More information on whole genetic content of an organism and mechanism of behind expression of specific genes under different environmental conditions, developmental stages, and different tissue types need to be gathered. In terms of revenue, genomic studies are expensive and the resources to support such studies are limited. It is critical that societies and scientific communities with knowledge of the scientific and economic importance of particular groups of organisms, such as the algae, make informed choices as to which organisms would be of most benefit for genomic examination, whether involving whole genome or cDNA projects. It would be most efficient to solicit the aid of large, well-equipped centers that have an expert staff to complete the required sequencing tasks efficiently. However, the first important step for the scientific community with a working knowledge of the field is to define the organisms for which full genome and cDNA sequences should be obtained, to develop collaborations to facilitate the generation and analysis of genomic information, to petition various agencies for the funds required to obtain the sequence information, and to train the community, either through courses or workshops and tutorials over the internet; in turn, switching on the ways in which the genomic information could be used and extended.

KEYWORDS

- **algal-based nanotechnology**
- **algal classification**
- **algal genome sequencing**
- **algal genomic model**
- **diverse application of algae**
- **genetic transformation methods**
- **genetically engineered algae**

- **genomic research development**
- **novel metabolites and pathways**
- **risk assessment**
- **vectors and gene selectivity**

REFERENCES

Abbasi, T., & Abbasi, S. A., (2010). Biomass energy and the environmental impacts associated with its production and utilization. *Renew. Sustain. Energy Rev., 4,* 919–937.

Ahmad, I., Sharma, A. K., Daniell, H., & Kumar, S., (2015). Altered lipid composition and enhanced lipid production in green microalga by introduction of brassica diacylglycerol acyltransferase 2. *Plant Biotechnol. J., 13,* 540–550.

Angaji, S. A., Hedayati, S. S., Poor, R. H., Poor, S. S., Shiravi, S., & Madani, S., (2010). Application of RNA interference in plants. *Plant Omics, 3,* 77–84.

Anila, N., Chandrashekar, A., Ravishankar, G. A., & Sarada, R., (2011). Establishment of *Agrobacterium tumefaciens*-mediated genetic transformation in *Dunaliella bardawil. Eur. J. Phycol., 46,* 36–44.

Annamalai, J., & Nallamuthu, T., (2014). Antioxidant potential phytochemicals from methanol extract of *Chlorella vulgaris* and *Chlamydomonas reinhardtii. J. Algal Biomass. Utln., 5,* 60–67.

Annamalai, J., Shanmugam, J., & Nallamuthu, T., (2012). Phytochemical screening and antimicrobial activity of *Chlorella vulgaris* Beijerinck. *Int. J. Curr. Res. Rev., 4,* 33–38.

Apt, K. E., Kroth-Pancic, P. G., & Grossman, A. R., (1996). Stable nuclear transformation of the diatom *Phaeodactylum tricornutum. Mol. Gen. Genet., 252,* 572–579.

Armbrust, E. V., Berges, J. A., Bowler, C., Green, B. R., Martinez, D., Putnam, N. H., et al., (2004). The genome of the diatom *Thalassiosira pseudonana*: Ecology, evolution, and metabolism. *Science, 306,* 79–86.

Bai, L. L., Yin, W. B., Chen, Y. H., Niu, L. L., Sun, Y. R., Zhao, S. M., et al., (2013). A new strategy to produce a defensin: Stable production of mutated NP-1 in nitrate reductase-deficient *Chlorella ellipsoidea. PLoS One, 8*(1), E54966. https://doi.org/10.1371/journal. pone.0054966 (accessed on 26 May 2020).

Baulcombe, D., (2004). RNA silencing in plants. *Nature, 431,* 356–363.

Baweja, P., Sahoo, D., Garcia-Jimenez, P., & Robaina, R. R., (2009). Seaweed tissue culture as applied to biotechnology: Problems, achievements and prospects. *Phycol. Res., 57,* 45–58.

Bennoun, P., & Levine, R. P., (1967). Detecting mutants that have impaired photosynthesis by their increased level of fluorescence. *Plant Physiol., 42,* 1284–1287.

Berkelmans, R., & Van, O. M. J., (2006). The role of zooxanthellae in the thermal tolerance of corals: A 'nugget of hope' for coral reefs in an era of climate change. *Proc. R. Soc. Lond. B. Biol. Sci., 273,* 2305–2312.

Bertagnolli, C., Uhart, A., Dupin, J. C., Da Silva, M. G. C., Guibal, E., & Desbrieres, J., (2014). Biosorption of chromium by alginate extraction products from *Sargassum filipendula*: Investigation of adsorption mechanisms using x-ray photoelectron spectroscopy analysis. *Bioresour. Technol., 164,* 264–269.

Bertalan, I., Munder, M. C., Weiß, C., Kopf, J., Fischer, D., & Johanningmeier, U., (2015). A rapid, modular and marker-free chloroplast expression system for the green alga *Chlamydomonas reinhardtii*. *J. Biotechnol., 195,* 60–66.

Bharathiraja, B., Chakravarthy, M., Kumar, R. R., Yogendran, D., Yuvaraj, D., Jayamuthunagai, J., et al., (2015). Aquatic biomass (algae) as a future feed stock for bio-refineries: A review on cultivation, processing and products. *Renew. Sustain, Energy Rev., 47,* 634–653.

Biegala, I. C., Not, F., Vaulot, D., & Simon, N., (2003). Quantitative assessment of picoeukaryotes in the natural environment by using taxon-specific oligonucleotide probes in association with tyramide signal amplification-fluorescence *in situ* hybridization and flow cytometry. *Appl. Environ. Microbiol., 69,* 5519–5529.

Boynton, J. E., Gillham, N. W., Harris, E. H., Hosler, J. P., Johnson, A. M., Jones, A. R., et al., (1988). Chloroplast transformation in *Chlamydomonas* with high velocity micro projectiles. *Science, 240,* 1534–1538.

Brennan, L., & Owende, P., (2010). Biofuels from microalgae-a review of technologies for production, processing, and extractions of biofuels and co-products. *Renew. Sustain. Energy Rev., 14,* 557–577.

Broadbent, F. E., & Clark, F. E., (1965). Dinitrification. In: Bartholomew, M. V., & Clark, F. E., (eds.), *Soil Nitrogen* (pp. 344–359). American Society of Agronomy, Madison, Wisconsin, USA.

Brown, L. E., Sprecher, S. L., & Keller, L. R., (1991). Introduction of exogenous DNA into *Chlamydomonas reinhardtii* by electroporation. *Mol. Cell Biol., 11,* 2328–2332.

Buchel, C., (2003). Fucoxanthin-chlorophyll proteins in diatoms: 18 and 19 kDa subunits assemble into different oligomeric states. *Biochem., 42,* 13027–13034.

Cavalier-Smith, T., (2000). Membrane heredity and early chloroplast evolution. *Trends Plant Sci., 5,* 174–182.

Cerutti, H., Ma, X. R., Msanne, J., & Repas, T., (2011). RNA-mediated silencing in algae: Biological roles and tools for analysis of gene function. *Eukaryot. Cell, 10,* 1164–1172.

Cha, T. S., Chen, C. F., Yee, W., Aziz, A., & Loh, S. H., (2011). Cinnamic acid, coumarin and vanillin: Alternative phenolic compounds for efficient *Agrobacterium*-mediated transformation of the unicellular green alga, *Nannochloropsis* sp. *J. Microbiol. Methods, 84,* 430–434.

Chen, H., Zhou, D., Luo, G., Zhang, S., & Chen, J., (2015). Macroalgae for biofuels production: Progress and perspectives. *Renew. Sustain. Energy Rev., 47,* 427–437.

Chen, Y. C., (2003). Immobilized *Isochrysis galbana* (Haptophyta) for long-term storage and applications for feed and water quality control in clam (*Meretrix lusoria*) cultures. *J. Appl. Phycol., 15,* 439–444.

Cheney, D., Metz, B., & Stiller, J., (2001). *Agrobacterium*-mediated genetic transformation in the macroscopic marine red alga *Porphyra yezoensis*. *J. Phycol., 37*(s3), 11. https://doi.org/10.1111/j.1529-8817.2001.jpy37303-22.x.

Ciniglia, C., Yoon, H. S., Pollio, A., Pinto, G., & Bhattacharya, D., (2004). Hidden biodiversity of the extremophilic cyanidiales red algae. *Mol. Ecol., 13,* 1827–1838.

Cock, J. M., Sterck, L., Rouze, P., Scornet, D., Allen, A. E., Amoutzias, G., et al., (2010). The *Ectocarpus* genome and the independent evolution of multicellularity in brown algae. *Nature, 465,* 617–621.

Coll, J. M., (2006). Methodologies for transferring DNA into eukaryotic microalgae. *Span. J. Agric. Res., 4,* 316–330.

Colla, L. M., Reinehr, C. O., Reichert, C., & Costa, J. A. V., (2007). Production of biomass and nutraceutical compounds by *Spirulina platensis* under different temperature and nitrogen regimes. *Bioresour. Technol., 98* (7), 1489–1493.

Da Silva, R. L., & Barbosa, J. M., (2008). Seaweed meal as a protein source for the white shrimp *Lipopenaeus vannamei*. *J. Appl. Phycol., 21*, 193–197.

Dauvillee, D., Delhaye, S., Gruyer, S., Slomianny, C., Moretz, S. E., D'Hulst, C., et al., (2010). Engineering the chloroplast targeted malarial vaccine antigens in *Chlamydomonas* starch granules. *PLoS One, 5*, e15424. doi: 10.1371/journal.pone.0015424.

Davis, T. A., Volesky, B., & Mucci, A., (2003). A review of the biochemistry of heavy metal biosorption by brown algae. *Water Res., 37*(18), 4311–4330.

De Almeida, C. L. F., De S. Falcao, H., De, M. Lima, G. R., De, A. Montenegro, C., Lira, N. S., De Athayde-Filho, P. F., et al., (2011). Bioactivities from marine algae of the genus *Gracilaria. Int. J. Mol. Sci., 12*, 4550–4573.

De Riso, V., Raniello, R., Maumus, F., Rogato, A., Bowler, C., & Falciatore, A., (2009). Gene silencing in the marine diatom *Phaeodactylum tricornutum. Nucleic Acids Res., 37*(14), e96, doi: 10.1093/nar/gkp448.

Delalat, B., Sheppard, V. C., Ghaemi, S. R., Rao, S., Prestidge, C. A., McPhee, G., et al., (2015). Targeted drug delivery using genetically engineered diatom biosilica. *Nat. Commun., 6*(8791), 1–11. doi: 10.1038/ncomms9791.

Demirbas, A., (2010). Use of algae as biofuel sources. *Energy Convers. Manage, 51*, 2738–2749.

Demirbas, M. F., (2011). Biofuels from algae for sustainable development. *Appl. Energy, 88*, 3473–3480.

Depicker, A., & Van Montagu, M., (1997). Post-transcriptional gene silencing in plants. *Curr. Opin. Cell Biol., 9*, 373–382.

Doron, L., Segal, N., & Shapira, M., (2016). Transgene expression in microalgae from tools to applications. *Front. Plant Sci., 7*, 505. doi: 10.3389/fpls.2016.00505.

Dufresne, A., Salanoubat, M., Partensky, F., Artiguenave, F., Axmann, I. M., Barbe, V., et al., (2003). Genome sequence of the cyanobacterium *Prochlorococcus marinus* SS120, a nearly minimal oxyphototrophic genome. *Proc. Natl. Acad. Sci. USA., 100*, 10020–10025.

Dunahey, T. G., Jarvis, E. E., & Roessler, P. G., (1995). Genetic transformation of the diatoms *Cyclotella cryptica* and *Navicula saprophila. J. Phycol., 31*, 1004–1012.

Dyo, Y. M., & Purton, S., (2018). The algal chloroplast as a synthetic biology platform for production of therapeutic proteins. *Microbiol., 164*, 113–121. doi: 10.1099/mic.0.000599.

Falciatore, A., Casotti, R., Leblanc, C., Abrescia, C., & Bowler, C., (1999). Transformation of non-selectable reporter genes in marine diatoms. *Mar. Biotechnol., 1*, 239–251.

Feng, S. Y., Xue, L. X., Liu, H. T., & Lu, P. J., (2009). Improvement of efficiency of genetic transformation for *Dunaliella salina* by glass beads method. *Mol. Biol. Rep., 36*, 1433–1439.

Feng, S., Feng, W., Zhao, L., Gu, H., Li, Q., Shi, K., et al., (2014). Preparation of transgenic *Dunaliella salina* for immunization against white spot syndrome virus in crayfish. *Arch. Virol., 159*, 519–525.

Flynn, K. J., Greenwell, H. C., Lovitt, R. W., & Shields, R. J., (2010). Selection for fitness at the individual or population levels: Modeling effects of genetic modifications in microalgae on productivity and environmental safety. *J. Theor. Biol., 263*, 269–280.

Fourest, E., & Volesky, B., (1995). Contribution of sulfonate groups and alginate to heavy metal biosorption by the dry biomass of *Sargassum fluitans. Environ. Sci. Technol., 30*(1), 277–282.

Geng, D., Wang, Y., Wang, P., Li, W., & Sun, Y., (2003). Stable expression of hepatitis B surface antigen gene in *Dunaliella salina* (Chlorophyta). *J. Appl. Phycol., 15*, 451–456.

Georgianna, D. R., Hannon, M. J., Marcuschi, M., Wu, S., Botsch, K., et al., (2013). Production of recombinant enzymes in the marine alga *Dunaliella tertiolecta. Algal Res., 2*, 2–9.

Ghosh, M., Shen, J., & Rosen, B. P., (1999). Pathways of As(III) detoxification in *Saccharomyces cerevisiae. Proc. Natl. Acad. Sci. USA., 96*(9), 5001–5006.

Gierz, S. L., Foret, S., & Leggat, W., (2017). Transcriptomic analysis of thermally stressed *Symbiodinium* reveals differential expression of stress and metabolism genes. *Front. Plant Sci., 8,* 271. doi: 10.3389/fpls.2017.00271.

Givan, A. L., & Levine, R. P., (1967). The photosynthetic electron transport chain of *Chlamydomonas reinhardtii*. VII. Photosynthetic phosphorylation by a mutant strain of *Chlamydomonas reinhardtii* deficient in active P700. *Plant Physiol., 42,* 1264–1268.

Gorman, D. S., & Levine, R. P., (1966). Cytochrome f and plastocyanin: Their sequence in the photosynthetic electron transport chain of *Chlamydomonas reinhardtii. Proc. Natl. Acad. Sci. USA., 54,* 1665–1669.

Goyen, S., Pernice, M., Szabó, M., Warner, M. E., Ralph, P. J., & Suggett, D. J., (2017). A molecular physiology basis for functional diversity of hydrogen peroxide production amongst *Symbiodinium* spp. (Dinophyceae). *Mar. Biol., 164,* 46. doi: 10.1007/s00227-017-3073-5.

Gressel, J., (2008). Transgenics are imperative for biofuel crops. *Plant Sci., 174,* 246–263.

Grossman, A. R., (2005). Update on genomic studies of algae: Paths toward algal genomics. *Plant Physiol., 137,* 410–427.

Guedes, E. A. C., Dos, S. A. M. A., Souza, A. K. P., De Souza, L. I. O., Barros, L. D., De Albuquerque, M. F. C., et al., (2012). Antifungal activities of different extracts of marine macroalgae against dermatophytes and *Candida* species. *Mycopathologia, 174*(3), 223–232.

Gust, D., Kramer, D., Moore, A., Moore, T. A., & Vermaas, W., (2008). Engineered and artificial photosynthesis: Human ingenuity enters the game. *MRS. Bull., 33,* 383–387.

Gutierrez, C. L., Gimpel, J., Escobar, C., Marshall, S. H., & Henríquez, V., (2012). Chloroplast genetic tool for the green microalgae *Haematococcus pluvialis* (Chlorophyceae, Volvocales). *J. Phycol., 48,* 976–983.

Haapa, S., Taira, S., Heikkinen, E., & Savilahti, H., (1999). An efficient and accurate integration of mini-Mu transposons *in vitro*: A general methodology for functional genetic analysis and molecular biology applications. *Nucleic Acids Res., 27,* 2777–2784.

Harris, E. H., (2001). *Chlamydomonas* as a model organism. *Annu. Rev. Plant Physiol. Plant Mol. Biol., 52,* 363–406.

He, D. M., Qian, K. X., Shen, G. F., Zhang, Z. F., Yi-Nu, L., Su, Z. L., et al., (2007). Recombination and expression of classical swine fever virus (CSFV) structural protein E2 gene in *Chlamydomonas reinhardtii* chroloplasts. *Colloids Surf. B. Biointerfaces, 55,* 26–30.

Hempel, F., Lau, J., Klingl, A., & Maier, U. G., (2011). Algae as protein factories: Expression of a human antibody and the respective antigen in the diatom *Phaeodactylum tricornutum. PLoS One, 6*(12), E28424. doi: 10.1371/journal.pone.0028424.

Hirakawa, Y., Kofuji, R., & Ishida, K., (2008). Transient transformation of a chlorarachniophyte alga, *Lotharella amoebiformis* (Chlorarachniophyceae), with uidA and *egfp* reporter genes. *J. Phycol., 44,* 814–820.

Hirata, R., Takahashi, M., Saga, N., & Mikami, K., (2011). Transient gene expression system established in *Porphyra yezoensis* is widely applicable in Bangiophycean algae. *Mar. Biotechnol., 13,* 1038–1047.

Hoegh-Guldberg, O., Mumby, P. J., Hooten, A. J., Steneck, R. S., Greenfield, P., Gomez, E., et al., (2007). Coral reefs under rapid climate change and ocean acidification. *Science, 318,* 1737–1742. doi: 10.1126/science.1152509.

Huang, X., Weber, J. C., Hinson, T. K., Mathieson, A. C., & Minocha, S. C., (1996). Transient expression of the GUS reporter gene in the protoplasts and partially digested cells of *Ulva lactuca* L. (Chlorophyta). *Bot. Marina., 39,* 467–474.

Huber, G. W., Iborra, S., & Corma, A., (2006). Synthesis of transportation fuels from biomass: Chemistry, catalysts, and engineering. *Chem. Rev., 106,* 4044–4098.

Hughes, T. P., Kerry, J. T., Álvarez-Noriega, M., Álvarez-Romero, J. G., Anderson, K. D., Baird, A. H., et al., (2017). Global warming and recurrent mass bleaching of corals. *Nature, 543,* 373–377.

Ishikawa, C., Tafuku, S., Kadekaru, T., Sawada, S., Tomita, M., Okudaira, T., et al., (2008). Anti- adult T-cell leukemia effects of brown algae fucoxanthin and its deacetylated product, fucoxanthinol. *Int. J. Cancer., 123*(11), 2702–2712.

Jia, Y. L., Xue, L. X., Liu, H. T., & Li, J., (2009). Characterization of the glyceraldehyde-3-phosphate dehydrogenase (GAPDH) gene from the halotolerant alga *Dunaliella salina* and inhibition of its expression by RNAi. *Curr. Microbiol., 58,* 426–431.

Jiang, P., Qin, S., & Tseng, C. K., (2003). Expression of the lacZ reporter gene in sporophytes of the seaweed *Laminaria japonica* (Phaeophyceae) by gametophyte-targeted transformation. *Plant Cell. Rep., 21,* 1211–1216.

Jones, C. S., Luong, T., Hannon, M., Tran, M., Gregory, J. A., Shen, Z., et al., (2013). Heterologous expression of the C-terminal antigenic domain of the malaria vaccine candidate Pfs48/45 in the green algae *Chlamydomonas reinhardtii. Appl. Microbiol. Biotechnol., 97,* 1987–1995.

Jung, K. A., Lim, S. R., Kim, Y., & Park, J. M., (2013). Potentials of macroalgae as feedstocks for biorefinery. *Bioresour. Technol., 135,* 182–190.

Kawata, Y., Yano, S., Kojima, H., & Toyomizu, M., (2004). Transformation of *Spirulina platensis* strain C1 (*Arthrospira* sp. PCC9438) with Tn5 transposase-transposon DNA-cation liposome complex. *Mar. Biotechnol., 6,* 355–363.

Kilian, O., Benemann, C. S. E., Niyogi, K. K., & Vick, B., (2011). High-efficiency homologous recombination in the oil-producing alga *Nannochloropsis* sp. *Proc. Natl. Acad. Sci. USA., 108,* 21265–21269.

Kim, D. H., Kim, Y. T., Cho, J. J., Bae, J. H., Hur, S. B., Hwang, I., et al., (2002). Stable integration and functional expression of flounder growth hormone gene in transformed microalga, *Chlorella ellipsoidea. Mar. Biotechnol., 4,* 63–73.

Kindle, K. L., (1990). High-frequency nuclear transformation of *Chlamydomonas reinhardtii. Proc. Natl. Acad. Sci. USA., 87,* 1228–1232.

Kirk, N. L., & Weis, V. M., (2016). In: Hurst, C. J., (ed.), *Animal-Symbiodinium Symbioses: Foundations of Coral Reef Ecosystems in the Mechanistic Benefits of Microbial Symbionts* (pp. 269–294). Springer, Cham, Switzerland. doi: 10.1007/978-3-319-28068-4_10.

Knowlton, N., Brainard, R. E., Fisher, R., Moews, M., Plaisance, L., & Caley, M. J., (2010). Coral reef biodiversity. In: Mcintyre, A. D., (ed.), *Life in the World's Oceans: Diversity Distribution and Abundance* (pp. 65–74). John Wiley and Sons, Hoboken, New Jersey. doi: 10.1002/9781444325508.ch4.

Kuang, M., Wang, S. J., Li, Y., Shen, D. L., & Zeng, C. K., (1998). Transient expression of exogenous gus gene in *Porphyra yezoensis* (Rhodophyta). *Chin. J. Oceanol. Limnol., 16,* 56–61.

Kuhlbusch, T. A., Lobert, J. M., Crutzen, P. J., & Warneck, P., (1991). Molecular nitrogen emission. Trace nutrification during biomass burning. *Nature, 351,* 135–137.

Kumar, S. V., Misquitta, R. W., Reddy, V. S., Rao, B. J., & Rajam, M. V., (2004). Genetic transformation of the green alga-*Chlamydomonas reinhardtii* by *Agrobacterium tumefaciens*. *Plant Sci., 166,* 731–738.

Kuroiwa, T., Kuroiwa, H., Sakai, A., Takahashi, H., Toda, K., & Itoh, R., (1998). The division apparatus of plastids and mitochondria. *Int. Rev. Cytol., 181,* 1–41.

Lavorel, J., & Levine, R. P., (1968). Fluorescence properties of wild-type *Chlamydomonas reinhardtii* and three mutant strains having impaired photosynthesis. *Plant Physiol., 43,* 1049–1055.

Lebeau, T., & Robert, J. M., (2003). Diatom cultivation and biotechnologically relevant products. Part II: Current and putative products. *Appl. Microbiol. Biotechnol., 60,* 624–632.

Leon, R., & Fernandez, E., (2007). Nuclear transformation of eukaryotic microalgae: Historical overview, achievements, and problems. *Adv. Exp. Med. Biol., 616,* 1–11.

Levin, R. A., Beltran, V. H., Hill, R., Kjelleberg, S., McDougald, D., Steinberg, P. D., et al., (2016). Sex, scavengers, and chaperones: Transcriptome secrets of divergent *Symbiodinium* thermal tolerances. *Mol. Biol. Evol., 33,* 2201–2215. doi: 10.1093/molbev/msw119.

Levin, R. A., Voolstra, C. R., Agrawal, S., Steinberg, P. D., Suggett, D. J., & Van, O. M. J. H., (2017). Engineering strategies to decode and enhance the genomes of coral symbionts. *Front. Microbiol., 8,* 1220, doi: 10.3389/fmicb.2017.01220.

Levine, R. P., & Goodenough, U. W., (1970). The genetics of photosynthesis and of the chloroplast in *Chlamydomonas reinhardii*. *Annu. Rev. Genet.*, *4,* 397–408.

Li, J. B., Gerdes, J. M., Haycraft, C. J., Fan, Y., Teslovich, T. M., May-Simera, H., et al., (2004). Comparative genomics identifies a flagellar and basal body proteome that includes the BBS5 human disease gene. *Cell, 117,* 541–552.

Liang, C. W., Zhang, X. W., Zou, J., Xu, D., Su, F., & Ye, N. H., (2010). Identification of miRNA from *Porphyra yezoensis* by high-throughput sequencing and bioinformatics analysis. *PLoS One, 19,* 5(5), e10698. doi: 10.1371/journal.pone.0010698.

Liu, H. Q., Yu, W. G., Dai, J. X., Gong, Q. H., Yang, K. F., & Zhang, Y. P., (2003). Increasing the transient expression of GUS gene in *Porphyra yezoensis* by 18S rDNA targeted homologous recombination. *J. Appl. Phycol., 15,* 371–377.

Lohuis, M. R., & Miller, D. J., (1998). Genetic transformation of dinoflagellates (*Amphidinium* and *Symbiodinium*): Expression of GUS in microalgae using heterologous promoter constructs. *Plant, J., 13,* 427–435.

Lowe, B. A., Prakash, N. S., Way, M., Mann, M. T., Spencer, T. M., & Boddupalli, R. S., (2009). Enhanced single copy integration events in corn via particle bombardment using low quantities of DNA. *Transgenic Res., 18,* 831–840.

Lu, Y. M., Li, J., Xue, L. X., Yan, H. X., Yuan, H. J., & Wang, C., (2011). A duplicated carbonic anhydrase 1 (DCA1) promoter mediates the nitrate reductase gene switch of *Dunaliella salina*. *J. Appl. Phycol., 23,* 673–680.

Madhyastha, H. K., & Vatsala, T. M., (2007). Pigment production in *Spirulina fussiformis* in different photophysical conditions. *Biomol. Eng.*, *24*(3), 301–305.

Maier, U. G., Douglas, S. E., & Cavalier-Smith, T., (2000). The nucleomorph genomes of cryptophytes and chlorarachniophytes. *Protist., 151,* 103–109.

Mani, D., & Kumar, C., (2014). Biotechnological advances in bioremediation of heavy metals contaminated ecosystems: An overview with special reference to phytoremediation. *Int. J. Environ. Sci. Technol., 11*(3), 843–872.

Marenkova, T. V., & Deineko, E. V., (2010). Transcriptional gene silencing in plants. *Russ. J. Genet., 46,* 511–520.

Matsunaga, T., & Takeyama, H., (1995). Genetic-engineering in marine cyanobacteria. *J. Appl. Phycol., 7,* 77–84.

Matsunaga, T., Sakaguchi, T., & Tadokoro, F., (1991). Magnetite formation by a magnetic bacterium capable of growing aerobically. *Appl. Microbiol. Biotechnol., 35,* 651–655.

Matsuzaki, M., Misumi, O., Shini, T., Maruyama, S., Takahara, M., Miyagishima, S., et al., (2004). Genome sequence of the ultrasmall unicellular red alga *Cyanidioschyzon merolae* 10D. *Nature, 428,* 653–657.

Mayfield, S. P., Franklin, S. E., & Lerner, R. A., (2003). Expression and assembly of a fully active antibody in algae. *Proc. Natl. Acad. Sci. USA, 100,* 438–442.

Meyers, B., Zaltsman, A., Lacroix, B., Kozlovsky, S. V., & Krichevsky, A., (2010). Nuclear and plastid genetic engineering of plants: Comparison of opportunities and challenges. *Biotechnol. Adv., 28,* 747–756.

Miyagawa, A., Okami, T., Kira, N., Yamaguchi, H., Ohnishi, K., & Adachi, M., (2009). High efficiency transformation of the diatom *Phaeodactylum tricornutum* with a promoter from the diatom *Cylindrotheca fusiformis. Phycol. Res., 57,* 142–146.

Miyagawa-Yamaguchi, A., Okami, T., Kira, N., Yamaguchi, H., Ohnishi, K., & Adachi, M., (2011). Stable nuclear transformation of the diatom *Chaetoceros* sp. *Phycol. Res., 59,* 113–119.

Mohan, D., Pittman, J. C. U., & Steele, P. H., (2006). Pyrolysis of wood/biomass for bio-oil: A critical review. *Energy Fuels, 20*(3), 848–889.

Moll, B., & Levine, R. P., (1970). Characterization of a photosynthetic mutant strain of *Chlamydomonas reinhardii* deficient in phosphoribulokinase activity. *Plant Physiol., 46,* 576–580.

Mora, C., Graham, N. A., & Nyström, M., (2016). Ecological limitations to the resilience of coral reefs. *Coral Reefs., 35,* 1271–1280. doi: 10.1007/s00338-016–1479-z.

Mulbry, W., Kondrad, S., Pizarro, C., & Kebede-Westhead, E., (2008). Treatment of dairy manure effluent using freshwater algae: Algal productivity and recovery of manure nutrients using pilot-scale algal turf scrubbers. *Bioresour. Technol., 99,* 8137–8142.

Murray, S. A., Suggett, D. J., Doblin, M. A., Kohli, G. S., Seymour, J. R., Fabris, M., et al., (2016). Unravelling the functional genetics of dinoflagellates: A review of approaches and opportunities. *Perspect. Phycol., 3,* 37–52. doi: 10.1127/pip/2016/0039.

Neuhaus, G., & Spangenberg, G., (1990), Plant transformation by microinjection techniques. *Physiol. Plant, 79,* 213–217.

Neuhaus, G., Neuhausurl, G., Degroot, E. J., & Schweiger, H. G., (1986). High-yield and stable transformation of the unicellular green-alga *Acetabularia* by microinjection of SV40 and PSV2NEO. *EMBO J., 5,* 1437–1444.

Neumann, E., Schaeferridder, M., Wang, Y., & Hofschneider, P. H., (1982). Gene-transfer into mouse lyoma cells by electroporation in high electric-fields. *EMBO J., 1,* 841–845.

Noraini, M. Y., Ong, H. C., Badrul, M. J., & Chong, W. T., (2014). A review on potential enzymatic reaction for biofuel production from algae. *Renew. Sustain. Energy Rev., 39,* 24–34.

Nozaki, H., Matsuzaki, M., Misumi, O., Kuroiwa, H., Hasegawa, M., Higashiyama, T., et al., (2004). Cyanobacterial genes transmitted to the nucleus before divergence of red algae in the *Chromista. J. Mol. Evol., 59,* 103–113.

Oeltjen, A., Marquardt, J., & Rhiel, E., (2004). Differential circadian expression of genes fcp2 and fcp6 in *Cyclotella cryptica. Int. Microbiol., 7,* 127–131.

Ogbonda, K. H., Aminigo, R. E., & Abu, G. O., (2007). Influence of temperature and pH on biomass production and protein biosynthesis in a putative *Spirulina* sp. *Bioresour. Technol., 98,* 2207–2211.

Ohta, N., Matsuzaki, M., Misumi, O., Miyagishima, S. Y., Nozaki, H., Tanaka, K., et al., (2003). Complete sequence and analysis of the plastid genome of the unicellular red alga *Cyanidioschyzon merolae. DNA Res., 10,* 67–77.

Pandolfi, J. M., Connolly, S. R., Marshall, D. J., & Cohen, A. L., (2011). Projecting coral reef futures under global warming and ocean acidification. *Science, 333,* 418–422.

Pfeiffer, A., Kunkel, T., Hiltbrunner, A., Neuhaus, G., Wolf, I., Speth, V., et al., (2009). A cell-free system for light-dependent nuclear import of phytochrome. *Plant J., 57,* 680–689.

Piaggio, A. J., Segelbacher, G., Seddon, P. J., Alphey, L., Bennett, E. L., Carlson, R. H., et al., (2016). Is it time for synthetic biodiversity conservation? *Trends Ecol. Evol., 32,* 97–107. doi: 10.1016/j.tree.2016.10.016.

Potvin, G., & Zhang, Z. S., (2010). Strategies for high-level recombinant protein expression in transgenic microalgae: A review. *Biotechnol. Adv.,* 28, 91–98.

Poulsen, N., & Kroger, N., (2005). A new molecular tool for transgenic diatoms-control of mRNA and protein biosynthesis by an inducible promoter-terminator cassette. *FEBS. J., 272,* 3413–3423.

Poulsen, N., Chesley, P. M., & Kroger, N., (2006). Molecular genetic manipulation of the diatom *Thalassiosira pseudonana* (Bacillariophyceae). *J. Phycol., 42,* 1059–1065.

Priyadarshani, I., & Rath, B., (2012). Commercial and industrial applications of micro algae: A review. *J. Algal Biomass Utln., 3*(4), 89–100.

Pulz, O., & Gross, W., (2004). Valuable products from biotechnology of microalgae. *Appl. Microbiol. Biotechnol., 65,* 635–648.

Purton, S., (2007). Tools and techniques for chloroplast transformation of *Chlamydomonas. Adv. Exp. Med. Biol., 616,* 34–45.

Purton, S., Szaub, J. B., Wannathong, T., Young, R., & Economou, C. K., (2013). Genetic engineering of algal chloroplasts: Progress and prospects. *Rus. J. Plant Physiol., 60,* 491–499.

Qin, S., Jiang, P., & Tseng, C. K., (2004). Molecular biotechnology of marine algae in China. *Hydrobiologia, 512,* 21–26.

Qin, S., Jiang, P., Li, X. P., Wang, X. H., & Zeng, C. K., (1998). A transformation model for *Laminaria japonica* (Phaeophyta, Laminariales). *Chin. J. Oceanol. Limnol., 16,* 50–55.

Qin, S., Zhang, J., Li, W. B., Wang, X. H., Tong, S., Sun, Y. R., et al., (1994). Transient expression of GUS gene in Phaeophytes using biolistic particle delivery system. *Oceanol. Limnol. Sin., 25,* 353–356.

Raheem, A., Wan, A. W. A. K. G., Taufiq, Y. Y. H., Danquah, M. K., & Harun, R., (2015). Thermochemical conversion of microalgal biomass for biofuel production. *Renew. Sustain. Energy Rev., 49,* 990–999.

Ramundo, S., & Rochaix, J. D., (2015). Controlling expression of genes in the unicellular alga *Chlamydomonas reinhardtii* with a vitamin repressible riboswitch. *Methods Enzymol., 550,* 267–281.

Rangel-Yagui, C. O., Danesi, E. D. G., Carvalho, J. C. M., & Sato, S., (2004). Chlorophyll production from *Spirulina platensis*: Cultivation with urea addition by fed-batch process. *Bioresour. Technol., 92*(2), 133–141.

Rania, M. A., & Hala, M. T., (2008). Antibacterial and antifungal activity of Cynobacteria and green Microalgae evaluation of medium components by Plackett-Burman design for antimicrobial activity of *Spirulina platensis. Glob. J. Biotechnol. Biochem., 3*(1), 22–31.

Raslavicius, L., Semenov, V. G., Chernova, N. I., Kersys, A., & Kopeyka, A. K., (2014). Producing transportation fuels from algae: In search of synergy. *Renew. Sustain. Energy Rev.*, *40,* 133–142.

Reddy, C., Gupta, M., Mantri, V., & Jha, B., (2008). Seaweed protoplasts: Status, biotechnological perspectives and needs. *J. Appl. Phycol., 20,* 619–632.

Reinfelder, J. R., Milligan, A. J., & Morel, F. M., (2004). The role of the C4 pathway in carbon accumulation and fixation in a marine diatom. *Plant Physiol.*, *135,* 2106–2111.

Rinkevich, B., (2014). Rebuilding coral reefs: Does active reef restoration lead to sustainable reefs? *Curr. Opin. Environ. Sustain, 7,* 28–36.

Rocap, G., Larimer, F. W., Lamerdin, J., Malfatti, S., Chain, P., Ahlgren, N. A., et al., (2003). Genome divergence in two *Prochlorococcus* ecotypes reflects oceanic niche differentiation. *Nature, 424,* 1042–1047.

Rosen, B. P., (2002). Transport and detoxification systems for transition metals, heavy metals, and metalloids in eukaryotic and prokaryotic microbes. *Comp. Biochem. Physiol. A. Mol. Integr. Physiol., 133*(3), 689–693.

Sager, R., (1960). Genetic systems in *Chlamydomonas*. *Science, 132,* 1459–1465.

Sajilata, M. G., Singhal, R. S., & Kamat, M. Y., (2008). Fractionation of lipids and purification of γ-linolenic acid (GLA) from *Spirulina platensis*. *Food Chem., 109*(3), 580–586.

Salvador, N., Gomez-Garreta, A., Lavelli, L., & Ribera, M. A., (2007). Antimicrobial activity of Iberian Macroalgae. *Scientia. Marina, 71,* 101–113.

Sambusiti, C., Bellucci, M., Zabaniotou, A., Beneduce, L., & Monlau, F., (2015). Algae as promising feedstocks for fermentative biohydrogen production according to a biorefinery approach: A comprehensive review. *Renew. Sustain. Energy Rev., 44,* 20–36.

Sato, V., Levine, R. P., & Neumann, J., (1971). Photosynthetic phosphorylation in *Chlamydomonas reinhardtii*. Effects of a mutation altering an ATP synthesizing enzyme. *Biochim. Biophys. Acta., 253,* 437–448.

Schnorf, M., Neuhaus-Url, G., Galli, A., Iida, S., Potrykus, I., & Neuhaus, G., (1991). An improved approach for transformation of plant cells by microinjection: Molecular and genetic analysis. *Transgenic Res., 1,* 23–30.

Shimogawara, K., Fujiwara, S., Grossman, A., & Usuda, H., (1998). High-efficiency transformation of *Chlamydomonas reinhardtii* by electroporation. *Genetics, 148,* 1821–1828.

Singh, A., Nigam, P. S., & Murphy, J. D., (2011). Renewable fuels from algae: An answer to debatable land based fuels. *Bioresour. Technol., 102,* 10–16.

Snell, W. J., Pan, J., & Wang, Q., (2004). Cilia and flagella revealed: From flagellar assembly in *Chlamydomonas* to human obesity disorders. *Cell, 117,* 693–697.

Sode, K., Tatara, M., Takeyama, H., Burgess, J. G., & Matsunaga, T., (1992). Conjugative gene-transfer in marine cyanobacteria-*Synechococcus* sp., *Synechocystis* sp. and *Pseudanabeaena* sp. *Appl. Microbiol. Biotechnol., 37,* 369–373.

Song, T., Martensson, L., Eriksson, T., Zheng, W., & Rasmussen, U., (2005). Biodiversity and seasonal variation of the cyanobacterial assemblage in a rice paddy field in Fujian, China. *FEMS. Microbiol. Ecol., 54,* 131–140.

Specht, E. A., & Mayfield, S. P., (2014). Algae-based oral recombinant vaccines. *Front. Microbiol., 5,* 60.

Spolaore, P., Joannis-Cassan, C., Duran, E., & Isambert, A., (2006). Commercial applications of microalgae. *J. Biosci. Bioeng., 101,* 87–96.

Stolz, P., & Obermayer, B., (2005). Manufacturing microalgae for skin care. *Cosmetics Toiletries*, *120,* 99–106.

Sun, M., Qian, K., Su, N., Chang, H., Liu, J., & Shen, G., (2003). Foot-and-mouth disease virus VP1 protein fused with cholera toxin B subunit expressed in *Chlamydomonas reinhardtii* chloroplast. *Biotechnol. Lett., 25,* 1087–1092.

Surzycki, R., Greenham, K., Kitayama, K., Dibal, F., Wagner, R., Rochaix, J. D., et al., (2009). Factors effecting expression of vaccines in microalgae. *Biologicals, 37,* 133–138.

Takahashi, M., Uji, T., Saga, N., & Mikami, K., (2010). Isolation and regeneration of transiently transformed protoplasts from gametophytic blades of the marine red alga *Porphyra yezoensis. Electron J. Biotechnol., 13*(2), 1–4.

Tan, C. P., Qin, S., Zhang, Q., Jiang, P., & Zhao, F. Q., (2005). Establishment of a micro-particle bombardment transformation system for *Dunaliella saliva. J. Microbiol., 43,* 361–365.

Ting, C. S., Rocap, G., King, J., & Chisholm, S. W., (2001). Phycobiliprotein genes of the marine photosynthetic prokaryote *Prochlorococcus*: Evidence for rapid evolution of genetic heterogeneity. *Microbiol (UK), 147,* 3171–3182.

Tisch, R., Yang, X. D., Singer, S. M., Liblau, R. S., Fugger, L., & McDevitt, H. O., (1993). Immune response to glutamic acid decarboxylase correlates with insulitis in non-obese diabetic mice. *Nature, 366,* 72–75.

Tolonen, A. C., Liszt, G. B., & Hess, W. R., (2006). Genetic manipulation of *Prochlorococcus* strain MIT9313: Green fluorescent protein expression from an RSF1010 plasmid and Tn5 transposition. *Appl. Environ. Microbiol., 72,* 7607–7613.

Tonon, T., Harvey, D., Qing, R., Li, Y., Larson, T. R., & Graham, I. A., (2004). Identification of a fatty acid Delta11-desaturase from the microalga *Thalassiosira pseudonana. FEBS Lett., 563,* 28–34.

Tran, M., Henry, R. E., Siefker, D., Van, C., Newkirk, G., Kim, J., et al., (2013a). Production of anti-cancer immunotoxins in algae: Ribosome inactivating proteins as fusion partners. *Biotechnol. Bioeng., 110,* 2826–2835.

Tran, M., Van, C., Barrera, D. J., Pettersson, P. L., Peinado, C. D., Bui, J., et al., (2013b). Production of unique immunotoxin cancer therapeutics in algal chloroplasts. *Proc. Natl. Acad. Sci. USA., 110,* E15–E22.

Tzfira, T., & Citovsky, V., (2006). *Agrobacterium*-mediated genetic transformation of plants: Biology and biotechnology. *Curr. Opin. Biotechnol., 17,* 147–154.

Ullah, K., Ahmad, M., Sofia, S. V. K., Lu, P., Harvey, A., Zafar, M., et al., (2015). Assessing the potential of algal biomass opportunities for bioenergy industry: A review. *Fuel, 143,* 414–423.

Vaishampayan, A., Sinha, R. P., & Hader, D. P., (1998). Use of genetically improved nitrogen-fixing cyanobacteria in rice paddy fields: Prospects as a source material for engineering herbicide sensitivity and resistance in plants. *Bot. Acta., 111,* 176–190.

Van, E. J. L., & Meints, R. H., (1999). Giant viruses infecting algae. *Annu. Rev. Microbiol., 53,* 447–494.

Van, H. M. C., Witteveldt, J., Peters, S., Kloosterboer, N., Tarchini, R., Fiers, M., et al., (2001). The white spot syndrome virus DNA genome sequence. *Virology, 286,* 7–22.

Van, O. M. J. H., Oliver, J. K., Putnam, H. M., & Gates, R. D., (2015). Building coral reef resilience through assisted evolution. *Proc. Natl. Acad. Sci. USA., 112,* 2307–2313.

Van, O. M. J., Gates, R. D., Blackall, L. L., Cantin, N., Chakravarti, L. J., Chan, W. Y., et al., (2017). Shifting paradigms in restoration of the world's coral reefs. *Glob. Chang. Biol., 23*(9), 3437–3448.

Wang, J. F., Jiang, P., Cui, Y. L., Guan, X. Y., & Qin, S., (2010). Gene transfer into conchospores of *Porphyra haitanensis* (Bangiales, Rhodophyta) by glass bead agitation. *Phycologia, 49,* 355–360.

Wang, X., Brandsma, M., Tremblay, R., Maxwell, D., Jevnikar, A. M., Huner, N., & Ma, S., (2008). A novel expression platform for the production of diabetes-associated autoantigen human glutamic acid decarboxylase (hGAD65). *BMC. Biotechnol., 8*, 87. doi: 10.1186/1472-6750-8-87.

Wannathong, T., Waterhouse, J. C., Young, R. E., Economou, C. K., & Purton, S., (2016). New tools for chloroplast genetic engineering allow the synthesis of human growth hormone in the green alga *Chlamydomonas reinhardtii. Appl. Microbiol. Biotechnol., 100*, 5467–5477.

Watanabe, S., Ohnuma, M., Sato, J., Yoshikawa, H., & Tanaka, K., (2011). Utility of a GFP reporter system in the red alga *Cyanidioschyzon merolae. J. Gen. Appl. Microbiol., 57*, 69–72.

Wen, Z. Y., & Chen, F., (2003). Heterotrophic production of eicosapentaenoic acid by microalgae. *Biotechnol. Adv., 21*, 273–294.

Xie, W. H., Zhu, C. C., Zhang, N. S., Li, D. W., Yang, W. D., Liu, J. S., et al., (2014). Construction of novel chloroplast expression vector and development of an efficient transformation system for the diatom *Phaeodactylum tricornutum. Mar Biotechnol., 16*, 538–546.

Yan, N., Fan, C., Chen, Y., & Hu, Z., (2016). The potential for microalgae as bioreactors to produce pharmaceuticals. *Int. J. Mol. Sci., 17*(6), E962. doi: 10.3390/ijms17060962.:962.

Yao, Q., Cong, L., Chang, J. L., Li, K. X., Yang, G. X., & He, G. Y., (2006). Low copy number gene transfer and stable expression in a commercial wheat cultivar via particle bombardment. *J. Exp. Bot., 57*, 3737–3746.

Yi, Z. W., Huang, H., Kuang, T. Y., & Sui, S. F., (2005). Three-dimensional architecture of phycobilisomes from *Nostoc flagelliforme* revealed by single particle electron microscopy. *FEBS. Lett., 579*, 3569–3573.

Young, R. E., & Purton, S., (2016). Codon reassignment to facilitate genetic engineering and biocontainment in the chloroplast of *Chlamydomonas reinhardtii. Plant Biotechnol. J., 14*, 1251–1260.

Zaslavskaia, L. A., Lippmeier, J. C., Kroth, P. G., Grossman, A. R., & Apt, K. E., (2000). Transformation of the diatom *Phaeodactylum tricornutum* (Bacillariophyceae) with a variety of selectable marker and reporter genes. *J. Phycol., 36*, 379–386.

Zeraatkar, A. K., Ahmadzadeh, H., Talebi, A. F., Moheimani, N. R., & McHenry, M. P., (2016). Potential use of algae for heavy metal bioremediation, a critical review. *J. Environ. Manage, 181*, 817–831.

Zhang, Z., Jiang, P., Yang, R., Fei, X., Tang, X., & Qin, S., (2010). Establishment of a concocelis-mediated transformation system for *Porphyra* by biolistic PDS-1000/He. *Chin. High Technol. Lett., 20*, 204–207.

Zienkiewicz, M., Krupnik, T., Drozak, A., Golke, A., & Romanowska, E., (2017). Transformation of the *Cyanidioschyzon merolae* chloroplast genome: Prospects for understanding chloroplast function in extreme environments. *Plant Mol. Biol., 93*, 171–183.

Ziolkowska, J. R., & Simon, L., (2014). Recent developments and prospects for algae-based fuels in the US. *Renew. Sustain. Energy Rev., 29*, 847–853.

CHAPTER 14

ALGAL PATHWAYS AND METABOLIC ENGINEERING FOR ENHANCED PRODUCTION OF LIPIDS, CARBOHYDRATES, AND BIOACTIVE COMPOUNDS

MOHD. AZMUDDIN ABDULLAH,[1] HANN LING WONG,[2]
SYED MUHAMMAD USMAN SHAH,[3] and PEK CHIN LOH[4]

[1]*Institute of Marine Biotechnology, University Malaysia Terengganu, 21030 Kuala Nerus, Terengganu, Malaysia*

[2]*Department of Biological Science, Faculty of Science, Universiti Tunku Abdul Rahman, Jalan Universiti, Bandar Barat, 31900 Kampar, Perak, Malaysia*

[3]*Department of Biosciences, COMSATS University Islamabad, Park Road, 44000, Islamabad, Pakistan*

[4]*Department of Biomedical Science, Faculty of Science, Universiti Tunku Abdul Rahman, Jalan Universiti, Bandar Barat, 31900 Kampar, Perak, Malaysia*

14.1 INTRODUCTION

During the first oil crisis in 1970, biofuel technologies based on corn for ethanol and soybeans for biodiesel have started to be developed. Corn and sugarcane are the two major crops extensively utilized for bioethanol. In 2008, USA produces 34 billion liters of fuel grade bioethanol from corn, while Brazil comes second at 27.5 billion liters of fuel grade bioethanol from sugarcane, and together, these account for 89% of global production (Chiu et al., 2009). To satisfy the US demand for transportation fuel, the biodiesel industry would need to produce 530 billion liters annually (Chisti, 2007). To address this, the Aquatic Species Program is initiated to produce biodiesel from high lipid content algae (Sheehan et al., 1998). However, the program has been discontinued in 1996, due to funding shortages which focus the shifts back towards corn and soybean technologies which are considered as proven technologies. A total of 14.3% of corn and 1.5% of the soybeans harvested in the US in 2005 have been channeled towards ethanol production or to make biodiesel. This merely offset 1.72% and 0.09% of the US gasoline and diesel demand, respectively. Even if 100% of corn and soybeans harvested were to be used for biofuels, these will only meet 12% of the gasoline demand and 6% of the diesel demand (Jason et al., 2006). Fuel crops take up a significant amount of land, leading to a significant ecological impact. On top of that, the fossil fuels are used for farming and in the production of the 'green' fuels. While carbon dioxide may be utilized for growth, corn, and soybean farming require the use of fertilizers, resulting in the bioethanol and biodiesel giving only a 12% and 41% reduction in greenhouse gases (GHGs) (Jason et al., 2006). There are also issues about "food" versus "fuel" as corn, soybean crops or oil palm are major crops for edible oil.

It seems that the only feasible source of biodiesel is microalgae as algae do not compromise a food stock or deplete nutrients in the soil though it still requires a nutrient source. In fact, microalgae as oil-producing organisms for biodiesel production are well recognized. Lipids from microalgae can be turned into biodiesel through transesterification and used for energy development including simple combustion in boiler or in diesel engine (Schenk et al., 2008; Meng et al., 2009). The entire process ranging from the cultivation of high-lipid microalgae to the production of biodiesel from microalgal oil has been explored. A typical algal culture system can generate 150 to 400 barrels of oil per acre per year, which is 30 times more than those produced via typical oilseed crops. The production cost of algal oil depends on many factors, such as yield of biomass from the culture system, oil content, scale of production systems, and the cost of recovering oil. Currently, algal-oil production is still far more expensive than petroleum diesel fuels. The petroleum-diesel price is at $3.80–$4.50 per gallon while the production cost of algae oil from a photobioreactor with an annual production capacity of 10,000 tons per year, assuming the oil content of approximately 30%, is estimated to be $2.80 per liter ($10.50 per gallon) (Chisti, 2007). This does not include the cost of converting algal oil to biodiesel, distribution, and marketing costs for biodiesel and taxes.

The algal productivity can be enhanced by efficient cultivation method and by genetic and metabolic engineering (Abdullah et al., 2015–2017).

Applying environmental stressors during algal cultivation is one of the effective strategies to enhance lipid for biodiesel (Shah et al., 2014; Shah and Abdullah, 2018). However, the major challenge will be to adapt the transformed and induced cultures in a large-scale integrated biorefinery set-up (Abdullah et al., 2017; Shah and Abdullah, 2017). To cope with unfavorable conditions such as lack of nitrogen, some species develop a number of adaptations, including alteration of lipid metabolism and synthesis of non-membranous lipids such as triacylglycerols (TAG) and carotenoids (Solovchenko et al., 2008). Although nitrogen deficiency inhibits the cell cycle and the production of almost all cellular components, the rate of lipid synthesis remains high, which leads to the accumulation of neutral lipids in starved cells (Rosenberg et al., 2008). The TAG accumulation which is often deposited in cytoplasmic lipid globules referred to as oil bodies (OB), increase vastly in sizes and numbers under mineral nutrition deficiency, high salinity, and high irradiances (Merzlyak et al., 2007). Lipids, particularly TAG, are thought to be a storehouse for the excessive photosynthates, which could not be utilized under unfavorable conditions. These adaptive responses help to ensure the survival of cells during times of stress, while lipids serve as energy stores (Bigogno et al., 2007). These have been described in chlorophytes such as *Dunaliella salina*, *Dunaliella bardawil*, *Haematococcus pluvialis*, and *Parietochloris incise* (Rabbani et al., 1998; Mendoza et al., 1999; Wang et al., 2003; Merzlyak et al., 2007).

There are research efforts aimed at increasing and modifying the accumulation of lipids, alcohols, hydrocarbons, polysaccharides, and other bio-based compounds in algae through genetic and metabolic engineering which is likely to have the greatest impact on improving the economics of biodiesel production from microalgae (Hankamer et al., 2007). During the aquatic species program, acetyl-CoA carboxylase (AACase) activity is observed to increase two-fold in the diatoms during silica deficiency. Later studies suggest that the AACase enzyme plays a key role in lipid accumulation. The enzyme is then isolated and cloned for a study on gene expression. The studies conclude that algae alter enzyme activity during silica deprivation for increased synthesis of lipids. Genetic engineering of an algal cell with increased activity of AACase during optimized cell growth produces a strain of algae efficient for bio-oil production. Although the Aquatic Species Program does make strides in genetic manipulations and mutations, the efforts are not successful at manipulating the algal strain for biofuel production (Sheehan et al., 1998). Algal biotechnology breakthroughs directly enable new approaches to generate algae with desirable properties for the production of biofuels and bioproducts. The strategies to introduce genes, to delete or disrupt genes, and to modify genes or gene expression in particular algal species could enhance physiological properties and optimize the production. Though the potential is great, to date, no commercial application of genetically manipulated algae has been reported. The limitation is due to the inability to silence foreign genes causing instability in the gene expression. If the regulation of the gene expression for algae is better understood, controlling the gene expression may be possible (Rosa et al., 2004).

In this chapter, an overview of biochemical pathways, biosynthesis, and metabolism of microalgae-based lipids, carbohydrates, and high-value products are given with special emphasis on the pathways and metabolic engineering considerations for enhanced productivity.

14.2 BIOCHEMICAL PATHWAYS AND BIOSYNTHESIS

Algae are one of the most abundant and primitive life on earth, categorized as plants for many years because of their photosynthetic ability, but with less complex structure which consists of one or more of the eukaryotic chlorophyll-containing cells. It may be a single cell, colonies, filament of cells, or as simple as in the kelp tissues. Of about 10 million estimated algal species in nature, only a small portion has been identified taxonomically. Algae are now placed within the diverse kingdom Protista of eukaryotic, predominantly single-celled microscopic organisms (Hollar, 2012), and can be classified into two main groups: (1) microalgae, which includes diatoms (Bacillariophyceae), green algae (Chlorophyceae), red algae (Rhodophyceae), yellow-green algae (Xanthophyceae), golden algae (Chrysophyceae), brown algae (Phaeophyceae), and Euglenoids; and (2) macroalgae (seaweeds or multicellular algae). Microalgae can further be categorized into three main sub-groups: green (Chlorophyta), brown (Phaeophyta), and red (Rhodophyta) algae. Most of the red and brown algae are marine species, though there are some rare freshwater species in each group. Blue-green algae (BGA), a type of chlorophyll-containing bacteria called cyanobacteria (CB) which do not have eukaryotic cells and therefore are not classified as algae; make up another group of single-celled organisms that use photosynthesis to create food (Hollar, 2012). In the tropics and subtropics, algae may be found on leaves, on woods and stones and some live within or on plants and animals. Because algae inhabit diverse ecological habitats ranging from seawater, freshwater or brackish water, in the oceans, rivers, lakes, streams, ponds, swamps, and the soils (Rosenberg et al., 2008), and can survive and flourish in dry conditions for a long time or different extreme temperatures such as at 80°C, in, and around hot springs, in the snow and ice of the Arctic and Antarctic regions, and extreme pHs, algae have developed defense strategies that result in a significant level of structural-chemical diversity from different metabolic pathways to survive in a competitive environment (Puglisi et al., 2004). These compounds include triglycerides, fatty acids, polysaccharides, pigments, antioxidants, β-carotenes, vitamins, and the biomass.

High productivity can be accomplished using genetic engineering tools in which crucial enzymes can be targeted for fatty acid and lipid metabolic pathways (Radakovits et al., 2010; Williams, 2010) or enhancing the photosynthetic efficiency and consequently the lipid production, by redirecting the lipid biosynthesis based on a fully annotated pathway (Armbrust et al., 2004). Despite the recent advances in recombinant DNA technologies which enable the manipulation of the function of key enzymes in algae, the major challenges to be overcome are the technical limitations with regards to the efficiency of genetic transformation tools. When lipids or carbohydrates are accumulated under stress conditions, the total proteins may decrease significantly (Stehfest et al., 2005). Under nutrient starvation, the protein pattern also changes (Tran et al., 2009a). Microalgae lipid metabolic network is highly complex. Understanding the biochemical pathways and biosynthesis is therefore of paramount importance to generate high yielding microalgal strains. Full characterization of the metabolic pathways is required before meaningful metabolic manipulation can be performed to enhance productivity (Radakovits et al., 2010; Williams, 2010).

14.3 LIPIDS

14.3.1 POLAR GLYCEROLIPIDS

The major components of polar glycerolipids are phosphoglycerides, glycosylglycerides (glycolipids), and betaine lipids. Phosphoglycerides possess the basic structure of phospholipids, where the glycerol backbones are metabolically derived from glycerol-3-phosphate (G3P). The hydrophobic acyl groups are esterified at the 1- and 2-positions, and the phosphate is esterified at the sn-3 position, with a link to a hydrophilic base group. The major phosphoglycerides identified in most algae species are phosphatidylethanolamine (PE), phosphatidylcholine (PC), and phosphatidylglycerol (PG) (Guschina and Harwood, 2009). Phosphatidic acid (PA), a minor component, serves as an important metabolic intermediate, possibly as a signaling compound. The phospholipids are mainly localized in the extrachloroplast membranes. PG which accounts for 10–20% of the total polar glycerolipids in eukaryotic green algae, present in high quantity in the thylakoid membranes. In general, fish phosphoglycerides; contain 50% of their total fatty acids as n-3 PUFA with a ratio of 22:6 (n-3) to 20:5 (n-3) of 2:1 and commonly found in fish eggs (Hemaiswarya et al., 2011). The δ-3-transhexadecenoic acid [16:1(3t)], which is an unusual fatty acid, is present in all eukaryotic photosynthetic organisms, and is highly enriched at the sn-2 position in PG. Its d-3 position and trans-configuration of the double bond are rare for naturally-occurring fatty acids (El Maanni et al., 1998).

Glycosylglycerides (glycolipids) possess a 1,2-diacyl-sn-glycerol moiety with a mono- or oligosaccharide linked to the glycerol backbone at the sn-3 position. Major plastid lipids, such as galactosylglycerides possess one or two galactose molecules linked to the glycerol at the sn-3 position, corresponding to monogalactosyldiacylglycerol, MGDG) or (1,2-diacyl-3-O-(β-D-galactopyranosyl)-sn-glycerol) and digalactosyldiacylglycerol (DGDG), or (1,2-diacyl-3-O-(α-D-galactopyranosyl)-(1→6)-O-β-D-galactopyranosyl-sn-glycerol). MGDG and DGDG constitute 40–55% and 15–35% of the total lipids in the thylakoid membranes of plants, respectively (Harwood and Okanenko, 2003). Glycosylglycerides, which belong to the class of sulfolipid, are either sulfoquinovosyldiacylglycerol (SQDG) or 1,2-diacyl-3-O-(6-deoxy-6-sulfo-α-D-glucopyranosyl)-sn-glycerol. They are found in both photosynthetic and non-photosynthetic membranes of algae and may constitute up to 30% of the total lipids, as reported in the raphidophycean alga *Chattonella antiqua* (Harwood and Okanenko, 2003). SQDG is uncommon due to its sulfonic acid linkage. The sulfonoglucosidic moiety (6-deoxy-6-sulfono-glucoside) is termed as sulfoquinovosyl and at physiological pH; its sulfonic residue bears a full negative charge (Harwood and Okanenko, 2003).

Betaine lipids are commonly found in algae, but not in higher plants such as gymnosperms and angiosperms. Lipids with polar betaine moiety are linked to glycerol at the sn-3 position by an ether bond. Betaine lipids possesses neither carbohydrate nor phosphorus groups and found in algae as 1,2-diacylglyceryl-3-O-carboxy-(hydroxymethyl)-choline (DGCC), 1,2-diacylglyceryl-3-O-2'-(hydroxymethyl)-(N,N,N-trimethyl)-β-alanine (DGTA), and 1,2-diacylglyceryl-3-O-4'-(N,N,N-trimethyl)-homoserine (DGTS) (Dembitsky, 1996). All are zwitterionic at neutral pH because of their negatively charged carboxyl group and positively charged trimethylam-monium group (Dembitsky, 1996).

14.3.2 FATTY ACIDS AND THE POLAR GLYCEROLIPID BIOSYNTHESIS

In plants and eukaryotic algae, fatty acids and glycerolipids biosynthesis may involve two subcellular organelles-the plastid and the endoplasmic reticulum (ER). In the ER, glycerolipids are synthesized via the core glycerol 3-phosphate pathway with triacylglycerols (TAG) and phosphoglycerides as major products. In higher plants, palmitate, oleate, and stearate are synthesized via a pathway localized in the plastid. This is one of the major pathways for lipid metabo-lism and the main *de novo* source where the acyl chains of complex lipids are derived. This pathway begins with acetyl-CoA derived from photosynthesis and the malonyl-acyl carrier protein (ACP) which serves as the two-carbon donor. For the *de novo* synthesis of long-chain saturated fatty acids, fatty acid synthase (FAS) and acetyl-CoA carboxylase (ACC) are involved. In most plants, the chloroplastic ACC is a multiprotein complex comprising several functional proteins (a biotin carboxylase, a biotin carboxyl carrier protein, and two different subunits of the carboxyltransferase (Guschina and Harwood, 2013a).

FAS is the second major enzyme complex that is involved in *de novo* biosynthesis of fatty acid. Plant FAS is a Type II dissociable multiprotein complex, similar to that the *E. coli* system, which is different from animals (Guschina and Harwood, 2013a). The individual subunit proteins that consti-tute FAS complex can be isolated. The catalytic function of FAS complex in fatty acid biosynthesis is contributed by the subunit protein β-ketoacyl-ACP synthase III (KAS III) which uses the acetyl-CoA and malonyl-ACP substrates to generate a 4C-keto-intermediate. Successive reduction, dehydration, and a second reduction produce a 4C fatty acid butyrate and all reactions occur while the intermediate is esterified to ACP. The next six condensations are catalyzed by KAS I, which produces 6–16C fatty acids. The final reaction which involves palmitoyl-ACP and malonyl-ACP is catalyzed by KAS II and this reaction results in the synthesis of a stearate. The remaining enzymes of FAS are β-hydroxylacyl-ACP dehydrase, β-ketoacyl-ACP reductase, and enoyl-ACP reductase. Numerous enzymes are involved in fatty acid biosyn-thesis, i.e., β-ketoacyl-ACP reductase, β-ketoacyl-ACP synthase, acyl-ACP thioesterase (TE), β-ketoacyl-CoA synthase, and β-ketoacyl-CoA reductase, and they are either up- or down-regulated in higher plants (Guschina and

Harwood, 2013a). Plastidic fatty acids may be assimilated into the plastidic phosphatidate pool, and then subsequently be converted into chloroplast lipids, such as MGDG, DGDG, SQDG, and PG. Similar to CB, algal glycerolipids are synthesized via this pathway in plastids and contain C16 fatty acids esterified to the *sn*-2 position of glycerol, and either C16 or C18 fatty acids, to the *sn*-1 position of their glycerol skeleton. Plastid galactolipids are characterized by a high PUFA content. Therefore, in freshwater algae, their MGDG contains α-linolenic (C18:3n-3) as the major fatty acid. Linolenic and palmitic acid (C16:0) are dominant in DGDG and SQDG. 1,2-di-O-acyl-3-O-(acyl-6'-galactosyl)-glycerol (GL 1a), the sulfonoglycolipid 2-O-palmitoyl-3-O-(6'-sulfoquinovopyranosyl)-glycerol and its ethyl ether derivative are three new minor glycolipids identified in the crude methanolic extracts of the red alga *Chondria armata* (Al-Fadhli et al., 2006). GL 1a has been reported as the first example of a naturally occurring glycolipid acylated at the 6' position of galactose (Al-Fadhli et al., 2006).

14.3.3 NON-POLAR STORAGE LIPIDS

Triacylglycerols (TAG) are found as storage products in many algae species. The accumulated TAG varies and may be elevated by environmental factors. Significant accumulation of neutral lipids in microalgae cells, i.e., 20 to 50% dry cell weight, mainly in the form of TAG, has been reported under unfavorable environmental or stress conditions, such as nutrition limitation. Under nitrogen deprivation, TAG synthesis may be stimulated, where a two- to three-fold increase in algal lipid content has been reported, predominantly as TAG (Shah, 2014). When growth of algae slows down, the demand for new membrane compounds biosynthesis dwindles, and fatty acids may be diverted into TAG synthesis before conditions improve, when they are needed again for further growth. TAGs are the key carbon and energy storage in the form of dense cellular lipid bodies and considered as the ideal source for biodiesel (Hu et al., 2008). In general, the light period favors TAG synthesis, where it is stored in cytosolic lipid bodies before they are reused for polar lipid synthesis in the dark. Algal TAG are generally characterized by the presence of saturated and monounsaturated fatty acids. However, in some oleaginous species may contain high levels of long-chain PUFA in TAG (Khozin-Goldberg et al., 2005). Therefore, quantitative evaluation of cellular TAG content in different cell growth conditions is essential for bioprocess monitoring and engineering of efficient and scalable biofuel production (Wang et al., 2014).

14.3.4 BIOSYNTHESIS OF TAG

In eukaryotic microalgae, the key regulatory steps identified for fatty acid biosynthesis and translocation of fatty acids to the ER for further glycerolipid biosynthesis, are basically similar to those in higher plant plastids (Riekhof et al., 2005; Sato et al., 2007; Moellering et al., 2009b). The first two reactions in the ER involving the core glycerol 3-phosphate pathway leading to TAG biosynthesis, are the generation of PA by the stepwise acylation of glycerol 3-phosphate. Two distinct acyltransferases catalyze these reactions which are specific for the positions sn-1 and sn-2. Firstly, the membrane-bound glycerol 3-phosphate acyltransferase (GPAT) initiates the reaction by transferring the acyl chain from acyl-CoA to the sn-1 position of glycerol 3-phosphate with the generation of lysophosphatidic acid (monoacylglycerol 3-phosphate) (Weselake et al., 2009). Then, the lysophosphatidic acid acyltransferase (LPAAT) catalyzes the transfer of acyl chains from acyl-CoAs to the sn-2 position to generate PA. PA is then dephosphorylated to form diacylglycerol (DAG). Finally, diacylglycerol acyltransferase (DGAT), an enzyme unique to TAG biosynthesis, catalyzes the final step in the pathway, where a final fatty-acyl group is added to the sn-3 position of DAG to generate TAG (Weselake et al., 2009).

In plants, two unrelated genes have been found to code for DGAT enzymes. One form (DGAT1) is related to acyl-CoA: cholesterol acyltransferase, while the second form (DGAT 2) does not resemble any other known genes (Weselake et al., 2009). Preliminary studies in plants show that alternative reactions for TAG synthesis are found in plants. In one of these reactions, a fatty acid residue is directly transferred from the sn-2 position of PC to DAG, generating lyso-PC and TAG. The enzyme involved is referred to as phospholipid:diacylglycerol acyltransferase (PDAT). Another reaction involving the transfer of acyl group between two molecules of DAG, i.e., DAG:DAG transacylase. In addition, another enzyme, which probably play a key role in the exchange of diacylglycerol from PC for the bulk pool and hence, facilitating the entry of PUFA into TAG synthesis, known as phosphatidylcholine:diacylglycerol cholinephosphotransferase (PDCT). DGAT, being a key mediator of plant TAG biosynthesis, the over expression of DGAT genes has been proposed as a promising strategy to enhance TAG production (Li et al., 2010). DGAT is known to be an integral ER protein found in OB and plastids. DGAT1 appears to be quantitatively most important for the biosynthesis of TAG, when common fatty acids are esterified. On the other hand, DGAT2 is important for TAG production in several cases that involve unusual fatty acids (Li et al., 2010). In tobacco leaves, over expression of acyl-CoA:diacylglycerol acyltransferase (*AtDGAT1*) produces up to seven-fold increase in TAG content. In *Arabidopsis*, seed-specific expression of *AtDGAT1* leads to an increase of 11–28% in seed oil content. Similar results have been obtained with experiments in oilseed rape (Weselake et al., 2009). These studies confirmed that DGAT plays an important role of in modulating the TAGs content in seeds.

One GPAT and two DGATs have been identified and characterized in algae (Xu et al., 2009; Wagner et al., 2010; Guihéneuf et al., 2011). A marine diatom *Thalassiosira pseudonana* membrane-bound GPAT is isolated and found to prefer saturated C16 fatty acid as a substrate. Apparently, it plays a significant function in determining the fatty acid profile of glycerolipids (Xu et al., 2009). In the green alga *Osteococcus tauri*, three putative DGAT23 genes are identified in the genome through database search (Wagner et al., 2010). Two of these DGAT cDNA sequences (OtDGAT2A and B) have been cloned and expression of these sequences in *Saccharomyces cerevisiae* mutant strains had impaired TAG metabolism (Wagner et al., 2010). In the diatom microalga *Phaeodactylum tricornutum*, the isolated PtDGAT1 shows a high homology to several higher plant DGAT1 proteins which have been functionally characterized. Functional study of PtDGAT1 is achieved via heterologous expression of in *S. cerevisiae* (Guihéneuf et al., 2011). *Chlamydomonas reinhardtii* has been shown to use a unique TAG biosynthesis pathway that uses DAG that is derived almost exclusively from the chloroplast (Fan et al., 2011).

14.4   HYDROCARBONS

Some algae are well known for their ability to synthesize and amass a significant amount of hydrocarbons and thus, exhibit good potential for biodiesel production. One of the most promising species is *Botryococcus braunii*. This freshwater microalga species forms green colonies, and has been recognized for its excellent potential for liquid hydrocarbons production. Indeed, in geochemical analysis of petroleum suggests that microalgae ancestral to *B. braunii*, may have generated botryococcene and methylated squalene type of hydrocarbons, which may be the source of today's petroleum deposits (Eroglu and Melis, 2010). The structure of hydrocarbons produced by *B. braunii* varies depending on its race. *B. braunii* is classified into the races A, B, and L, based on the type of hydrocarbons they synthesize. For instances, race A produces up to 61% (based on a dry biomass) of non-isoprenoid dienic and trienic hydrocarbons, odd numbered n-alkadienes, mono-, tri, tetra-, and pentaenes, from C25 to C31, which are derived from fatty acids, while race B produces C30–C37, which are highly unsaturated isoprenoid hydrocarbons, known as botryococcenes and small amounts of methyl branched squalenes and race L produces lycopadiene, a single tetraterpenoid hydrocarbon (Rao et al., 2007).

14.5 CARBOHYDRATES

14.5.1 COMPOSITION

Microalgae cell walls are primarily composed of an inner and an outer cell wall layers. The latter can be further classified into: (1) a trilaminar outer layer, (2) a thin outer monolayer, and (3) without an outer layer (Yamada and Sakaguchi, 1982). The outer cell wall composition differs from species to species, but often contains specific polysaccharides, such as agar, pectin, and alginate. The inner cell wall layer is mainly consisting of cellulose and other materials, such as hemicellulose and glycoprotein. For some micoralgae, the glucose polymers generated from cellulose/starch are the main component in the cell walls stored with starch products (Metting, 1996). Except for the starch in plastids, algal extracellular coverings, e.g., cell wall, are rich in carbohydrate, which can be converted to biofuel (Harun et al., 2010). However, the compositions of microalgal extracellular coverings vary according species (Domozych et al., 2012). Most cell wall polysaccharides and starch can be converted into fermentable sugars for subsequent bioethanol production via microbial fermentation (Wang et al., 2011).

14.5.2 BIOSYNTHESIS OF STARCH/CARBOHYDRATES

Carbohydrates serve as energy sources and as structural elements where the monosaccharide glucose is a prominent energy source in almost all living cells. Carbohydrate metabolism can occur via gluconeogenesis and glycolysis (McKee and McKee, 2015). During glycolysis, a small amount of energy is captured as a glucose molecule and converted to two molecules of pyruvate. Glycogen, a storage form of glucose in vertebrates, is synthesized by glycogenesis when glucose levels are high and degraded by glycogenolysis when glucose is in short supply. Glucose can also be synthesized from noncarbohydrate precursors by reactions referred to as gluconeogenesis (McKee and McKee, 2015). The glycolysis reactions occur only in cytoplasm. In gluconeogenesis, which occurs when sugar levels are low and liver glycogen is depleted in higher animals, 7 of the 10 reactions of glycolysis are reversed. The major substrates for gluconeogenesis are certain amino acids (derived from muscle), lactate (formed in muscle and red blood cells), and glycerol (produced from the degradation of TAGs) (McKee and McKee, 2015).

In microalgae, the accumulated carbohydrate is derived from CO_2 fixation during the photosynthetic process. Photosynthesis is a biological process that utilizes ATP/NADPH to fix and convert CO_2 captured from the environment to produce glucose and other sugars via a metabolic pathway called the Calvin-Benson cycle or the pentose phosphate cycle (Lehninger et al., 2005). The metabolic pathways of energy-rich molecules (e.g., carbohydrate, and lipid) are closely linked. Lipid and starch synthesis may be a competitive due to G3P, the major precursor for TAG synthesis, is produced via glucose catabolism (glycolysis) (Rismani-Yazdi et al., 2011; Ho et al., 2012). The gluconeogenesis reactions catalyzed by pyruvate carboxylase and, in some species, phosphoenolpyruvate (PEP) carboxykinase, occur within the mitochondria. The reaction catalyzed by glucose-6-phosphatase takes place in the ER (McKee and McKee, 2015).

Starch, which forms around a crystallizing nucleus, is present in an amorphous starch grain. When a chloroplast accumulates sufficient starch, it may convert into an amyloplast (Radakovits et al., 2010). In most of green algae, cellulose is one of the main fermentable carbohydrates (Radakovits et al., 2010). Cellulose synthesis involves many enzymatic reactions and thus, is a complicated process. Uridine diphosphate (UDP)-glucose, is the starting substrate for cellulose synthesis and its formation is catalyzed by sucrose synthase using UDP and fructose (Kimura and Kondo, 2002). It is vital to understand and manipulate the related metabolisms to achieve higher carbohydrate accumulation via strategies, such as increasing glucan storage and decreasing starch degradation for the purpose of enhancing biofuels production (Radakovits et al., 2010).

14.6 CAROTENOIDS

Carotenoids belong to the class of terpenoids, which are derived from a 40-carbon polyene chain, which may be considered as their molecular backbones. It gives carotenoids their distinctive structures, and the associated chemical properties, including light-absorption properties (Del Campo et al., 2007). The molecular backbone may be accompanied by oxygen-containing functional groups and cyclic groups (rings). In general, carotenoids are denoted as carotenes, but their oxygenated derivatives are distinguished as cantaxanthin (with oxi-groups), or xanthophylls (with -OH groups) (e.g., lutein) or as a combination of both (e.g., astaxanthin) (Del Campo et al., 2007). Primary carotenoids (i.e., xanthophylls) form part of the functional and structural components of the cellular photo-synthetic machinery, and therefore essential for cell survival (Jin et al., 2003). Secondary carotenoids however are produced by microalgae in large amounts in response to specific environmental conditions (through carotenogenesis). Xanthophylls are typically found in the thylakoid membrane, while secondary carotenoids are located in lipid vesicles, e.g., the cytosol or plastid stroma. In most green microalgae, xanthophylls, and carotenes are produced and accu-mulate only within plastids. However, in some green microalgae, secondary xanthophylls, such as astaxanthin in *Haematococcus* sp., are accumulated in the cytoplasm. This may suggest the possibility of an extra-plastidic site of carotenoid biosynthesis in *Haematococcus* sp. It is possible that the chloro-plast-produced xanthophylls may be exported, and consequently accumulated in the cytoplasm, and thus found in all cellular compartments (Tardy et al., 1996; Rabbani et al., 1998; Jin et al., 2003).

All xanthophylls derived from higher plants, (e.g., antheraxanthin, lutein, violaxanthin, neoxanthin, zeaxanthin) may be produced by green microalgae but more specifically the astaxanthin, canthaxanthin, and loroxanthin. Brown algae or diatoms may produce diadinoxanthin, diatoxanthin, and fucoxanthin (Jin et al., 2003).

Xanthophylls, which are relatively hydrophobic, usually involve in non-covalent binding to specific proteins and/or associate with membranes. In oxygenic photosynthetic bacteria and CB, xanthophylls usually associate with chlorophyll-binding polypeptides of photosynthetic machinery (Grossman et al., 1995). In microalgae, carotenoids play key roles in harvesting light, but they also function in aiding the function of photosynthetic complexes and stabilizing the structure. They scavenge reactive oxygen species (ROS), thus quenching chlorophyll triplet states, and dissipating excess energy (Demming-Adams et al., 2002). The antioxidant activity of carotenoids is limited by their capability in scavenging free radicals and ROS as a defense against oxidative stress.

14.7   PATHWAYS, ENZYMES, AND GENES

Among oxygenic phototrophs, CB, and land plants have been the focus of studies on carotenogenesis pathways and their enzymes (Britton et al.,

1998; Takaichi et al., 2007). In land plants, carotenogenesis pathways and characteristics of enzymes have been studied in detail but though sharing the pathways with land plants, algae possess additional algae-specific pathways, which are proposed entirely based on the chemical structures of carotenoids. The carotenogenesis genes and enzymes, whose functions are confirmed, in algae, have been reported (Takaichi, 2011). Some common carotenogenesis genes in algae show homology to known genes (Frommolt et al, 2008; Bertrand, 2010). However, most enzymes and genes involved in algae-specific pathways remain unknown. Isopentenyl pyrophosphate (IPP), a C5-compound, is the source of terpenes, isoprenoids, and phytol of chlorophylls, quinones, sterols, and carotenoids. Two independent pathways of IPP synthesis are known-the classic mevalonate (MVA) pathway and the alternative, non-MVA, 1-deoxy-D-xylulose-5-phosphate (DOXP) pathway (Lichtenthaler et al., 1999; Eisenreich et al., 2004). In the MVA pathway, acetyl-CoA is converted to IPP via MVA, and the enzymes and genes involved are well known (Miziorko et al., 2011). This pathway is found in animals, plant cytoplasm, and some bacteria (Lichtenthaler et al., 1999; Miziorko et al., 2011). The DOXP pathway, discovered in the 1990s, is found in the plastids of algae, CB, land plants, and some bacteria. In this pathway, glycelaldehyde, and pyruvate are converted to IPP. The carotenoid biosynthesis pathway has been corroborated by the molecular cloning of the numerous relevant genes from oxygenic phototrophs, including CB and green algae (Misawa et al., 1990; Chamovitz et al., 1992; Martínez-Férez et al., 1994; Sandmann et al., 1994; Armstrong et al., 1997; Ohto et al., 1999; Steinbrenner et al., 2001; McCarthy et al., 2004; Steiger et al., 2005). *crtE* and *crtB* genes are highly conserved among oxygenic phototrophs from bacteria to land plants. Most Euglenophyceae only possess the MVA pathway, while Chlorophyceae only possess the DOXP pathway (Lichtenthaler et al., 1999).

The carotenogenesis pathways and enzymes in CB (Takaichi, 2011). Carotenoids are produced in the plastids. Most carotenoids are composed of eight IPP units. Initially, farnesyl pyrophosphate (C15) is produced from three IPPs, subsequently one IPP is added to farnesyl pyrophosphate by geranylgeranyl pyrophosphate synthase (CrtE, GGPS) to generate geranylgeranyl pyrophosphate (GGPP) (C20). Next, phytoene (C40) is formed by phytoene synthase (CrtB, Pys, Psy) via a head-to-head condensation of the two GGPS using ATP (adenosine triphosphate) (Sandmann et al., 1994; Armstrong et al., 1997). Conversion from phytoene to lycopene involves four desaturation steps. Oxygenic phototrophs employ three enzymes, i.e., phytoene desaturase (CrtP, PDS), ζ-carotene desaturase (CrtQ, ZDS) and *cis*-carotene isomerase (CrtH, CrtISO). The first two desaturation steps CrtP, where phytoene is converted to ζ-carotene through phytofluene. Next, the two following desaturation steps are catalyzed by CrtQ from ζ-carotene to lycopene through neurosporene. During the desaturation step catalyzed by CrtQ, lycopene, and neurosporene are isomerized to poly-*cis* forms, and subsequently, CrtH isomerizes them into all-*trans* forms. The functions of these enzymes have been mainly confirmed in CB, green algae and land plants (Chamovitz et al., 1992; Martínez-Férez et al., 1992; Linden et al., 1993; Breitenbach et al., 1998, 2001; Masamoto et al., 2001; Huang et al., 2008; Vila et al., 2008; Liu et al., 2010).

CB produce two ketocarotenoids, canthaxanthin, and 4-ketomyxol. Two classes of β-carotene ketolases, CrtO, and CrtW, are known. However, only seven functional enzymes from four species of CB have been reported (Takaichi et al., 2007). CrtO catalyzes β-carotene to form echinenone, and the final product is canthaxanthin (Steiger et al., 2005). CrtW attaches a keto group to β-carotene, myxol, and zeaxanthin to generate canthaxanthin, 4-ketomyxol, and astaxanthin, respectively (Steiger et al., 2004; Mochimaru et al., 2005; Steiger et al., 2005; Makino et al., 2008). Consequently, these ketolases are used in two pathways involving β-carotene and myxol, depending on the species (Takaichi et al., 2007). The pathway and the enzymes that produce the right half of myxol 2'-pentoside remains unknown (Takaichi et al., 2007).

14.8 GENETIC ENGINEERING

14.8.1 MODEL ORGANISMS

Genetic engineering has the potential to improve high oil-yielding microalgal strains, based on knowledge on lipid metabolic pathway of microalgae, whereby critical enzymes, for instance, those involved in TAG accumulation, can be targeted for modification (Radakovits et al., 2010; Williams, 2010). However, details for metabolic flux of carbohydrate biosynthesis and enzymatic activity in microalgae are not well studied. Therefore, genetic engineering may be used to manipulate carbohydrate metabolisms of microalgae to gain a more understanding of the biochemistry of the relevant-metabolic pathways, before superior strains are engineered (Radakovits et al., 2010). Since the early 2000s, *C. reinhardtii* has became emerged a key model organism for studying photosynthesis. Thus, substantial advances such as the development of genetic tools and a fully annotated genome have been made. On the basis of these advances, molecular toolkits for a variety of algal species are now being developed and these include strains are used for bioenergy production (Georgianna and Mayfield, 2012). Genetically modified superior microorgansims (GMSMs) present the opportunity for breaking the genetic bottlenecks in algae by providing a basis for laboratory and computational experiments (Lehr and Posten, 2009; Chisti, 2013). Genetic transformation and stable heterologous gene expression in algae facilitate the development of economically viable algal strains. In this aspect, *C. reinhardtii* being a relatively simple, single-celled organism that thrives as a haploid or diploid, with sexual and asexual lifecycles that can be easily manipulating using nutrient deprivation and light cycles, is a remarkable choice as a model organism. Furthermore, *C. reinhardtii* can be easily grown photoautotrophically, mixotrophically, or heterotrophically, with a doubling time (DT) of 5–8 hrs in to density above 10^7 cells mL^{-1}. The three *C. reinhardtii* genomes, i.e., nuclear, chloroplast, and mitochondria, have been sequenced (Popescu and Lee, 2007). *C. reinhardtii* has been used to express a wide range of recombinant proteins, for instances protein therapeutics, such as antibodies, and reporters, such as GFP, GUS, luciferase (Tran et al., 2009b), vaccines (Surzycki et al., 2009), hormones, growth factors (Rasala et al., 2010), and industrial enzymes. The recombinant proteins produced may accumulate to a level as high as 10% of total soluble protein (TSP), properly folded, functional active (Tran et al., 2009b; Rasala et al., 2010).

In general, for expressing a recombinant gene construct, the transgene need to be placed under the control of a promoter, and flanked by 5' and 3' untranslated regions (UTRs). The promoter region is required for initiating gene transcription, while the 5' UTR regulates ribosome association and translation rates (Marin-Navarro et al., 2007). Apparently, the 3' UTR modulates mRNA stability and may associate with the 5' UTR. Nuclear transformation of *C. reinhardtii* occurs mainly by random insertion via non-homologous end-joining. Transformation with linear DNA fragments foster insertion of multiple DNA copies in one locus, thereby leading to fewer large rearrangements or deletions. In targeted gene insertion, homologous recombination (HR) is employed. However, such event occurs at a low, but measurable frequency. The use of single-stranded DNA in transformation improves the ratio of HR to non-homologous integration events, thereby enabling the generation of homologous recombinants (Zorin et al., 2009). However, besides *Chlamydomonas*, few algal species, including chlorella, diatoms, dinoflagellates, and red algae, have been successfully transformed (Walker et al., 2005).

Further efforts are required to understand direct HR for targeted gene replacement and mutagenesis in *C. reinhardtii* (Zorin et al., 2009). Generally, the regulation of endogenous chloroplast and heterologous gene expression from the chloroplast genome appears to be mainly regulated at the translation level (Rasala et al., 2011). Although it is relatively easy to transfer DNA into the chloroplast genome, *Chlamydomonas* maintain a tight control on plastidic gene expression. The use of synthetic or heterologous promoters/UTRs produces little to no recombinant protein accumulation in *Chlamydomonas*. Exceptionally, the use of viral elements for heterologous gene expression is an unlikely option because there is no known virus that infects *C. reinhardtii* chloroplasts. Therefore, much work focused on understanding endogenous transcriptional and translational regulatory elements to facilitate improving heterologous gene expression in algae (Harris et al., 2009).

14.8.2   VECTOR SYSTEM

In constructing an algal expression systems vector, it is important to determine the stability and frequency of exogenous DNA expression. In marine CB, the backbone of shuttle vectors may be constructed from a segment of cyanobacterial chromosome and cyanophage DNA. In eukaryotic algae, vectors are usually constructed based on segment of their own chromosome. In addition, in marine algal transformations, some vectors are derived the vectors that are used for *E. coli* and high plants. The choice of promoter and its availability is an important factor in genetic transformation. The SV40 and CaMV35S promoters from viruses are generally used, because essential genetic information of the endogenous promoters in marine algae, in general, is lacking (Liu et al., 2003; Qin et al., 2004; Wang et al., 2010; Anila et al., 2011). In vector construction, besides viral universal promoters, endogenous promoters from specific marine algae are thought to be the more effective. The use of the promoter of duplicated carbonic anhydrase 1 (DCY1) is generated stable *D. salina* nuclear transformants (Li et al., 2010; Lu et al., 2011). The promoter of diatom fucoxanthin-chlorophyll a/c binding protein (fcp) gene is active in marine algae and diatoms (Apt et al., 1996; Zaslavskaia et al., 2000; Qin et al., 2004; Li et al., 2009; Miyagawa-Yamaguchi et al., 2011). The promoter, including the 5′ UTR of the actin1 (PyAct1) gene from *Pyropia yezoensis* also is effective for the transient gene expression in 12 red seaweed species (Takahashi et al., 2010; Hirata et al., 2011). In *Nannochloropsis* sp., the endogenous promoters from violaxanthin/chlorophyll α-binding protein (VCP) genes, VCP1, and VCP2, were developed successfully developed (Kilian et al., 2011). In addition, the promoter of nuclear-encoded plastid-targeted protein Rubisco SSU (rbcS) of chlorarachniophyte *Lotharella amoebiformis* has been cloned and used transgenic studies (Hirakawa et al., 2008).

Heterologous protein expression often poses a challenge. This may be due to a difference in codon preference of the hosts, where the exogenous gene may carry codons that are rarely used in the desired host, non-canonical codons or regulatory elements that impose limitation on expression (Gustafsson et al., 2004). Other problems associated with transgene expression include codon bias and the increased susceptibility to gene silencing (Heitzer et al., 2007; Potvin and Zhang, 2010). Codon bias is one of the most critical factors for protein expression in prokaryotic or prokaryotic-derived genomes, such as chloroplasts from eukaryotic algae (Surzycki et al., 2009). Recently, codon optimization for marine algal genetic transformation are receiving increasing attention in transgenic research (Lerche and Hallmann, 2009; Takahashi et al., 2010). Software and web applications are freely available to predict and optimize the codon usage of sequences (Potvin and Zhang, 2010). Today, the rapidly growing field of synthetic biology, codon optimization is becoming a necessity, particularly for *de novo* DNA synthesis and generation of gene design (Welch et al., 2009; McArthur and Fong, 2010).

For the ease of monitoring protein expression, reporter genes that are highly sensitive for detection, provide a better measure for transformation frequency, the stability of the transgene, the expression efficiency and the protein localization in the transformed cells. The reporter genes that widely used in marine algal transformation include *GFP*, *GUS*, *lacZ*, and Luc. The *GUS* gene, which encodes a β-glucuronidase, has been used as the reporter in transient and stable transfromation of marine algae, e.g., *Amphidinium* sp., *D. salina*, *Ectocarpus* sp., *Laminaria japonica*, *P. yezoesis*, *Symbiodinium microadriaticum*, *T. weissflogii*, *Undaria pinnatifida*, and *Ulva lactuca* (Cheney and Kurtzman, 1992; Qin et al., 1994; Huang et al., 1996; Kuang et al., 1998; Ten Lohuis and Miller, 1998; Falciatore et al., 1999; Liu et al., 2003; Tan et al., 2005; Li et al., 2009; Hirata et al., 2011). However, due to the impermeability of the substrate of this gene product to cell membrane, the substrate is toxic to the cell and damages the cellular ultrastructure during the staining process. Furthermore, in marine seaweeds and higher plants, a weak background of the GUS activity has been found and, thus the negative control included in the experimental design to eliminate false positive caused by the background signal. The homolog of the *lacZ* gene is found in several bacteria, as in the case of the *GUS* gene, negative and blank controls are required to eliminate the background problem. The lacZ assay is used in many marine algae, such as brown alga *L. japonica* (Jiang et al., 2003; Qin et al., 1998) and red alga *Porphyra haitanensis* (Zhang et al., 2010). The *Luc* gene, which encodes a luciferase, is another commonly used reporter gene in a freshwater microalgae and the marine diatom *P. tricornutum* (Falciatore et al., 1999). In addition, the green fluorescent protein (GFP) from the jellyfish *Aequorea victoria* has also been used as a reporter of gene expression and protein subcellular localization in various marine algae (Zaslavskaia et al., 2000; Poulsen et al., 2006; Hirakawa et al., 2008; Takahashi et al., 2010; Wang et al., 2010; Miyagawa-Yamaguchi et al., 2011; Watanabe et al., 2011). However, due to the endogenous photosynthetic pigments and other fluorescent substances that is commonly found in algae, in the GFP expression, a strong promoter used to reduce the interference by autofluorescence.

In any transformation protocols, an effective selection marker is required to distinguish successful transformants from non-transformed cells. Most selectable markers belong to two types; genes that confer resistance to antibiotics or herbicide and those that complement a metabolic deficiency. Selectable marker genes that confer antibiotics or herbicides are most widely used for the selecting for marine algal transformants. Marine algae differ from higher plants in that marine algae are not susceptible to neomycin and kanamycin, but are typically susceptible to hygromycin, chloromycetin, and herbicide glufosinate. The second type of marker or auxotrophy marker is based on the homologous complementation of metabolic mutants. This approach may be used for chloroplast transformations. However, its usage is limited by the availability of suitable mutants. Commonly used selectable markers in microalgae have been compiled in past reviews (León-Bañares et al., 2004; Walker et al., 2005; Griesbeck et al., 2006), but recently novel markers, such as ARG9 genes (Remacle et al., 2009) and PDS (Steinbrenner and Sandmann, 2006), have been developed. Concerns have been raised on the biosafety of antibiotic- and herbicide-resistant genes during the release of the transgenic plants to the environment. Efforts have also been on the development of marker-free selection systems and alternative marker systems (Manimaran et al., 2011). The removal of the selectable marker may quell some public concern and promote the acceptance for genetically modified organisms (Miki and McHugh, 2004). Several approaches for marker elimination have been developed in higher plants, including co-transformation, which is usually used in *Agrobacterium*-mediated transformation (Sripriya et al., 2008) and several site-specific recombination approaches that remove the selection marker by deleting or inverting the marker gene aided by an recombinase (Cotsaftis et al., 2002; Ow, 2002; Kopertekh et al., 2004; Darbani et al., 2007). Due to the differences in the growth and development mode, propagation, genetic background, and breeding type between high plants and marine algae, and particularity of the marine environments, on the basis of the marker elimination approaches used in higher plants, modifications need to made for their application in marine algae (Qin et al., 2012).

Transgenes expression in marine algae is variable and difficult to predict. The key factor for this unpredictability comes from the uncertain copy number of inserted transgene and the subsequent gene silencing. Transformants with single-copy insertion are desirable and preferred due to their higher expression level and their more predictable nature. Gene silencing, which occurs at the transcriptional or post-transcriptional level in general, increases with the increasing number of inserted gene copies (Depicker and VanMontagu, 1997; Baulcombe, 2004; Marenkova and Deineko, 2010; Potvin and Zhang, 2010). Transformation by electroporation typically results in transformants with highly variable copy number of inserted transgene. In contrast, *Agrobacterium*-mediated transformation typically leads to low copy numbers and higher single-copy transformants. The direct DNA-transfer methods such as glass beads and biolistic bombardment usually produce to a large number of inserted gene copies in algal transformants, which may increase the silencing effects. However, transgenic lines with single or low number copy can be obtained using reduced amount of cassette DNA during transformation cassettes by the biolistic approach (Yao et al., 2006; Lowe et al., 2009).

14.8.3   TRANSFORMATION METHODS

Until now, various transformation approaches have been successfully used to transform microalgae. The biolistic and electroporation approaches are popular. In some instances, stable transgenes expression in either the nucleus or the plastid is obtained, but in most cases, the only transient expression is observed. To date, efforts main focus on increasing lipid yield of by optimizing growth and induction conditions, such as salinity, temperature, light, and nutrient depletion (Horvatic and Persic, 2007; Liu et al., 2008; Yang et al., 2013). However, genetic manipulation of microalgae to alter either lipid quality or quantity or are rarely reported. The root cause for this probably due to the lack of a universally applicable transformation protocol for microalgae. Microalgae, being a diverse group of organisms, a method that works for one species may not be applicable in another. For instance, species such as *D. salina* do not possess a rigid cell wall, thereby making it easier to be transformed. In contrast, diatoms, such as *P. triconutum*, possess a very rigid silicate shell making it very difficult recalcitrant to transform. In addition, the availability of selective marker is another problem. Although auxotrophy markers may be used for some species such as *C. reinhardtii*, unfortunately stable transformation of other species still depends on selectable markers that confer antibiotics resistance. Antibiotics that are routinely used in the transformation of plants, such as zeocin and kanamycin, may not work well in microalgae. Furthermore, heterologous gene expression may call modification on gene sequence to comply with codon usage preference and appropriate promoter sequences for expression algal hosts. Undoubtedly, this would be ameliorated by a growing number of fully sequenced and annotated microalgal genomes. Nevertheless, genetic engineering has good potential as a key tool for studying the metabolic pathways in microalgae and contributes towards economically sustainable production of biofuels from microalgae (Georgianna and Mayfield, 2012; Schuhmann et al., 2012; Blatti et al., 2013).

14.8.4 *ELECTROPORATION*

Electroporation has been widely used for genetic transformation of various cell types in the last 40 years due to its simplicity and the high efficiency of its protocols involving only a small amount of DNA (Zimmermann et al., 1975; Neumann et al., 1982). Theoretically, the principles of electroporation for transferring exogenous DNA apply to all kinds of cells. The first application of this approach in cyanobacterium is performed in the unicellular *Synechococcus* sp. NKBG042902-YG 1116 (Matsunaga et al., 1990). The cyanobateria have lower electric field strength than that of freshwater strains. Efficient electroporation-mediated transformation is obtained in both wild-type and cell wall-less eukaryotic *C. reinhardtti* strains (Brown et al., 1991). The transformation efficiency obtained using of electroporation is two orders of magnitude higher than that obtained with the glass beads approach (Shimogawara et al., 1998). Today, the electroporation transformation approach has been established in many marine algae, including from prokaryotic cells to eukaryotic green algae, red algae, and diatoms. *D. salina* has been successfully transformed by electroporation (Sun et al., 2005). Recently, *Nannochloropsis* sp. and *C. vulgaris*, which are the most popular microalgae for their potentials in biofuels production, using electroporation, and several genes involved in the nitrogen metabolism, are modified by the HR method (Kilian et al., 2011; Niu et al., 2011). Electroporation has no need for special consumables, thus making it affordable. However, the application of this technology in brown algae is limited by the lack of protoplast preparation and regeneration technologies (Niu et al., 2012, 2013).

14.8.5   BIOLISTIC TRANSFORMATION

The biolistic approach, also known as the microprojectile or particle bombardment method, has been established as the most efficient method and is highly reproducible in transferring exogenous DNA into algal cells. The protocol has been successfully used for the transformation of many types of microalgal chloroplast and nuclear expression systems. Biolistic transformation is one of the most useful tools in transgenic research of microalgae, irrespective of their cell wall rigidity or life cycle. Biolistic transformation possesses several advantages. First of all, various types of vectors can be used in biolistic transformation. Therefore, even when understanding of the genetic background of microalgae is limited, biolistic transformation can be used to overcome difficulty by allowing flexibility in the choice of vectors. For instance, sometimes *E. coli* vectors are used in algal biolistic transformations. Furthermore, exogenous DNA can be introduced into a variety of tissues and cells, including microbes, animals, and plants. More importantly, this approach remains the only effective method for transforming nucleus, mitochondria, and chloroplasts. The biolistic transformation protocol is well-established, albeit the requirement for expensive equipment (gene gun) and special pretreatment. To date, this approach is the only efficient tool for genetic transformation of marine diatoms (Dunahay et al., 1995; Apt et al., 1996). At present, several selectable markers and reporter genes have been used in diatom *Phaeodactylum tricornutum* (Falciatore et al., 1999; Zaslavskaia et al., 2000; Bozarth et al., 2009). Biolistic transformation have been established for various species of diatoms, including the pinnate diatoms *Navicula saprophila* (Dunahay et al., 1995), *Cylindrotheca fusiformis* (Poulsen and Kroger, 2005), and *Phaeodactylum tricornutum* (Miyagawa et al., 2009), and the centric diatom *Cyclotella cryptic* (Dunahay et al., 1995), *Thalassiosira weissflogii* (Falciatore et al., 1999), *Thalassiosira pseudonana* (Poulsen et al., 2006), and *Chaetoceros* sp. (Miyagawa-Yamaguchi et al., 2011).

Successful of biolistic transformations have been report for red algae: plastic transformation of *Porphyridium* spp. and nuclear transformation of *C. merolae* (Minoda et al., 2004). The primary strategy for the transformation of *V. carteri* is through particle bombardment. *Haematococcus pluvialis,* an important producer of the keto-carotenoid astaxanthin, has been transiently and stably transformed by biolistic transformation (Steinbrenner and Sandmann, 2006). The halotolerant green microalga *D. salina,* which accumulates β-carotene, has been successfully transformed by biolistic transformation (Tan et al., 2005). Stable nuclear transformation of the volvocine alga *Gonium pectoral* is achieved by biolistic transformation (Lerche and Hallmann, 2009).

14.9 GENE SILENCING

Although gene silencing can pose a serious problem to heterologous gene expression, in *C. reinhardtii,* efforts have been made to use this phenomenon for gene function discovery and metabolic engineering (Molnar et al., 2009; Zhao et al., 2009). Gene silencing or RNA silencing occurs via RNA interference (RNAi). Due to direct gene knockout through HR is rare, suppression of gene expression or gene knockdown, presents a valuable reverse genetics tool for studying gene function and harbor good potential to be used for algal genetic engineering for biofuels production (Cerutti et al., 2011). *C. reinhardtii* possesses Dicer and Argonaute (AGO) proteins, which are the essential machinery in RNAi (Casas-Mollano et al., 2008). It facilitates the knockdown of specific genes. In most eukaryotes, naturally occurring microRNAs (miRNAs), which are small double-stranded RNAs (dsRNAs), function to suppress the expression of endogenous genes by mRNA cleavage. Initially, this process is thought to exist only in multicellular organisms, and that unicellular organisms do not produce miRNAs. Now, the unicellular *C. reinhardtii* has been shown to possess miRNAs that are similar to those of land plants (Molnar et al., 2009; Zhao et al., 2009). The endogenous miRNAs in land plants differ from those in animals, in that they typically have few offsite targets and cleave only a limited number of mRNAs. In addition, in *C. reinhardtii,* the miRNA precursors produce only one miRNA and are also not normally associated with transcriptional silencing (Molnar et al., 2009). Characteristics of these miRNAs enable them to be highly specific for their mRNA target, making this process amenable to the generation of an artificial miRNA system for use as a genetic tool. Recently, artificial miRNA (amiRNA)-mediated gene silencing systems have been developed for *C. reinhardtii* using amiRNAs based on endogenous miRNA precursors (Molnar et al., 2009; Zhao et al., 2009). Continuous improvements in gene silencing technology has open up a new era in genetics of *C. reinhardtii*, particularly in targeted knockdown specific genes, thereby allowing the manipulation of their biological functions.

RNAi has opened the way for reverse genetics in diatoms, such as *P. tricornutum* (De Riso et al., 2009). As new algae genomes are sequenced, RNAi should be applicable to more species. In addition, in some cases, gene transcript regulatable at the translational level through riboswitches. In *C. reinhardtii*, the thiamine pyrophosphate (TPP), riboswitch has been found functional for regulating the luciferase reporter gene, in the presence or absence of thiamine. Similar regulatory sequences have been found in *Volvox carteri*, suggesting that this mechanism may be applicable in other algae species (Croft et al., 2007).

14.10   METABOLIC ENGINEERING

14.10.1   *MUTAGENESIS*

Mutagenesis has long been used to generate genetic variants and under-standing gene function. Mutant strains can be selected based on phenotype and reproduced asexually, giving rise to a culture expressing that phenotype. Most algae can exist as haploids, and even recessive mutants can be propa-gated in this way, and thus algae are amenable to mutagenesis.

Chemical and physical mutagens are commonly used, both in basic and applied science, particularly because they can be applied at desirable doses and their mutagenic potentials are well characterized. The wild type algal cells are mutagenized to improve or obtain new useful biotechnological properties. Alkylating agents, such as methyl nitro nitroso guanidine (MNNG) and ethyl methanesulfonate (EMS) are among the most commonly used chemical muta-gens. *Nannochloropsis oculata* has been mutagenized to screen for mutants with increased EPA production (Chaturvedi and Fujita, 2006) and *Chlorella* for enhanced growth properties (Ong et al., 2010). The most widely physical mutagens include various types of irradiation, such as gamma, UV, or heavy ion beams. The mode of action and mutagenic potential of each type of radia-tion on cells depends on the energy input, while the frequency of their use is based on the ease of application. UV Mutagenesis relatively simple, because it requires neither chemicals nor specialized equipment and can be readily performed, basically by exposing cells to germicidal UV lamps in a sterile hood. Due to its simplicity and potential, this method has been widely used in basic research to generate algal mutants with specific features (Neupert et al., 2009), and in applied science to generate mutants with increased oil production (Vigeolas et al., 2012; De Jaeger et al., 2014). Gamma and particularly, heavy ion-beam irradiation need specific equipment, are thus not commonly used. However, the applicability of gamma irradiation to mutagenesis has been proven in improving the productivity of astaxanthin (Najafi et al., 2011; Hlavova et al., 2015).

When essential genes are inactivated by mutation, the mutant cannot be recovered due to its lethality, and thus essential genes are less amenable to a classical genetic approach. To circumvent this problem, point mutations which may affect only the activity or behavior of a gene product without inactivating the genes, are induced to allow the isolation of essential gene mutants. For ensuring the survival of essential gene mutants, additional precautions are performed in the conditional mutant screens. Conditional mutants show the phenotype only under specific, restrictive conditions, whereas they are similar to the wild type under other, permissive conditions. Temperature-sensitive mutants are the most widely used type of conditional mutants. Typically, temperature-sensitive mutants carry mutations in cell cycle regulators, e.g., genes essential for processes related to nuclear and/or cell division. Such mutants will grow and divide normally at a permissive (usually lower) temperature, but will stop growing and cell dividing at restrictive (higher) temperatures (Hartwell et al., 1974; Nurse et al., 1976; Thuriaux et al., 1978; Harper et al., 1995). Such mutants are found in two common algal species, *C. reinhardtii* and *C. vulgaris*, where they were tested for lipid production at a restrictive temperature (Yao et al., 2012). Interestingly, approximately 20% of the mutants in both species showed increased amounts of neutral lipids when the mutants are shifted to restrictive temperature and some of them showed changes in lipid composition. However, a temperature switching is only possible in a small scale indoor photobioreactor, but is impractical in large scale outdoor algal cultivations due to costing. Thus in the best-case scenario, such mutants and the temperature switch could only be used with temperature control. However, such tunable metabolic production suggests that it may be exploited using synthetic biology for manipulating quantitative and qualitative biofuel production (Hlavova et al., 2015).

Mutants can also be generated by random or targeted insertional mutagenesis, is based on the integration of a foreign DNA fragment(s) into the genome. This can be accomplished using DNA transformation. As in the case of genetic transformation, insertional mutagenesis, also a selective marker to distinguish transformed (mutated) lines from non-transformed ones. The insertion of a large DNA fragment may produce detrimental effects, and thus only mutations in non-essential genes are usually recovered. In addition, mutation site needs to be identified and this is usually based on the inserted DNA fragment, which as serves as a tag. The inserted DNA fragment may be naturally occurring transposons or viruses, or artificially engineered DNA vectors, such as modified viruses, transposons, T-DNA (transfer DNA), antibiotic resistance cassettes, or artificial DNA plasmids. In plant biology, T-DNA delivered by *Agrobacterium tumefaciens* is the commonly used as an insertional mutagen (Krysan et al., 1999; Jeon et al., 2000; Alonso et al., 2003). Using these DNA mutagens, mutant libraries may generate for fundamental study of gene function by identifying the phenotype of knock-out, or knock-down mutants. Insertional mutagenesis is also widely used for studying gene function in microalgae (Colombo et al., 2002; Dent et al., 2005; Fang et al., 2006; Galván et al., 2007; Umen and Goodenough, 2010; Gonzalez-Ballester et al., 2011; Jungnick et al., 2014). Recently, in *C. reinhardtii*, a tagged mutant library has been generating in a manner similar to that with *A. thaliana* T-DNA insertional mutagenesis (Zhang et al., 2014). This mutant library enabled the isolation of mutants with increased lipid production (Terashima et al., 2014) and with appropriate screening procedure/s it should serve as a basis for mutants with diverse phenotypes. Besides basic research, these newly developed tools may be useful in algal biotechnology (Hlavova et al., 2015).

14.10.2 LIPID PATHWAYS ENGINEERING

As algal biomass and TAGs compete for photosynthetic assimilation, pathway engineering is needed to elevate the metabolic flux into lipid biosynthesis (Sharma et al., 2012). PEP is the common substrate for fatty acid and protein biosynthetic pathways. When PEP carboxylase (PEPC) converts PEP to oxaloacetate (OAA), it enters into protein synthesis. When PEP is transformed to malonyl-CoA, it is directed towards fatty acid synthesis. Conversion of PEP to pyruvate by pyruvate kinase, followed by pyruvate dehydrogenase in a second reaction, forms acetyl-coenzyme A (acetyl-CoA). Pyruvate can be converted into alanine to participate in protein metabolism. Acetyl-CoA can be converted to malonyl-CoA in a rate-limiting reaction catalyzed by ACC leading towards fatty acid synthesis (Quintana et al., 2011).

In the metabolic engineering of microalgae, several strategically important steps aimed at increasing the production of TAG have been proposed for sustainable biodiesel production (Rosenberg et al., 2008; Beer et al., 2009; Radakovits et al., 2010). Some of these strategies include increasing the availability of acetyl-CoA precursors by over expression of these enzymes involved in its production via over expression of FAS and ACCase enzymes, interfering with the competing pathways that lead to TAG degradation, e.g., inhibition of lipolysis and β-oxidation, and downregulation of PEP conversion to oxalo-acetate. The *Cyclotella cryptica acc*1 gene, which codes for the ACCase, has been cloned and characterized. However, over expression of ACCase may not lead to increased oil production (Hu et al., 2008). Chain length and saturation levels of fatty acid may be modified to generate monounsaturated and saturated TAG, which is ideal for biodiesel. It may be possible through the expression of TEs and gene silencing of desaturases. Such strategies are made possible by recent success in modification of lipid biosynthesis pathways using molecular tools and discoveries on lipid metabolism of higher plants (Wallis, 2010). It is plausible that algae, which are more amenable to genetic transformation and produce TAG under conditions of nutrient deficiency or environmental stress, can be subjected to similar engineering strategies. In addition, it may also be feasible to modulate algal lipid metabolism by regulating the gene expression using specific transcription factors identified in genomics and bioinformatics (Moreno-Risueno et al., 2007; Courchesne et al., 2009).

Diatoms possess novel biological processes not present in other more well-studied organisms and are thus of major interest (Siaut et al., 2007). *Phaeodactylum tricornutum* is a potential producer of biodiesel, due to its established genetic tools, rapid growth, and lipid accumulation capability (Yang et al., 2013; Xie et al., 2014). Traits, such as high lipid production and high biomass of *P. tricornutum* may be genetically manipulated to improve for industrial production. Genetic manipulation of *P. tricornutum* for modifying PUFA and fatty acid chain length has been reported (Niu et al., 2013; Hamilton et al., 2014). Malic enzyme (ME) catalyzes the irreversible oxidative decarbooxylation of malate to pyruvate, generating pyruvate, NADH, and CO_2 (Vongsangnak et al., 2012). NADH production, which provides the requisite reducing power for cell metabolism, is vital for biosynthesis of fatty acid. ME is involved in diverse metabolic pathways, such as lipogenesis, energy metabolism and photosynthesis. ME over expression under by the constitutive promoter *gpd1* yields a 2- to 3-fold increase in ME activity and a 2.5-fold increase in lipid accumulation in the fungus *Mucor circinelloides* (Zhang et al., 2007).

14.10.3   LIGHT-HARVESTING PROTEINS

Hydrogen production by algae has been studied as an alternative for source of biofuel (Rupprecht, 2009). *C. reinhardtii*, *Chlamydomonas moewusii*, *Chlorella fusca* and *Lobochlamys culleus* and *Scenedesmus obliquus*, produce hydrogen in the presence of light under anaerobic conditions. However, this process is not efficient and difficult to be reproduced at a large scale (Meuser et al., 2009). Problems such as anaerobiosis induction, oxygen sensitivity of hydrogenase, avoiding competition for electrons from other pathways and increasing the sources of electrons, may need be tackled using genetic engineering (Beer et al., 2009). Another system for achieving anaerobiosis based on a copper-responsive nuclear transgene that requires expression of protein D2 of PSII, has been reported. In this system, a short anaerobiosis period and hydrogen production is achieved, but this productive period is much shorter than the standard method through sulfur deprivation (Surzycki et al., 2007).

Light-harvesting antennae are important in microalgae for maximizing light absorption under low-light conditions. However, under artificial culture or high light conditions, light absorption is in excess, thus, the excess energy needs to be dissipated through heat and fluorescence quenching, otherwise, it results in photodamage. Furthermore, light penetration into the culturing media is limited by the large size of the light-harvesting complexes (LHC), thereby lowering the maximum cell density in a bioreactor (Mussgnug et al., 2007). To overcome this, all 20 LHC protein isoforms in *C. reinhardtii* has been effectively silenced using a single RNAi construct. In these transformants, fluorescence quenching is less, leading to an increase in quantum yield of photosynthetic. Under high light conditions, transformed cells are less susceptible to photoinhibition (measured as the decrease in oxygen evolution) and grow at faster rate (Mussgnug et al., 2007). However, the transformed cells do not reach higher cell density, which is considered a highly desirable trait for biofuel production. To overcome the problem of poor light penetration in dense algal culture, *C. reinhardtii* mutants with truncated light-harvesting chlorophyll antenna size (*tla*) have used to improve photosynthetic productivity and solar energy conversion efficiency in mass culture and bright light (Kirst et al., 2012). In another study, different pathways for the process of carbon fixation have modeled as a mean to overcome the low oxygenase activity of Rubisco (Bar-Even et al., 2010; Whitney et al., 2011). Using an algal GMSM to study these pathways would help in understanding how these predictions may affect biomass and product synthesis in microalgae.

14.10.4 CARBOHYDRATE PATHWAYS ENGINEERING

The most common source of metabolic energy among living organisms is sugar. The glyoxylate cycle is used primarily by plants to manufacture carbohydrate from fatty acids and photosynthesis is a process in which light energy is captured to drive carbohydrate synthesis (McKee and McKee, 2015). In glycolysis, pyruvate is converted either to acetyl-CoA or to lactate. The major route of glucose degradation is the oxidative pentose phosphate pathway (OPP) cycle and is considered as the main CO_2 fixation mechanism in CB. The OPP pathway enables cells to convert glucose-6-phosphate (G6P), a derivative of glucose, to ribose-5-phosphate (the sugar used to synthesize nucleotides and nucleic acids), and other types of monosaccharides and also NADPH, an important cellular reducing agent (McKee and McKee, 2015). Sugar catabolic pathways are active mainly during the dark phase of the light-dark cycle. During the first step of Calvin cycle (dark phase of photosynthesis), CO_2 is assimilated through the carboxylation of ribulose-1,5-biphosphate (RuBP), to form 3-phosphoglycerate (3PG), which then forms G6P through gluconeogenesis (Quintana et al., 2011). The key enzymes in the oxidation of G6P through the OPP cycle are G6P dehydrogenase and 6-phosphogluconate dehydrogenase. For metabolic regulation, G6P dehydrogenase controlled at the level of gene expression can be targeted as low RuBP levels significantly reduce this enzyme activity (Kaplan et al., 2008).

14.10.5 CAROTENOIDS PATHWAYS ENGINEERING

The carotenoid biosynthesis in all organisms emerges directly from the central isoprenoid pathway via prenyl diphosphate, the precursor of all isoprenoids in plants, which is synthesized by either the MVA pathway in the cytoplasm or the 2-C-methyl-D-erythritol 4-phosphate (MEP) pathway in plastids (Miziorko, 2011). Due to the potential for producing carotenoids in a heterologous organism, manipulating carotenogenic genes expression attracted much interest (Wang et al., 2012). Early attempts lead to the production of astaxanthin, β-carotene, and lycopene, in *Saccharomyces cerevisiae* and *Candida utilis,* by expressing *Pantoea ananatis* carotenogenic enzymes. *Corynebacterium glutamicum,* a native producer of decaprenoxanthin and its glucosides, has been engineered to produce sarcinaxanthin and C50 carotenoids C.p.450 (Heider et al., 2014). Carotenogenic enzymes derived from different sources show different capacities in carotenoid biosynthesis.

Expression of carotenogenic enzymes from *P. agglomerans* and *P. ananatis* in *E. coli* produces 27 mg/L and 12 mg/L of lycopene, respectively; a two-fold difference (Yoon et al., 2007). New carotenoids can be produced via metabolic engineering to allow genes from different organisms to be assembly for novel carotenoids production (Misawa, 2011). Supplying a sufficient precursor is a prerequisite for higher yield. For instance, over expression of the rate-limiting enzyme 3-hydroxy-3-methyl-glutaryl-coenzyme A (HNG-CoA) reductase from *Xanthophyllomyces dendrorhous*, which is found in the MVA pathway, has significantly increased the production of β-carotene in *S. cerevisiae* (Verwaal et al., 2007).

Introducing a hybrid MVA pathway from *Streptococcus pneumonia* and *Enterococcus faecalis* into *E. coli* has resulted in 465 mg/L of β-carotene produced (Yoon et al., 2009). These demonstrate that with the availability of more genetic tools, microbial organisms such as *Bacillus subtilis* and *Pseudomonas putida*, can be developed into effective platform hosts for carotenoid production (Beuttler et al., 2011). In *C. reinhardtii*, production of ketocarotenoids, such as astaxanthin, has been reported, by over-expressing β-carotene ketolases from *C. reinhardtii* itself (*CRBKT*), *H. pluvialis* (*bkt1* and *bkt3*) (Wong, 2006; Leon et al., 2007). In *H. pluvialis*, expression of a herbicide-resistant version of its PDS gene has led to 26% increase in astaxanthin accumulation (Steinbrenner and Sandmann, 2006). Isoprenoids, such as carotenoids, can also be considered as potential fuel molecules, where highly valuable carotenoids may be recovered as co-products in storage lipids, thus contributing to the production costs for microalgae biofuels. Expression of a thermostable archeal geranylgeranyl-pyrophosphate (GGPP) synthase gene from the isoprenoid biosynthesis pathway in the chloroplast of *C. reinhardtii* has been reported. However, no effect on isoprenoid biosynthesis is observed (Fukusaki et al., 2003).

Owing to the involvement of multiple enzymes in carotenoids biosynthesis, expression level of all the components of a multigene circuit should be coordinated to optimize metabolic flux to increase yield (Ye and Bhatia, 2012). Several advanced assembly methods based on HR, such as Gibson DNA assembly, sequence and ligation-independent cloning (SLIC), and reiterative recombination, have been used in constructing multigene circuits (Wingler and Cornish, 2011). Such tools allow a randomized combination of all genetic components, such as ribosome binding sites, promoters, coding region, and other control modules to assemble a large number of individual genetic circuits for screening purposes. A "randomized BioBrick assembly" approach has been used to optimize the lycopene biosynthesis pathway, resulting in a 30% increase in lycopene production (Sleight and Sauro, 2013). For engineering a longer and more complicated pathway, the component gene cassettes can be modularized into subsets, thereby making the assemblage becoming more efficient to improve production (Yadav et al., 2012). Using this multivariate modular metabolic engineering (MMME) approach, the production of taxadiene, a precursor of the anti-cancer drug taxol, attains an astounding 15,000-fold increase (Ajikumar et al., 2010). A variety of promising approaches, such as ribosome binding site design, tunable promoters, and tunable intergenic regions, can be used to fine-tune the expression of modules (Temme et al., 2012). In the other approach, a multigenic operon is transcribed into a single large polycistronic mRNA. This large transcript is spliced into small monocistronic transcripts via post-transcriptional RNA processing, such as ribozyme cleavage and clustered regularly interspaced short palindromic repeats (CRISPR) editing. As such, the stability of the monocistronic transcripts can be independently modulated to distinguish at the expression level for each enzyme, even when it is derived from a multigene operon (Qi et al., 2012). With the aid of these tools, metabolic engineering of carotenoids in algae is expected to become increasingly important in research field, as it is in higher plants.

14.10.6   THE OMICS

Today, DNA sequencing has become a routine procedure to determine the sequence of a small DNA fragment in many research laboratories. Laboratories without the necessary equipment would outsource the service from a commercial laboratory at an affordable cost. The availability of rapid large-scale sequencing technology has facilitated the whole genome sequencing of microalgae, thereby opening wider interest for academic or industrial applications of algal genetic engineering. Significant advances in microalgal genomics have been achieved during the last decade. Nuclear, mitochondrial, and chloroplast genomes, and expressed sequence tag (EST) from several microalgae have been sequenced and established. Many tools for the transgenes expression and gene knockdown have been established for the model organism *Chlamydomonas reinhardtii*. However, tools for other algal species are scarce, including diatoms, thus attracting great interest for industrial development (Radakovits et al., 2010). Several nuclear genome sequencing projects have been completed including *C. merolae* (Matsuzaki et al., 2040), *Chlamydomans reinhardtii* (Shrager et al., 2003), *Ostreococcus lucimarinus* (Palenik et al., 2007), *Ostreococcus tauri* (Derelle et al., 2006), *Phaeodactylum tricornutum* (Bowler et al., 2008), *Micromonas pusilla* (Worden et al., 2009), and *Thalassiosira pseudonana* (Armbrust et al., 2008).

DNA sequencing of plastid and mitochondrial genomes, and transcriptomes from many different microalgae also have been reported (Crepineau et al., 2000; Archibald et al., 2003; Bachvaroff et al., 2004; Henry et al., 2004; Hackett et al., 2005; Maheswari et al., 2005; O'Brien et al., 2007). Transcriptomic analysis of microalagae is expected to provide clues for targeted manipulation of metabolic pathways. Many metabolic pathways occur in the chloroplast, but knowledge in chloroplast gene regulation is lacking. Therefore, plastid transcriptome analysis may reveal promoter/UTRs that bear potential for achieving higher expression levels of transgenes in the plastid (Rasala et al., 2012) and provide pointers to designing novel synthetic UTRs that evade negative regulatory mechanisms (Specht et al., 2012). Taking advantage of the chloroplast's minimal genome, an *ex vivo* genome assembly has been used to transfer genes for core photosystem subunits from *Scenedesmus* into multiple loci in the *Chlamydomonas* plastid genome (O'Neill et al., 2012). This demonstrates a synthetic biology approach for engineering complex photosynthetic traits from diverse algae into more tractable production strains.

The growing genome information has enabled the analysis of fatty acid and glycerolipid metabolism pathways *in silico* and the prediction of genes encoding proteins that are involved in storage lipid and membrane biogenesis in some microalgae (Riekhof et al., 2005; Sato et al., 2007; Moellering et al., 2009a, b). Major fatty acid synthesis pathways have been reconstructed for *C. reinhardtii* (Riekhof et al., 2005; Moellering et al., 2009a, b). Based on their protein sequence similarities with proteins with experimentally verified functions in higher plants, all components of the plastidial multisubunit acetyl coenzyme A (acetyl-CoA) carboxylase (ACCase) and FAS complexes, have been identified. Orthologs of higher plant key enzymes for the plastidial FAS complex have been identified in the genomic and EST databases of other algae (Weber et al., 2004; Sato et al., 2007). Analysis of EST sequences of the thermo-acidophilic red microalga *Galdieria sulphuraria* (Weber et al., 2004) and of red microalga *C. merolae* genome (Sato et al., 2007) has revealed that their fatty acid biosynthesis is likely to be similar to that of the green lineage, which includes higher plants and green algae.

GFP-fusion experiments confirm the plastid localization of the nuclear-encoded subunit of a multisubunit ACCase. Based on sequence similarity to those of higher plants, genes coding for enzymes involved in acetyl-CoA production in the plastid, such as pyruvate dehydrogenase complex (PDC) and pyruvate kinase, have been identified in the *Chlamydomonas* genome. Web-based protein-targeting algorithms, such as TargetP (http://www. cbs.dtu.dk/services/TargetP), is commonly used to predict the cellular localization of the putative plastid proteins. However, these predictions could produce uncertain probabilities of the protein being targeted either to mitochondria or chloroplasts. This may be due to the dual-targeting nature some proteins in both organelles (Moellering et al., 2009a). Understanding the protein-protein interactions occurring between both autologous and heterologous FAS enzymes is critical for metabolic engineering. Using *C. reinhardtii* as a model, interaction between TE and its ACP has been shown to control fatty acid hydrolysis in the algal chloroplast (Blatti et al., 2012). *In silico* structural docking simulation and *in vitro* protein activity-based, crosslinking experiments for CrTE and CrACP of *C. reinhardtii* provide insights into the importance of protein-protein interactions in manipulating algal fatty acid biosynthesis (Worthington et al., 2010; Blatti et al., 2012). The use of activity-based chemical probes also presents a novel approach towards understanding functional protein-protein interactions in the algal chloroplast.

Besides protein-protein interactions, apparently, the tight regulation of these metabolic systems makes fatty-acid biosynthesis pathways difficult to be modified. The minimal increases of lipid production in microalgae resulting from ACCase over expression, as compared to that in plants, suggests the presence of regulatory control. In *Brassica napus*, feedback inhibition of the ACCase has been reported (Andre et al., 2012). Other studies that focus on differential gene and protein expressions, including that of transcription factors, lipid production is increased when microalgae are subjected to different environmental stressors and stimuli (Georgianna and Mayfield, 2012). In eukaryotic algae, the process of protein secretion is well studied and by combining these protein secretion signals with the synthetic biology approach, studies have shown that such combination can increase the recombinant protein abundance for nuclear gene expression with extracellular localization (Rasala et al., 2012; Lauersen et al., 2013). Promising strategies for the construction of more efficient algal strains in the future will be based on the availability of advanced synthetic biological tools for metabolic engineering via transformation of heterologous (prokaryotic and eukaryotic) genes in established eukaryotic hosts such as *C. reinhardtii* (Niu et al., 2013).

14.11   BIOINFORMATICS

The rapidly increasing arrays of new online bioinformatic tools are becoming essential to researchers who wish to characterize, forecast outcomes of metabolic shifts, gene regulatory pathways and functionally annotate genomes from diverse algal species. Metabolic networks that can be presented in detailed *in silico* models, have been useful for identifying rate-limiting steps in starch metabolism (Nag et al., 2011) and for predicting the light-induced metabolic response to various wavelengths in a metabolic reconstruction that includes over 1000 genes across ten cellular compartments (Chang et al., 2011). The BioModels Database hosts these algal metabolic models, which can be readily accessed (www.ebi.ac.uk/biomodels-main). Predictions of biosynthetic pathways solely based on genome sequence, can now be performed using new approaches for assembling and annotating *de novo* algal genomes (Rismani-Yazdi et al., 2011) based KEGG assignments (www.genome.jp/kegg). This greatly increases the utility of diverse algal genomic datasets for identifying lipid synthesis or metabolic pathways of interest for biofuel production (Georgianna and Mayfield, 2012). NCBI provides access nearly 189 plant genomes either in full, as scaffolds, or with ongoing sequencing (www.ncbi.nlm.nih.gov/genomes), while Phytozome hosts 65 sequenced and annotated green plant genomes (www.phytozome.net).

DNA sequences such as 18S ribosomal RNA gene, internal transcribed spacer (ITS) region of ribosomal RNA transcription unit and *rbcL* gene have been widely used to identify species (Li et al., 2011). A species in genus *Nannochloropsis* has been identified based on 18S ribosomal RNA gene, ITS region of ribosomal RNA transcription unit and *rbcL* gene (Shah, 2014). Table 14.1 and Figure 14.1 show 18S ribosomal RNA gene, *rbcL* gene and ITS region of ribosomal RNA. The phylogenetic analysis carried out with the neighbor-joining tree of other species in genus *Nannochloropsis*. The consensus tree was calculated by 1000 permutation with bootstrap values of >50% (Figures 14.2–14.4). The bootstrap consensus maximum parsimony tree of algae in class Eustigmatophyceae based on 18S rRNA sequences shows identical phylogenetic relationship brought out by the Neighbor-Joining tree, which proves the accuracy of the phylogenetic analysis. Figure 14.2 suggests that *Nannochloropsis* sp. is grouped with a branch containing *N. oceanica*, *N. granulata* and *N. oculata* with the outgroups *Coccoid pelagophyte* and *Chrysosaccus* sp. Figure 14.3 suggests that *Nannochloropsis* sp. is grouped with a branch containing *N. garditana*, *N. salina*, *N. oceanica*, *N. maritima* and *N.*

oculata with the outgroups *Vischeria helvetica, Eustigmatos magna, N. oceanica* (LAMB0001), and *Nannochloropsis* (EU165325). Figure 14.4 suggests that *Nannochloropsis* sp. is grouped with a branch containing *Nannochloropsis* (JX 913539) and *N. oceanica* with the outgroups *Rhizosolenia setigera* and *N. oceanica* (LAMB0001). Based on a BLAST search of GenBank and complete sequences, the 18S rRNA of *Nannochloropsis* sp. is 99% identical to that of *N. oculata* (AF045044) and *N. oceanica* (LAMB0001). The partial sequence of *rbcL* gene of *Nannochloropsis* sp. is 97% similar to that of *N. oculata* (AB052286) and the ITS region of *Nannochloropsis* sp. gives the identification until genus level. The phylogenetic tree based on the ITS region of rRNA shows that the species belong to *Nannochloropsis* while the Neighbor-Joining tree based on partial sequence of 18S rRNA and *rbcL* gene confirms that the species is *Nannochloropsis oculata. Nannochloropsis* is a genus of marine eukaryotic unicellular algae, generally regarded as picoeukaryotic plankton which belongs to class Eustigmatophyceae. Traditional morphological observation is difficult as the sizes are usually 2–5 μm in size and morphologically similar. Hence, bioinformatics has greatly speed up the characterization and identification of unknown species.

TABLE 14.1   18S Ribosomal RNA Gene, *rbcL* Gene and ITS Region of Ribosomal RNA

Oligo Name	Primer Sequence (5'– 3')	GenBank Accession Number	Position
F298-18S rRNA	CAAGTTTCTGCCCTATCAGCT	AF045045	298–318
R948-18S rRNA	GCTTTCGCAGTAGTTCGTCTT	AF045045	928–948
F1-*rbcL*	GATGCAAACTACACAATTAAAGATACTG	AB280614	1–28
R1486-*rbcL*	ATTTTGTTCGTTTGTTAAATCCG	AB280614	1464–1486
F3-ITS	GTCGCACCTACCGATTGA	DQ069777	3–20
R1049-ITS	CGGGTAGCCTTGCTTGAT	DQ069777	1032–1049

The 18S rRNA has been used in the phylogenetic analysis of Eustigmatophyceae (Andersen et al., 1998). The *rbcL* gene which encodes the large unit of ribulose-1,5-bisphate carboxylase/oxygenase (RUBISCO) has been cloned from a large number of plant species (Chase et al., 1993). The ITS of ribosomal RNA transcription unit is a desirable nuclear marker due to its high evolution rate and moderate length (Adachi et al., 1996).

For higher plants, chloroplast DNA (cpDNA) variation is commonly used for molecular phylogenic analysis (Despres et al., 2003). However, the low evolutionary rate of cpDNA limits its use in identifying an organism to genus or species. In our study *rbcL* gene is able to distinguish micro-algal species. However, the ITS region is more varied than 18S rRNA (Mai and Coleman, 1997). The combination of 18S rRNA and ITS region could identify an organism to the taxa below species, which has been applied in identification of two populations of *Ditylum brightwellii* (Rynearson et al., 2006).

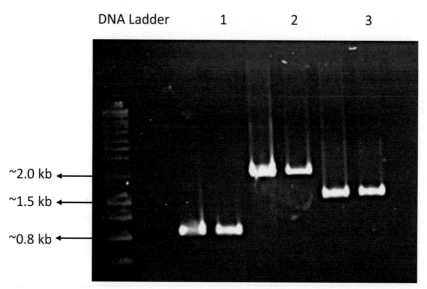

FIGURE 14.1   Genomic DNA with Lane 1: 18S ribosomal RNA gene (~0.8 kb), Lane 2: *rbcL* gene (~2.0 kb) and Lane 3: ITS region of ribosomal RNA (~1.5 kb).

The molecular markers need to be developed so that the strains or the varieties of a species could be distinguished further. The DNA finger-printing technique based on molecular markers such as microsatellite DNA and single nucleotide polymorphism (SNP) are among the alternatives. Using microsatellite DNA, molecular identification can tell the differences among the varieties or strains within a species. The toxic dinoflagellate *Alexandrium minutum* has been investigated with microsatellite DNA where Global and Pacific clades are distinguished clearly. SNP has been applied in deepening the identification of species in habitat types as well as in eelgrass (Oetjen et al., 2010). The drawback is the development of both microsatellite DNA and SNP markers is time-consuming and labor-intensive. Microsatellite DNA can be amplified only in a specific species or its closely related species (cross species amplification). SNP markers are powerful only when they are associated with high throughput-genotyping techniques (e.g., DNA chips) (Li et al., 2011).

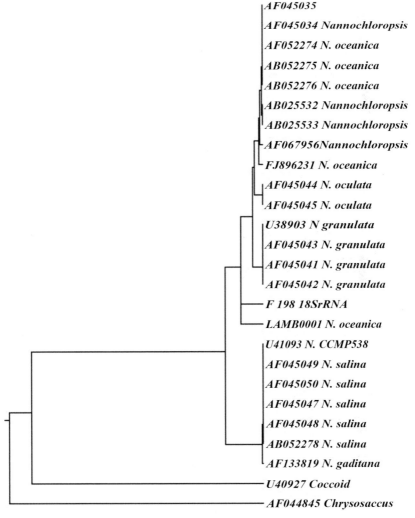

FIGURE 14.2 Neighbor-Joining tree of Eustigmatophyceae and other chromophyte algae based on 18S rRNA sequence analysis.

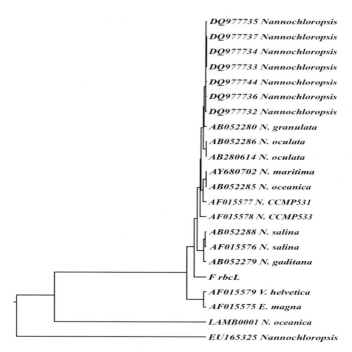

FIGURE 14.3 Neighbor-joining tree of algae in class Eustigmatophyceae based on *rbcL* gene sequence analysis.

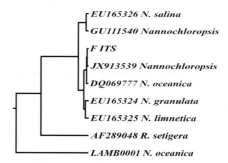

FIGURE 14.4 Neighbor-joining tree of algae in class Eustigmatophyceae based on ITS region of ribosomal RNA sequence analysis.

14.12 CONCLUSION AND FUTURE OUTLOOK

Algae naturally produce storage lipids, carbohydrates, and value-added biocompounds and tend to increase accumulation when placed under light, nutrient or temperature stress. The ability of eukaryotic microalgae to accumulate and store secondary metabolites in cellular compartments and to produce and secrete functionally active protein products make them ideal hosts and powerful alternatives to CB in industrial biotechnology. Understanding the molecular events that result in lipid, carbohydrate, and value-added compounds accumulation under stress should pave the way for engineering algae that constitutively accumulate large amounts of desired products. Likewise, engineering the metabolic enzymes should alter the product composition within algae. However, it is not as simple as inserting a single biosynthetic enzyme, as there is a global metabolic network control of carbon flux that has to be unmasked. Although many advances have been made in studying lipid metabolism in algae, a more detailed understanding of lipid (especially TAG) formation, as well as regulation of the carbon flux in general, is necessary to optimize TAG biosynthesis in algae. Identification of all genes encoding the key enzymes (ACCases, DGAT) controlling TAG synthesis is of great importance. Understanding the fundamentals underlying the carbohydrate metabolism of microalgae is a prerequisite to develop effective strategies to increase the carbohydrates productivity apart from manipulation of the key operating factors.

The efforts to genetically engineer eukaryotic microalgae to realize their full potential as biofactories hinge upon the types of tools that may be developed, taking cues from other genetic and model systems. These include fluorescent or luminescent reporter systems, tags that promote easy protein purification, improved selection markers, improved expression and inducible systems, and improved homologous gene replacement, and vectors that work in a broad range of algae species. Mutagenesis and enzyme modulators may be useful to understand the primary regulatory mechanisms for lipid pathways in different algae species. However, engineering strategy in suitable eukaryotic hosts for efficient hydrocarbon secretion mechanisms, has not been well studied, and thus remains as one of the most challenging future tasks for algal biotechnology research. More effective saccharification processes should also be developed to enhance the efficiency of biofuel conversion through the microalgae biomass. Economic assessment and life cycle analyses of the microalgae-based biofuels producing system based on engineered strains should also be conducted to assess the commercial feasibility for conversion into various bioproducts.

KEYWORDS

- **bioactive compounds**
- **biodiesel**
- **bioenergy**
- **carbohydrates**
- **carotenoids**
- **fatty acids**
- **genetic engineering**
- **glycolysis**
- **lipids**
- **metabolic engineering**
- **metabolic pathways**
- **microalgae**

REFERENCES

Abalde, J., Fabregas, J., & Herrero, C., (1991). β-Carotene, vitamin C and vitamin E content of the marine microalga *Dunaliella tertiolecta* cultured with different nitrogen sources. *Bioresour. Technol.*, *38*, 121–125.

Abd El-Baky, H. H., El Baz, F. K., & El-Baroty, G. S., (2004). Production of lipids rich in omega-3 fatty acids from the halotolerant alga *Dunaliella salina. Biotechnology*, *3*, 102–108.

Abd El-Baky, H., El Baz, F. K., & El-Baroty, G., (2009). Enhancement of antioxidant production in *Spirulina platensis* under oxidative stress. *Acta Physiol. Plant*, *31*, 623–631.

Abdullah, M. A., Ahmad, A., Shah, S. M. U., Shanab, S. M. M., Ali, H. E. A., Abo-State, M. A. M., & Othman, M. F., (2016). Integrated algal engineering for bioenergy generation, effluent remediation and production of high-value bioactive compounds. *Biotechnol. Bioproc. Eng., 21*, 236–249.

Abdullah, M. A., Shah, S. M. U., Ahmad, A., & El-Sayed, H., (2015). Algal biotechnology for bioenergy, environmental remediation, and high value biochemicals. In: Thangadurai, D., & Sangeetha, J., (eds.), *Biotechnology and Bioinformatics: Advances and Applications for Bioenergy, Bioremediation, and Biopharmaceutical Research* (pp. 301–344). Apple Academic Press, New Jersey, USA.

Abdullah, M. A., Shah, S. M. U., Shanab, S. M. M., & Ali, H. E. A., (2017). Integrated algal bioprocess engineering for enhanced productivity of lipid, carbohydrate, and high-value bioactive compounds. *Research and Review: Journal of Microbiology and Biotechnology*, *6*(2), 61–92.

Abe, K., Hattor, H., & Hiran, M., (2005). Accumulation and antioxidant activity of secondary carotenoids in the aerial microalga *Coelastrella striolata* var. *multistriata*. *Food Chem.*, *100*, 656–661.

Adachi, M., Sako, Y., & Ishida, Y., (1996). Analysis of *Alexandrium* (Dinophyceae) species using sequences of the 5.8S ribosomal DNA and internal transcribed spacer regions. *J. Phycol.*, *32*(3), 424–432.

Aflalo, C., Meshulam, Y., Zarka, A., & Boussiba, S., (2007). On the relative efficiency of two vs. one-stage production of astaxanthin by the green alga *Haematococcus pluvialis*. *Biotechnol. Bioeng.*, *98*, 300–305.

Afolayan, A. F., Mann, M. G., Lategan, C. A., Smith, P. J., Bolton, J. J., et al., (2009). Antiplasmodial halogenated monoterpenes from the marine red alga *Plocamium cornutum*. *Phytochemistry*, *70*, 597–600.

Afolayana, A. F., Boltonb, J. J., Lateganc, C. A., Smithc, P. J., & Beukesa, D. R., (2008). Fucoxanthin, tetraprenylated toluquinone and toluhydroquinone metabolites from *Sargassum heterophyllum* inhibit the *in vitro* growth of the malaria parasite *Plasmodium falciparum*. *Z. Naturforsch. C, 63*, 848–852.

Ahmad, A., Shah, S. M. U., Othman, M. F., & Abdullah, M. A., (2014). Enhanced palm oil mill effluent treatment and biomethane production by co-digestion of oil palm empty fruit bunches with *Chlorella* sp. *Can. J. Chem. Eng., 92,* 1636–1642.

Ajikumar, P. K., Xiao, W. H., Tyo, K. E., Wang, Y., Simeon, F., Leonard, E., Mucha, O., Phon, T. H., Pfeifer, B., & Stephanopoulos, G., (2010). Isoprenoid pathway optimization for taxol precursor overproduction in *Escherichia coli*. *Science*, *330*, 70–74.

Al-Fadhli, A., Wahidulla, S., & D'Souza, L., (2006). Glycolipids from the red alga *Chondria armata* (Kütz.) Okamura. *Glycobiology*, *16*, 902–915.

Alfafara, C. G., Nakano, K., Nomura, N., Igarashi, T., & Matsumura, M., (2002). Operating and scale-up factors for the electrolytic removal of algae from eutrophied lake water. *J. Chem. Technol. Biotechnol.*, *77*, 871–876.

Alonso, J. M., Stepanova, A. N., Leisse, T. J., Kim, C. J., Chen, H., Shinn, P., et al., (2003). Genome-wide insertional mutagenesis of *Arabidopsis thaliana*. *Science*, *301*, 653–657.

Amin, S., (2009). Review on biofuel oil and gas production processes from microalgae. *Energ. Convers. Manag.*, *50*, 1834–1840.

Andersen, R. A., Brett, R. W., Potter, D., & Sexton, J. P., (1998). Phylogeny of the Eustigmatophyceae based upon 18S rDNA, with emphasis on *Nannochloropsis*. *Protist, 149*(1), 61–74.

Andre, C., Haslam, R. P., & Shanklin, J., (2012). Feedback regulation of plastidic acetyl-CoA carboxylase by 18:1-acyl carrier protein in *Brassica napus*. *Proc. Natl. Acad. Sci. USA*, *109*, 10107–10112.

Anila, N., Chandrashekar, A., Ravishankar, G. A., & Sarada, R., (2011). Establishment of *Agrobacterium tumefaciens*-mediated genetic transformation in *Dunaliella bardawil*. *Eur. J. Phycol.*, *46*, 36–44.

Apt, K. E., Kroth-Pancic, P. G., & Grossman, A. R., (1996). Stable nuclear transformation of the diatom *Phaeodactylum tricornutum*. *Mol. Gen. Genet.*, *252*, 572–579.

Arab-Tehrani, K., Colteu, A., Rasoanarivo, I., & Michel-Sargos, F., (2010). Design a new high intensity magnetic separator with permanent magnets for industrial applications. *Int. J. Appl. Electromagnet.*, *32*, 237–248.

404 Phycobiotechnology

Araujo, S. D., & Garcia, V. M. T., (2005). Growth and biochemical composition of the diatom *Chaetoceros cf. wighamii* bright well under different temperature, salinity, and carbon dioxide levels I. Protein, carbohydrates, and lipids. *Aquaculture, 246*, 405–412.
Archibald, J. M., Rogers, M. B., Toop, M., Ishida, K., & Keeling, P. J., (2003). Lateral gene transfer and the evolution of plastid-targeted proteins in the secondary plastid-containing alga *Bigelowiella natans*. *Proc. Natl. Acad. Sci. USA, 100*, 7678–7683.
Armbrust, E. V., Berges, J. A., Bowler, C., Green, B. R., Martinez, D., et al., (2004). The genome of the diatom *Thalassiosira pseudonana*: Ecology, evolution, and metabolism. *Science, 306*, 79–86.
Armstrong, G. A., (1997). Genetics of eubacterial carotenoid biosynthesis: A colorful tale. *Annu. Rev. Microbiol., 51*, 629–659.
Azachi, M., Sadka, A., Fisher, M., Goldshlag, P., Gokhman, I., & Zamir, A., (2002). Salt induction of fatty acid elongase and membrane lipid modifications in the extreme halotolerant alga *Dunaliella salina*. *Plant Physiol., 129*, 1320–1329.
Azamai, E. S. M., Sulaiman, S., Habib, S. H. M., Looi, M. L., Das, S., Hamid, N. A. A., Ngah, W. Z. W., & Yusof, Y. A. M., (2009). *Chlorella vulgaris* triggers apoptosis in hepatocarcinogenesis-induced rats. *J. Zhejiang Univ. Sci. B, 10*, 14–21.
Bachvaroff, T. R., Concepcion, G. T., Rogers, C. R., Herman, E. M., & Delwiche, C. F., (2004). Dinoflagellate expressed sequence tag data indicate massive transfer of chloroplast genes to the nuclear genome. *Protist, 155*, 65–78.
Bar-Even, A., Noora, E., Lewis, N. E., & Milo, R., (2010). Design and analysis of synthetic carbon fixation pathways. *Proc. Natl. Acad. Sci. USA, 107*, 8889–8894.
Barros, M. P., Pinto, E., Colepicolo, P., & Pedersén, M., (2001). Astaxanthin and peridinin inhibit oxidative damage in Fe^{2+} loaded liposomes: Scavenging oxyradicals or changing membrane permeability? *Biochem. Biophys. Res. Commun., 288*, 225–232.
Baulcombe, D., (2004). RNA silencing in plants. *Nature, 431*, 356–363.
Beer, L. L., Boyd, E. S., Peters, J. W., & Posewitz, M. C., (2009). Engineering algae for biohydrogen and biofuel production. *Curr. Opin. Biotechnol., 20*, 264–271.
Bertrand, M., (2010). Carotenoid biosynthesis in diatoms. *Photosynth. Res., 106*, 89–102.
Beuttler, H., Hoffmann, J., Jeske, M., Hauer, B., Schmid, R. D., Altenbuchner, J., & Urlacher, V. B., (2011). Biosynthesis of zeaxanthin in recombinant *Pseudomonas putida*. *Appl. Microbiol. Biotechnol., 89*, 1137–1147.
Bhadury, P., & Wright, P. C., (2004). Exploitation of marine algae: Biogenic compounds for potential antifouling applications. *Planta, 219*, 561–578.
Bigogno, C., Khozin-Goldberg, I., Boussiba, S., Vonshak, A., & Cohen, Z., (2002). Lipid and fatty acid composition of the green oleaginous alga *Parietochloris incisa*, the richest plant source of arachidonic acid. *Phytochemistry, 60*, 497–503.
Blatti, J. L., Beld, J., Behnke, C., Mendez, M., Mayfield, S., & Burkart, M. D., (2012). Manipulating fatty acid biosynthesis in microalgae for biofuel through protein-protein interactions. *PLoS One, 7*, e42949. doi: 10.1371/journal.pone.0042949.
Blatti, J. L., Michaud, J., & Burkart, M. D., (2013). Engineering fatty acid biosynthesis in microalgae for sustainable biodiesel. *Curr. Opin. Chem. Biol., 17*, 496–505.
Bligh, E. G., & Dyer, W. J., (1959). A rapid method for total lipid extraction and purification. *Can. J. Biochem. Physiol., 67*, 911–917.
Blunt, J. W., Copp, B. R., Munro, M. H. G., Northcote, P. T., & Prinsep, M. R., (2005). Marine natural products. *Nat. Prod. Rep., 22*, 15–61.

Bowler, C., Allen, A. E., Badger, J. H., Grimwood, J., Jabbari, K., et al., (2008). The *Phaeodactylum* genome reveals the evolutionary history of diatom genomes. *Nature, 456,* 239–244.

Bozarth, A., Maier, U. G., & Zauner, S., (2009). Diatoms in biotechnology: Modern tools and applications. *Appl. Microbiol. Biotechnol., 82,* 195–201.

Breitenbach, J., Fernández-González, B., Vioque, A., & Sandmann, G., (1998). A higher-plant type δ-carotene desaturase in the cyanobacterium *Synechocystis* PCC6803. *Plant Mol. Biol., 36,* 725–732.

Breitenbach, J., Vioque, A., & Sandmann, G., (2001). Gene sll0033 from *Synechocystis* 6803 encodes a carotene isomerase involved in the biosynthesis of all-Elycopene. *Z. Naturforsch., 56,* 915–917.

Brennan, L., & Owende, P., (2010). Biofuels from microalgae: a review of technologies for production, processing, and extractions of biofuels and co-products. *Renew. Sust. Energ. Rev., 14,* 557–577.

Britton, G., (1998). Overview of carotenoid biosynthesis. In: Britton, G., Liaaen-Jensen, S., & Pfander, H., (eds.), *Carotenoids: Biosynthesis and Metabolism* (Vol. 3, pp. 13–147). Birkhäuser, Basel, Switzerland.

Brown, G. N., Müller, C., Theodosiou, E., Franzreb, M., & Thomas, O. R., (2013). Multi-cycle recovery of lactoferrin and lactoperoxidase from crude whey using fimbriated high-capacity magnetic cation exchangers and a novel 'rotor-stator' high gradient magnetic separator. *Biotechnol. Bioeng., 110,* 1714–1725.

Brown, L. E., Sprecher, S. L., & Keller, L. R., (1991). Introduction of exogenous DNA into *Chlamydomonas reinhardtii* by electroporation. *Mol. Cell Biol., 11,* 2328–2332.

Brown, M. R., Jeffrey, S. W., Volkman, J. K., & Dunstan, G. A., (1997). Nutritional properties of microalgae for mariculture. *Aquaculture, 151,* 315–331.

Cai, M., Li, Z., & Qi, A., (2009). Effects of iron electrovalence and species on growth and astaxanthin production of *Haematococcus pluvialis. Chin. J. Oceanol. Limnol., 27,* 370–375.

Cara, C., Moya, M., Ballesteros, I., Negro, M. J., Gonzalez, A., & Ruiz, E., (2007). Influence of solid loading on enzymatic hydrolysis of steam exploded or liquid hot water pretreated olive tree biomass. *Proc. Biochem., 42,* 1003–1009.

Carvalho, A. P., Monteiro, C. M., & Malcata, F. X., (2009). Simultaneous effect of irradiance and temperature on biochemical composition of the microalga *Pavlova lutheri. J. Appl. Phycol., 21,* 543–552.

Casas-Mollano, J. A., Rohr, J., Kim, E. J., Balassa, E., Van, D. K., & Cerutti, H., (2008). Diversification of the core RNA interference machinery in *Chlamydomonas reinhardtii* and the role of DCL1 in transposon silencing. *Genetics, 179,* 69–81.

Catchpole, O., Ryan, J., Zhu, Y., Fenton, K., Grey, J., Vyssotski, M., MacKenzie, A., et al., (2010). Extraction of lipids from fermentation biomass using near-critical dimethylether. *J. Supercritical Fluids, 53,* 34–41.

Catchpole, O., Tallon, S., Dyer, P., Montanes, F., Moreno, T., Vagi, E., Eltringham, W., & Billakanti, J., (2012). Integrated supercritical fluid extraction and bioprocessing. *Am. J. Biochem. Biotech., 8,* 263–287.

Cerutti, H., Ma, X., Msanne, J., & Repas, T., (2011). RNA-mediated silencing in algae: Biological roles and tools for analysis of gene function. *Eukaryot. Cell, 10,* 1164–1172.

Cha, K. H., Koo, S. Y., & Lee, D. U., (2008). Antiproliferative effects of carotenoids extracts from *Chlorella ellipsoidea* and *Chlorella vulgaris* on human colon cancer cells. *J. Agric. Food. Chem., 56,* 10521–10526.

Chamovitz, D., Misawa, N., Sandmann, G., & Hirschberg, J., (1992). Molecular cloning and expression in *Escherichia coli* of a cyanobacterial gene coding for phytoene synthase, a carotenoid biosynthesis enzyme. *FEBS Lett.*, *296*, 305–310.

Champoux, J. J., (2001). DNA topoisomerases: Structure, function, and mechanism. *Annu. Rev. Biochem.*, *70*, 369–413.

Chan, M. C., Ho, S. H., Lee, D. J., Chen, C. Y., Huang, C. C., & Chang, J. S., (2013). Characterization, extraction and purification of lutein produced by an indigenous microalga *Scenedesmus obliquus* CNW-N. *Biochem. Eng. J.*, *78*, 24–31.

Chang, R. L., Ghamsari, L., Manichaikul, A., Hom, E. F., Balaji, S., Fu, W., Shen, Y., Hao, T., Palsson, B. O., Salehi-Ashtiani, K., et al., (2011). Metabolic network reconstruction of *Chlamydomonas* offers insight into light-driven algal metabolism. *Mol. Syst. Biol.*, *7*, 518.

Chase, M. W., Soltis, D. E., Olmstead, R. G., Morgan, D., Les, D. H., & Mishler, B. D., (1993). Phylogenetics of seed plants: An analysis of nucleotide sequences from the plastid gene rbcL. *Annals of the Missouri Botanical Garden*, *80*(3), 528–580.

Chaturvedi, R., & Fujita, Y., (2006). Isolation of enhanced eicosapentaenoic acid-producing mutants of *Nannochloropsis oculata* ST-6 using ethyl methanesulfonate induced mutagenesis techniques and their characterization at mRNA transcript level. *Phycol. Res.*, *54*, 208–219.

Chen, C. Y., Zhao, X. Q., Yen, H. W., Ho, S. H., Cheng, C. L., Lee, D. J., & Chang, J. S., (2013). Microalgae-based carbohydrates for biofuel production. *Biochem. Eng. J.*, *78*, 1–10.

Chen, L., Liu, T., Zhang, W., Chen, X., & Wang, J., (2012). Biodiesel production from algae oil high in free fatty acids by two-step catalytic conversion. *Bioresour. Technol.*, *111*, 208–214.

Chen, W., Zhang, C., Song, L., Sommerfeld, M., & Hu, Q., (2009). A high throughput Nile red method for quantitative measurement of neutral lipids in microalgae. *J. Microb. Meth.*, *77*, 41–47.

Cheney, D., & Kurtzman, A., (1992). Progress in protoplast fusion and gene transfer in red algae. *Proceedings of the XIV International Seaweed Symposium* (p. 68). Brittany, France.

Chisti, Y., (2007). Biodiesel from microalgae. *Biotechnol. Adv.*, *25*, 294–306.

Chisti, Y., (2013). Constraints to commercialization of algal fuels. *J. Biotechnol.*, *167*, 201–214.

Chiu, S. Y., Kao, C. Y., Tsai, M. T., Ong, S. C., Chen, C. H., & Lin, C. S., (2009). Lipid accumulation and CO_2 utilization of *Nannochloropsis oculata* in response to CO_2 aeration. *Bioresour. Technol.*, *100*, 833–838.

Choi, S. P., Nguyen, M. T., & Sim, S. J., (2010). Enzymatic pretreatment of *Chlamydomonas reinhardtii* biomass for ethanol production. *Bioresour. Technol.*, *101*, 5330–5336.

Choi, Y. E., Yun, Y. S., & Park, J. M., (2002). Evaluation of factors promoting astaxanthin production by a unicellular green alga, *Haematococcus pluvialis*, with fractional factorial design. *J. Biotechnol.*, *18*, 1170–1175.

Christenson, L., & Sims, R., (2011). Production and harvesting of microalgae for wastewater treatment, biofuels, and bioproducts. *Biotechnol. Adv.*, *29*, 686–702.

Chung, J. H., Kang, S., Varani, J., Lin, J., Fisher, G. J., & Voorhees, J. J., (2000). Decreased extracellular-signal-regulated kinase and increased stress-activated MAP kinase activities in aged human skin *in vivo. J. Invest. Dermatol.*, *115*, 177–182.

Colombo, S. L., Pollock, S. V., Eger, K. A., Godfrey, A. C., Adams, J. E., Mason, C. B., et al., (2002). Use of the bleomycin resistance gene to generate tagged insertional mutants of *Chlamydomonas reinhardtii* that require elevated CO_2 for optimal growth. *Funct. Plant Biol.*, *29*, 231–241.

Cotsaftis, O., Sallaud, C., Breitler, J. C., Meynard, D., Greco, R., Pereira, A., et al., (2002). Transposon-mediated generation of T-DNA and marker-free rice plants expressing a Bt endotoxin gene. *Mol. Breed., 10*, 165–180.

Courchesne, N. M., Parisien, A., Wang, B., & Lan, C. Q., (2009). Enhancement of lipid production using biochemical, genetic and transcription factor engineering approaches. *J. Biotechnol., 141*, 31–41.

Crepineau, F., Roscoe, T., Kaas, R., Kloareg, B., & Boyen, C., (2000). Characterization of complementary DNAs from the expressed sequence tag analysis of life cycle stages of *Laminaria digitata* (Phaeophyceae). *Plant Mol. Biol., 43*, 503–513.

Croft, M. T., Moulin, M., Webb, M. E., & Smith, A. G., (2007). Thiamine biosynthesis in algae is regulated by riboswitches. *Proc. Natl. Acad. Sci. USA, 104*, 20770–20775.

Cunningham, F. X. Jr., & Gantt, E., (2001). One ring or two? Determination of ring number in carotenoids by lycopene ε-cyclases. *Proc. Natl. Acad. Sci. USA, 98*, 2905–2910.

Cunningham, F. X. Jr., Lee, H., & Gantt, E., (2007). Carotenoid biosynthesis in the primitive red alga *Cyanidioschyzon merolae*. *Eukaryot. Cell, 6*, 533–545.

Cunningham, F. X. Jr., Sun, Z., Chamovitz, D., Hirschberg, J., & Gantt, E., (1994). Molecular structure and enzymatic function of lycopene cyclase from the cyanobacterium *Synechococcus* sp. strain PCC7942. *Plant Cell, 6*, 1107–1121.

D'Souza, F. M. L., & Kelly, G. J., (2000). Effects of a diet of a nitrogen-limited alga (*Tetraselmis suecica*) on growth, survival, and biochemical composition of tiger prawn (*Penaeus semisulcatus*) larvae. *Aquaculture, 181*, 311–329.

Damonte, E. B., Pujol, C. A., & Coto, C. E., (2004). Prospects for the therapy and prevention of Dengue virus infections. *Adv. Virus Res., 63*, 239–285.

Darbani, B., Eimanifar, A., Stewart, Jr. C. N., & Camargo, W. N., (2007). Methods to produce marker-free transgenic plants. *Biotechnol. J., 2*, 83–90.

Darzins, A., Pienkos, P., & Edye, L., (2010). Current status and potential for algal biofuels production. Bioindustry Partners and NREL. *Bioenergy Task, 39*, 131.

Das, B. K., Pradhan, J., Pattnaik, P., Samantaray, B. R., & Samal, S. K., (2005). Production of antibacterials from the freshwater alga *Euglena viridis* (Ehren). *World J. Microbiol. Biotechnol., 21*, 45–50.

De Jaeger, L., Verbeek, R., Draaisma, R., Martens, D., Springer, J., Eggink, G., et al., (2014). Superior triacylglycerol (TAG) accumulation in starchless mutants of *Scenedesmus obliquus*: (I) mutant generation and characterization. *Biotechnol. Biofuels, 7*, 69. doi: 10.1186/1754-6834-7-69.

De Oliveira, M. A. C. L., Monteiro, M. P. C., Robbs, P. G., & Leite, S. G. F., (1999). Growth and chemical composition of *Spirulina maxima* and *Spirulina platensis* biomass at different temperatures. *Aquacult. Int., 7*, 261–275.

De Riso, V., Raniello, R., Maumus, F., Rogato, A., Bowler, C., & Falciatore, A., (2009). Gene silencing in the marine diatom *Phaeodactylum tricornutum*. *Nucleic Acids Res., 37*, e96. doi: 10.1093/nar/gkp448.

De Souza, S. A. P. F., Costa, M. C., Lopes, A. C., Neto, E. F. A., Leitão, R. C., Mota, C. R., & Dos, S. A. B., (2014). Comparison of pretreatment methods for total lipids extraction from mixed microalgae. *Renew. Energ., 63*, 762–766.

Del Campo, A. J., García-González, M., & Guerrero, M. G., (2007). Outdoor cultivation of microalgae for carotenoid production: Current state and perspectives. *Appl. Microb. Biotechnol., 74*, 1163–1174.

Dembitsky, V. M., (1996). Betaine ether-linked glycerolipids: Chemistry and biology. *Prog. Lipid Res.*, *35*, 1–51.

Demming-Adams, B., & Adams, W. W., (2002). Antioxidants in photosynthesis and human nutrition. *Science*, *298*, 2149–2153.

Dent, R. M., Haglund, C. M., Chin, B. L., Kobayashi, M. C., & Niyogi, K. K., (2005). Functional genomics of eukaryotic photosynthesis using insertional mutagenesis of *Chlamydomonas reinhardtii*. *Plant Physiol.*, *137*, 545–556.

Depicker, A., & Van, M. M., (1997). Post-transcriptional gene silencing in plants. *Curr. Opin. Cell Biol.*, *9*, 373–382.

Derelle, E., Ferraz, C., Rombauts, S., Rouze, P., Worden, A. Z., Robbens, S., & Partensky, F., (2006). Genome analysis of the smallest free-living eukaryote *Ostreococcus tauri* unveils many unique features. *Proc. Natl. Acad. Sci. USA*, *103*, 11647–11652.

Desbois, A. P., Mearns-Spragg, A., & Smith, V. J., (2009). A fatty acid from the diatom *Phaeodactylum tricornutum* is antibacterial against diverse bacteria including multi-resistant *Staphylococcus aureus* (MRSA). *Mar. Biotechnol.*, *11*, 45–52.

Despres, L., Gielly, L., Redoutet, B., & Taberlet, P., (2003). Using AFLP to resolve phylogenetic relationships in a morphologically diversified plant species complex when nuclear and chloroplast sequences fail to reveal variability. *Molecular Phylogenetics and Evolution*, *27*(2), 185–196.

Dismukes, G. C., Carrieri, D., Bennette, N., Ananyev, G. M., & Posewitz, M. C., (2008). Aquatic phototrophs: Efficient alternatives to land based crops for biofuels. *Curr. Opin. Biotechnol.*, *19*, 235–240.

Domozych, D. S., Ciancia, M., Fangel, J. U., Mikkelsen, M. D., Ulvskov, P., & Willats, W. G. T., (2012). The cell walls of green algae: A journey through evolution and diversity. *Front Plant Sci.*, *3*, 82.

Dos Santos, M. D., Guaratini, T., Lopes, J. L. C., Colepicolo, P., & Lopes, N. P., (2005). Plant cell and microalgae culture. In: *Modern Biotechnology in Medicinal Chemistry and Industry*. Research Signpost, Kerala, India.

Dragone, G., Fernandes, B. D., Abreu, A. P., Vicente, A. A., & Teixeira, J. A., (2011). Nutrient limitation as a strategy for increasing starch accumulation in microalgae. *Appl. Energ.*, *88*, 3331–3335.

Dunahay, T. G., Jarvis, E. E., & Roessler, P. G., (1995). Genetic transformation of the diatoms *Cyclotella cryptica* and *Naviculas aprophila*. *J. Phycol.*, *31*(6), 1004–1012.

Efremenko, E. N., Nikolskaya, A. B., Lyagin, I. V., Senko, O. V., Makhlis, T. A., Stepanov, N. A., et al., (2012). Production of biofuels from pretreated microalgae biomass by anaerobic fermentation with immobilized *Clostridium acetobutylicum* cells. *Bioresour. Technol.*, *114*, 342–348.

Eisenreich, W., Bacher, A., Arigoni, D., & Rohdich, F., (2004). Biosynthesis of isoprenoids via the non-mevalonate pathway. *Cell. Mol. Life Sci.*, *61*(12), 1401–1426.

El Maanni, A., Dubertret, G., Delrieu, M. J., Roche, O., & Trémolières, A., (1998). Mutants of *Chlamydomonas reinhardtii* affected in phosphatidylglycerol metabolism and thylakoid biogenesis. *Plant Physiol. Biochem.*, *36*(8), 609–619.

Eroglu, E., & Melis, A., (2010). Extracellular terpenoid hydrocarbon extraction and quantitation from the green microalgae *Botryococcus braunii* var. showa. *Bioresour. Technol.*, *101*(7), 2359–2366.

Fábregas, J., Garcıa, D., Fernandez-Alonso, M., Rocha, A. I., Gomez-Puertas, P., Escribano, J. M., et al., (1999). *In vitro* inhibition of the replication of haemorrhagic septicaemia virus (VHSV) and African swine fever virus (ASFV) by extracts from marine microalgae. *Antivir. Res.*, *44*(1), 67–73.

Fábregas, J., Maseda, A., Domínguez, A., & Otero, A., (2004). The cell composition of *Nannochloropsis* sp. changes under different irradiances in semi continuous culture. *World J. Microbiol. Biotechnol.*, *20*(1), 31–35.

Fajardo, A. R., Cerdan, L. E., Medina, A. R., Fernandez, F. G. A., Moreno, P. A. G., & Grima, E. M., (2007). Lipid extraction from the microalga *Phaeodactylum tricornutum*. *Eur. J. Lipid Sci. Tech.*, *109*(2), 120–126.

Falciatore, A., Casotti, R., Leblanc, C., Abrescia, C., & Bowler, C., (1999). Transformation of nonselectable reporter genes in marine diatoms. *Mar. Biotechnol.*, *1*(3), 239–251.

Fan, J., Andre, C., & Xu, C., (2011). A chloroplast pathway for the *de novo* biosynthesis of triacylglycerol in *Chlamydomonas reinhardtii*. *FEBS Lett.*, *585*(12), 1985–1991.

Fang, S. C., De Los, R. C., & Umen, J. G., (2006). Cell size checkpoint control by the retinoblastoma tumor suppressor pathway. *PLoS Genet.*, *2*(10), 1565–1579.

Fernandes, B. D., Dragone, G. M., Teixeira, J. A., & Vicente, A. A., (2010). Light regime characterization in an airlift photobioreactor for production of microalgae with high starch content. *Appl. Biochem. Biotechnol.*, *161*(1–8), 218–226.

Frommolt, R., Werner, S., Paulsen, H., Goss, R., Wilhelm, C., Zauner, S., et al., (2008). Ancient recruitment by chromists of green algal genes encoding enzymes for carotenoid biosynthesis. *Mol. Biol. Evol.*, *25*(12), 2653–2667.

Fujii, K., (2012). Process integration of supercritical carbon dioxide extraction and acid treatment for astaxanthin extraction from a vegetative microalga. *Food Bioprod. Process.*, *90*(4), 762–766.

Fukusaki, E. I., Nishikawa, T., Kato, K., Shinmyo, A., Hemmi, H., Nishino, T., & Kobayashi, A., (2003). Introduction of the archaebacterial geranylgeranyl pyrophosphate synthase gene into *Chlamydomonas reinhardtii* chloroplast. *J. Biosci. Bioeng.*, *95*(3), 283–287.

Galvan, A., Gonzalez-Ballester, D., & Fernandez, E., (2007). Insertional mutagenesis as a tool to study genes/functions in *Chlamydomonas*. In: *Transgenic Microalgae as Green Cell Factories* (pp. 77–89). Springer, New York.

Gao, S., Yang, J., Tian, J., Ma, F., Tu, G., & Du, M., (2010). Electro-coagulation–flotation process for algae removal. *J. Hazard. Mater*, *177*(1–3), 336–343.

García-Malea, M. C., Acién, F. G., Del, R. E., Fernández, J. M., Cerón, M. C., Guerrero, M. G., & Molina-Grima, E., (2009). Production of astaxanthin by *Haematococcus pluvialis*: Taking the one-step system outdoors. *Biotechnol. Bioeng.*, *102*(2), 651–657.

Georgianna, D. R., & Mayfield, S. P., (2012). Exploiting diversity and synthetic biology for the production of algal biofuels. *Nature*, *488*(7411), 329.

Gerber, G. B., Leonard, A., & Hantson, P. H., (2002). Carcinogenicity, mutagenicity and teratogenicity of manganese compounds. *Crit. Rev. Oncol. Hematol.*, *42*(1), 25–34.

Ghasemi, Y., Moradian, A., Mohagheghzadeh, A., Shokravi, S., & Morowvat, M. H., (2007). Antifungal and antibacterial activity of the microalgae collected from paddy fields of Iran: Characterization of antimicrobial activity of *Chroococcus dispersus*. *J. Biol. Sci.*, *7*, 904–910.

Gheshlaghi, R. E. Z. A., Scharer, J. M., Moo-Young, M., & Chou, C. P., (2009). Metabolic pathways of clostridia for producing butanol. *Biotechnol. Adv.*, *27*(6), 764–781.

Giordano, M., (2001). Interactions between C and N metabolism in *Dunaliella salina* cells cultured at elevated CO_2 and high N concentrations. *J. Plant Physiol.*, *158*(5), 577–581.

Gonzalez-Ballester, D., Pootakham, W., Mus, F., Yang, W., Catalanotti, C., Magneschi, L., et al., (2011). Reverse genetics in *Chlamydomonas*: A platform for isolating insertional mutants. *Plant Methods*, *7*(1), 24. https://doi.org/10.1186/1746-4811-7-24 (accessed on 26 May 2020).

Gouveia, L., Marques, A. E., Da Silva, T. L., & Reis, A., (2009). *Neochloris oleabundans* UTEX# 1185: A suitable renewable lipid source for biofuel production. *J. Ind. Microbiol. Biotechnol.*, *36*(6), 821–826.

Graham-Rowe, D., (2011). Beyond food versus fuel. *Nature*, *474*(7352), S6–S8.

Granado-Lorencio, F., Herrero-Barbudo, C., Acién-Fernández, G., Molina-Grima, E., Fernández-Sevilla, J. M., Pérez-Sacristán, B., & Blanco-Navarro, I., (2009). *In vitro* bioaccesibility of lutein and zeaxanthin from the microalgae *Scenedesmus almeriensis*. *Food Chem.*, *114*(2), 747–752.

Greenwell, H. C., Laurens, L. M. L., Shields, R. J., Lovitt, R. W., & Flynn, K. J., (2009). Placing microalgae on the biofuels priority list: A review of the technological challenges. *J. R. Soc. Interface*, *7*(46), 703–726.

Griesbeck, C., Kobl, I., & Heitzer, M., (2006). *Chlamydomonas reinhardtii. Mol. Biotechnol.*, *34*(2), 213–223.

Grima, E. M., Belarbi, E. H., Fernández, F. A., Medina, A. R., & Chisti, Y., (2003). Recovery of microalgal biomass and metabolites: Process options and economics. *Biotechnol. Adv.*, *20*(7/8), 491–515.

Grimi, N., Dubois, A., Marchal, L., Jubeau, S., Lebovka, N. I., & Vorobiev, E., (2014). Selective extraction from microalgae *Nannochloropsis* sp. using different methods of cell disruption. *Bioresour. Technol.*, *153*, 254–259.

Grossman, A. R., Bhaya, D., Apt, K. E., & Kehoe, D. M., (1995). Light-harvesting complexes in oxygenic photosynthesis: Diversity, control, and evolution. *Annu. Rev. Genet.*, *29*(1), 231–288.

Guedes, A. C., Amaro, H. M., & Malcata, F. X., (2011). Microalgae as sources of high added-value compounds: A brief review of recent work. *Biotechnol. Progr.*, *27*(3), 597–613.

Guerin, M., Huntley, M. E., & Olaizola, M., (2003). *Haematococcus astaxanthin*: Applications for human health and nutrition. *Trends Biotechnol.*, *21*(5), 210–216.

Guihéneuf, F., Leu, S., Zarka, A., Khozin-Goldberg, I., Khalilov, I., & Boussiba, S., (2011). Cloning and molecular characterization of a novel acyl-CoA: Diacylglycerol acyltransferase 1-like gene (PtDGAT1) from the diatom *Phaeodactylum tricornutum. FEBS, J., 278*(19), 3651–3666.

Gunnlaugsdottir, H., & Ackman, R. G., (1993). Three extraction methods for determination of lipids in fish meal: Evaluation of a hexane/isopropanol method as an alternative to chloroform-based methods. *J. Sci. Food Agric.*, *61*(2), 235–240.

Guschina, I. A., & Harwood, J. L., (2009). Algal lipids and effect of the environment on their biochemistry. In: *Lipids in Aquatic Ecosystems* (pp. 1–24). Springer, New York, NY.

Guschina, I. A., & Harwood, J. L., (2013a). Chemical diversity of lipids. *Enc. Biophys.*, 268–279.

Guschina, I. A., & Harwood, J. L., (2013b). Algal lipids and their metabolism. In: *Algae for Biofuels and Energy* (pp. 17–36). Springer, Dordrecht.

Gustafsson, C., Govindarajan, S., & Minshull, J., (2004). Codon bias and heterologous protein expression. *Trends Biotechnol.*, *22*(7), 346–353.

Gutiérrez, R., Ferrer, I., García, J., & Uggetti, E., (2015). Influence of starch on microalgal biomass recovery, settleability and biogas production. *Bioresour. Technol.*, *185*, 341–345.

Guzman, S., Gato, A., Lamela, M., Freire-Garabal, M., & Calleja, J. M., (2003). Anti-inflammatory and immunomodulatory activities of polysaccharide from *Chlorella stigmatophora* and *Phaeodactylum tricornutum*. *Phytother. Res.*, *17*(6), 665–670.

Hackett, J. D., Scheetz, T. E., Yoon, H. S., Soares, M. B., Bonaldo, M. F., Casavant, T. L., & Bhattacharya, D., (2005). Insights into a dinoflagellate genome through expressed sequence tag analysis. *BMC Genom.*, *6*(1), 80. https://doi.org/10.1186/1471-2164-6-80.

Hamilton, M. L., Haslam, R. P., Napier, J. A., & Sayanova, O., (2014). Metabolic engineering of *Phaeodactylum tricornutum* for the enhanced accumulation of omega-3 long chain polyunsaturated fatty acids. *Metab. Eng.*, *22*, 3–9.

Hankamer, B., Lehr, F., Rupprecht, J., Mussgnug, J. H., Posten, C., & Kruse, O., (2007). Photosynthetic biomass and H_2 production by green algae: From bioengineering to bioreactor scale-up. *Physiol. Plantarum*, *131*(1), 10–21.

Harker, M., & Hirschberg, J., (1998). Molecular biology of carotenoid biosynthesis in photosynthetic organisms. In: *Methods in Enzymology* (Vol. 297, pp. 244–263). Academic Press, USA.

Harris, E. H., (2009). *The Chlamydomonas Sourcebook: Introduction to Chlamydomonas and its Laboratory Use* (Vol. 1, p. 480). Academic Press, Oxford.

Hartwell, L. H., Culotti, J., Pringle, J. R., & Reid, B. J., (1974). Genetic control of the cell division cycle in yeast. *Science*, *183*(4120), 46–51.

Harun, R., & Danquah, M. K., (2011). Enzymatic hydrolysis of microalgal biomass for bioethanol production. *Chem. Eng. J.*, *168*(3), 1079–1084.

Harun, R., Danquah, M. K., & Forde, G. M., (2010). Microalgal biomass as a fermentation feedstock for bioethanol production. *J. Chem. Technol. Biotechnol.*, *85*(2), 199–203.

Harwood, J. L., & Guschina, I. A., (2013). Regulation of lipid synthesis in oil crops. *FEBS Lett.*, *587*(13), 2079–2081.

Harwood, J. L., & Okanenko, A. A., (2003). Sulphoquinovosyl diacylglycerol (SQDG)-the sulfolipid of higher plants. In: *Sulphur in Plants* (pp. 189–219). Springer, Dordrecht.

Heider, S. A., Peters-Wendisch, P., Netzer, R., Stafnes, M., Brautaset, T., & Wendisch, V. F., (2014). Production and glucosylation of C50 and C40 carotenoids by metabolically engineered *Corynebacterium glutamicum*. *Appl. Microbiol. Biotechnol.*, *98*(3), 1223–1235.

Heitzer, M., Eckert, A., Fuhrmann, M., & Griesbeck, C., (2007). Influence of codon bias on the expression of foreign genes in microalgae. In: *Transgenic Microalgae as Green Cell Factories* (pp. 46–53). Springer, New York.

Hemaiswarya, S., Raja, R., Kumar, R. R., Ganesan, V., & Anbazhagan, C., (2011). Microalgae: A sustainable feed source for aquaculture. *World J. Microbiol. Biotechnol.*, *27*(8), 1737–1746.

Hemmi, H., Ikejiri, S., Nakayama, T., & Nishino, T., (2003). Fusion-type lycopene β-cyclase from a thermoacidophilic archaeon *Sulfolobus solfataricus*. *Biochem. Biophys. Res. Commun.*, *305*(3), 586–591.

Henry, I. M., Wilkinson, M. D., Hernandez, J. M., Schwarz-Sommer, Z., Grotewold, E., & Mandoli, D. F., (2004). Comparison of ESTs from juvenile and adult phases of the giant unicellular green alga *Acetabularia acetabulum*. *BMC Plant Biol.*, *4*(1), 3. https://doi.org/10.1186/1471-2229-4-3 (accessed on 26 May 2020).

Herrero, M., & Ibáñez, E., (2015). Green processes and sustainability: An overview on the extraction of high added-value products from seaweeds and microalgae. *J. Supercrit. Fluid*, *96*, 211–216.

Hildebrand, M., (2005). Prospects of manipulating diatom silica nanostructure. *J. Nanosci. Nanotechnol.*, *5*(1), 146–157.

Hirakawa, Y., Kofuji, R., & Ishida, K. I., (2008). Transient transformation of a chlorarachniophyte alga, *Lotharella amoebiformis* (Chlorarachniophyceae), with *uid A* and *egfp* reporter genes. *J. Phycol.*, *44*(3), 814–820.

Hirano, A., Ueda, R., Hirayama, S., & Ogushi, Y., (1997). CO_2 fixation and ethanol production with microalgal photosynthesis and intracellular anaerobic fermentation. *Energy*, *22*(2, 3), 137–142.

Hirata, R., Takahashi, M., Saga, N., & Mikami, K., (2011). Transient gene expression system established in *Porphyra yezoensis* is widely applicable in bangiophycean algae. *Mar. Biotechnol.*, *13*(5), 1038–1047.

Hlavova, M., Turoczy, Z., & Bisova, K., (2015). Improving microalgae for biotechnology: From genetics to synthetic biology. *Biotech. Adv., 33,* 1194–1203.

Ho, S. H., Chen, C. Y., & Chang, J. S., (2012). Effect of light intensity and nitrogen starvation on CO_2 fixation and lipid/carbohydrate production of an indigenous microalga *Scenedesmus obliquus* CNW-N. *Bioresour. Technol.*, *113*, 244–252.

Ho, S. H., Chen, C. Y., Lee, D. J., & Chang, J. S., (2011). Perspectives on microalgal CO_2-emission mitigation systems: A review. *Biotechnol. Adv.*, *29*(2), 189–198.

Ho, S. H., Huang, S. W., Chen, C. Y., Hasunuma, T., Kondo, A., & Chang, J. S., (2013). Characterization and optimization of carbohydrate production from an indigenous microalga *Chlorella vulgaris* FSP-E. *Bioresour. Technol.*, *135*, 157–165.

Hollar, S., (2011). *A Closer Look at Bacteria, Algae, and Protozoa* (p.88). Britannica Educational Publishing, New York.

Horvatić, J., & Persić, V., (2007). The effect of Ni^{2+}, Co^{2+}, Zn^{2+}, Cd^{2+} and Hg^{2+} on the growth rate of marine diatom *Phaeodactylum tricornutum* Bohlin: Microplate growth inhibition test. *Bull. Environ. Contam. Toxicol.*, *79*, 494–498.

Hsieh, C. H., & Wu, W. T., (2009). Cultivation of microalgae for oil production with a cultivation strategy of urea limitation. *Bioresour. Technol.*, *100*(17), 3921–3926.

Hu, Q., Sommerfeld, M., Jarvis, E., Ghirardi, M., Posewitz, M., Seibert, M., & Darzins, A., (2008). Microalgal triacylglycerols as feedstocks for biofuel production: Perspectives and advances. *Plant, J., 54*(4), 621–639.

Hu, Q., Zheungu, H., Cohen, Z., & Richond, A., (1997). Enhancement of eicosapentaenoic acid (EPA) and γ-linolenic acid (GLA) production by manipulating algal density of outdoor cultures of *Monodus subterraneus* (Eustigmatophyta) and *Spirulina platensis* (Cyanobacteria). *Eur. J. Phycol.*, *32*(1), 81–86.

Hu, Y. R., Guo, C., Wang, F., Wang, S. K., Pan, F., & Liu, C. Z., (2014). Improvement of microalgae harvesting by magnetic nanocomposites coated with polyethylenimine. *Chem. Eng. J.*, *242*, 341–347.

Hu, Y. R., Guo, C., Xu, L., Wang, F., Wang, S. K., Hu, Z., & Liu, C. Z., (2014). A magnetic separator for efficient microalgae harvesting. *Biores. Technol.*, *158*, 388–391.

Huang, J., Liu, J., Li, Y., & Chen, F., (2008). Isolation and characterization of the phytoene desaturase gene as a potential selective marker for genetic engineering of the astaxanthin-producing green alga *Chlorella zofingiensis* (Chlorophyta). *J. Phycol., 44* (3), 684–690.

Huang, X., Weber, J. C., Hinson, T. K., Mathieson, A. C., & Minocha, S. C., (1996). Transient expression of the GUS reporter gene in the protoplasts and partially digested cells of *Ulva lactuca* L. (Chlorophyta). *Bot. Marina*, *39*(1–6), 467–474.

Huleihel, M., Ishanu, V., Tal, J., & Arad, S. M., (2001). Antiviral effect of red microalgal polysaccharides on Herpes simplex and Varicella zoster viruses. *J. Appl. Phycol.*, *13*(2), 127–134.

Huntley, M. E., & Redalje, D. G., (2007). CO_2 mitigation and renewable oil from photosynthetic microbes: A new appraisal. *Mitig. Adapt. Strat. Gl.*, *12*(4), 573–608.

Huo, Y. X., Cho, K. M., Rivera, J. G. L., Monte, E., Shen, C. R., Yan, Y., & Liao, J. C., (2011). Conversion of proteins into biofuels by engineering nitrogen flux. *Nature Biotechnol.*, *29*(4), 346–351.

Ike, A., Toda, N., Tsuji, N., Hirata, K., & Miyamoto, K., (1997). Hydrogen photoproduction from CO_2-fixing microalgal biomass: Application of halotolerant photosynthetic bacteria. *J. Ferment. Bioeng.*, *84*(6) 606–609.

Illman, A. M., Scragg, A. H., & Shales, S. W., (2000). Increase in *Chlorella* strains calorific values when grown in low nitrogen medium. *Enzyme Microb. Technol.*, *27*(8), 631–635.

Imamoglu, E., Dalay, M. C., & Sukan, F. V., (2009). Influences of different stress media and high light intensities on accumulation of astaxanthin in the green alga *Haematococcus pluvialis*. *New Biotechnol.*, *26*(3, 4), 199–204.

Jaime, L., Rodríguez-Meizoso, I., Cifuentes, A., Santoyo, S., Suarez, S., Ibáñez, E., & Señorans, F. J., (2010). Pressurized liquids as an alternative process to antioxidant carotenoids' extraction from *Haematococcus pluvialis* microalgae. *LWT Food Sci. Technol.*, *43*(1), 105–112.

Jason, H., Nelson, E., Tilman, D., Polasky, S., & Tiffany, D., (2006). Environmental, economic, and energetic costs and benefits of biodiesel and ethanol biofuels. Proceedings *Proc. Natl. Acad. Sci. USA*, *103*(30), 11206–11210.

Jazzar, S., Olivares-Carrillo, P., De Los, R. A. P., Marzouki, M. N., Acién-Fernández, F. G., Fernández-Sevilla, J. M., et al., (2015). Direct supercritical methanolysis of wet and dry unwashed marine microalgae (*Nannochloropsis gaditana*) to biodiesel. *Appl. Energ.*, *148*, 210–219.

Jeon, J. S., Lee, S., Jung, K. H., Jun, S. H., Jeong, D. H., Lee, J., et al., (2000). T-DNA insertional mutagenesis for functional genomics in rice. *Plant J.*, *22*(6), 561–570.

Jiang, P., Qin, S., & Tseng, C. K., (2003). Expression of the *lacZ* reporter gene in sporophytes of the seaweed *Laminaria japonica* (Phaeophyceae) by gametophyte-targeted transformation. *Plant Cell Rep.*, *21*(12), 1211–1216.

Jin, E. S., Polle, J. E., Lee, H. K., Hyun, S. M., & Chang, M., (2003). Xanthophylls in microalgae: From biosynthesis to biotechnological mass production and application. *J. Microbiol. Biotechnol.*, *13*(2), 165–174.

John, R. P., Anisha, G. S., Nampoothiri, K. M., & Pandey, A., (2011). Micro and macroalgal biomass: A renewable source for bioethanol. *Bioresour. Technol.*, *102*(1), 186–193.

Jungnick, N., Ma, Y., Mukherjee, B., Cronan, J. C., Speed, D. J., Laborde, S. M., et al., (2014). The carbon concentrating mechanism in *Chlamydomonas reinhardtii*: Finding the missing pieces. *Photosynth. Res.*, *121*(2/3), 159–173.

Kaplan, A., Hagemann, M., Bauwe, H., Kahlon, S., & Ogawa, T., (2008). Carbon acquisition by cyanobacteria: Mechanisms, comparative genomics and evolution. In: Antonia, H., & Enrique, F., (eds.), *The Cyanobacteria: Molecular Biology, Genomics and Evolution* (pp. 305–334). Caister Academic Press, UK.

Kawaguchi, H., Hashimoto, K., Hirata, K., & Miyamoto, K., (2001). H_2 production from algal biomass by a mixed culture of *Rhodobium marinum* A-501 and *Lactobacillus amylovorus*. *J. Biosci. Bioeng.*, *91*(3), 277–282.

Keidan, M., Friedlander, M., & Arad, S. M., (2009). Effect of brefeldin A on cell-wall polysaccharide production in the red microalga *Porphyridium* sp. (Rhodophyta) through its effect on the Golgi apparatus. *J. Appl. Phycol.*, *21*(6), 707–717.

Keris-Sen, U. D., Sen, U., Soydemir, G., & Gurol, M. D., (2014). An investigation of ultrasound effect on microalgal cell integrity and lipid extraction efficiency. *Bioresour. Technol.*, *152*, 407–413.

Khalil, Z. I., Asker, M. M., El-Sayed, S., & Kobbia, I. A., (2010). Effect of pH on growth and biochemical responses of *Dunaliella bardawil* and *Chlorella ellipsoidea*. *World J. Microbiol. Biotechnol.*, *26*(7), 1225–1231.

Khanavi, M., Toulabi, P. B., Abai, M. R., Sadati, N., Hadjiakhoondi, F., Hadjiakhoondi, A., & Vatandoost, H., (2011). Larvicidal activity of marine algae, *Sargassum swartzii* and *Chondria dasyphylla*, against malaria vector *Anopheles stephensi*. *J. Vec. Born. Dis.*, *48*(4), 241–245.

Khotimchenko, S. V., & Yakovleva, I. M., (2005). Lipid composition of the red alga *Tichocarpus crinitus* exposed to different levels of photon irradiance. *Phytochem.*, *66*(1), 73–79.

Khozin-Goldberg, I., & Cohen, Z., (2011). Unraveling algal lipid metabolism: Recent advances in gene identification. *Biochimie*, *93*(1), 91–100.

Kilian, O., Benemann, C. S., Niyogi, K. K., & Vick, B., (2011). High-efficiency homologous recombination in the oil-producing alga *Nannochloropsis* sp. *Proc. Natl. Acad. Sci. USA*, *108*(52), 21265–21269.

Kim, J., Ryu, B. G., Kim, B. K., Han, J. I., & Yang, J. W., (2012a). Continuous microalgae recovery using electrolysis with polarity exchange. *Bioresour. Technol.*, *111*, 268–275.

Kim, J., Ryu, B. G., Kim, K., Kim, B. K., Han, J. I., & Yang, J. W., (2012b). Continuous microalgae recovery using electrolysis: Effect of different electrode pairs and timing of polarity exchange. *Bioresour. Technol.*, *123*, 164–170.

Kim, J., Yoo, G., Lee, H., Lim, J., Kim, K., Kim, C. W., et al., (2013). Methods of downstream processing for the production of biodiesel from microalgae. *Biotechnol. Adv.*, *31*(6), 862–876.

Kim, M. S., Baek, J. S., Yun, Y. S., Sim, S. J., Park, S., & Kim, S. C., (2006). Hydrogen production from *Chlamydomonas reinhardtii* biomass using a two-step conversion process: Anaerobic conversion and photosynthetic fermentation. *Int. J. Hydrogen Energ.*, *31*(6), 812–816.

Kimura, S., & Kondo, T., (2002). Recent progress in cellulose biosynthesis. *J. Plant Res.*, *115*(4), 297–302.

Kirst, H., Garcia-Cerdan, J. G., Zurbriggen, A., Ruehle, T., & Melis, A., (2012). Truncated photosystem chlorophyll antenna size in the green microalga *Chlamydomonas reinhardtii* upon deletion of the *TLA3-CpSRP43* gene. *Plant Physiol.*, *160*(4), 2251–2260.

Knoshaug, E. P., & Zhang, M., (2009). Butanol tolerance in a selection of microorganisms. *Appl. Biochem. Biotechnol.*, *153*(1–3), 13–20.

Kopertekh, L., Jüttner, G., & Schiemann, J., (2004). PVX-Cre-mediated marker gene elimination from transgenic plants. *Plant Mol. Biol.*, *55*(4), 491–500.

Krubasik, P., & Sandmann, G., (2000). Molecular evolution of lycopene cyclases involved in the formation of carotenoids with ionone end groups. *Biochem. Soc. Trans.*, *28*, 806–810.

Krysan, P. J., Young, J. C., & Sussman, M. R., (1999). T-DNA as an insertional mutagen in Arabidopsis. *Plant Cell*, *11*(12), 2283–2290.

Kuang, M., Su-Juan, W., Yao, L., Da-leng, S., & Cheng-Kui, Z., (1998). Transient expression of exogenous *gus* gene in *Porphyra yezoensis* (Rhodophyta). *Chin. J. Oceanol. Limnol.*, *16*(1), 56–61.

Lagarde, D., & Vermaas, W., (1999). The zeaxanthin biosynthesis enzyme β-carotene hydroxylase is involved in myxoxanthophyll synthesis in *Synechocystis* sp. PCC 6803. *FEBS Lett.*, *454*(3), 247–251.

Lauersen, K. J., Berger, H., Mussgnug, J. H., & Kruse, O., (2013). Efficient recombinant protein production and secretion from nuclear transgenes in *Chlamydomonas reinhardtii*. *J. Biotechnol.*, *167*(2), 101–110.

Lee, A. K., Lewis, D. M., & Ashman, P. J., (2009). Microbial flocculation, a potentially low-cost harvesting technique for marine microalgae for the production of biodiesel. *J. Appl. Phycol.*, *21*(5), 559–567.

Lee, A. K., Lewis, D. M., & Ashman, P. J., (2013). Harvesting of marine microalgae by electroflocculation: The energetics, plant design, and economics. *Appl. Energ.*, *108*, 45–53.

Lee, J. B., Hayashi, K., Hirata, M., Kuroda, E., Suzuki, E., Kubo, Y., & Hayashi, T., (2006). Antiviral sulfated polysaccharide from *Navicula directa*, a diatom collected from deep-sea water in Toyama Bay. *Biol. Pharm. Bull.*, *29*(10), 2135–2139.

Lee, S. Y., Park, J. H., Jang, S. H., Nielsen, L. K., Kim, J., & Jung, K. S., (2008). Fermentative butanol production by *Clostridia*. *Biotechnol. Bioeng.*, *101*(2), 209–228.

Lee, Y. C., Lee, H. U., Lee, K., Kim, B., Lee, S. Y., Choi, M. H., et al., (2014). Aminoclay conjugated TiO_2 synthesis for simultaneous harvesting and wet-disruption of oleaginous *Chlorella* sp. *Chem. Eng. J.*, *245*, 143–149.

Lehninger, A. L., Nelson, D. L., Cox, M. M., & Cox, M. M., (2005). *Lehninger Principles of Biochemistry* (4th edn., p. 1216). W.H. Freeman and Company, New York.

Lehr, F., & Posten, C., (2009). Closed photo-bioreactors as tools for biofuel production. *Curr. Opin. Biotechnol.*, *20*(3), 280–285.

León, R., Couso, I., & Fernández, E., (2007). Metabolic engineering of ketocarotenoids biosynthesis in the unicelullar microalga *Chlamydomonas reinhardtii*. *J. Biotechnol.*, *130*(2), 143–152.

León-Bañares, R., González-Ballester, D., Galván, A., & Fernández, E., (2004). Transgenic microalgae as green cell-factories. *Trends Biotechnol.*, *22*(1), 45–52.

León-Deniz, L. V., Dumonteil, E., Moo-Puc, R., & Freile-Pelegrin, Y., (2009). Antitrypanosomal *in vitro* activity of tropical marine algae extracts. *Pharm. Biol.*, *47*(9), 864–871.

Lerche, K., & Hallmann, A., (2009). Stable nuclear transformation of *Gonium pectorale*. *BMC Biotechnol.*, *9*(1), 64. https://doi.org/10.1186/1472-6750-9-64.

Letelier-Gordo, C. O., Holdt, S. L., De Francisci, D., Karakashev, D. B., & Angelidaki, I., (2014). Effective harvesting of the microalgae *Chlorella protothecoides* via bioflocculation with cationic starch. *Bioresour. Technol.*, *167*, 214–218.

Leung, D. Y., Wu, X., & Leung, M. K. H., (2010). A review on biodiesel production using catalyzed transesterification. *Appl. Energ.*, *87*(4), 1083–1095.

Li, F., Qin, S., Jiang, P., Wu, Y., & Zhang, W., (2009). The integrative expression of GUS gene driven by FCP promoter in the seaweed *Laminaria japonica* (Phaeophyta). *J. Appl. Phycol.*, *21*(3), 287–293.

Li, H. B., Cheng, K. W., Wong, C. C., Fan, K. W., Chen, F., & Jiang, Y., (2007). Evaluation of antioxidant capacity and total phenolic content of different fractions of selected microalgae. *Food Chem.*, *102*(3), 771–776.

Li, J., Lu, Y., Xue, L., & Xie, H., (2010). A structurally novel salt-regulated promoter of duplicated carbonic anhydrase gene 1 from *Dunaliella salina*. *Mol. Biol. Rep.*, *37*(2), 1143–1154.

Li, R., Yu, K., & Hildebrand, D. F., (2010). DGAT1, DGAT2, and PDAT expression in seeds and other tissues of epoxy and hydroxy fatty acid accumulating plants. *Lipids, 45*(2), 145–157.

Li, X., Xu, H., & Wu, Q., (2007). Large scale biodiesel production from microalga *Chlorella protothecoides* through heterotrophic cultivation in bioreactors. *Biotechnol. Bioeng.*, *98*(4), 764–771.

Li, Y., Han, D., Hu, G., Dauvillee, D., Sommerfeld, M., et al., (2010). *Chlamydomonas* starchless mutant defective in ADP-glucose pyrophosphorylase hyper-accumulates triacylglycerol. *Metab. Eng.*, *12*, 387–391.

Li, Y., Han, D., Hu, G., Sommerfeld, M., & Hu, Q., (2010). Inhibition of starch synthesis results in overproduction of lipids in *Chlamydomonas reinhardtii*. *Biotechnol. Bioeng.*, *107*, 258–268.

Li, Y., Horsman, M., Wang, B., Wu, N., & Lan, C. Q., (2008). Effects of nitrogen sources on cell growth and lipid accumulation of green alga *Neochloris oleoabundans*. *Appl. Microbiol. Biotechnol.*, *81*(4), 629–636.

Liang, Y., Sarkany, N., & Cui, Y., (2009). Biomass and lipid productivities of *Chlorella vulgaris* under autotrophic, heterotrophic and mixotrophic conditions. *Biotechnol Lett.*, *31*, 1043–1049.

Lichtenthaler, H. K., (1999). The 1-deoxy-D-xylulose-5-phosphate pathway of isoprenoid biosynthesis in plants. *Annu. Rev. Plant Physiol. Plant Mol. Biol.*, *50*, 47–65.

Lin, Z., Xu, Y., Zhen, Z., Fu, Y., Liu, Y., Li, W., et al., (2015). Application and reactivation of magnetic nanoparticles in *Microcystis aeruginosa* harvesting. *Bioresour. Technol.*, *190*, 82–88.

Linden, H., (1999). Carotenoid hydroxylase from *Haematococcus pluvialis*: CDNA sequence, regulation and functional complementation. *Biochim. Biophys. Acta, 1446*, 203–212.

Linden, H., Vioque, A., & Sandmann, G., (1993). Isolation of a carotenoid biosynthesis gene coding for δ-carotene desaturase from *Anabaena* PCC 7120 by heterologous complementation. *FEMS Microbiol. Lett., 106*, 99–104.

Liu, H. Q., Yu, W. G., Dai, J. X., Gong, Q. H., Yang, K. F., & Zhang, Y. P., (2003). Increasing the transient expression of GUS gene in *Porphyra yezoensis* by 18S rDNA targeted homologous recombination. *J. Appl. Phycol., 15*, 371–377.

Liu, J., Zhong, Y., Sun, Z., Huang, J., Sandmann, G., & Chen, F., (2010). One amino acid substitution in phytoene desaturase makes *Chlorella zofingiensis* resistant to norflurazon and enhances the biosynthesis of astaxanthin. *Planta, 232*, 61–67.

Liu, X., Sheng, J., & Curtiss, R. III., (2011). Fatty acid production in genetically modified cyanobacteria. *Proc. Natl. Acad. Sci. USA, 108*, 6899–6904.

Liu, Z. Y., Wang, G. C., & Zhou, B. C., (2008). Effect of iron on growth and lipid accumulation in *Chlorella vulgaris*. *Bioresour. Technol., 99*, 4717–4722.

Lowe, B. A., Prakash, N. S., Way, M., Mann, M. T., Spencer, T. M., & Boddupalli, R. S., (2009). Enhanced single copy integration events in corn via particle bombardment using low quantities of DNA. *Transgenic Res., 18*, 831–840.

Lu, Y. M., Li, J., Xue, L. X., Yan, H. X., Yuan, H. J., & Wang, C., (2011). A duplicated carbonic anhydrase 1 (DCA1) promoter mediates the nitrate reductase gene switch of *Dunaliella salina*. *J. Appl. Phycol., 23*, 673–680.

Maheswari, U., Montsant, A., Goll, J., Krishnasamy, S., Rajyashri, K. R., Patell, V. M, et al., (2005). The diatom EST database. *Nucleic Acids Res., 33,* D344–D347.

Mai, J. C., & Coleman, A. W., (1997). The internal transcribed spacer 2 exhibits a common secondary structure in green algae and flowering plants. *J. Mol. Evol., 44*(3), 258–271.

Makino, T., Harada, H., Ikenaga, H., Matsuda, S., Takaichi, S., Shindo, K., et al., (2008). Characterization of cyanobacterial carotenoid ketolase CrtW and hydroxylase CrtR by complementation analysis in *Escherichia coli. Plant Cell Physiol., 49,* 1867–1878.

Mandal, S., & Mallick, N., (2009). Microalga *Scenedesmus obliquus* as a potential source for biodiesel production. *Appl. Microbiol. Biotechnol., 84,* 281–291.

Manimaran, P., Ramkumar, G., Sakthivel, K., Sundaram, R. M., Madhav, M. S., & Balachandran, S. M., (2011). Suitability of non-lethal marker and marker-free systems for development of transgenic crop plants: Present status and future prospects. *Biotechnol. Adv., 29,* 703–714.

Maresca, J. A., Frigaard, N. U., & Bryant, D. A., (2005). Identification of a novel class of lycopene cyclases in photosynthetic organisms. In: Van, D. E. A., & Bruce, D., (eds.), *Photosynthesis: Fundamental Aspects to Global Perspectives* (pp. 884–886). Allen Press, Lawrence, Kansas.

Maresca, J. A., Graham, J. E., Wu, M., Eisen, J. A., & Bryant, D. A., (2007). Identification of a fourth family of lycopene cyclases in photosynthetic bacteria. *Proc. Natl. Acad. Sci. USA, 104,* 11784–11789.

Marin-Navarro, J., Manuell, A. L., Wu, J., & Mayfield, S. P., (2007). Chloroplast translation regulation. *Photosynth. Res., 94,* 359–374.

Martínez-Férez, I. M., & Vioque, A., (1992). Nucleotide sequence of the phytoene desaturase gene from *Synechocystis* sp. PCC 6803 and characterization of a new mutation which confers resistance to the herbicide norflurazon. *Plant Mol. Biol., 18,* 981–983.

Martínez-Férez, I., Fernández-González, B., Sandmann, G., & Vioque, A., (1994). Cloning and expression in *Escherichia coli* of the gene coding for phytoene synthase from the cyanobacterium *Synechocystis* sp. PCC6803. *Biochim. Biophys. Acta, 1218,* 145–152.

Masamoto, K., Misawa, N., Kaneko, T., Kikuno, R., & Toh, H., (1998). β-Carotene hydroxylase gene from the cyanobacterium *Synechocystis* sp. PCC6803. *Plant Cell Physiol., 39,* 560–564.

Masamoto, K., Wada, H., Kaneko, T., & Takaichi, S., (2001). Identification of a gene required for *cis*-to-*trans* carotene isomerization in carotenogenesis of the cyanobacterium *Synechocystis* sp. PCC 6803. *Plant Cell Physiol., 42,* 1398–1402.

Masojodek, J., & Torzillo, G., (2003). Mass cultivation of freshwater microalgae. *Encycl. Ecol., 3,* 2226–2235.

Mata, M. T., Martins, A. A., & Caetano, N. S., (2010). Micaroalgae for biodiesel production and other applications: A review. *Renew. Sust. Energy Rev., 14,* 217–232.

Matsui, M. S., Muizziddin, A. S., & Marenuss, K., (2003). Sulfated polysaccharides from red microalgae have antiinflammatory properties *in vitro* and *in vivo. Appl. Biochem. Biotechnol., 104,* 13–22.

Matsunaga, T., Matsumoto, M., Maeda, Y., Sugiyama, H., Sato, R., & Tanaka, T., (2009). Characterization of marine microalga, *Scenedesmus* sp. strain JPCC GA0024 towards biofuel production. *Biotechnol. Lett., 31,* 1367–1372.

Matsunaga, T., Takeyama, H., & Nakamura, N., (1990). Characterization of cryptic plasmids from marine cyanobacteria and construction of a hybrid plasmid potentially capable of transformation of marine cyanobacterium, *Synechococcus* sp., and its transformation. *Appl. Biochem. Biotechnol., 24,* 151–160.

Matsuzaki, M., Misumi, O., Shin-I, T., Maruyama, S., Takahara, M., Miyagishima, S. Y., et al., (2004). Genome sequence of the ultrasmall unicellular red alga *Cyanidioschyzon merolae* 10D. *Nature, 428,* 653–657.

Mayer, A. M. S., & Hamann, M. T., (2005). Marine pharmacology in 2001–2002: Marine compounds with anthelmintic, antibacterial, anticoagulant, antidiabetic, antifungal, anti inflammatory, antimalarial, antiplatelet, antiprotozoal, antituberculosis, and antiviral activities; affecting the cardiovascular, immune and nervous systems and other miscellaneous mechanisms of action. *Comp. Biochem. Physiol. C. Toxicol. Pharmacol., 140,* 265–286.

McArthur, G. H., & Fong, S. S., (2010). Toward engineering synthetic microbial metabolism. *J. Biomed. Biotechnol.,* 459760. http://dx.doi.org/10.1155/2010/459760 (accessed on 26 May 2020).

McCarthy, S. S., Kobayashi, M. C., & Niyogi, K. K., (2004). White mutants of *Chlamydomonas reinhardtii* are defective in phytoene synthase. *Genetics, 168,* 1249–1257.

McKee, T., & McKee, J. R., (2015). *Biochemistry: The Molecular Basis of Life* (6th edn., p. 928). Oxford University Press.

McNulty, H. P., Byun, J., Lockwood, S. F., Jacob, R. F., & Mason, R. P., (2007). Differential effects of carotenoids on lipid peroxidation due to membrane interactions: X-ray diffraction analysis. *Biochim. Biophys. Acta, 1768,* 167–174.

Mendiola, J. A., Torres, C. F., Martın-Alvarez, P. J., Santoyo, S., Tore, A., Arredondo, B. O., et al., (2007). Use of supercritical CO_2 to obtain extracts with antimicrobial activity from *Chaetoceros muelleri* microalga. A correlation with their lipidic content. *Eur. Food Res. Technol., 224,* 505–510.

Mendoza, H., Martel, A., Jimenez, D. R. M., & Reina, G. G., (1999). Oleic acid is the main fatty acid related with carotenogenesis in *Dunaliella salina*. *J. Appl. Phycol., 11,* 15–19.

Meng, X., Yang, J., Xu, X., Zhang, L., Nie, Q., & Xian, M., (2009). Biodiesel production from oleaginous microorganisms. *Renew. Energ., 34,* 1–5.

Merzlyak, M. N., Chivkunova, O. B., Gorelova, O. A., Reshetnikova, I. V., Solovchenko, A. E., Khozin-Goldberg, I., et al., (2007). Effect of nitrogen starvation on optical properties, pigments and arachidonic acid content of the unicellular green alga *Parietochloris incise* (Trebouxiophyceae, Chlorophyta). *J. Phycol., 43,* 833–843.

Metting, F. B., (1996). Biodiversity and application of microalgae. *J. Ind. Microbiol. Biotech., 17,* 477–489.

Meuser, J. E., Ananyev, G., Wittig, L. E., Kosourov, S., Ghirardi, M. L., Seibert, M., et al., (2009). Phenotypic diversity of hydrogen production in chlorophycean algae reflects distinct anaerobic metabolisms. *J. Biotechnol., 142,* 21–30.

Miki, B., & McHugh, S., (2004). Selectable marker genes in transgenic plants: Applications, alternatives and biosafety. *J. Biotechnol., 107,* 193–232.

Minoda, A., Sakagami, R., Yagisawa, F., Kuroiwa, T., & Tanaka, K., (2004). Improvement of culture conditions and evidence for nuclear transformation by homologous recombination in a red alga, *Cyanidioschyzon merolae* 10D. *Plant Cell Physiol., 45,* 667–671.

Misawa, N., (2011). Pathway engineering for functional isoprenoids. *Curr. Opin. Biotechnol., 22,* 627–633.

Misawa, N., Nakagawa, M., Kobayashi, K., Yamano, S., Izawa, Y., Nakamura, K., & Harashima, K., (1990). Elucidation of the *Erwinia uredovora* carotenoid biosynthetic pathway by functional analysis of gene products expressed in *Escherichia coli. J. Bacteriol., 172,* 6704–6712.

Misra, R., Guldhe, A., Singh, P., Rawat, I., Stenström, T. A., & Bux, F., (2015). Evaluation of operating conditions for sustainable harvesting of microalgal biomass applying electrochemical method using non sacrificial electrodes. *Bioresour. Technol., 176,* 1–7.

Miyagawa, A., Okami, T., Kira, N., Yamaguchi, H., Ohnishi, K., et al., (2009). High efficiency transformation of the diatom *Phaeodactylum tricornutum* with a promoter from the diatom *Cylindrotheca fusiformis. Phycol. Res., 57,* 142–146.

Miyagawa-Yamaguchi, A., Okami, T., Kira, N., Yamaguchi, H., Ohnishi, K., & Adachi, M., (2011). Stable nuclear transformation of the diatom *Chaetoceros* sp. *Phycol. Res., 59,* 113–119.

Miziorko, H. M., (2011). Enzymes of the mevalonate pathway of isoprenoid biosynthesis. *Arch Biochem. Biophys., 505,* 131–143.

Mochimaru, M., Msukawa, H., & Takaichi, S., (2005). The cyanobacterium *Anabaena* sp. PCC 7120 has two distinct β-carotene ketolase: CrtO for echinenone and CrtW for ketomyxol synthesis. *FEBS Lett., 579,* 6111–6114.

Mochimaru, M., Msukawa, H., Maoka, T., Mohamed, H. E., Vermaas, W. F. J., & Takaichi, S., (2008). Substrate specificities and availability of fucosyltransferase and β-carotene hydroxylase for myxol 2-fucoside synthesis in *Anabaena* sp. strain PCC 7120 compared with *Synechocystis* sp. strain PCC 6803. *J. Bacteriol., 190,* 6726–6733.

Moellering, E. R., & Benning, C., (2009). Glycerolipid biosynthesis. In: Stern, D., & Harris, E. H., (eds.), *The Chlamydomonas Sourcebook: Organellar and Metabolic Processes* (Vol. 2, pp. 41–68). Elsevier, Dordrecht.

Moellering, E. R., & Benning, C., (2010). RNA interference silencing of a major lipid droplet protein affects lipid droplet size in *Chlamydomonas reinhardtii. Eukaryot. Cell, 9,* 97–106.

Moellering, E. R., Miller, R., & Benning, C., (2009). Molecular genetics of lipid metabolism in the model green alga *Chlamydomonas reinhardtii.* In: Wada, H., & Murata, N., (eds.), *Lipids in Photosynthesis: Essential and Regulatory Functions* (pp. 139–155). Springer Science, Netherlands.

Moeser, G. D., Roach, K. A., Green, W. H., Hatton, T. A., & Laibinis, P. E., (2004). High gradient magnetic separation of coated magnetic nanoparticles. *AIChE, J., 50,* 2835–2848.

Molnar, A., Bassett, A., Thuenemann, E., Schwach, F., Karkare, S., Ossowski, S., et al., (2009). Highly specific gene silencing by artificial microRNAs in the unicellular alga *Chlamydomonas reinhardtii. Plant, J., 58,* 165–174.

Moreno-Risueno, M. A., Martínez, M., Vicente-Carbajosa, J., & Carbonero, P., (2007). The family of DOF transcription factors: From green unicellular algae to vascular plants. *Mol. Genet. Genomics, 277,* 379–390.

Mussatto, S. I., Dragone, G., Guimaraes, P. M. R., Silva, J. P. A., Carneiro, L. M., Roberto, I. C., et al., (2010). Technological trends, global market, and challenges of bioethanol production. *Biotechnol. Adv., 28,* 817–830.

Mussgnug, J. H., Klassen, V., Schluter, A., & Kruse, O., (2010). Microalgae as substrates for fermentative biogas production in a combined biorefinery concept. *J. Biotechnol., 150,* 51–56.

Mussgnug, J. H., Thomas-Hall, S., Rupprecht, J., Foo, A., Klassen, V., McDowall, A., et al., (2007). Engineering photosynthetic light capture: Impacts on improved solar energy to biomass conversion. *Plant Biotechnol. J., 5,* 802–814.

Nag, A., Lunacek, M., Graf, P. A., & Chang, C. H., (2011). Kinetic modeling and exploratory numerical simulation of chloroplastic starch degradation. *BMC Syst. Biol., 5,* 94, https://doi.org/10.1186/1752-0509-5-94 (accessed on 26 May 2020).

Najafi, N., Ahmadi, A. R., Hosseini, R., & Golkhoo, S., (2011). Gamma irradiation as a useful tool for the isolation of astaxanthin-overproducing mutant strains of *Phaffia rhodozyma. Can. J. Microbiol., 57,* 730–734.

Nakashima, Y., Ohsawa, I., Konishi, F., Hasegawa, T., Kumamoto, S., Suzuki, Y., & Ohta, S., (2009). Preventive effects of *Chlorella* on cognitive decline in age-dependent dementia model mice. *Neurosci. Lett., 464,* 193–198.

Naviner, M., Berge, J. P., Durand, P., & Le Bris, H., (1999). Antibacterial activity of the marine diatom *Skeletonema costatum* against aquacultural pathogens. *Aquaculture, 174,* 15–24.

Naylor, R. L., Liska, A. J., Burke, M. B., Falcon, W. P., Gaskell, J. C., Rozelle, S. D., & Cassman, K. G., (2007). The ripple effect: Biofuels, food security, and the environment. *Environment, 49,* 30–43.

Ndikubwimana, T., Zeng, X., Liu, Y., Chang, J. S., & Lu, Y., (2014). Harvesting of microalgae *Desmodesmus* sp. F51 by bioflocculation with bacterial bioflocculant. *Algal Res., 6,* 186–193.

Neti, N. R., & Misra, R., (2012). Efficient degradation of reactive blue 4 in carbon bed electrochemical reactor. *Chem. Eng. J., 184,* 23–32.

Neupert, J., Karcher, D., & Bock, R., (2009). Generation of *Chlamydomonas* strains that efficiently express nuclear transgenes. *Plant, J., 57,* 1140–1150.

Newman, D. J., & Cragg, G. M., (2006). Natural products from marine invertebrates and microbes as modulators of antitumor targets. *Curr. Drug Targets, 7,* 279–304.

Nguyen, M. T., Choi, S. P., Lee, J., Lee, J. H., & Sim, S. I., (2009). Hydrothermal acid pretreatment of *Chlamydomonas reinhardtii biomass* for ethanol production. *J. Microbiol. Biotechnol., 19,* 161–166.

Niehaus, T. D., Okada, S., Devarenne, T. P., Watt, D. S., Sviripa, V., & Chappell, J., (2011). Identification of unique mechanisms for triterpene biosynthesis in *Botryococcus braunii. Proc. Natl. Acad. Sci. USA, 108,* 12260–12265.

Nishida, I., & Murata, N., (1996). Chilling sensitivity in plants and cyanobateria: The crucial contribution of membrane lipids. *Annu. Rev. Plant. Physiol. Plant Mol. Biol. 47,* 541–568.

Niu, Y. F., Yang, Z. K., Zhang, M. H., Zhu, C. C., Yang, W. D., Liu, J. S., et al., (2012). Transformation of diatom *Phaeodactylum tricornutum* by electroporation and establishment of inducible selection marker. *Biotechniques, 52*(6), 379–386.

Niu, Y. F., Zhang, M. H., Li, D. W., Yang, W. D., Liu, J. S., Bai, W., & Li, H. Y., (2013). Improvement of neutral lipid and polyunsaturated fatty acid biosynthesis by over expressing a type 2 diacylglycerol acyltransferase in marine diatom *Phaeodactylum tricornutum. Mar. Drugs, 11,* 4558–4569.

Niu, Y. F., Zhang, M. H., Xie, W. H., Li, J. N., Gao, Y. F., et al., (2011). A new inducible expression system in a transformed green alga, *Chlorella vulgaris. Genet. Mol. Res., 10,* 3427–3434.

Nurse, P., Thuriaux, P., & Nasmyth, K., (1976). Genetic control of the cell division cycle in the fission yeast *Schizosaccharomyces pombe. Mol. Gen. Genet., 146,* 167–178.

O'Brien, E. A., Koski, L. B., Zhang, Y., Yang, L. S., Wang, E., Gray, M. W., et al., (2007). TBestDB: A taxonomically broad database of expressed sequence tags (ESTs). *Nucleic Acids Res., 35,* D445–D451.

O'Neill, B. M., Mikkelson, K. L., Gutierrez, N. M., Cunningham, J. L., Wolff, K. L., Szyjka, S. J., et al., (2012). An exogenous chloroplast genome for complex sequence manipulation in algae. *Nucleic Acids Res., 40,* 2782–2792.

Oetjen, K., Ferber, S., Dankert, I., & Reusch, T. B. H., (2010). New evidence for habitat-specific selection in Wadden Sea *Zostera marina* populations revealed by genome scanning using SNP and microsatellite markers. *Mar. Biol., 157*(1), 81–89.

Oh, S. H., Han, J. G., Kim, Y., Ha, J. H., & Kim, S. S., (2009). Lipid production in *Porphyridium cruentum* grown under different culture conditions. *Jpn. J. Biosci. Bioeng., 108,* 429–434.

Ohta, S., Ono, F., Shiomi, Y., Nakao, T., Aozasa, O., Nagate, T., et al., (1998). Anti-herpes simplex virus substances produced by the marine green alga. *Dunaliella primolecta. J. Appl. Phys., 10,* 349–356.

Ohto, C., Ishida, C., Nakane, H., Muramatsu, M., Nishino, T., & Obata, S., (1999). A thermophilic cyanobacterium *Synechococcus elongatus* has three different Class I prenyltransferase genes. *Plant Mol. Biol., 40,* 307–321.

Okuda, K., Oka, K., Onda, A., Kajiyoshi, K., Hiraoka, M., & Yanagisawa, K., (2008). Hydrothermal fractional pretreatment of sea algae and its enhanced enzymatic hydrolysis. *J. Chem. Technol. Biotechnol., 83,* 836–841.

Olaizola, M., (2003). Commercial development of microalgal biotechnology: From the test tube to the marketplace. *Biomol. Eng., 20,* 459–466.

Ong, S. C., Kao, C. Y., Chiu, S. Y., Tsai, M. T., & Lin, C. S., (2010). Characterization of the thermal-tolerant mutants of *Chlorella* sp. with high growth rate and application in outdoor photobioreactor cultivation. *Bioresour. Technol., 101,* 2880–2883.

Oono, M., Kikuchi, K., Oonishi, S., Nishino, H., & Tsushima, Y., (1995). Anti-cancer agents containing carotenoids. *Japan Kokai Tokkyo Koho,* p. 5, Japanese Patent: JP 07101872 A2 19950418 Heisei. Appl: JP 93-248267 19931004.

Orhan, I., Sener, B., Atici, T., Brun, R., Perozzo, R., & Tasdemir, D., (2006). Turkish freshwater and marine macrophyte extracts show *in vitro* antiprotozoal activity and inhibit FabI, a key enzyme of *Plasmodium falciparum* fatty acid biosynthesis. *Phytomed., 13,* 388–393.

Ota, M., Kato, Y., Watanabe, H., Watanabe, M., Sato, Y., Smith, R. L. Jr., & Inomata, H., (2009). Fatty acid production from a highly CO_2 tolerant alga, *Chlorococcum littorale,* in the presence of inorganic carbon and nitrate. *Bioresour. Technol., 100,* 5237–5242.

Ow, D. W., (2002). Recombinase-directed plant transformation for the post-genomic era. *Plant Mol. Biol., 48,* 183–200.

Palenik, B., Grimwood, J., Aerts, A., Rouze, P., Salamov, A., Putnam, N., et al., (2007). The tiny eukaryote *Ostreococcus* provides genomic insights into the paradox of plankton speciation. *Proc. Natl. Acad. Sci. USA, 104,* 7705–7710.

Palozza, P., Serini, S., Maggiano, N., Tringali, G., Navarra, P., Ranelletti, F. O., et al., (2005). Beta-carotene downregulates the steady state and heregulin-α-induced COX-2 pathways in colon cancer cells. *J. Nutr., 135,* 129–136.

Palozza, P., Torelli, C., Boninsegna, A., Simone, R., Catalano, A., Mele, M. C., & Picci, N., (2009). Growth-inhibitory effects of the astaxan-thin-rich alga *Haematococcus pluvialis* in human colon cancer cells. *Cancer Lett., 283,* 108–117.

Park, P. K., Kim, E. Y., & Chu, K. H., (2007). Chemical disruption of yeast cells for the isolation of carotenoid pigments. *Separation and Purification Technology, 53,* 148–152.

Pashkow, F., Watumull, D., & Campbell, C., (2008). Astaxanthin: A novel potential treatment for oxidative stress and inflammation in cardiovascular disease. *Am. J. Cardiol., 101,* S58–S68.

Paudel, A., Jessop, M. J., Stubbins, S. H., Champagne, P., & Jessop, P. G., (2015). Extraction of lipids from microalgae using CO_2-expanded methanol and liquid CO_2. *Bioresour. Technol., 184*, 286–290.

Pavoni, J. L., Tenney, M. W., & Echelberger, Jr. W. F., (1972). Bacterial exocellular polymers and biological flocculation. *J. Water Pollut. Control Fed., 44*, 414–431.

Peer, M. S., Skye, R., Hall, T., Marx, E. S. U., Mussgnug, J. H., Posten, C., et al., (2008). Second generation biofuels: High-efficiency microalgae for biodiesel production. *Bioenerg. Res., 1*, 20–43.

Perez-Garcia, O., Escalante, F. M. E., de-Bashan, L. E., & Bashan, Y., (2011). Heterotrophic cultures of microalgae: Metabolism and potential products. *Water Res., 45*, 11–36.

Pienkos, P. T., & Darzins, A., (2009). The promise and challenges of micro-algal derived biofuels. Biofuels, *Bioproducts and Biorefining, 3*, 431–440.

Plaza, M., Herrero, M., Cifuentes, A., & Ibanez, E., (2009). Innovative natural functional ingredients from microalgae. *J. Agric. Food. Chem., 57*, 7159–7170.

Pollet, P., Davey, E. A., Urena-Benavides, E. E., Eckert, C. A., & Liotta, C. L., (2014). Solvents for sustainable chemical processes. *Green Chem., 16*, 1034–1055.

Popescu, C. E., & Lee, R. W., (2007). Mitochondrial genome sequence evolution in *Chlamydomonas. Genetics, 175*, 819–826.

Potvin, G., & Zhang, Z. S., (2010). Strategies for high-level recombinant protein expression in transgenic microalgae: A review. *Biotechnol. Adv., 28*, 910–918.

Poulsen, N., & Kröger, N., (2005). A new molecular tool for transgenic diatoms: Control of mRNA and protein biosynthesis by an inducible promoter-terminator cassette. *FEBS J., 272*, 3413–3423.

Poulsen, N., Chesley, P. M., & Kröger, N., (2006). Molecular genetic manipulation of the diatom *Thalassiosira pseudonana* (Bacillariophyceae). *J. Phycol., 42*, 1059–1065.

Pruvost, J., Cornet, J. F., & Legrand, J., (2008). Hydrodynamics influence on light conversion in photobioreactors: An energetically consistent analysis. *Chem. Eng. Sci., 63*, 3679–3694.

Puglisi, M. P., Tan, L. T., Jensen, P. R., & Fenical, W., (2004). Capisterones A and B from the tropical green alga *Penicillus capitatus*: Unexpected anti-fungal defenses targeting the marine pathogen *Lindra thallasiae. Tetrahedron, 60*, 7035–7039.

Qi, L., Haurwitz, R. E., Shao, W., Doudna, J. A., & Arkin, A. P., (2012). RNA processing enables predictable programming of gene expression. *Nat. Biotechnol., 30*, 1002–1006.

Qin, S., Jiang, P., & Tseng, C. K., (2004). Molecular biotechnology of marine algae in China. *Hydrobiologia, 512*, 21–26.

Qin, S., Jiang, P., Li, X. P., Wang, X. H., & Zeng, C. K., (1998). A transformation model for *Laminaria japonica* (Phaeophyta, Laminariales). *China J. Oceanol. Limnol., 16*, 50–55.

Qin, S., Zhang, J., Li, W. B., Wang, X. H., Tong, S., Sun, Y. R., et al., (1994). Transient expression of GUS gene in Phaeophytes using biolistic particle delivery system. *Oceanol. Limnol. Sin., 25*, 353–356.

Quintana, N., Van, D. K. F., Van, D. R. M. D., Voshol, G. P., & Verpoorte, R., (2011). Renewable energy from cyanobacteria: Energy production optimization by metabolic pathway engineering. *Appl. Microbiol. Biotechnol., 91*, 471–490.

Rabbani, S., Beyer, P., Von, L. J., Hugueney, P., & Kleinig, H., (1998). Induced β-carotene synthesis driven by triacylglycerol deposition in the unicellular alga *Dunaliella bardawil. Plant Physiol., 116*, 1239–1248.

Radakovits, R., Eduafo, P. M., & Posewitz, M. C., (2011). Genetic engineering of fatty acid chain length in *Phaeodactylum tricornutum*. *Metab. Eng., 13,* 89–95.

Radakovits, R., Jinkerson, R. E., Darzins, A., & Posewitz, M. C., (2010). Genetic engineering for enhanced biofuel production. *Eukaryot. Cell, 9,* 486–501.

Raja, R., Hemaiswarya, S., Ashok, Kumar, N., Sridhar, S., & Rengasamy, R., (2008). A perspective on the biotechnological potential of microalgae. *Crit. Rev. Microbiol., 34,* 77–88.

Ramos, A., Coesel, S., Marques, A., Rodrigues, M., Baumgartner, A., Noronha, J., et al., (2008). Isolation and characterization of a stress-inducible *Dunaliella salina Lyc-β* gene encoding a functional lycopene β-cyclase. *Appl. Microbiol. Biotechnol., 79,* 819–828.

Rangel-Yagui, C. D., Danesi, E. D. G., De Carvalho, J. C. M., & Sato, S., (2004). Chlorophyll production from *Spirulina platensis*: Cultivation with urea addition by fed-batch process. *Bioresour. Technol., 92,* 133–141.

Rao, A. R., Dayananda, C., Sarada, R., Shamala, T. R., & Ravishankar, G. A., (2007). Effects of salinity on growth of green alga *Botryococcus braunii* and its constituents. *Bioresour. Technol., 98,* 560–564.

Rasala, B. A., Lee, P. A., Shen, Z., Briggs, S. P., Mendez, M., & Mayfield, S. P., (2012). Robust expression and secretion of Xylanase1 in *Chlamydomonas reinhardtii* by fusion to a selection gene and processing with the FMDV 2A peptide. *PLoS One, 7*(8), e43349. https://doi.org/10.1371/journal.pone.0043349.

Rasala, B. A., Muto, M., Lee, P. A., Jager, M., Cardoso, R. M., Behnke, C. A., et al., (2010). Production of therapeutic proteins in algae, analysis of expression of seven human proteins in the chloroplast of *Chlamydomonas reinhardtii*. *Plant Biotechnol. J., 8,* 719–733.

Rasala, B. A., Muto, M., Sullivan, J., & Mayfield, S. P., (2011). Improved heterologous protein expression in the chloroplast of *Chlamydomonas reinhardtii* through promoter and 50 untranslated region optimization. *Plant Biotechnol. J., 9,* 674–683.

Remacle, C., Cline, S., Boutaffala, L., Gabilly, S., Larosa, V., Barbieri, M. R., et al., (2009). The ARG9 gene encodes the plastid-resident n-acetyl ornithine aminotransferase in the green alga *Chlamydomonas reinhardtii*. *Eukaryot. Cell, 8,* 1460–1463.

Renaud, S. M., Thinh, L. V., Lambrinidis, G., & Parry, D. L., (2002). Effect of temperature on growth, chemical composition, and fatty acid composition of tropical Australian microalgae grown in batch cultures. *Aquaculture, 211,* 195–214.

Richardson, J. W., Johnson, M. D., & Outlaw, J. L., (2012). Economic comparison of open pond raceways to photo bio-reactors for profitable production of algae for transportation fuels in the Southwest. *Algal Res., 1,* 93–100.

Richmond, A., (2004). Biological principles of mass cultivation. In: Richmond, A., (ed.), *Handbook of Microalgal Cultures, Biotechnology and Applied Phycology* (pp. 125–177). Blackwell, Oxford.

Riekhof, W. R., Sears, B. B., & Benning, C., (2005). Annotation of genes involved in glycerolipid biosynthesis in *Chlamydomonas reinhardtii*: Discovery of the betaine lipid synthase BTA1Cr. *Eukaryot. Cell, 4,* 242–252.

Rio, E., Acien, F. G., Rivas, J., Molina-Grima, E., & Guerrero, M. G., (2008). Efficiency assessment of the one-step production of astaxanthin by the microalga *Haematococcus pluvialis*. *Biotechnol. Bioeng. 100,* 397–402.

Rismani-Yazdi, H., Haznedaroglu, B. Z., Bibby, K., & Peccia, J., (2011). Transcriptome sequencing and annotation of the microalgae *Dunaliella tertiolecta*: Pathway description and gene discovery for production of next-generation biofuels. *BMC Genomics 12,* 148. doi: 10.1186/1471-2164-12-148.

Rodolfi, L., Zittelli, C. G., Bassi, N., Padovani, G., Biondi, N., Bonini, G., & Tredici, M. R., (2009). Microalgae for oil: Strain selection, induction of lipid synthesis and outdoor mass cultivation in a low-cost photobioreactor. *Biotechnol. Bioeng., 102,* 100–112.

Rosa, L., Gonzalez-Ballester, D., Galvan, A., & Fernandez, E., (2004). Transgenic microalgae as green cell-factories. *Trends in Biotechnology, 22,* 45–52.

Rosenberg, J. N., Oyler, G. A., Wilkinson, L., & Betenbaugh, M. J., (2008). A green light for engineered algae: Redirecting metabolism to fuel a biotechnology revolution. *Curr. Opin. Biotechnol., 19,* 430–436.

Rupprecht, J., (2009). From systems biology to fuel: *Chlamydomonas reinhardtii* as a model for a systems biology approach to improve biohydrogen production. *J. Biotechnol., 142,* 10–20.

Rynearson, T. A., Newton, J. A., & Armbrust, E. V., (2006). Spring bloom development, genetic variation, and population succession in the planktonic diatom *Ditylum brightwellii*. *Limnol. Oceanography, 51*(3), 1249–1261.

Sanchez, J., Fernandez-Sevilla, J. M., Acien, F. G., Ceron, M. C., Perez-Parra, J., & Molina-Grima, E., (2008). Biomass and lutein productivity of *Scenedesmus almeriensis*: Influence of irradiance, dilution rate and temperature. *Appl. Microbiol. Biotechnol., 79,* 719–729.

Sandmann, G., (1994). Carotenoid biosynthesis in microorganisms and plants. *Eur. J. Biochem., 223,* 7–24.

Sandmann, G., (2002). Molecular evolution of carotenoid biosynthesis from bacteria to plants. *Physiol. Plant, 116,* 431–440.

Sangha, A. K., Petridis, L., Smith, J. C., Ziebell, A., & Parks, J. M., (2012). Molecular simulation as a tool for studying lignin. *Environ. Prog. Sustain., 31,* 47–54.

Sato, N., & Moriyama, T., (2007). Genomic and biochemical analysis of lipid biosynthesis in the unicellular rhodophyte *Cyanidioschyzon merolae*: Lack of a plastidic desaturation pathway results in the coupled pathway of galactolipid synthesis. *Eukaryot. Cell 6,* 1006–1017.

Savage, N., (2011). The ideal biofuel. *Nature, 474,* S9–S11.

Schenk, P. M., Thomas-Hall, S. R., Stephens, E., Marx, U. C., Mussgnug, J. H., Posten, C., Kruse, O., & Hankamer, B., (2008). Second generation biofuels: High-efficiency microalgae for biodiesel production, *Bioenerg. Res., 1,* 20–43.

Schmollinger, S., Strenkert, D., & Schroda, M., (2010). An inducible artificial microRNA system for *Chlamydomonas reinhardtii* confirms a key role for heat shock factor 1 in regulating thermotolerance. *Curr. Genet., 56,* 383–389.

Schuhmann, H., Lim, D. K., & Schenk, P. M., (2012). Perspectives on metabolic engineering for increased lipid contents in microalgae. *Biofuels, 3,* 71–86.

Shah, S. M. U., & Abdullah, M. A., (2017). *Nannochloropsis oculata* and integrated biorefinery based on palm oil milling. In: Marcel, J., & Przemek, K., (eds.), *Nannochloropsis Biology, Biotechnological Potential and Challenges* (pp. 135–180). Nova Science Publisher, New York, USA.

Shah, S. M. U., & Abdullah, M. A., (2018). Effects of macro/micronutrients on green and brown microalgal cell growth and fatty acids in photobioreactor and open-tank systems. *Biocatalysis and Agricultural Biotechnology, 14,* 10–17.

Shah, S. M. U., (2014). *Cell Culture Optimization and Reactor Studies of Green and Brown Microalgae for Enhanced Lipid Production*. PhD Thesis, Universiti Teknologi Petronas, Tronoh, Malaysia.

Shah, S. M. U., Che Radziah, C. M., Ibrahim, Z. S., Latif, F., Othman, M. F., & Abdullah, M. A., (2014). Effects of photoperiod, salinity, and pH on cell growth and lipid content of *Pavlova lutheri*. *Annals Microbiol., 64*, 157–164.

Sharma, K. K., Schuhmann, H., & Schenk, P. M., (2012). High lipid induction in microalgae for biodiesel production. *Energies, 5*, 1532–1553.

Sheehan, J., Dunahay, T., Benemann, J., & Roessler, P., (1998). *A Look Back at the U.S. Department of Energy's Aquatic Species Program-Biodiesel from Algae*. Golden, Colorado, USA, National Renewable Energy Laboratory.

Shih, S. R., Ho, M. S., Lin, K. H., Wu, S. L., Chen, Y. T., Wu, C. N., et al., (2000). Genetic analysis of enterovirus 71 isolated from fatal and non-fatal cases of hand, foot, and mouth disease during an epidemic in Taiwan 1998. *Virus Res., 68*, 127–136.

Shrager, J., Hauser, C., Chang, C. W., Harris, E. H., Davies, J., McDermott, J., Tamse, R., Zhang, Z., & Grossman, A. R., (2003). *Chlamydomonas reinhardtii* genome project, a guide to the generation and use of the cDNA information. *Plant Physiol., 131*, 401–408.

Siaut, M., Cuine, S., Cagnon, C., Fessler, B., Nguyen, M., Carrier, P., et al., (2011). Oil accumulation in the model green alga *Chlamydomonas reinhardtii*: Characterization, variability between common laboratory strains and relationship with starch reserves. *BMC Biotechnol., 11*, 7.

Siaut, M., Heijde, M., Mangogna, M., Montsant, A., Coesel, S., Allen, A. E., et al., (2007). Molecular toolbox for studying diatom biology in *Phaeodactylum tricornutum. Gene, 406*, 23–35.

Sing, S. F., Isdepsky, A., Borowitzka, M. A., & Moheimani, N. R., (2013). Production of biofuels from microalgae. *Mitigation and Adaptation Strategies for Global Change, 18*, 47–72.

Singh, J., & Gu, S., (2010). Commercialization potential of microalgae for biofuels production. *Renew. Sust. Energy, 14*, 2596–2610.

Singh, S., Kate, B. N., & Banerjee, U. C., (2005). Bioactive compounds from cyanobacteria and microalgae: An overview. *Crit. Rev. Biotechnol., 25*, 73–95

Sleight, S. C., & Sauro, H. M., (2013). Randomized biobrick assembly: A novel DNA assembly method for randomizing and optimizing genetic circuits and metabolic pathways. *ACS Synth. Biol., 2*, 506–518.

Solana, M., Rizza, C. S., & Bertucco, A., (2014). Exploiting microalgae as a source of essential fatty acids by supercritical fluid extraction of lipids: Comparison between *Scenedesmus obliquus, Chlorella protothecoides* and *Nannochloropsis salina. J. Supercritical Fluids, 92*, 311–318.

Song, Q. S., Lin, H., & Jiang, P., (2012). Advances in genetic engineering of marine algae. *Biotech. Adv., 30*, 1602–1613.

Soni, R., Nazir, A., & Chadha, B. S., (2010). Optimization of cellulase production by a versatile *Aspergillus fumigatus* Fresenius strain (AMA) capable of efficient deinking and enzymatic hydrolysis of Solka floc and bagasse. *Ind. Crop Prod., 31*, 277–283.

Spavieri, J., Kaiser, M., Casey, R., Hingley-Wilson, S., & Lalvani, A., (2010). Antiprotozoal, antimycobacterial and cytotoxic potential of some british green algae. *Phytother. Res., 24*, 1095–1098.

Specht, E. A., & Mayfield, S. P., (2012). Synthetic oligonucleotide libraries reveal novel regulatory elements in *Chlamydomonas* chloroplast mRNAs. *ACS Synth. Biol., 1*, 34–36.

Spolaore, P., Joannis-Cassan, C., Duran, E., & Isambert, A., (2006). Commercial applications of microalgae. *J. Biosci. Bioeng., 101*, 87–96.

Sripriya, R., Raghupathy, V., & Veluthambi, K., (2008). Generation of selectable marker-free sheath blight resistant transgenic rice plants by efficient co-transformation of a cointegrate vector T-DNA and a binary vector T-DNA in one *Agrobacterium tumefaciens* strain. *Plant Cell Rep., 27*, 1635–1644.

Stehfest, K., Toepel, J., & Wilhelm, C., (2005). The application of micro-FTIR spectroscopy to analyze nutrient stress-related changes in biomass composition of phytoplankton algae. *Plant Physiol. Biochem., 43*, 717–726.

Steiger, S., & Sandmann, G., (2004). Cloning of two carotenoid ketolase genes from *Nostoc punctiforme* for the heterologous production of canthaxanthin and astaxanthin. *Biotechnol. Lett., 26*, 813–817.

Steiger, S., Jackisch, Y., & Sandmann, G., (2005). Carotenoid biosynthesis in *Gloeobacter violaceus* PCC4721 involves a single crtI-type phytoene desaturase instead of typical cyanobacterial enzymes. *Arch. Microbiol., 184*, 207–214.

Steinbrenner, J., & Linden, H., (2001). Regulation of two carotenoid biosynthesis genes coding for phytoene synthase and carotenoid hydroxylase during stress-induced astaxanthin formation in the green alga *Haematococcus pluvialis*. *Plant Physiol., 125*, 810–817.

Steinbrenner, J., & Linden, H., (2003). Light induction of carotenoid biosynthesis genes in the green alga *Haematococcus pluvialis*: Regulation by photosynthetic redox control. *Plant Mol. Biol., 52*, 343–356.

Steinbrenner, J., & Sandmann, G., (2006). Transformation of the green alga *Haematococcus pluvialis* with a phytoene desaturase for accelerated astaxanthin biosynthesis. *Appl. Environ. Microbiol., 72*, 7477–7484.

Stickforth, P., Steiger, S., Hess, W. R., & Sandmann, G., (2003). A novel type of lycopene ε-cyclase in the marine cyanobacterium *Prochlorococcus marinus* MED4. *Arch. Microbiol., 179*, 409–415.

Sukenik, A., (1991). Ecophysiological considerations in the optimization of eicosapentaenoic acid production by *Nannochloropsis* sp. (Eustigmatophyceae). *Bioresource Technology, 35*, 263–269.

Sun, G., Zhang, X., Sui, Z., & Mao, Y., (2008). Inhibition of pds gene expression via the RNA interference approach in *Dunaliella salina* (Chlorophyta). *Marine Biotechnology, 10*, 219–226.

Surzycki, R., Greenham, K., Kitayama, K., Dibal, F., Wagner, R., Rochaix, J. D., Ajam, T., & Surzycki, S., (2009). Factors effecting expression of vaccines in microalgae. *Biologicals, 37*, 133–138.

Suzuki, Y., Ohgami, K., Shiratori, K., Jin, X. H., Ilieva, I., Koyama, Y., Yazawa, K., Yoshida, K., Kase, S., & Ohno, S., (2006). Suppressive effects of astaxanthin against rat endotoxin-induced uveitis by inhibiting the NF-kB signaling pathway. *Exp. Eye Res., 82*, 275–281.

Takagi, M., Karseno, Y., & Yoshida, T., (2006). Effect of salt concentration on intracellular accumulation of lipids and triacylglycerols in marine microalgae *Dunaliella* cells. *J. Biosci. Bioeng., 3*, 223–226.

Takahashi, M., Uji, T., Saga, N., & Mikami, K., (2010). Isolation and regeneration of transiently transformed protoplasts from gametophytic blades of the marine red alga *Porphyra yezoensis*. *Electron J. Biotechnol., 13*, 7. doi: 10.2225/vol13-issue2-fulltext-7.

Takaichi, S., & Mochimaru, M., (2007). Carotenoids and carotenogenesis in cyanobacteria: Unique ketocarotenoids and carotenoid glycosides. *Cell Mol. Life Sci., 64*, 2607–2619.

Takaichi, S., (2009). Distribution and biosynthesis of carotenoids. In: Hunter, C. N., Daldal, F., Thurnauer, M. C., & Beatty, J. T., (eds.), *The Purple Phototrophic Bacteria* (pp. 97–117). Springer: Dordrecht, the Netherlands.

Takaichi, S., (2011). Carotenoids in algae: Distributions, biosyntheses and functions. *Mar. Drugs, 9*, 1101–1118.

Tan, C. P., Qin, S., Zhang, Q., Jiang, P., & Zhao, F. Q., (2005). Establishment of a microparticle bombardment transformation system for *Dunaliella salina*. *J. Microbiol., 43*, 361–365.

Tannin-Spitz, T., Bergman, M., Van-Moppes, D., Grossman, S., & Arad, S., (2005). Antioxidant activity of the polysaccharide of the red microalgae *Porphyridium* sp. *J. Appl. Phycol., 17*, 215–222.

Tanoi, T., Kawachi, M., & Watanabe, M. M., (2011). Effects of carbon source on growth and morphology of *Botryococcus braunii*. *J. Appl. Phycol., 23*, 25–33.

Taraldsvik, M., & Myklestad, S. M., (2000). The effect of pH on growth rate, biochemical composition and extracellular carbohydrate production of the marine diatom *Skeletonema costatum*. *Eur. J. Pharmacol., 35*, 189–194.

Tardy, F., & Havaux, M., (1996). Photosynthesis, chlorophyll fluorescence, light-harvesting system, and photoinhibition resistance of a zeaxanthin-accumulating mutant of *Arabidopsis thaliana*. *J. Photochem. Photobiol. B, 34(1)*, 87–94.

Temme, K., Hill, R., Segall-Shapiro, T. H., Moser, F., & Voigt, C. A., (2012). Modular control of multiple pathways using engineered orthogonal T7 polymerases. *Nucleic Acids Res., 40*, 8773–8781.

Ten, L. M. R., & Miller, D. J., (1998). Genetic transformation of dinoflagellates (*Amphidinium* and *Symbiodinium*): Expression of GUS in microalgae using heterologous promoter constructs. *Plant J., 13*, 427–435.

Terashima, M., Freeman, E. S., Jinkerson, R. E., & Jonikas, M. C., (2014). A fluorescence-activated cell sorting-based strategy for rapid isolation of high-lipid *Chlamydomonas* mutants. *Plant J., 81*, 147–159.

Terekhova, V. E., Aizdaicher, N. A., Buzoleva, L. S., & Somov, G. P., (2009). Influence of extrametabolites of marine microalgae on the reproduction of the bacterium *Listeria monocytogenes*. *Russ. J. Mar. Biol., 35*, 355–358.

Thuriaux, P., Nurse, P., & Carter, B., (1978). Mutants altered in the control co-ordinating cell division with cell growth in the fission yeast *Schizosaccharomyces pombe*. *Mol. Gen. Genet., 161*, 215–220.

Toh, P. Y., Yeap, S. P., Kong, L. P., Ng, B. W., Chan, D. J. C., Ahmad, A. L., & Lim, J. K., (2012). Magnetophoretic removal of microalgae from fish pond water: Feasibility of high gradient and low gradient magnetic separation. *Chem. Eng. J., 211*, 22–30.

Tran, M., Zhou, B., Pettersson, P. L., Gonzalez, M. J., & Mayfield, S. P., (2009b). Synthesis and assembly of a full-length human monoclonal antibody in algal chloroplasts. *Biotechnol. Bioeng., 104*, 663–673.

Tran, N. P., Park, J. K., Hong, S. J., & Lee, C. G., (2009a). Proteomics of proteins associated with astaxanthin accumulation in the green algae *Haematococcus lacustris* under the influence of sodium ortho-vanadate. *Biotechnol. Lett. 31*, 1917–1922.

Tsuchiya, T., Takaichi, S., Misawa, N., Maoka, T., Miyashita, H., & Mimuro, M., (2005). The cyanobacterium *Gloeobacter violaceus* PCC 7421 uses bacterial-type phytoene desaturase in carotenoid biosynthesis. *FEBS Lett., 579*, 2125–2129.

Turpin, D. H., (1991). Effects of inorganic N availability on algal photosynthesis and carbon metabolism. *J. Phycol., 27*, 14–20.

Umen, G. J., & Goodenough, U. W., (2010). Control of cell division by a retinoblastoma protein homolog in *Chlamydomonas*. *Genes Dev., 15*, 1652–1661.

Verwaal, R., Wang, J., Meijnen, J. P., Visser, H., Sandmann, G., Van, D. B. J. A., & Van, O. A. J., (2007). High-level production of beta-carotene in *Saccharomyces cerevisiae* by successive transformation with carotenogenic genes from *Xanthophyllomyces dendrorhous*. *Appl. Environ. Microbiol., 73*, 4342–4350.

Vigeolas, H., Duby, F., Kaymak, E., Niessen, G., Motte, P., Franck, F., et al., (2012). Isolation and partial characterization of mutants with elevated lipid content in *Chlorella sorokiniana* and *Scenedesmus obliquus*. *J. Biotechnol., 162*, 3–12.

Vila, M., Couso, I., & León, R., (2008). Carotenoid content in mutants of the chlorophyte *Chlamydomonas reinhardtii* with low expression levels of phytoene desaturase. *Process Biochem., 43*, 1147–1152.

Vongsangnak, W., Zhang, Y., Chen, W., Ratledge, C., & Song, Y., (2012). Annotation and analysis of malic enzyme genes encoding for multiple isoforms in the fungus *Mucor circinelloides* CBS 277.49. *Biotechnol. Lett., 34*, 941–947.

Vonthron-Sénécheau, C., Kaiser, M., Devambez, I., Vastel, A., & Mussio, I., (2011). Antiprotozoal activities of organic extracts from French marine seaweeds. *Mar. Drugs, 9*, 922–933.

Wagner, M., Hoppe, K., Czabany, T., Heilmann, M., Daum, G., Feussner, I., & Fulda, M., (2010). Identification and characterization of an acyl-CoA: diacylglycerol acyltransferase 2 (DGAT2) gene from the microalga *O. tauri*. *Plant Physiol. Biochem., 48*, 407–416.

Walker, T. L., Collet, C., & Purton, S., (2005). Algal transgenics in the genomic era. *J. Phycol., 41*, 1077–1093.

Wallis, J. G., & Browse, J., (2010). Lipid biochemists salute the genome. *Plant J., 61*, 1092–1106.

Wang, G. S., Grammel, H., Abou-Aisha, K., Sagesser, R., & Ghosh, R., (2012). High-level production of the industrial product lycopene by the photosynthetic bacterium *Rhodospirillum rubrum*. *Appl. Environ. Microbiol., 78*, 7205–7215.

Wang, G. Y., Wang, X., & Liu, X. H., (2011). Two-stage hydrolysis of invasive algal feedstock for ethanol fermentation. *J. Integr. Plant Biol., 53*, 246–252.

Wang, J. F., Jiang, P., Cui, Y. L., Guan, X. Y., & Qin, S., (2010). Gene transfer into conchospores of *Porphyra haitanensis* (Bangiales, Rhodophyta) by glass bead agitation. *Phycologia, 49*, 355–360.

Wang, M., Yuan, W., Jiang, X., Jing, Y., & Wang, Z., (2014). Disruption of microalgal cells using high-frequency focused ultrasound. *Bioresour. Technol., 153*, 315–321.

Wang, Q., Guan, Y., Ren, X., Cha, G., & Yang, M., (2011). Rapid extraction of low concentration heavy metal ions by magnetic fluids in high gradient magnetic separator. *Sep. Purif. Technol., 82*, 185–189.

Wang, T., Ji, Y., Wang, Y., Jia, J., Li, J., Huang, S., Danxiang, H., Qiang, H., Wei, E. H., & Xu, J., (2014). Quantitative dynamics of triacylglycerol accumulation in microalgae populations at single-cell resolution revealed by Raman micro spectroscopy. *Biotech. Biofuels, 7*, 58.

Ward, O. P., & Singh, A., (2005). Omega-3/6 fatty acids: Alternative sources of production. *Process Biochem., 40*, 3627–3652.

Washida, K., Koyama, T., Yamada, K., Kitab, M., & Uemura, D., (2006). Karatungiols A and B, two novel antimicrobial polyol compounds, from the symbiotic marine dinoflagellate *Amphidinium* sp. *Tetrahedron Lett., 47*, 2521–2525.

Watanabe, S., Ohnuma, M., Sato, J., Yoshikawa, H., & Tanaka, K., (2011). Utility of a GFP reporter system in the red alga *Cyanidioschyzon merolae. J. Gen. Appl. Microbiol., 57*, 69–72.

Weber, A., Oesterhelt, C., Gross, W., Bräutigam, A., Imboden, L., Krassovskaya, I., Linka, N., et al., (2004). EST-analysis of the thermoacidophilic red microalga *Galdieria sulphuraria* reveals potential for lipid A biosynthesis and unveils the pathway of carbon export from rhodoplasts. *Plant Mol. Biol., 55*, 17–32.

Welch, M., Villalobos, A., Gustafsson, C., & Minshull, J., (2009). You're one in a googol: Optimizing genes for protein expression. *J. R. Soc. Interface, 6*, S467–S476.

Weselake, R. J., Taylor, D. C., Rahman, H., Laroche, A., McVetty, P. B. E., & Harwood, J. L., (2009). Increasing the flow of carbon into seed oil. *Biotechnol. Adv., 27*, 866–878.

Whitney, S. M., Houtz, R. L., & Alonso, H., (2011). Advancing our understanding and capacity to engineer nature's CO_2-sequestering enzyme, Rubisco. *Plant Physiol. 155*, 27–35.

Williams, N., (2010). New biofuel questions. *Curr. Biol., 20*, 219–220.

Wingler, L. M., & Cornish, V. W., (2011). Reiterative recombination for the *in vivo* assembly of libraries of multigene pathways. *Proc. Natl. Acad. Sci. USA., 108*, 15135–15140.

Worden, A. Z., Lee, J. H., Mock, T., Rouze, P., Simmons, M. P., Aerts, A. L., Allen, A. E., & Cuvelier, M. L., (2009). Green evolution and dynamic adaptations revealed by genomes of the marine picoeukaryotes *Micromonas. Science, 324*, 268–272.

Worthington, A. S., Porter, D. F., & Burkart, M. D., (2010). Mechanism-based crosslinking as a gauge for functional interaction of modular synthases. *Org. Biomol. Chem., 8*, 1769–1772.

Wu, Z., Wu, S., & Shi, X., (2007). Supercritical fluid extraction and determination of lutein in heterotrophically cultivated *Chlorella pyrenoidosa. J. Food Process Eng., 30*, 174–185.

Xie, W. H., Zhu, C. C., Zhang, N. S., Li, D. W., Yang, W. D., Liu, J. S., & Li, H. Y., (2014). Construction of novel chloroplast expression vector and development of high efficient and economical transformation system for diatom *Phaeodactylum tricornutum. Mar. Biotechnol., 16*, 538–546.

Xu, J., Zheng, Z., & Zou, J., (2009). A membrane-bound glycerol-3-phosphate acyltransferase from *Thalassiosira pseudonana* regulates acyl composition of glycerolipids. *Botany, 87*, 544–551.

Xu, L., Brilman, D. W. F., Withag, J. A. M., Brem, G., & Kersten, S., (2011). Assessment of a dry and a wet route for the production of biofuels from microalgae: Energy balance analysis. *Biores. Technol., 102*, 5113–5122.

Yadav, V. G., De Mey, M., Lim, C. G., Ajikumar, P. K., & Stephanopoulos, G., (2012). The future of metabolic engineering and synthetic biology: Towards a systematic practice. *Metab. Eng., 14*, 233–241.

Yamada, T., & Sakaguchi, K., (1982). Comparative studies on *Chlorella* cell walls-induction of protoplast formation. *Arch. Microbiol., 132*, 10–13.

Yang, Z. K., Niu, Y. F., Ma, Y. H., Xue, J., Zhang, M. H., Yang, W. D., Liu, J. S., Lu, S. H., Guan, Y., & Li, H. Y., (2013). Molecular and cellular mechanisms of neutral lipid accumulation in diatom following nitrogen deprivation. *Biotechnol. Biofuels, 6*, 67.

Yao, Q., Cong, L., Chang, J. L., Li, K. X., Yang, G. X., & He, G. Y., (2006). Low copy number gene transfer and stable expression in a commercial wheat cultivar via particle bombardment. *J. Exp. Bot., 57*, 3737–3746.

Yao, S., Brandt, A., Egsgaard, H., & Gjermansen, C., (2012). Neutral lipid accumulation at elevated temperature in conditional mutants of two microalgae species. *Plant Physiol. Biochem., 61*, 71–79.

Ye, V. M., & Bhatia, S. K., (2012). Pathway engineering strategies for production of beneficial carotenoids in microbial hosts. *Biotechnol. Lett., 34*, 1405–1414.

Yim, J. H., Kim, S. J., Ahn, S. H., Lee, C. K., Rhie, K. T., & Lee, H. K., (2004). Antiviral effects of sulfated exopolysacchride from the marine microalga *Gyrodinium impudicum* strain KG03. *Mar. Biotechnol., 6*, 17–25.

Yingying, S., Changhai, W., & Jing, C., (2008). Growth inhibition of the eight species of microalgae by growth inhibitor from the culture of *Isochrysis galbana* and its isolation and identification. *J. Appl. Phycol. 20*, 315–321.

Yoon, S. H., Kim, J. E., Lee, S. H., Park, H. M., Choi, M. S., Kim, J., et al., (2007). Engineering the lycopene synthetic pathway in *E. coli* by comparison of the carotenoid genes of *Pantoea agglomerans* and *Pantoea ananatis*. *Appl. Microbiol. Biotechnol., 74*, 131–139.

Yoon, S. H., Lee, S. H., Das, A., Ryu, H. K., Jang, H. J., & Kim, J. Y., (2009). Combinatorial expression of bacterial whole mevalonate pathway for the production of β-carotene in *E. coli. J. Biotechnol. 140*, 218–226.

Zaslavskaia, L. A., Lippmeier, J. C., Kroth, P. G., Grossman, A. R., & Apt, K. E., (2000). Transformation of the diatom *Phaeodactylum tricornutum* (Bacillariophyceae) with a variety of selectable marker and reporter genes. *J. Phycol. 36*, 379–386.

Zhang, R., Patena, W., Armbruster, U., Gang, S. S., Blum, S. R., & Jonikas, M. C., (2014). High-throughput genotyping of green algal mutants reveals random distribution of mutagenic insertion sites and endonucleolytic cleavage of transforming DNA. *Plant Cell, 26*, 1398–1409.

Zhang, Y., Adams, I. P., & Ratledge, C., (2007). Malic enzyme: The controlling activity for lipid production? Over expression of malic enzyme in *Mucor circinelloides* leads to a 2.5-fold increase in lipid accumulation. *Microbiol., 153*, 2013–2025.

Zhang, Z., Jiang, P., Yang, R., Fei, X., Tang, X., & Qin, S., (2010). Establishment of a concocelis-mediated transformation system for *Porphyra* by biolistic PDS-1000/He. *Chin High Technol. Lett., 20*, 204–207.

Zhao, T., Wang, W., Bai, X., & Qi, Y., (2009). Gene silencing by artificial microRNAs in *Chlamydomonas. Plant J., 58*, 157–164.

Zorin, B., Lu, Y., Sizova, I., & Hegemann, P., (2009). Nuclear gene targeting in *Chlamydomonas* as exemplified by disruption of the PHOT gene. *Gene, 432*, 91–96.

INDEX

Ketones, 25
Kinetic parameters, 272, 276, 279, 282, 301

L

Lactate dehydrogenase (LDH), 160
Lactic acid bacteria (LAB), 193
Lactose, 118
Lagoons, 2
Large scale
  cultivation, 28
  investments, 99
  microalgal biomass production, 94
Lectins, 191, 192, 194, 205
Lessoniopsis littoralis, 93
Leukemia, 147, 148, 329
Ligands, 151
Light-harvesting complex, 390
Lignocelluloses, 28, 82, 86
Linear
  chain, 37
  polymer, 86
  polyuronic acid, 56
Linoleic acid, 50
Linolenic acid, 53
Lipid, 24, 25, 29, 30, 47, 50, 64–66, 72–74,
    81, 89, 92, 93, 95, 112–114, 122–126,
    128, 131, 173, 174, 216, 217, 221, 222,
    226–240, 271, 280–282, 317, 320, 328,
    348, 365–370, 387, 392, 400, 402
  accumulation, 29, 30, 280, 317, 320, 365,
    389
  compounds, 24
  extraction, 123–125, 216, 218, 226,
    228–231, 233–240
    cell wall disruption methods, 229
  metabolism, 228, 365, 369, 389, 401
  molecules, 24, 28, 227
  peroxidation, 71, 174
Lipopeptides, 69, 188, 201, 203
Lipophilic compounds, 52
Lipopolysaccharide (LPS), 156, 164
Liver fibrosis, 163, 164
Long chain
  aliphatic compounds, 31
  complex organic compounds, 23
  PUFA (LC-PUFA), 307–311, 313–315,
    317, 320, 321
  steroid compounds, 25

Low
  density lipoprotein (LDL), 69, 311
  molecular weight fucoidan (LMWF), 161,
    165, 166
Luminescent reporter systems, 401
Lycopadiene, 31–33, 35, 36, 39, 373
Lycopene, 48, 376, 377, 391–393
Lymphocytes, 159, 190, 346
Lyngbya majuscula, 10, 13, 195, 198, 199,
    203
Lyngbyatoxin, 49
Lyophilization, 218, 235, 240
Lysophosphatidic acid acyltransferase
    (LPAAT), 371

M

Macroalgae, 1, 2, 6, 58, 64, 82, 111, 112,
    141, 325, 326, 329, 336, 344, 366
Macroalgal diversity, 16
Macrolactones, 188
Macromolecules, 50, 121, 174
Macrophytic algal resources, 3
Macular degeneration, 48
Magnetic microparticles, 96
Major surface protein, 346
Malic enzyme (ME), 389
Malnutrition, 264, 308, 321
Malondialdehyde (MDA), 152, 155, 164, 165
Mammalian cell cycle regulation, 149
Mannose, 118, 192
Marine
  algae, 3, 4, 6, 53, 81, 168, 169, 173, 340,
    379–383
  microalgae, 85, 86, 100, 101, 220
Mass
  culture, 47, 95, 390
  spectrometry, 126
Mathematical models, 272, 273, 302
Matrix metalloproteinases (MMPs), 148,
    150, 156, 170, 171
  inhibition activities, 170, 176
Maximum specific growth rate, 274, 279,
    289, 294, 296, 299
Melanogenesis, 176
Membrane-bound enzyme, 36
Metabolic
  engineering, 91, 93, 97, 320, 364–366,
    385, 389, 392, 393, 395, 402

444 Index

recombination, 257, 259, 262, 265, 381
Skin matrix enzymes, 170
Sodium hydroxide (NaOH), 52, 127, 221,
   233, 318
Solar
   drying, 218, 235, 240
   energy, 30, 90, 112, 348, 390
Solid
   liquid extraction, 238
   phase extraction (SPE), 52
Solvent-lipid complex, 237
Soxhlet equipment, 238
Specific growth rate, 53, 274, 279, 289, 294,
   296, 299, 301
Spectrometers, 24
Spectrophotometer, 126
Spirulina
   maxima, 300, 327
   platensis, 113, 119, 273, 327, 328
Spliceosomes, 332
Spray-drying, 120, 235
Spyridia filamentosa, 12, 15
Squalene, 34, 36–38, 372
   synthase, 36, 37, 38
      like (SSL), 38
Stearidonic acid (SDA), 307, 313, 315
Steatohepatitis, 163
Steroid, 25, 36, 71, 172, 327
   hormones, 53
Sterol, 25, 53, 58, 59, 227, 273, 376
Strain improvement, 83, 90, 91, 101
Stromal-derived factor 1 (SDF-1), 154, 160,
   161
Strong acid, 127
Substrate
   affinity constant, 274
   concentration, 274, 275, 302
Sucrose, 118, 374
Sulfated fucosyl units, 176
Sulfoglycolipids, 192
Sulfoquinovosyldiacylglycerol (SQDG),
   368, 370
Sulfuric acid, 54, 230
Supercritical
   carbon dioxide, 236
   fluid, 123, 131, 236
      extraction (SFE), 123, 124, 131
   state, 124, 128, 237

transesterification, 126, 128
Superoxide radicals, 174
Supply chains, 98, 99
Surfactants, 220, 233
Sustainability, 27, 59, 72, 85, 97, 99, 101,
   225, 320, 321
Switchgrass, 28
Synechococcus, 66, 190, 253, 254, 273, 331,
   334–336, 383
   adaptive sensor (SasA), 254
   strains, 335

T

Taxadiene, 393
Taxonomic
   groups, 89, 141
   study of seaweeds, 16
Terpenoids, 24, 25, 70, 71, 327, 374
Tetraterpenoids, 24, 25, 26
Thalassiosira
   pseudonana, 92, 336, 349, 372, 384, 394
   weissflogii, 336, 384
Therapeutic
   compounds, 189, 204
   effects, 173
Thermodynamic limits, 97
Thiamine pyrophosphate (TPP), 386
Thioesterase (TE), 369, 395
Third-generation biofuel, 27, 83, 99
Thraustochytriaceae, 317
Thromboxanes, 52
Thylakoid membrane, 331, 375
Timedomain nuclear magnetic resonance
   (TD-NMR), 126, 127
TNFR1-associated death domain (TRADD),
   165
Toll-like receptors (TLRs), 159, 160
Torbanite, 33, 39
Total soluble protein (TSP), 345–347, 378
Toxic chemical substance, 56
Transesterification, 73, 87, 126–128, 233,
   236, 364
Transforming growth factor beta 1 (TGF-
   β1), 164, 165
Trans-membrane receptors, 151
Triacylglyceride (TAG), 73, 91, 235, 365,
   369–372, 374, 377, 389, 401